Principles of Population Genetics

Daniel L. Hartl
Harvard University

Andrew G. Clark
Pennsylvania State University

Principles of
Population Genetics

THIRD EDITION

Sinauer Associates, Inc. Publishers
Sunderland, Massachusetts

THE COVER

This image represents data from the first study of nucleotide sequence variation in a natural population, conducted by Martin Kreitman (1983; *Nature* 304:412–417). Each of the 11 vertical bands represents the 43 varying nucleotide sites in one allele of the gene alcohol dehydrogenase taken from a global distribution of the fruit fly *Drosophila melanogaster*. The colors correspond to the bases at each site: adenine = green, cytosine = yellow, guanine = blue, and thymine = red. Note that at each position only two different bases were observed. Blocks of sites in strong linkage disequilibrium can also be seen as repeated patterns of color. The sequences are oriented with the 5' end at the top, and the nucleotide corresponding to the fast/slow difference in the ADH protein is the twelfth one from the bottom.

PRINCIPLES OF POPULATION GENETICS, Third Edition
Copyright © 1997 by Sinauer Associates, Inc. All rights reserved.
This book may not be reproduced in whole or in part without permission from the publisher.

For information or to order, address:
    Sinauer Associates, Inc., PO Box 407
    23 Plumtree Road, Sunderland, MA, 01375 U.S.A.
    FAX: 413-549-1118
    Internet: publish@sinauer.com; http://www.sinauer.com

Library of Congress Cataloging-in-Publication Data
Hartl, Daniel L.
    Principles of population genetics / Daniel L. Hartl, Andrew G. Clark. — 3rd ed.
            p.   cm.
    Includes bibliographical references and index.
    ISBN 0-87893-306-9 (hardcover)
    1. Population genetics. 2. Quantitative genetics. 3. Population genetics—Problems, exercises, etc. I. Clark, Andrew G., 1954–   .
II. Title.
QH455.H36 1997   97-34505
576.5'8—dc21   CIP

Printed in Canada

5 4 3 2 1

To Barbara and Christine

Table of Contents

Preface

Thanks in part to the power of molecular methods, population genetics has been reinvigorated. As some genome projects are approaching closure and methods of "functional genomics" are scaling up to identify the roles of novel genes, inevitably increasing attention is being paid to the significance of genetic variation in populations. Nowhere is this more evident than in medical genetics. Within a decade we can expect that all major single-gene inherited disorders will be identified, genetically mapped, cloned, and characterized at a fine molecular level. Health professionals realize that this impressive feat will have an impact only on a small minority of individuals. Most of the genetic variation in disease risk is multifactorial, which means that the risk is determined by multiple genetic and environmental factors acting together. Killer diseases such as familial forms of cancer, diabetes, and cardiovascular disease fall into this category. The fact that these diseases aggregate in families implies that there is probably a genetic component, but the genetic component may differ from one family or ethnic group to another. Prompted by the high incidence of multifactorial diseases as a group, the medical community has become acutely aware of the need to understand the basic structure of genetic variation in populations in order to determine what aspects of the variation cause disease.

The exciting practical applications of population genetics to the analysis of multifactorial diseases have received great atten-

tion, but the scope of population genetics actually is much broader. Population genetics provides the genetic underpinning for all of evolutionary biology. By "evolution" we mean descent with modification. Species undergo progressive genetic modification as they adapt to their environments, and new species arise as a by-product of this process. The intellectual excitement of biological evolution arises from the fact that it addresses the fundamental questions, "What are we?" and "Where did we come from?"

Patterns of evolutionary history are recorded in DNA sequences, and the application of population genetics to interpreting DNA sequences is revealing many secrets about the evolutionary past, including the history of our own species. But population genetics embraces much more than the analysis of evolutionary relationships. It is particularly concerned with the processes and mechanisms by which evolutionary changes are made. The field is inherently multidisciplinary, cutting across molecular biology, genetics, ecology, evolutionary biology, systematics, natural history, plant breeding, animal breeding, conservation and wildlife management, human genetics, sociology, anthropology, mathematics, and statistics.

Students taking population genetics are usually expected to have completed, or to be taking concurrently, a course in differential calculus. While this book assumes a familiarity with the elementary notation for differentials and integrals, it does not require great mathematical proficiency. We have kept the mathematics to a minimum. On the other hand, some of the most important models in population genetics require quite advanced mathematics. Rather than ignore these approaches, we have made a concerted effort to present these models in such a way that the assumptions can be understood and the main results appreciated without much mathematics. References are provided for the interested reader to learn more about the details.

Several important changes distinguish the third edition of *Principles* from the second edition. The level of the treatment is more tailored to the needs of a one-semester or one-quarter course, with the intended audience being third- and fourth-year undergraduates as well as beginning graduate students. Population genetics is not only an experimental science but also a theoretical one. Special care has been taken to explain the biological motivation behind the theoretical models so that the models do not simply materialize out of thin air, and to explain in plain English the implications of the results. Many concepts are illustrated by numerical examples, using actual data wherever possible. Special topics and examples are often set off from the text as boxed problems whose solutions are explained step by step. Every chapter ends with about 20 problems, graded in difficulty, and solutions worked in full appear at the end of the text.

This edition of *Principles* is organized into nine chapters that gradually build concepts from measuring variation and the various forces that influence genetic variation through a sequential progression to concepts of molecular population genetics and quantitative genetics. The first chapter provides a background in basic genetic and statistical principles. We discuss the fundamental concepts of allelism, dominance, segregating,

recombination, and population frequencies. The role of model building and testing in population genetics is emphasized. Chapter 2 introduces the student to the primary data of population genetics, namely, the many levels of genetic variation. Chapter 3 is concerned with the organization of genetic variation into genotypes in populations. Here the Hardy-Weinberg principle gets very thorough coverage, including the cases of X-linkage and multiple alleles. Chapter 4 widens the perspective and considers the organization of genetic variation among spatially structured populations. Population substructure is measured by Wright's *F* statistics, and is presented in a way that conveys their biological meaning. The Wahlund principle and inbreeding are also covered in Chapter 4.

The goal of population genetics is to understand the forces that have an impact on levels of genetic variation. The forces of mutation, recombination, and migration are outlined in Chapter 5. Darwinian selection is the topic of Chapter 6, including both the theoretical foundations and empirical observations of the dynamics of gene-frequency change under the action of selection. Haploid and diploid cases are developed, as are the concepts of equilibrium, stability, and context dependence. After classical models of mutation-selection balance are developed, a series of more complex scenarios of natural selection are presented.

Chapter 7 deals with random genetic drift. In the absence of other forces, allele and genotype frequencies change as a result of random sampling from one generation to another. The Wright-Fisher model and diffusion approximations are presented in such a way that the student gains an appreciation for the importance of random genetic drift. The process of the coalescence of genealogies is an important innovation in theoretical population genetics, and some of the basic concepts of coalescence are presented in Chapter 7.

In Chapter 8 we cover the rapidly expanding data on molecular evolutionary genetics. The unifying theme in the study of molecular evolution is Kimura's neutral theory, and a close examination is made of the correspondence between the data and theory. This is a field in which advances in our empirical database and statistical tools for quantifying and manipulating the data are growing at a dizzying pace. Our goal is to give the student a firm grasp of the fundamentals, and a deep enough understanding of the principles to identify important gaps in our knowledge. One intriguing aspect of molecular evolutionary genetics is the discovery of new phenomena and forces taking place at the molecular level that go beyond the realm of classical population genetics. Multigene families and organelle genomes are described in some detail to illustrate these uniquely molecular phenomena.

Chapter 9 covers the problem of quantitative genetics from an evolutionary perspective. A compelling argument for using quantitative genetics for the study of evolution is that adaptive evolution takes place at the level of the phenotype, and quantitative genetics provides the tools for understanding transmission of phenotypic traits. Theoretical quantitative genetics is given special importance by the paradoxes it raises in contrasting evolution at the levels of the phenotype and of the DNA sequence. Our understanding of the correspondence between phenotypic and molecular differentiation is very incomplete, and our understanding of the correspondence between the rates of morphological and molecular evolution is even less well developed. As in the preceding chapters, we hope that the student is left with a feeling that there is plenty of room for imaginative

work in this area. Population genetics is a field with a bright and expanding future.

ACKNOWLEDGMENTS

This book was greatly improved by the efforts of many people. The staff at Sinauer Associates did a splendid job assisting us with the revision. Nan Sinauer kept us on track, collecting and assembling dozens of computer files, revisions, FAXes, phone and email messages. Chris Small oversaw the page layout and managed the art program. Andy Sinauer played an essential role in having the book reviewed and in giving helpful advice as to level and length. We are grateful to Chip Aquadro, James Jacobson, Trudy Mackay, Roger Milkman, Tim Prout, Glenys Thomson, and Ken Weiss for their comments on the previous edition. Their insights greatly improved the presentation in this one.

Neither author could participate in writing a book such as this without a supportive, patient, sympathetic laboratory staff, able and willing to keep things running smoothly while the boss is at his word-processor doing a neuronal fusion with a silicon chip. We are grateful to all of them. In Dan Hartl's laboratory, the list includes Lara Forde, Elena Lozovskaya, Dmitry Nurminsky, E. Fidelma Boyd, Allan Lohe, Javare Nagaraju, David Sullivan, Charles Hill, Dmitri Petrov, Mark Siegal, Daniel De Aguiar, Carlos Bustamante, Jeffrey Townsend, Jorges Vieira, Christina Vieira, Isabel Beerman, Yunsun Nam, Elizabeth Stover, and Susan Yuknis. In Andy Clark's laboratory, the acknowledgments include Michael Abraham, Joe Canale, Manolis Dermitzakis, Chris Fucito, Cristina Gonzalez, Jen Jordan, Angela Lambert, Brian Lazzaro, J.P. Masly, Hamish Spencer, Sarah Tishkoff, Bridget Todd, Carrie Tupper, and Lei Wang.

CHAPTER 1

Genetic and Statistical Background

GENES • GENE EXPRESSION • PROBABILITY • ALLELE FREQUENCY ESTIMATES
STANDARD ERROR • MODELS • POPULATION GROWTH

T HE SCIENCE OF **population genetics** deals with Mendel's laws and other genetic principles as they affect entire populations of organisms. The organisms may be human beings, animals, plants, or microbes. The populations may be natural, agricultural, or experimental. The environment may be city, farm, field, or forest. The habitat may be soil, water, or air. Because of its wide-ranging purview, population genetics cuts across many fields of modern biology. A working knowledge has become essential in genetics, evolutionary biology, systematics, plant breeding, animal breeding, ecology, natural history, forestry, horticulture, conservation, and wildlife management. A basic understanding of population genetics is also useful in medicine, law, biotechnology, molecular biology, cell biology, sociology, and anthropology.

Population genetics also includes the study of the various forces that result in evolutionary changes in species through time. By defining the framework within which evolution takes place, the principles of population genetics are basic to a broad evolutionary perspective on biology. From an experimental point of view, evolution provides a wealth of testable hypotheses for all other branches of biology. Many oddities in biology become comprehensible in the light of evolution: they result from shared ancestry among organisms, and they attest to the unity of life on earth.

Practical applications of population genetics are extensive. Many applications, particularly those relevant to human beings, also have important implications in ethics and social policy. Among the applications of population genetics in medicine, agriculture, conservation, and research are:

- Genetic counseling of parents and other relatives of patients with hereditary diseases.
- Genetic mapping and identification of genes for disease susceptibility in human beings, including breast cancer, colon cancer, diabetes, schizophrenia, and so forth.
- Implications of population screening for carriers of disease genes, confidentiality of results, and maintenance of health insurability.
- Studies of the heritability of IQ score and its implications for affirmative action, welfare, and other social programs.
- Statistical interpretation of the significance of matching DNA types found between a suspect and a blood or semen sample from the scene of a crime.
- Design of studies to sample and preserve a record of genetic variation among human populations throughout the world.
- Improvement in the performance of domesticated animals and crop plants.
- Organization of mating programs for the preservation of endangered species in zoos and wildlife refuges.
- Sampling and preservation of germ plasms of potentially beneficial plants and animals that may soon vanish from the wild.
- Interpretation of differences in the nucleotide sequences of genes or amino acid sequences of proteins among members of the same or closely related species.

The genetic and statistical principles underlying population genetics are for the most part simple and straightforward, but it may be helpful to preface the discussion with a few key definitions and concepts.

GENE EXPRESSION AND GENE INTERACTION

Gene is a general term meaning, loosely, the physical entity transmitted from parent to offspring in reproduction that influences hereditary traits. Genes influence human traits such as hair color, eye color, skin color, height, weight, and various aspects of behavior—although most of these traits are also influenced more or less strongly by environment. Genes also determine the makeup of proteins such as hemoglobin, which carries oxygen in the red blood cells, or insulin, which is important in maintaining glucose balance in the blood. Genes can exist in different forms or states. For example, a gene for hemoglobin may exist in a normal form or in any one of a number of alternative forms that result in hemoglobin molecules that are more or less abnormal. These alternative forms of a gene are called **alleles**.

From a biochemical point of view, a gene corresponds to a region along a molecule of DNA (deoxyribonucleic acid). DNA is the genetic material. A molecule of DNA consists of two strands wound around each other in the form of a right-handed helix (the celebrated "double helix"). Each strand is a polymer of constituents called **nucleotides**, of which there are four, conventionally symbolized A, T, G, and C according to the nitrogen-rich base that each contains — either adenine (A), thymine (T), guanine (G), or cytosine (C). The paired strands are held together by weak chemical bonds (hydrogen bonds) that form between A and T at corresponding positions in opposite strands or between G and C at corresponding positions in opposite strands (Figure 1.1). Wherever one strand contains an A, the other across the way contains a T; and wherever one strand contains a G, the other across the way contains a C. Because of the pairing of complementary bases—A with T and G with C—a double-stranded DNA molecule contains an equal number of A and T nucleotides as well as an equal number of G and C nucleotides. DNA molecules can be very long. The DNA molecule in the bacterium *E. coli* is about 4.7 million base pairs, that in the largest chromosome in the fruit fly *Drosophila melanogaster* is about 65 million base pairs, and that in the largest human chromosome is about 230 million base pairs. Physical manipulation of such large molecules is impractical. In order to be studied, they must first be broken into smaller pieces.

Gene Expression

Most genes code for the polypeptide chains that constitute proteins. The code is the sequence of nucleotides along the DNA. In the decoding of the nucleotide sequence in DNA and also in the synthesis of proteins, several

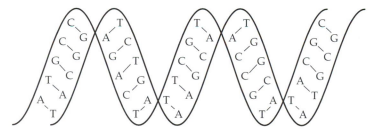

Figure 1.1   Genes are fundamental units of genetic information that correspond chemically to the sequence of nucleotides in a segment of DNA. A molecule of duplex DNA is composed of two intertwined strands, each of which consists of a long sequence of nucleotides. The strands are held together by pairing between the bases A and T in opposite strands and between the bases G and C in opposite strands. The short diagonal lines indicate the paired bases. There are 10 base pairs per turn of the double helix. A typical gene consists of hundreds of thousands of nucleotides, only a few of which are shown here.

types of RNA (ribonucleic acid) are essential. RNA is also a polymer of nucleotides, each of which carries a base. Three of the bases in RNA (A, C, and G) are the same as those in DNA. The fourth [uracil (U)] is different. When an RNA strand pairs with a complementary strand of DNA, U in the RNA pairs with A in the DNA. Hence, the base-pairing role of U in RNA is the same as that of T in DNA.

The essentials of gene expression in the cells of higher organisms (eukaryotes) are outlined in Figure 1.2. The coding regions of the DNA in a

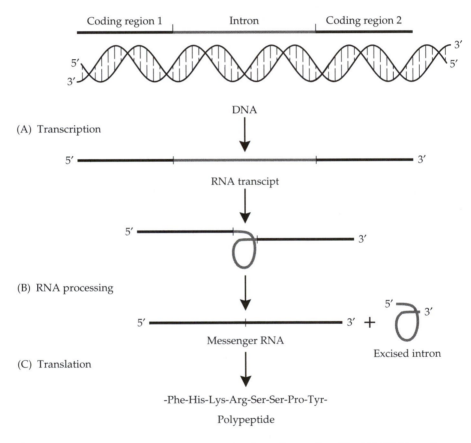

Figure 1.2   Processes in gene expression in eukaryotic cells. (A) DNA regions coding for the amino acids in a single polypeptide can be interrupted by non-coding regions (introns). (B) When the DNA is copied into RNA in transcription, both coding and noncoding regions are transcribed. However, the introns are removed from the transcript by processing. (C) In the messenger RNA, the coding regions are contiguous. The messenger RNA is translated to form the chain of linked amino acids constituting the polypeptide.

gene, which code for amino acids, are often interrupted by one or more non-coding regions known as intervening sequences or **introns**. In the first step in gene expression (transcription), a molecule of RNA is produced that is complementary in base sequence to one of the strands of DNA (Figure 1.2A). Every gene includes a regulatory region (sometimes more than one) that determines when transcription takes place, the types of cells in which it takes place, and the strand that is to be transcribed. Because of the base pairing rules, a DNA sequence—say, 3'–ATCG–5'—results in a complementary RNA sequence—in this example, 5'–UAGC–3'. Note that the DNA and RNA strands each have a **polarity** or directionality. The terms 5' and 3' refer to the polarity of the strands. The 5' end typically terminates with a free phosphate group and the 3' end typically terminates with a free hydroxyl group (—OH). When two strands of nucleic acid are paired, the polarity of each strand is opposite to that of the other. In the duplex DNA in Figure 1-2, for example, the left-to-right polarity of one strand is 5'-to-3', whereas the left-to-right polarity of the partner strand is 3'-to-5'. Similarly, in transcription, the template DNA strand has a left-to-right polarity of 3'-to-5', whereas the RNA transcript has the left-to-right polarity of 5'-to-3'. Because of the complementary base pairing between DNA and RNA nucleotides, the base-sequence code in DNA becomes converted into a base-sequence code in RNA. In transcription, the base sequence present in the introns is also faithfully copied into the base sequence of the RNA transcript.

The second step in gene expression in eukaryotes is RNA processing (Figure 1.2B). The beginning and end of the RNA transcript are chemically modified and the introns are removed by splicing (cutting and rejoining). RNA processing results in a molecule called **messenger RNA (mRNA)**, in which the coding regions have been made contiguous. The regions in the original RNA transcript that are retained in the mature mRNA are called **exons**. The central part of the mRNA contains the spliced exons that code for the amino acid sequence of a polypeptide chain. The mRNA also includes exons upstream and downstream from the protein-coding region. The upstream region is the *5' untranslated region* and the downstream region is the *3' untranslated region*.

The final step in gene expression is **translation**, in which the mRNA molecule combines with ribosomes and other types of RNA molecules in the cytoplasm to produce the final polypeptide (Figure 1.2C). In the coding region of the mRNA, each adjacent group of three nucleotides constitutes a separate coding group or **codon** that specifies which amino acid is to be incorporated into the polypeptide chain. The ribosome moves along the mRNA in steps of three nucleotides (codon by codon). As each new codon comes into place, the correct amino acid is brought into line and attached to the end of the growing chain of amino acids. New amino acids are added to the growing chain until a codon specifying "stop" is encountered. At this point synthesis of the chain of amino acids is finished and the polypeptide is released from the ribosome.

In prokaryotes, which includes bacteria and other organisms lacking a nucleus, gene expression is essentially identical to that in eukaryotes except for the absence of RNA processing. Genes in prokaryotes do not contain introns and so splicing is unnecessary. In prokaryotes, the original RNA transcript is used immediately as mRNA and translated into a polypeptide. Because there is no separate nucleus, translation in prokaryotes often begins immediately when the 5′ end of an RNA transcript comes off the DNA and even before transcription of the 3′ end of the same molecule has been completed.

The central role of RNA in gene expression is one of the oddities of biology that makes sense in the light of evolution. That gene expression is configured around RNA is a legacy of the earliest forms of life when RNA molecules served both as carriers of genetic information and as catalytic molecules. The role of RNA as carrier of genetic information was gradually replaced by DNA, and the role of RNA as catalytic molecules was gradually replaced by proteins. At every step along the way, as the RNA world evolved into the DNA world, the role of RNA was indispensable in the processes of information transfer and protein synthesis, and so the RNA intermediates became locked in place.

The Genetic Code

The **genetic code** is the list of all codons showing which amino acid each codon specifies. Table 1.1 shows the standard genetic code used in nuclear genes in most organisms. A few organisms and some cellular organelles, such as mitochondria, use slightly altered codes. The codons in Table 1.1 are those found in the mRNA. The amino acids are given by three-letter abbreviations as well as by conventional single-letter abbreviations. Codon AUG is the start codon in polypeptide synthesis; it specifies methionine (Met) at the beginning of the polypeptide as well as at internal positions. Three codons are stops that result in termination of polypeptide synthesis: UAA, UAG, and UGA. The genetic code is redundant in that most amino acids are specified by more than one codon. Most of the redundancy is in the third codon position.

A code for an amino acid is *twofold degenerate* if either of two sequences specifies the same amino acid. Twofold degenerate codes have the pattern ··Y or ··R, where ·· stands for the bases in codon positions 1 and 2. The symbol Y stands for any pyrimidine base (either U or C); the symbol R stands for any purine base (either A or G). For example, CAU and CAC both code for histidine (His), fitting the pattern CAY; and CAA and CAG both code for glutamine (Gln), fitting the pattern CAR. A code for an amino acid is *fourfold degenerate* if any of four sequences specifies the same amino acid; fourfold degenerate codes have the form ··N, where N means any nucleotide (U, C, A, or G). For example, GUU, GUC, GUA, and GUG all code for valine (Val),

---

TABLE 1.1 **THE STANDARD GENETIC CODE**

		Second nucleotide in codon		
	U	**C**	**A**	**G**
U	UUU UUC }Phe (F) UUA UUG }Leu (L)	UCU UCC UCA UCG }Ser (S)	UAU UAC }Tyr (Y) UAA Stop UAG Stop	UGU UGC }Cys (C) UGA Stop UGG Trp (W)
C	CUU CUC CUA CUG }Leu (L)	CCU CCC CCA CCG }Pro (P)	CAU CAC }His (H) CAA CAG }Gln (Q)	CGU CGC CGA CGG }Arg (R)
A	AUU AUC AUA }Ile (L) AUG Met (M*)	ACU ACC ACA ACG }Thr (T)	AAU AAC }Asn (N) AAA AAG }Lys (K)	AGU AGC }Ser (S) AGA AGG }Arg (R)
G	GUU GUC GUA GUG }Val (V)	GCU GCC GCA GCG }Ala (A)	GAU GAC }Asp (D) GAA GAG }Glu (E)	GGU GGC GGA GGG }Gly (G)

(*First nucleotide in codon* is the row label, *Second nucleotide in codon* is the column heading)

Note: Codons are nonoverlapping three-base sequences present in mRNA, each of which specifies an amino acid in a polypeptide chain or terminates synthesis ("Stop"). The full names of the amino acids are phenylalanine (Phe), leucine (Leu), isoleucine (Ile), methionine (Met), valine (Val), serine (Ser), proline (Pro), threonine (Thr), alanine (Ala), tyrosine (Tyr), histidine (His), glutamine (Gln), asparagine (Asn), lysine (Lys), aspartic acid (Asp), glutamic acid (Glu), cysteine (Cys), tryptophan (Trp), arginine (Arg), and glycine (Gly).

which fits the pattern GUN. Note in Table 1.1 that the code for isoleucine is *threefold degenerate* and those for leucine, arginine, and serine are each *sixfold degenerate*.

The codons for amino acids are not used randomly in proteins. There are preferred codons for amino acids that differ from one gene to the next and from one organism to another. Codon preferences exist even within redundancy classes. In *Drosophila*, for example, among codons for histidine, CAC is used more than CAU in a ratio of about 2 : 1. Similarly, among codons for glutamine, CAG is used more than CAA in a ratio of about 3 : 1. Another example of nonrandom codon usage is the AUA codon for isoleucine, which tends to be avoided in most proteins in most organisms. In *Drosophila*, AUU and AUC are used more than AUA in a ratio of about 10 : 1. One evolutionary hypothesis that explains the avoidance of AUA is that, because of the degeneracy of the genetic code, the AUA codon might sometimes be translated as AUG, which codes for methionine. Because methionine is likely to change protein structure radically, the mistranslation would be a costly mistake. Through evolutionary time, one by one, the AUA codons in a messenger RNA become replaced with AUU or AUC, minimizing this type of misincorporation error. This misincorporation hypothesis for AUA codon avoidance has not been tested, but it is testable.

Alleles

Alternative alleles of a gene differ in their sequence of nucleotides (Figure 1.3). For example, where one allele has a T-A base pair in the DNA, another may have a C-G base pair at the same position. Because of redundancy in the code, not all nucleotide substitutions result in a replacement of one amino acid for another. In Figure 1.3B, for example, if a mutation at the third position in the second codon (asterisk) changes one pyrimidine into the other, the new codon still codes for histidine. On the other hand, some nucleotide substitutions at the third position do result in amino acid replacements. For example, in Figure 1.3C, if the third position in the second codon changes from a pyrimidine to a purine, the codon changes from one for histidine to one for glutamine. Most nucleotide substitutions at codon positions one and two result in amino acid replacements (Figure 1.2D).

Not all alleles differ by a mere nucleotide substitution. Relative to the typical or **wildtype** allele, some alleles may have a deletion of a number of nucleotide pairs or an insertion into the DNA molecule. The number of nucleotides deleted or inserted may be small (as few as one nucleotide pair) or large. Some insertions are thousands of nucleotide pairs in size. Many large insertions result from the activity of **transposable elements**, which are specialized sequences of DNA able to replicate and insert at novel positions virtually anywhere in the DNA of the organism in which they are present. Alleles also may differ in the number of copies of short sequences present in tandem arrays in the DNA. For example, near many genes in human beings are tandem copies of dinucleotides, such as 5′–CACACACA . . . –3′. Such a repeating sequence is symbolized as (5′–CA–3′)n. The number of copies (n) of the dinucleotide repeat often range from fewer than ten to hundreds, and the number of copies may differ dramatically from one allele to the next. Some alleles even differ from wildtype in having an inversion of the nucleotide sequence in a region of DNA.

Genotype and Phenotype

Within a living cell, genes are arranged in linear order along microscopic threadlike bodies called **chromosomes**. A typical chromosome may contain

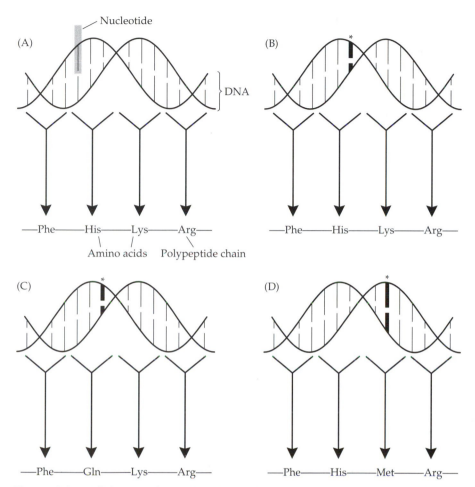

Nucleotide

(A)

DNA

—Phe——His——Lys——Arg—

Amino acids   Polypeptide chain

(B)

—Phe——His——Lys——Arg—

(C)

—Phe——Gln——Lys——Arg—

(D)

—Phe——His——Met——Arg—

Figure 1.3   Alleles are alternative forms of a gene. (A) The arrows show how the genetic information in a portion of the nucleotide sequence of DNA specifies the amino acid sequence in a portion of a polypeptide. Each group of three adjacent nucleotides corresponds to one amino acid in the polypeptide. (B, C, D) Substitution of one nucleotide for another in the DNA (indicated by the asterisks and heavy lines) can result in the replacement of one amino acid for another in the polypeptide.

several thousand genes. The position of a gene along a chromosome is called the **locus** of the gene. In most higher organisms, each cell contains two copies of each type of chromosome. Such organisms, in which the chromosomes are present in pairs, are said to be **diploid**. In each pair of chromosomes, one member is inherited from the mother through the egg and the other is inherited from the father through the sperm. At every locus, therefore, diploid organisms contain two alleles, one each at corresponding positions in the maternal and paternal chromosomes. If the two alleles at a locus are chemically identical (in the sense of having the same nucleotide sequence along the DNA), the organism is said to be **homozygous** at the locus under consideration; if the two alleles at a locus are chemically different, the organism is said to be **heterozygous** at the locus. The term *gene* is a general term usually used in the sense of *locus*.

Geneticists make a fundamental distinction between the genetic constitution of an organism and the physical or biochemical attributes of the organism. The genetic constitution of an organism is called the **genotype**; genotype thus refers to the particular alleles present in an organism at all loci that affect the trait in question. For example, if a trait is influenced by two genes, each with two alleles, then there are nine possible genotypes, as follows:

$$AA\,;BB \quad AA\,;Bb \quad AA\,;bb$$

$$Aa\,;BB \quad Aa\,;Bb \quad Aa\,;bb$$

$$aa\,;BB \quad aa\,;Bb \quad aa\,;bb$$

where A and a refer to the alleles of the first gene and B and b refer to the alleles of the second gene. In some cases when the genes are **linked** (located in the same chromosome), it is sometimes necessary to distinguish between the genotypes AB/ab and Ab/aB, in which case there are ten possible genotypes.

In contrast to genotype, the physical expression of a genotype is called the **phenotype**. Examples of phenotypes include hair color, eye color, height, weight, number of kernels on an ear of corn, number of eggs laid by a hen, and round versus wrinkled pea seeds. The distinction between the genetic constitution of an organism (genotype) and the physical or biochemical attributes of the organism (phenotype) is particularly important in cases in which the environment can affect the trait; in such cases, two organisms with the same genotype can nevertheless have different phenotypes because of differences in the environment. Conversely, two organisms with the same phenotype can have different genotypes.

PROBLEM 1.1   If a gene in a diploid organism has m alternative alleles, show that the number of possible genotypes equals $m(m + 1)/2$.

ANSWER: Consider first the heterozygotes. There are m ways of choosing the first allele and, having done that, there are $m-1$ ways of choosing a different second allele. Altogether, there are $m(m-1)/2$ different heterozygotes. The division by 2 is necessary because, for each heterozygote—say, A_iA_j—it makes no difference whether A_i was chosen first and A_j second or the other way around. In addition to the heterozygotes, there are m possible homozygotes. Hence, the total number of diploid genotypes equals $[m(m-1)/2] + m = m(m+1)/2$.

Dominance and Gene Interaction

Whether each genotype has a single, unique expression of the trait depends on the manner in which the alleles of a gene interact in development. For the alleles of one gene, **dominance** refers to the concealment of the presence of one allele by the strong phenotypic effects of another. For example, with two alleles there are three possible genotypes:

$$AA \quad Aa \quad aa$$

Several types of dominance are distinguished and exemplified in the following examples:

- *Complete dominance:* A is completely dominant to a if the phenotypes of AA and Aa cannot be distinguished.
- *Incomplete dominance:* A shows incomplete dominance with respect to a if the phenotype of Aa is intermediate between that of AA and that of aa. This situation is also referred to as partial dominance or intermediate dominance. When the phenotype can be measured on a quantitative scale, for example, the number of kernels on an ear of corn, and the phenotype of Aa is exactly the average between that of AA and that of aa, then the alleles are said to be additive alleles and the type of dominance is sometimes called semidominance.
- *Codominance:* A and a are codominant if the products of both alleles can be detected in Aa heterozygotes. Many alleles are codominant at the level of their protein products because two different forms of the polypeptide, encoded by A and a, can be detected in heterozygotes. At the level of the DNA sequences, all alleles differing in DNA sequence are codominant.

It is important to note that dominance is not a characteristic of alleles so much as a characteristic of the manner in which the phenotype is examined. An allele may show complete dominance if the phenotype is examined in one way, no dominance if examined in another, and codominance if examined in still another. For example, the allele for round pea seeds W studied by Gregor Mendel is completely dominant to that for wrinkled seeds w when the phenotype "round" versus "wrinkled" is examined. The genetic defect in wrinkled seeds is the absence of an enzyme needed for the synthesis of a branched-chain form of starch. Microscopic examination reveals subtle differences in the form of the starch grains in seeds of the three genotypes: WW seeds contain large, well-rounded starch grains, retain water and shrink uniformly as they ripen, so the seeds do not become wrinkled; ww seeds lack the branched-chain starch and are irregular in shape because the ripening seeds lose water more rapidly and shrink unevenly. However, heterozygous Ww seeds have starch grains that are intermediate in shape even though the seeds shrink uniformly and show no wrinkling. Therefore, at the level of the starch grains, there is incomplete dominance of W and w because the starch grains in the heterozygotes are intermediate between the two homozygotes. Furthermore, the difference in DNA sequence between W and w can readily be detected with modern methods, so that W and w are codominant at the level of DNA sequence.

For traits affected by more than one gene, the relation between genotype and phenotype depends not only on the degree of dominance of the alleles of each gene but also on the type of interaction between the genes in development. For example, suppose that the trait in question is degree of pigmentation and that pigmentation is determined by two alleles of each of two genes, say, A, a and B, b. Suppose further that the total amount of pigment in an organism results from the total number of A and B alleles present, each of which adds a single unit of pigmentation to the phenotype. Then, as shown in Table 1.2, there are only five possible levels of pigmentation (0 through 4) and genotypes $aa\ BB$, $Aa\ Bb$, and $AA\ bb$ all have the same phenotype. Because each uppercase allele adds the same quantity to the total phenotype, the type of gene interaction in Table 1.2 is said to be *additive*.

Segregation and Recombination

The essential mechanism of inheritance was established by Gregor Mendel (1822–1884) in experiments with garden peas carried out in the years 1856 to 1863 in a small garden plot next to the monastery in which he lived. Mendel showed that the alleles of each gene **segregate** from one another in the formation of reproductive cells or **gametes**. Because of segregation, heterozygous genotypes form equal numbers of gametes containing each allele.

TABLE 1.2   A MODEL OF THE ADDITIVE GENE ACTION[a]

Genotype			Amount of pigmentation[b]
AA BB			4
Aa BB,	AA Bb		3
aa BB,	Aa Bb,	AA bb	2
Aa bb,	aa Bb		1
aa bb			0

[a]At left are shown the nine possible genotypes of two genes with two alleles of each gene. At right is shown the amount of pigmentation expected in each genotype when it is assumed that each allele designated by an uppercase letter is responsible for producing a certain amount of pigment.
[b]Measured as an increase in pigmentation over that in *aa bb* genotypes.

Furthermore, because gametes unite at random in fertilization, the following are the results of simple Mendelian segregation:

- *AA* × *AA* matings produce all *AA* progeny.
- *AA* × *Aa* matings produce ½ *AA* and ½ *Aa* progeny.
- *AA* × *aa* matings produce all *Aa* progeny.
- *Aa* × *Aa* matings produce ¼ *AA*, ½ *Aa*, and ¼ *aa* progeny.
- *Aa* × *aa* matings produce ½ *Aa* and ½ *aa* progeny.
- *aa* × *aa* matings produce all *aa* progeny.

The physical basis of Mendelian segregation is that the maternal and paternal pairs of chromosomes are separated into different cells in the formation of gametes. Prior to their separation, the maternal and paternal chromosomes associate intimately all along their length and alleles may be interchanged in the process of **recombination** (Figure 1.4). The interchange of parts takes place after the chromosomes have replicated, and only two of the four chromosome strands participate in any one exchange. Recombination results in the creation of allele combinations different from either parental chromosome. In Figure 1.4, the *A b* and *a B* combinations are recombinant, whereas the *A B* and *a b* combinations are parental (nonrecombinant). Therefore, a single exchange between parental chromosomes results in two recombinant and two nonrecombinant gametes.

In organisms with an XX–XY chromosomal mechanism of sex determination, Mendelian segregation randomizes the sex ratio at fertilization. In mammals and many other animals, sex is determined by sex chromosomes: males have an X and a Y chromosome, and females have two X chromosomes. In males, the X and Y chromosomes segregate, yielding equal proportions of

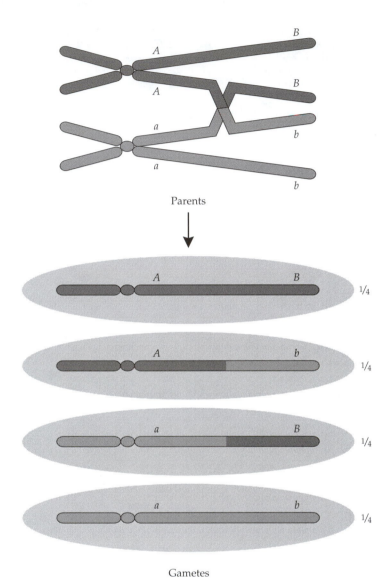

Parents

Gametes

Figure 1.4   Recombination results from a physical interchange of parts between chromosomes. New combinations of alleles are created that differ from either parental chromosome. The physical interchange of parts takes place in gamete formation after the chromosomes have replicated, and only two of the four chromosome strands participate in any one exchange.

X-bearing and Y-bearing sperm. If both types of sperm are equally able to fertilize eggs, then random union of sperm with eggs yields $\frac{1}{2}$ XX (female) and $\frac{1}{2}$ XY (male) chromosome constitutions.

PROBABILITY IN POPULATION GENETICS

The basic concepts of probability needed for elementary population genetics are quite straightforward. They will be introduced with the concrete example of genetic segregation in Figure 1.5, which deals with the progeny of the mating

(A) Addition rule

Mating: $Aa \times Aa$

Offspring: $\frac{1}{4}AA + \frac{1}{2}Aa + \frac{1}{4}aa$

$A-$ means "Offspring either AA or Aa"

$\Pr(A-) = \frac{1}{4}(AA) + \frac{1}{2}(Aa) = \frac{3}{4}$

(B) Multiplication rule

| | Birth Order | | | |
Sibship	1	2	3	Probability
1	$A-$	$A-$	$A-$	$\frac{3}{4} \times \frac{3}{4} \times \frac{3}{4} = \frac{27}{64}$
2	$A-$	$A-$	aa	$\frac{3}{4} \times \frac{3}{4} \times \frac{1}{4} = \frac{9}{64}$
3	$A-$	aa	$A-$	$\frac{3}{4} \times \frac{1}{4} \times \frac{3}{4} = \frac{9}{64}$
4	aa	$A-$	$A-$	$\frac{1}{4} \times \frac{3}{4} \times \frac{3}{4} = \frac{9}{64}$
5	$A-$	aa	aa	$\frac{3}{4} \times \frac{1}{4} \times \frac{1}{4} = \frac{3}{64}$
6	aa	$A-$	aa	$\frac{1}{4} \times \frac{3}{4} \times \frac{1}{4} = \frac{3}{64}$
7	aa	aa	$A-$	$\frac{1}{4} \times \frac{1}{4} \times \frac{3}{4} = \frac{3}{64}$
8	aa	aa	aa	$\frac{1}{4} \times \frac{1}{4} \times \frac{1}{4} = \frac{1}{64}$

Figure 1.5 Basic concepts of probability illustrated by Mendelian segregation in the mating $Aa \times Aa$. The elementary outcomes of the mating are the possible genotypes of each progeny—AA, Aa, and aa—and these are realized with probabilities $\frac{1}{4}$, $\frac{1}{2}$, and $\frac{1}{4}$, respectively. (A) The compound event $A-$ consists of the two elementary outcomes AA and Aa, and the probability of $A-$ is the sum of the probabilities of these elementary outcomes (addition rule). (B) The possible distributions of genotypes $A-$ and aa in sibships of size three offspring. Successive births are independent, and so the probability of any sibship equals the product of the probabilities for each birth separately (multiplication rule).

$Aa \times Aa$. Considerations in probability always begin with an experiment of some kind. The experiment may be either a real experiment or a conceptual experiment. In Figure 1.5, it is a conceptual experiment in which Aa is crossed with Aa. In probability calculations, it is also necessary to define all possible outcomes of the experiment. The outcomes are called **elementary outcomes** because they are defined in such a way that, in any repetition of the experiment, one and only one of the elementary outcomes must be realized. For example, if we are interested in the genotypes among the progeny of the mating Aa, the possible elementary outcomes for each offspring are either AA, Aa, or aa. (Note that, in defining these as the elementary outcomes, we are ignoring the possibility of either A or a mutating to a novel allele.) To proceed further, we must assign to each elementary outcome a *probability*, a number between 0 and 1 that measures how much confidence we have that the outcome will be realized. The probabilities assigned to the outcomes are based on genetic reasoning, intuition, or experience. One requirement of the assigned probabilities is that the probabilities of all the elementary outcomes must add to 1; this is the mathematical consequence of requiring that one of the elementary outcomes must be realized. For example, if there are three elementary outcomes, and all are equally probable, then each has a probability of $1/3$. In Figure 1.5, the probabilities assigned to the elementary outcomes AA, Aa, and aa are $1/4$, $1/2$, and $1/4$, respectively, because these are the relative proportions of the three progeny genotypes expected from Mendelian segregation.

The Addition Rule

An outcome of a conceptual experiment is an **event**. The distinction between an event and an elementary outcome is that an event can include more than one elementary outcome. For example, in Figure 1.5A, the event "the offspring has at least one copy of the dominant A allele" consists of two elementary outcomes, namely, genotypes AA and Aa. This event may be symbolized $A-$, where the dash indicates that the unspecified allele may be either A or a. For events defined in terms of elementary outcomes, the probability of an event equals the sum of the probabilities of the elementary outcomes included in the event. In the present example,

$$\Pr(A-) = \Pr(AA) + \Pr(Aa) = 1/4 + 1/2 = 3/4$$

More generally, two events are **mutually exclusive** if they cannot be realized simultaneously. The **addition rule** states that, for mutually exclusive events, the probability that either one or the other is realized equals the sum of the probabilities of the separate events.

The Multiplication Rule

Figure 1.5B shows all possible genotypes of sibships of three offspring from the mating $Aa \times Aa$, with each offspring classified as $A-$ versus aa. (A **sibship** is a group of brothers and sisters.) The probability of $A-$ in any particular birth is $3/4$ and that of aa is $1/4$. The probabilities at the right are the overall probabilities for each of the sibships. They are obtained by multiplication of the probability for each birth because successive births are **independent**, which means that the genotype of any birth has no effect on the genotype of any other birth. Because of the independence, among the $3/4$ of the sibships with $A-$ in the first birth, $3/4$ will have $A-$ in the second birth, and among the $3/4 \times 3/4$ of the sibships with $A-$ in the first two births, $3/4$ will have $A-$ in the third birth. Therefore, the overall probability of three $A-$ births is $3/4 \times 3/4 \times 3/4$. The reasoning for the other types of sibships is similar. More generally, the **multiplication rule** states that, whenever two events are independent, the probability of their joint realization is the product of the probabilities of their being realized separately.

Repeated Trials

The sibships in Figure 1.5B are an example of **repeated trials** of a conceptual experiment. Repeated trials are encountered frequently in probability. They govern tosses of a coin or dice, deals of cards, successive spins of a roulette wheel, and so forth. Repeated trials are also important in population genetics because successive offspring of a mating are independent events and thus repeated trials. Furthermore, it is apparent from Figure 1.5B that the different birth orders are mutually exclusive: any sibship can have one and only one birth order of $A-$ or aa. Because the birth orders are mutually exclusive, their probabilities may be combined by the addition rule. Hence, the composite events below have the following probabilities:

$$\text{Pr(two } A- \text{ and one } aa) = 9/64 + 9/64 + 9/64 = 27/64$$

$$\text{Pr(one } A- \text{ and two } aa) = 3/64 + 3/64 + 3/64 = 9/64$$

Note in Figure 1.5B that, when the sibships with the same number of $A-$ and aa genotypes are combined, the overall probabilities are given by successive terms in the expansion of:

$$(3/4\, A- + 1/4\, aa)^3 = 1 \times (3/4)^3 \qquad\qquad A-\ A-\ A-$$
$$+ 3 \times (3/4)^2 (1/4)^1 \qquad\qquad A-\ A-\ aa$$
$$+ 3 \times (3/4)^1 (1/4)^2 \qquad\qquad A-\ aa\ aa$$
$$+ 1 \times (1/4)^3 \qquad\qquad aa\ aa\ aa$$

The coefficients $1 : 3 : 3 : 1$ are the number of combinations in which each triad of genotypes can be born: 1 for $A-\ A-\ A-$, 3 for $A-\ A-\ aa$ (because the aa genotype can be born either first, second, or third), and so forth. Each power of $3/4$ and $1/4$ is the probability that any one of the birth orders will be realized;

for example, $(\frac{3}{4})^2(\frac{1}{4})^1$ is the probability that any sibship with two $A-$ and one aa genotype will be realized.

In all cases of repeated and independent trials, the overall probabilities are given by analogous expansions. Suppose that any one trial may result in either of two mutually exclusive events, A or B, and that the probability of event A is p and that of event B is q (with $p + q = 1$). Among a total of n independent trials, what is the probability that A is realized exactly r times and B is realized exactly $n - r$ times? By the multiplication rule, any particular combination of r As and $n - r$ Bs has a probability $p^r q^{n-r}$. Deducing the total number of combinations of r As and $n - r$ Bs is a little less obvious, but it is given by the coefficient of the term $p^r q^{n-r}$ in the expansion of $(p + q)^n$, which equals

$$\frac{n!}{r!(n-r)!} \qquad\qquad 1.1$$

where the exclamation point means the **factorial**, the product of all integers from 1 through the number in question. For example, $n! = 1 \times 2 \times 3 \times \cdots \times n$. For consistency, the number $0!$ is defined as $0! = 1$.

Equation 1.1 is often called a **binomial coefficient** because it arises in the expansion of the two terms $(p + q)^n$. To understand the reason why Equation 1.1 yields the correct number of combinations of r As and $(n - r)$ Bs, first consider what the $n!$ means. It is the total number of ways that any set of n objects can be arranged in order. There are n ways to choose the first object and, having chosen the first, $n - 1$ ways to choose the second and, having chosen the first two, $n - 2$ ways to choose the third, and so on, yielding $n \times (n - 1) \times (n - 2) \times \cdots \times 1 = n!$. Furthermore, for each arrangement of n objects of which r are As and $(n - r)$ are Bs, there are $r!$ ways to arrange the As among themselves and $(n - r)!$ ways to arrange the Bs among themselves, for a total of $r! \times (n - r)!$ arrangements. Because each of the $n!$ combinations of r As and $(n - r)$ Bs includes $r! \times (n - r)!$ equivalent arrangements of the As and Bs, the total number of different arrangements of r As and $(n - r)$ Bs equals the ratio given in Equation 1.1.

Equation 1.1 gives the number of different arrangements of r As and $(n - r)$ Bs. Each arrangement has a probability given by $p^r q^{n-r}$. Therefore, using the addition rule, the probability that n repeated trials yields r realizations of A and $(n - r)$ realizations of B equals

$$\frac{n!}{r!(n-r)!} p^r q^{n-r} \qquad\qquad 1.2$$

As an example of the use of Equation 1.2, consider the probability that a sibship of 12 offspring from the mating $Aa \times Aa$ perfectly matches the expected Mendelian ratio of 9 $A-$ and 3 aa. In this case, $p = \frac{3}{4}$, $q = \frac{1}{4}$, $n = 12$, $r = 9$, and $n - r = 3$. The required probability from Equation 1.2 is therefore

$$\frac{12!}{9!3!}\left(\frac{3}{4}\right)^9\left(\frac{1}{4}\right)^3 = 220 \times 0.0751 \times 0.0156 = 0.258$$

The implication of this calculation is that, whereas the "expected" ratio is 9 $A-$: 3 aa, only a little more than 25% of such sibships actually have the expected distribution.

PROBLEM 1.2   Suppose that a society decided to limit the number of males by passing a law denying further reproduction to any woman who gives birth to a male child. Given a ratio of males to females at birth of 1 : 1, how would such a law affect the sex ratio? Suppose further that, in practice, any woman who has a female child voluntarily terminates further reproduction with probability p. In this case, what is the proportion of males in sibships of size n?

ANSWER   The law would have no effect on the sex ratio. To understand why, consider the first birth across the entire population. The sex ratio among these offspring must be 50% males. Consider now the second birth. The sex ratio among these offspring must also be 50% males. Indeed, the sex ratio in any birth must be 50% males, and so this is the sex ratio in the population of births as a whole. In regard to the second part of the problem, note that sibships of size n can be separated into two classes: those in which the final birth is a male (and the mother's further reproduction is denied) and those in which the final birth is a girl (in which the mother voluntarily stops reproducing with probability p). These types of sibships occur in the ratio $\frac{1}{2}$: $p/2$, which means in the proportions $1/(1 + p)$ and $p/(1 + p)$, respectively. The first type of sibship has a proportion of males of $1/n$ and the second has a proportion of males of 0. Hence, the proportion of males as a function of sibship size equals $(1/n) \times [1/(1 + p)] + 0 \times [p/(1 + p)] = 1/n(1 + p)$. Note that, for $p = 0$, the proportion of males as a function of sibship size decreases according to the series 1, $\frac{1}{2}$, $\frac{1}{3}$, $\frac{1}{4}$, ... Nevertheless, the sex ratio in the population as a whole equals $\frac{1}{2}$ for this and any other value of p.

PHENOTYPIC DIVERSITY AND GENETIC VARIATION

One of the universal attributes of natural populations is that organisms differ in phenotype with respect to many traits. Phenotypic diversity in many traits is impressive even with the most casual observation. Among human beings, for example, there is diversity with respect to height, weight, body conformation, hair color and texture, skin color, eye color, and many other physical and psychological attributes or skills. Population genetics must deal with this phenotypic diversity, and especially with that portion of the diversity that is caused by differences in genotype. In particular, the field of population genetics has set for itself the tasks of determining how much genetic variation exists in natural populations and of explaining its origin, maintenance, and evolutionary importance. Genetic variation, in the form of multiple alleles of many genes, exists in most natural populations. In most sexually reproducing populations, no two organisms (barring identical twins or other multiple identical births) can be expected to have the same genotype for all genes. Thus, it becomes important to describe how alleles in natural populations are organized into genotypes—to determine, for example, whether alleles of the same or different genes are associated at random.

Allele Frequencies in Populations

Much of the phenotypic variation in natural populations does not yield simple Mendelian segregation ratios such as 1 : 1 or 3 : 1 in pedigrees. Some differences in phenotype are environmental in origin and so are not expected to show Mendelian segregation. However, simple Mendelian segregation is not usually observed even for traits whose expression is influenced more or less strongly by genetic factors. Although the underlying genetic factors do segregate in pedigrees in Mendelian fashion, the segregation is concealed by several complications. First, environmental effects on the trait may be strong enough to mask the genetic segregation. Second, genetic effects on many traits are determined by the joint effects of the alleles of two or more genes, and the segregation of any one gene in a pedigree may be obscured by the segregation of others.

On the other hand, some phenotypic diversity in populations does show simple Mendelian segregation. In the snapdragon *Antirrhinum majus,* for example, whether the flower color is red, pink, or white is determined by the alleles *I* and *i* of a single gene. The genotypes *II, Ii,* and *ii* have red, pink, and white flowers, respectively, an example of incomplete dominance.

Populations containing both the *I* and *i* alleles will include plants whose flowers are red (*II*), pink (*Ii*), or white (*ii*) in proportions determined by the allele frequencies of the *I* and *i* alleles in the population as well as by the manner in which the alleles are united in fertilization. By the **allele frequency** of a specified allele, we mean the proportion of all alleles of the gene that are of the specified type. To take a hypothetical example, suppose 400 members of a population were classified as to flower color and the finding was: 165 red, 190 pink, and 45 white. Because the flower color reveals the genotype, we may infer that the sample of 400 includes 165 *II*, 190 *Ii*, and 45 *ii* genotypes. The observed numbers of *I* and *i* alleles are therefore:

$$I: 2 \times 165 + 190 = 520$$

$$i: 190 + 2 \times 45 = 280$$

The factors of 2 are included for the homozygous genotypes because each *II* genotype contains two *I* alleles and each *ii* genotype contains two *i* alleles. The total number of alleles in the sample equals $2 \times 400 = 800$. Therefore, if we let p represent the frequency of the *I* allele and q represent the frequency of the *i* allele (with $p + q = 1$ because these are the only alleles of the gene in question), then we can estimate p and q from the observations as:

$$\hat{p} = 520/800 = 0.65$$

$$\hat{q} = 280/800 = 0.35$$

Note that, if the *I* and *i* alleles were combined into genotypes at random, the expected frequencies of three genotypes can be calculated from the rule for repeated trials by expanding the binomial $(p\,I + q\,i)^2 = p^2\,II + 2pq\,Ii + q^2\,ii$. Therefore, assuming random combination into genotypes, the expected numbers of the three genotypes are:

$$II: (0.65)^2 \times 400 = 169$$

$$Ii: 2 \times 0.65 \times 0.35 = 182$$

$$ii: (0.35)^2 \times 400 = 49$$

Hence, the observed numbers in this hypothetical population are very close to those expected with random combinations of alleles. The proportions p^2, $2pq$, and q^2 for the three genotypes when two alleles are combined at random constitutes the **Hardy-Weinberg principle**, which is one of the basic principles in population genetics. The Hardy-Weinberg principle is discussed in detail in Chapter 2.

PROBLEM 1.3   Suppose that a random sample of 400 snapdragons from a population includes 185 red, 150 pink, and 65 white. Estimate the allele frequency p of *I* and q of *i*. Assuming random combinations of alleles in the genotypes, what are the expected numbers of the three genotypes? Do the observed data seem to fit the expectations?

ANSWER   Among the total of 800 alleles, the observed number of I alleles is $2 \times 185 + 150 = 520$ and that of i alleles is $150 + 2 \times 65 = 280$. Therefore, $\hat{p} = 520/800 = 0.65$ and $\hat{q} = 280/800 = 0.35$. Note that the estimated allele frequencies are the same as above, even though the observed numbers of the genotypes are different. With random combinations of alleles in the genotypes, the expected numbers are again 169 red, 182 pink, and 49 white. Compared to the observations, there appear to be too many homozygous genotypes and too few heterozygous genotypes. (A statistical method for deciding whether the fit is satisfactory or not is discussed in Chapter 2.)

Parameters and Estimates

In the discussion of flower color in snapdragons, we made a subtle distinction between the actual allele frequency of the I allele (designated p) and the estimated allele frequency of the I allele (symbolized $\hat{p}$). The distinction is necessary whenever an experimenter makes inferences about an entire population from an examination of a random sample from the population. Quantities used in describing entire populations are **parameters**. In the snapdragon example, the parameter of interest is the allele frequency p of I in the entire population. Because we only have access to a sample of 400 organisms from the population, the true value of p is unknown. The best we can do is make an estimate of p based on a sample, hoping that the sample is representative of the population as a whole. The estimate obtained from the sample is designated $\hat{p}$ to emphasize that it is an estimate rather than the true value. In this book, whenever it is necessary to distinguish parameters from their estimates, we use unembellished symbols for parameters (for example p for the unknown frequency of an allele in a specified population) and the same symbol with a circumflex for the estimated value (in this example $\hat{p}$).

The Standard Error of an Estimate

The distinction between a parameter and an estimate is important because different samples may yield different values of the estimate for the same reason that different sibships may yield different segregation ratios, namely, chance variation from one repeated trial to the next. The estimation of an allele frequency can be treated as repeated trials by supposing that the alleles are sampled at random, one by one, from a very large population. In the snapdragon example, there are 800 alleles sampled. If the allele frequency of I has the true value $p = 0.65$, then the repeated-trials interpretation implies that all possible outcomes of 800 trials have probabilities given by successive terms in the expansion of $(0.65\ I + 0.35\ i)^{800}$. This is not an expansion that one would want to do by hand, but the binomial expression makes evident the underlying random-sampling process that accounts for variation in the estimate of $\hat{p}$ from one sample of 800 alleles to the next.

Unless p is quite close to 0 or quite close to 1, there is a convenient approximation to the binomial expansion $(p\ I + q\ i)^n$, where n is the number of alleles sampled. As n becomes large, the distribution of $\hat{p}$ approaches the familiar bell-shaped curve called the *normal distribution*. The normal distribution features prominently in the analysis of traits determined jointly by multiple genetic and environmental factors and it is discussed in detail in that context (Chapter 9). For present purposes, it is sufficient to note that the degree to which the values of $\hat{p}$ are clustered around the overall average depends on a quantity called the **standard error**:

$$s = \sqrt{\frac{\hat{p}\hat{q}}{n}} \qquad\qquad 1.3$$

where $\hat{q} = 1 - \hat{p}$. If the sampling and estimation of p were repeated many times using the same population, then the values of $\hat{p}$ would be expected to be clustered symmetrically around p according to the standard error as follows:

- Approximately 68% of the estimates $\hat{p}$ lie within plus or minus one standard error of p.
- Approximately 95% of the estimates $\hat{p}$ lie within two standard errors of p.
- Approximately 99.7% of the estimates $\hat{p}$ lie within three standard errors of p.

To put the matter in another way, with repeated sampling, 32% of the estimates would be expected to differ from the true value by more than one standard error, 5% by more than two standard errors, and only 0.3% by more than three standard errors.

As an illustration of the variation among repeated estimates of p, Figure 1.6 shows the values of $\hat{p}$ obtained in 100 repetitions of the experiment of sampling 800 alleles from a large population in which the true allele frequency is $p = 0.65$. Each of the 100 samples was created by computer simulation using a random-number generator that yielded a 1 with probability 0.65 and a 0 with probability 0.35. For each sample of 800, therefore, the estimate $\hat{p}$ equals the number of 1s in the sample divided by 800. As is evident in Figure 1.6, the distribution of $\hat{p}$ values is more or less bell-shaped but not

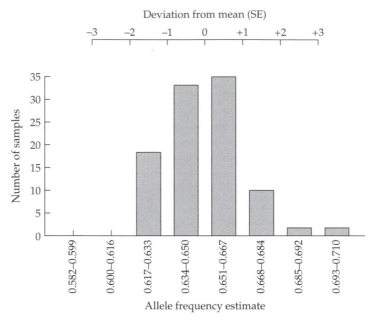

Figure 1.6   Estimates of allele frequency based on 100 samples, each of size 400 diploid organisms, from a population in which the actual allele frequency is 0.65. The standard error equals 0.017, and the distribution of the estimates is very close to the bell-shaped distribution expected theoretically. The scale across the top gives the ranges of the estimates as multiples of the standard error.

exactly so because it is based on only 100 samples rather than an infinite number. The overall mean $\hat{p}$ from all 100 samples combined (80,000 observations) equals 0.6492, which is very close to the true value of p. Furthermore, the distribution of the estimates fits the predictions based on the standard error quite well.

To apply Equation 1.3 to the data in Figure 1.6, note first that $p = 0.65$ with $n = 800$, and so s in Equation 1.3 equals $\sqrt{[(0.65 \times 0.35)/800]} = 0.017$. Because 68% of the samples are expected to yield values of $\hat{p}$ in the range $p \pm s$, and because the expected distribution is symmetrical, 34 of the values in Figure 1.6 are expected in the range $p - s$ to p (0.633 – 0.650) and 34 in the range p to $p + s$ (0.650 – 0.667); the actual numbers are 33 in the first interval and 35 in the second. By the same reasoning, 95% of the values should lie in the range $p \pm 2s$, or 47.5% on each side of the mean; because 34% of the values on either side of the mean are in the range $p \pm s$, the implication is that 47.5 – 34 or 13.5% of the values should lie in the range $p - 2s$ to $p - s$ and 13.5% should lie in the range $p + s$ to $p + 2s$. For the data in Figure 1.6, these ranges are 0.616 – 0.633 and 0.667 – 0.684; the actual number in each interval is 18 and 10, respectively, as against the theoretical 13.5 in each. Likewise, the standard error predicts that 0.3% of the samples will deviate by more than $3s$ from the mean, as compared with the observed 2.

Estimates and their standard errors are often presented as $\hat{p} \pm s$, or 0.65 ± 0.017 in the present example. The 68%, 95%, and 99.7% cutoffs for ± 1, ± 2, and ± 3 standard errors provide one manner in which the reliability of an estimate may be interpreted. Estimates may also be presented alternatively in terms of a range called a **confidence interval**, which expresses a degree of confidence that the true value of a parameter lies in some specified interval. The most frequently encountered confidence interval is the 95% confidence interval, defined as the interval $(\hat{p} - 2s, \hat{p} + 2s)$. Because 95% of repeated samples are expected to yield estimates in a range $\pm 2s$ around the true mean, then 95% of the time the interval $(\hat{p} - 2s) - (\hat{p} + 2s)$ is expected to include the true value of the parameter p. In the snapdragon example with $\hat{p} = 0.65$ and $s = 0.017$, the 95% confidence interval is 0.616–0.684.

PROBLEM 1.4   The MN blood groups in human beings are determined by two alleles of a single gene, designated M and N. Each allele results in the production of a different type of polysaccharide molecule on the surface of red blood cells, which can be distinguished by means of appropriate chemical reagents. The types of molecules corresponding to the M and N alleles are designated M and N, respectively. The M and N alleles are codominant; that is, genotype MM produces only the M substance and has blood group M, genotype NN produces only the N substance and has blood group N, and the heterozygous genotype MN produces both the M and N substances and has blood group MN. Among a sample of 1000 British people (Race and Sanger 1975), the observed numbers of each blood group were 298 M, 489 MN, and 213 N. Using these data, estimate the allele frequency p of the M allele and calculate its standard error. What are the 68%, 95%, and 99.7% confidence intervals for p?

ANSWER   Because each genotype has a unique phenotype, the sample contains $2 \times 298 + 489 = 1085$ M alleles, and so $\hat{p} = 1085/2000 = 0.5425$. The standard error $s = \sqrt{(0.5425)(10.5425)/2000} = 0.0111$. The 68%, 95%, and 99.7% confidence intervals for p are $\hat{p} \pm 1s$, $2s$, and $3s$, respectively, and so the confidence intervals are $0.5314 - 0.5536$ (68%), $0.5202 - 0.5647$ (95%), and $0.5092 - 0.5758$ (97.5%).

MODELS IN POPULATION GENETICS

Population geneticists must contend with factors such as population size, patterns of mating, geographical distribution of organisms, mutation, migration, and natural selection. Although we wish ultimately to understand the combined effects of all these factors and more, the factors are so numerous and interact in such complex ways that they cannot usually be grasped all at once. Simpler situations are therefore devised, situations in which a few identifiable factors are the most important ones and others can be neglected. An intentional simplification of a complex situation is a **model**. There are several types of models, each designed to eliminate extraneous detail in order to focus attention on the essentials. Some models are experimental. An experimental model may consist of a laboratory experiment with population cages of *Drosophila* or growing cultures of bacteria. An experimental model may also consist of observations of natural populations in particular locations or at particular times in which evolutionary forces of interest may be presumed to be present. Models of this type include the study of the origin and spread of insecticide resistance in insects or antibiotic resistance in bacteria.

A model may also be a conceptual simplification. Conceptual models have a number of uses. They require a concise statement of presumed mechanisms and interactions; they afford a framework for interpreting observations and setting research priorities; they enable extrapolation into the future or beyond the range of known parameters; and they suggest tests of consistency between theory and observation

A conceptual model may consist of verbal arguments logically linking a chain of hypothesis and deductions. Another type of conceptual model is a computer program that simulates the random component in a process or that calculates the values of changing quantities in a complex system based on prescribed numerical relations. An example of a computer model is the one for examining the result of repeated random sampling whose outcome is depicted in Figure 1.6. In population genetics, a kind of model frequently encountered is a **mathematical model**, which is a set of hypotheses that specifies the mathematical relations between measured or measurable quantities (the parameters) in a system or process. Mathematical models can be extremely useful:

- They express concisely the hypothesized quantitative relationships between parameters.
- They reveal which parameters are the most important in a system and thereby suggest critical experiments or observations.
- They serve as guides to the collection, organization, and interpretation of observed data.

- They make quantitative predictions about the behavior of a system that can, within limits, be confirmed or shown to be false.

The validity of any model must be tested by determining whether the hypotheses on which it is based and the predictions that grow out of it are consistent with observations.

A mathematical model is always simpler than the actual situation it is designed to elucidate. A model is supposed to be simple: If it is not simpler than the real situation, then it isn't a model. Models are simpler than real situations because many features of real life are intentionally ignored. To include every aspect of a complex system would make a model too complex and unwieldy. Construction of a model always requires a compromise between realism and manageability. A completely realistic model is likely to be too complex to handle mathematically, and a model that is mathematically simple may be so unrealistic as to be useless. Ideally, a model should include all essential features of the system and exclude all nonessential ones. How good or useful a model is often depends on how closely this ideal is approximated. In short, a model is a sort of metaphor or analogy. Like all analogies, it is valid only within certain limits but, when pushed beyond these limits, becomes misleading or even absurd.

In this book, we are going to take many liberties with mathematical rigor. Our excuse is that the basic ideas of a model are often obscured rather than illuminated by excessive attention to mathematical detail. Our authority for the approach is the great physicist Richard Feynman, who wrote in one of his papers:

> Mathematicians may be completely repelled by the liberties taken here. The liberties are taken not because the mathematical problems are considered unimportant. On the contrary, [I hope] to encourage the study of these forms from a mathematical standpoint. In the meantime, just as a poet has a license from the rules of grammar and pronunciation, we should like to ask for "physicists' license" from the rules of mathematics in order to express what we wish to say in as simple a manner as possible.

Exponential Population Growth

To illustrate the nature of mathematical models (as well as some of their limitations) we consider the dynamics of population growth, a subject of considerable interest in population genetics and population biology. In Figure 1.7, the solid dots show the increase in the number of cells of the yeast *Saccharomyces cerevisiae* in a defined quantity of culture medium. The number of cells increases slowly at first (0–4 hours), then more rapidly (hours 4–12), then more slowly again (hours 12–18). As a first approximation of the early stages of population growth, we may assume that a constant fraction

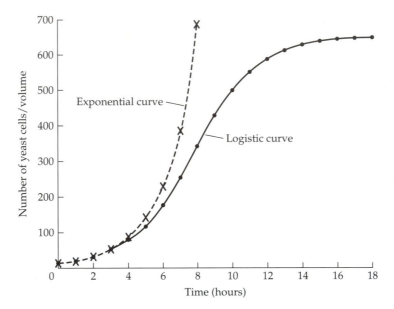

Figure 1.7   Increase in the number of cells of the yeast *Saccharomyces cerevisiae* in a defined quantity of culture medium (dots). The smooth curves are made from mathematical models of exponential growth or logistic growth. (Data from Pearl 1927.)

of the cells reproduces in each interval of time. To simplify matters further, we will assume that the population size does not change gradually but changes in a discrete and instantaneous "jump" at the end of each hour. A model of this type is a **discrete model** of population growth. Thus, we may write

$$N_t = N_{t-1} + rN_{t-1} \qquad\qquad 1.4$$

where N_t and N_{t-1} represent population size at the end of hours t and $t-1$ and where r is a constant called the **intrinsic rate of increase** equal to the fraction of cells that reproduce in each interval of time. This equation says that the population size at the end of hour t is the sum of two components: (1) all the cells present at the end of hour $t-1$ (which means that none of the cells die), and (2) the progeny of the rN_{t-1} cells that divided in the interval.

Equation 1.4 illustrates a feature of theoretical population genetics that sometimes leads to confusion: the same symbols are often used for different things. In this equation, r is the intrinsic rate of increase in population number. In other equations in population genetics, r is the recombination fraction between two genes linked in the same chromosome. The symbol r is used for still other parameters also. Any possible confusion could be avoided by indicating each parameter with a different letter; this solution is impractical because one quickly runs out of letters, even including Greek letters. Another way is to distinguish different meanings of the same letter by typography, the use of superscripts, subscripts, and so forth. The problem with this approach is that even simple equations get to look imposing. Still another solution, which is the one adopted in this book, is to ask the reader to play close attention to the context so that, for example, r as used in the context of population growth is not confused with r used in the context of genetic linkage and recombination.

The solution to Equation 1.4 is straightforward. Because $N_t = (1 + r)N_{t-1}$, it follows that $N_{t-1} = (1 + r)N_{t-2}$. Consequently, we can write $N_t = (1 + r)(1 + r)N_{t-2} = (1 + r)^2 N_{t-2}$. However, $N_{t-2} = (1 + r)N_{t-3}$, and so $N_t = (1 + r)^3 N_{t-3}$. Continuing in this manner, we eventually deduce that

$$N_t = (1+r)^t N_0 \qquad\qquad 1.5$$

For the data in Figure 1.7, if we set $N_0 = 10$ (the observed number) and $r = 0.7083$, the first few points from Equation 1.5 (indicated by crosses) fit very well — $N_0 = 10$, $N_1 = 17$, $N_2 = 29$, $N_3 = 50$. Then the model starts to break down: $N_4 = 85$, $N_5 = 145$, $N_6 = 249$, and thereafter the fit becomes very bad indeed. The lesson from this example is that many models have a range over which they are reasonable approximations to the real world, in this case, for a short time after a yeast culture is inoculated. If the model is extrapolated beyond its range of validity, it yields nonsense. The problem for many models in population genetics is that their range of validity is unknown.

In Equation 1.5, N is defined only for t equal to positive integers because of the discrete nature of the model. Population growth is actually a continuous process. Population size increases gradually rather than in jumps. The continuous-growth version of Equation 1.5, shown by the dashed line labeled "exponential curve" in Figure 1.7, is given by

$$N(t) = N(0)e^{r_0 t} \qquad\qquad 1.6$$

where $r_0 = \ln (1 + r)$. The rationale for Equation 1.6 is based on the same sort of argument as Equation 1.4 but compressing the time scale. Whereas Equation 1.4 assumes that each unit of time is one hour, suppose that each time unit were, say, one minute. In slowing down the time scale in this manner, we must also decrease the value of r, otherwise too many organisms would reproduce in each unit of time. Therefore, by analogy with Equation 1.4, we can write $N_t - N_{t-1} = r_0 N_{t-1}$, but here r_0 is the intrinsic rate of increase in the new time scale. If $N(t)$ is a smooth, continuous function and not changing too fast, then it is easy to convince yourself that $N_t - N_{t-1}$ should approximate the derivative of $N(t)$, which is the change in $N(t)$ in a small interval of time, and that N_{t-1} should be close to $N(t)$ because we have assumed that $N(t)$ is not changing very fast in the new time scale. Therefore, we can write

$$\frac{dN(t)}{dt} = r_0 N(t)$$ 1.7

or

$$\frac{dN(t)}{N(t)dt} = r_0$$ 1.8

Because $d\ln N(t) = d\,N(t)/N(t)dt$, where ln is the base of natural logarithms, the solution of Equation 1.8 is $\ln N(t) = r_0 t + C$, where C is a constant chosen so that $N(t) = N(0)$ when $t = 0$. (Hence, $C = \ln N(0)$.) Expressing the solution in terms of $N(t)$ rather than $\ln N(t)$ yields Equation 1.6. Furthermore, comparing Equation 1.6 with Equation 1.5, it is clear that

$$N(0)e^{r_0 t} = (1+r)^t N_0$$ 1.9

and therefore $r_0 = \ln(1 + r)$ is the relation between the parameter r_0 in the continuous model and the parameter r in the discrete model. Equation 1.6 is the exponential function plotted In Figure 1.7 with $N(0) = 10$ and $r_0 = 0.5355$.

PROBLEM 1.5   Under optimal culture conditions, the bacterium *Escherichia coli* can double in population size every 20 minutes. Because population growth is continuous, Equation 1.7 is the appropriate model. A single cell of *E. coli* is cylindrical in shape and has a volume of approximately 1.6 μm³ (1.6×10^{-12} cm³). A standard soccer ball has a diameter of 22 cm (roughly 9 inches) and a volume of approximately 5600 cm³.

(a) What intrinsic rate of increase r_0 per minute results in a doubling time of 20 minutes?

(b) Starting with a single cell of *E. coli* growing under optimal conditions, how long would it take to produce enough cells to fill one soccer ball?

(c) How many soccer balls could be filled with cells after 24 hours of unrestricted growth?

ANSWER   (a) Set $N(20) = 2N(0) = N(0) \exp (r_0 \times 20)$, where exp $(\cdot)$ stands for $e^{(\cdot)}$. Therefore, $r_0 = (\ln 2)/20 = 0.034657$. (b) One soccer ball full of cells equals $5600/(1.6 \times 10^{-12}) = 3.5 \times 10^{15}$ cells. The time needed to produce this many cells is given by $t = [\ln (3.5 \times 10^{15})]/r_0 = 1032.7$ minutes (17.2 hours). (c) After 24 hours (1440 minutes) of unrestricted growth, one cell yields $\exp (r_0 \times 1440) = 4.7 \times 10^{21}$ cells, which would fill more than 1.35 million soccer balls. (*Note:* If your answers to this problem are a little different from those given, it is probably because the numbers given were calculated to nine significant digits before rounding off.)

Logistic Population Growth

The calculations in Problem 1.5 indicate that no real population can grow exponentially for more than a relatively small number of generations without catastrophic consequences. In nature, although factors such as disease and predation often contribute to the control of population size, populations that grow too large ultimately must deplete the available resources. The kind of growth curve in Figure 1.7 is typical for populations expanding in a new environment: the initial population growth is exponential, but then the rate of growth gradually decreases.

A simple alternative to exponential growth is the **logistic model**; the term *logistic* refers to proportions and, in the logistic model, the rate of population growth is assumed to decrease in proportion to the population size. By analogy with Equation 1.4, the change in population size with a discrete model of population growth takes the form

$$N_t = N_{t-1} + rN_{t-1}\left(\frac{K - N_{t-1}}{K}\right) \qquad 1.10$$

In this equation, K is a constant known as the **carrying capacity of the environment**. Observe that, when N is very small compared with K, then $N_t \approx N_{t-1} + rN_{t-1}$, and so population growth is nearly exponential. On the other hand, when N is close to K, then $N_t \approx N_{t-1}$, and so population growth comes to a standstill.

Unlike Equation 1.4, Equation 1.10 does not have a simple solution for N_t in terms of N_0. However, if the population grows sufficiently slowly, then population growth can be treated as continuous, and Equation 1.10 yields the differential equation

$$\frac{dN(t)}{dt} = rN(t)\left(\frac{K - N(t)}{K}\right) \qquad 1.11$$

The solution of Equation 1.11 is given by

$$N(t) = \frac{K}{1 + Ce^{-rt}} \qquad 1.12$$

where the constant $C = (K - N_0)/N_0$. Equation 1.12 is called the **logistic growth curve** and it is derived in Problem 1.7 below. Logistic population growth results in a sort of S-shaped curve like that shown in Figure 1.7, where the parameters are $r = 0.5355$, $N_0 = 10$ and $K = 665$. (Note that the r and N_0 parameters are the same as in the exponential-growth model for the same data.) The fit is obviously very good indeed.

PROBLEM 1.6   Use Equation 1.12 with $N_0 = 10$, $r = 0.5355$, and $K = 665$ to calculate $N(t)$ for the times $t = 7$ and 8 and $t = 13$ and 14. What are the values of r in Equation 1.10 for $t = 8$ and $t = 14$? Why are they not equal to 0.5355? Why are they not equal to each other?

ANSWER   With the given parameters, $N(7) = 261.53$, $N(8) = 349.43$, $N(13) = 626.13$, and $N(14) = 641.68$. Solving Equation 1.10 for r and substituting $N(t)$ yields $r = 0.5540$ for $t = 8$ and $r = 0.425$ for $t = 14$. Neither of these values agrees with $r = 0.5355$, nor do they agree with each other, because Equation 1.10 pertains to a discrete model and Equation 1.12 to a continuous model. When the population grows continuously, the value of r needed to produce a given change in population size in some discrete interval of time differs according to the magnitude of the change in population size.

PROBLEM 1.7   Use the expression $\int [1/x(a + bx)]dx = -(1/a) \ln (a + bx)/x$ to derive the logistic growth curve from Equation 1.11.

ANSWER   Write Equation 1.11 as $dN(t)/\{N(t)[K- N(t)]\} = rK\, dt$ so that, comparing with the integral form, it is clear that $a = K$ and $b = -1$. Integrating both sides in accordance with the formula results in $-(1/K) \ln [K - N(t)]/N(t) = rt/K + cnst$, where $cnst$ is a constant of integration chosen so that $N(t) = N(0)$ when $t = 0$. Hence, $cnst = -(1/K) \ln [K - N(0)]/N(0) = -(1/K) \ln C$, where C is the constant appearing in Equation 1.12. Consequently, $\ln [K - N(t)]/N(t) = -rt + C$, and so $[K- N(t)]/N(t) = C \exp -rt$. Equation 1.12 follows after some simplification.

SUMMARY

Population genetics is the application of Mendel's laws and other genetic principles to entire populations of organisms. It includes the study of genetic variation within and between species and attempts to understand the processes resulting in adaptive evolutionary changes in species through time. Population genetics has many practical applications in medicine, agriculture, conservation, and other fields.

A gene is a hereditary determinant transmitted from parent to offspring that influences a hereditary trait, often in combination with other genes and also with the environment. Alleles are alternative forms of a gene. Genotypes are formed from pairs of alleles and are either homozygous (if the alleles in the genotype are the same) or heterozygous (if the alleles are different). The physical or biochemical characteristics of an organism constitute its phenotype. The essential mechanism of genetic transmission was established in experiments by Gregor Mendel in the years 1856 to 1863. Mendel showed that the alleles of each gene separate (segregate) from one another in the formation of reproductive cells or gametes. Genes are arranged in linear order along chromosomes. A chromosome may contain several thousand genes. Alleles of different genes present in the same chromosome tend to be inherited together (linkage), but the allele combinations can be broken up by recombination.

Chemically, a gene is a region of a DNA molecule. DNA is a metaphorical "twisted ladder" consisting of two paired strands composed of polymers of nucleotides (the sidepieces of the ladder) whose bases (either A, T, G, or C) jut inward from the sidepieces to form the rungs. Each rung of the ladder consists of either an A–T base pair or a G–C base pair. Most genes code for the polypeptide chains of proteins through a transcript of RNA that is processed into the messenger RNA (mRNA). The polypeptide is produced stepwise by translation of the mRNA according to a triplet genetic code, in which each nonoverlapping group of three adjacent bases (a codon) specifies the amino acid to be attached to the growing chain. Alleles differ in their sequence of nucleotides. A nucleotide substitution in the third position of a codon may not result in an amino replacement in the encoded polypeptide because of redundancy in the genetic code. However, most nucleotide substitutions in either of the first two positions do result in amino acid replacements.

A probability is a number between 0 and 1 that measures the likelihood of a particular event being realized in an actual or conceptual experiment. The addition rule applies to mutually exclusive events and states that the probability of one or the other event being realized equals the sum of the separate probabilities. The multiplication rule applies to independent events and states that the probability of both events being realized simultaneously equals the product of the separate probabilities. The probabilities of various outcomes of repeated and independent trials can be deduced by application of the addition and multiplication rules and conforms to successive terms in the binomial expansion $(p + q)^n$.

Natural populations contain genetic variation in the form of multiple alleles of many genes. For any specified allele, the allele frequency is the proportion of all alleles of the gene that are of the specified type. The allele frequency in a population must usually be estimated from a sample, and so there is variation in the estimate from one sample to the next. The variation is quantified by the standard error. If the distribution of the estimates conforms to a normal, bell-shaped distribution, then the proportions of the estimates lying within ± 1, ± 2, and ± 3 standard deviations of the true value of the parameter are 68%, 95%, and 99.7%, respectively. Estimates are also often presented as a confidence interval, which expresses the degree of confidence that the true value of a parameter lies in some specified interval.

A model is a deliberate simplification of a complex situation. Models may be experimental or conceptual. Conceptual models may be verbal, computational, or mathematical. Mathematical models are widely used in population genetics. They specify the mathematical relations between measured or measurable quantities that determine the changes in allele frequency in populations. Population growth affords an example of mathematical modeling. In the simplest model of discrete population growth, at discrete times a constant fraction of the population reproduces, and so the population jumps instantaneously from one size to the next. A more realistic model envisages continuous reproduction through time, in which case population growth is exponential. The exponential model often fits population growth in newly colonized environments when the population density is low. Population growth is ultimately limited by nutrients, space, or other resources. When population growth decreases in proportion to population size, the S-shaped logistic curve of population growth results; this curve is determined by the intrinsic rate of increase r and the carrying capacity of the environment K.

PROBLEMS

1. If you were to catch a collection of *Drosophila*, grind each one individually in a buffer solution, and measure the rate at which this crude whole-fly homogenate catalyzed the reaction for glucose-6-phosphate dehydrogenase, you would find that the activities would vary by more than four fold. Make a list of possible causes of this variation.

2. Given the complexity of causes of variation in Problem 1, how much variation would you expect to see in the underlying genetic cause of a human inborn error of metabolism such as phenylketonuria? This disorder is caused by insufficient activity of phenylalanine hydroxylase.

3. There are 64 codons in the genetic code, and each codon can undergo nine single-site mutations (each base can mutate to three other bases), for

a total of 576 mutations. How many of these result in no change in the "meaning" of the encoded sequence?

4. Assuming that all nucleotides in all codons mutate with equal frequency (i.e., that all 576 mutations in Problem 3 occur at the same rate), are mutations from one amino acid to another all equally likely?

5. The correspondence between genotype and phenotype is one of the most complex and difficult aspects of evolutionary genetics. Describe an example of a gene whose mutations cause more than one distinctly different phenotype that do not appear to be related.

6. A population cage of *Drosophila melanogaster* is started with 50 males and 50 females, all having the genotype $(e\ st)/(e^+\ st^+)$. This notation implies that one chromosome has the e and st mutations, and the other has the wild type allele at both loci. These two loci show a frequency of recombination in females of $r = 0.37$, and the males produce only non-recombinant gametes. Calculate the expected frequency of the gametes for both males and females and the expected offspring genotype frequencies.

7. In some human cultures it is very important to have a son and a daughter, and couples continue having offspring until they have one of each. If an entire population followed this rule, what would happen to the sex ratio in the population?

8. If two genes are on different chromosomes, the probability that a gamete has a particular allele of each of the two genes is the product of the probability of drawing each allele because the draws are independent of one another (see the multiplication rule). If each gene is on a different chromosome, what is the chance that genotype *Aa Bb CC Dd* produces two consecutive gametes that are *ABCD*?

9. If individual X has an autosomal recessive disease and both parents are unaffected, what is the chance that the sibling of X is a heterozygous carrier?

10. A line of mice seems to consistently produce 55% male and 45% female offspring. In order to test whether this deviation is significant, how many offspring would you have to could to be able to reject a 50 : 50 sex ratio at a probability of $\alpha = 0.05$? (Assume that the sex ratio of the mice remains 55 : 45.)

11. A species of butterflies occurs in two distinct morphs, A and B. You sample two areas and count 26 A and 28 B butterflies in one area, and 10 A and 21 B in another area. Is it possible that these two samples could come from a single homogeneous population, or are the frequencies of the two morphs significantly different from one another?

12. Levy and Levin (1975) used electrophoresis to study the phosphoglucose isomerase-2 gene in the evening primrose *Oenothera biennis,* a complex genomic heterozygote made true breeding by chromosomal translocations. They observed two alleles affecting electrophoretic mobility of the

enzyme, and among 57 strains they found 35 *PGI-2a/PGI-2a*, 19 *PGI-2a/PGI-2b*, and 3 *PGI-2b/PGI-2b* genotypes.

a. Calculate the allele frequencies of *PGI-2a* and *PGI-2b*.

b. With random mating, what would be the expected numbers?

13. The simple models of population growth fail to take into account many factors that affect rates of change. The global human population at 0 A.D., 200 A.D., and at intervals of 200 years up to the present has been estimated in millions of people as 200, 200, 200, 200, 250, 280, 350, 400, 550, 980, and 6000. If the population were growing exponentially, these points would fall on a straight line when plotted on a logarithmic scale. Draw this plot. What do you conclude?

14. A healthy pair of *Drosophila* can produce 500 offspring in 12 days, each adult fly weighing about 1 mg. Assume that the parental flies die after they finish reproducing. (Actually, they live about a month.) If all successive generations get enough to eat and remain this fecund, what will the mass of flies be in one year?

CHAPTER 2

Genetic and Phenotypic Variation

PHENOTYPIC VARIATION · NORMAL DISTRIBUTION · MENDELIAN VARIATION ·
PROTEIN POLYMORPHISMS · DNA POLYMORPHISMS · MULTIPLE-FACTOR MODEL

G ENETIC VARIATION IN POPULATIONS became a subject of scientific
inquiry in the late nineteenth century prior even to the rediscovery of Mendel's paper in 1900. The leading exponent of the study
of hereditary differences among human beings was Francis Galton
(1822–1911). Galton was a pioneer in the application of statistics to biology.
He used statistical methods to study physical traits such as eye color and fingerprint ridges as well as behavioral traits such as temperament and musical
ability. Galton was among the first to examine the statistical relations
between the distributions of phenotypic traits in successive generations. He
is regarded as the founder of biometry, the application of statistics to biological problems.

PHENOTYPIC VARIATION IN NATURAL POPULATIONS

Galton and Mendel exemplify opposite approaches to the study of inherited
traits. Mendel's point of departure in the study of genetics was **discrete variation**, in which phenotypic differences among organisms can be assigned to
a small number of clearly distinct classes, such as round versus wrinkled
peas. Galton's point of departure was **continuous variation**, in which the
phenotypes of organisms are measured on a quantitative scale, like height or
weight, and in which the phenotypes grade imperceptibly from one category into the next. As material for the study of phenotypic variation, Galton's
choice was good: most of the differences among normal people that are vis-

ible to the unaided eye are differences in continuous traits—height, weight, skin color, hair color, facial features, running speed, shoe size, and so forth. The same is true of phenotypic variation in other organisms. On the other hand, as material for the study of genetic variation, Mendel's choice was good: The pattern of segregation of alleles is revealed most clearly in pedigrees of discrete, simple Mendelian traits.

Continuous Variation: The Normal Distribution

With continuous traits, not only do the phenotypes grade into one another, but the traits also usually present difficulties for genetic analysis. The problems are of two principal types:

- Most continuous traits are influenced by the alleles of two or more genes, hence the segregation of any one gene in pedigrees is obscured by the segregation of other genes that affect the trait.
- Most continuous traits are influenced by environmental factors as well as by genes, and so genetic segregation is obscured by environmental effects.

These problems are not insurmountable in organisms with a sufficiently high density of genetic markers scattered throughout the genome (the complement of chromosomes) because the genetic markers can be tracked in pedigrees along with the continuous trait of interest. Organisms with sufficiently dense genetic maps include human beings, laboratory animals, and many domesticated animals and crop plants.

In Galton's time, however, studies of continuous traits based on genetic linkage were unknown. Why, then, did Galton focus on continuous traits? Because they have a sort of regularity—a statistical predictability—of their own. For many continuous traits, when the phenotypes are grouped into suitable intervals and plotted as a bar graph, the distribution of phenotypes conforms closely to the **normal distribution**, the symmetrical, bell-shaped curve discussed briefly in Chapter 1 in the section on phenotypic diversity and genetic variation. For example, a bar graph of Galton's data on the heights of 1329 men, rounded to the nearest inch, is plotted in Figure 2.1. The smooth curve is the normal distribution that best fits the data. The equation of the normal curve is:

$$f(x) = \frac{1}{\sqrt{2\pi}\sigma} e^{-\frac{(x-\mu)^2}{2\sigma^2}} \hspace{2cm} 2.1$$

where x ranges from $-\infty$ to $+\infty$, and $\pi = 3.14159$ and $e = 2.71828$ are constants. The location of the peak of the distribution along the x axis is determined by the parameter μ, which is the **mean**, or average, of the phenotypic values. The degree to which the phenotypes are clustered around the mean is deter-

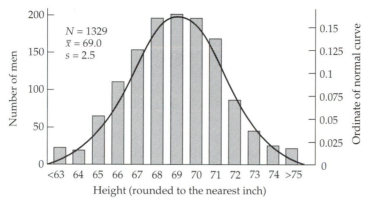

Figure 2.1   Distribution of height among 1329 British men. (Data from Galton 1889.)

mined by the parameter σ^2, which is the **variance** of the distribution. Mathematically, the variance is the average of the squared difference of each phenotypic value from the mean; that is, it is the average of the values of $(x - \mu)^2$. How μ and σ^2 are estimated from data is considered next.

Mean and Variance

Because μ and σ^2 are parameters, their values are unknown, and they must be estimated from the data themselves. The height data are tabulated in Table 2.1, in which f_i is the number of men whose height is x_i, rounded to the nearest inch. (The fact that the shortest and tallest men are grouped in the tails of the distribution makes no difference because these men account for only a small proportion of the total sample.) Also tabulated are the products $f_i \times x_i$ and $f_i \times x_i^2$ as well as their sums.

The mean μ of the distribution is estimated as the mean of the sample, which is conventionally denoted $\bar{x}$ (also sometimes as $\hat{\mu}$):

$$\bar{x} = \frac{\sum f_i x_i}{\sum f_i} \qquad\qquad 2.2$$

In this example, $\bar{x} = 91{,}639/1329 = 68.95$ inches.

Likewise, the variance σ^2 of the distribution is estimated as the variance of the sample, which is conventionally denoted s^2 (also sometimes as $\hat{\sigma}^2$):

$$s^2 = \frac{\sum f_i (x_i - \bar{x})^2}{\sum f_i} = \frac{\sum f_i x_i^2}{\sum f_i} - (\bar{x})^2 \qquad\qquad 2.3$$

The expression in the middle follows directly from the definition of the variance: it is the average of the squared deviations from the mean because, for each value of x_i, $(x_i - \bar{x})$ is the deviation of that value from the mean. The expression on the right is identical arithmetically but easier to apply in practice. In the example in Table 2.1, $s^2 = 6{,}326{,}939/1329 - (68.96)^2 = 6.11$. (This value may differ slightly from your own calculation according to the number of significant digits you carried along before rounding off.) If the sample size is small (say, less than 50), then a slightly better estimate of the variance is obtained by multiplying the expression in Equation 2.3 by $n/(n-1)$, where n is the total size of the sample (in this case, 1329).

Closely related to the variance is the **standard deviation** of the distribution, which is the square root of the variance. The standard deviation is a natural quantity to consider in view of the units of measurement. In Table 2.1, for example, each measurement is in inches. The mean is also in inches. However, the variance, being the average of squared deviations, has the units of squared inches—which seems more appropriate for an area than for a height. Taking the square root of the variance restores the correct unit of measure: in this example, inches. The estimate of the standard deviation is conventionally denoted s (also sometimes as $\hat{\sigma}$) and it is calculated as the square root of the quantity in Equation 2.3. In the height example, $s = 2.47$ (which may

TABLE 2.1   HEIGHTS OF 1329 MEN

Height interval (i)	Height range (in.)	Nearest inch (x_i)	Number of men (f_i)	$f_i \times x_i$	$f_i \times x_i^2$
1	<63.5	63	23	1,449	91,287
2	63.5–64.5	64	20	1,280	81,920
3	64.5–65.5	65	64	4,160	270,400
4	65.5–66.5	66	110	7,260	479,160
5	66.5–67.5	67	155	10,385	695,795
6	67.5–68.5	68	199	13,532	920,176
7	68.5–69.5	69	203	14,007	966,483
8	69.5–70.5	70	198	13,860	970,200
9	70.5–71.5	71	171	12,141	862,011
10	71.5–72.5	72	88	6,336	456,192
11	72.5–73.5	73	47	3,431	250,463
12	73.5–74.5	74	27	1,998	147,852
13	>74.5	75	24	1,800	135,000
Totals			1,329 (Σf_i)	91,639 ($\Sigma f_i x_i$)	6,326,939 ($\Sigma f_i x_i^2$)

Source: Data from Galton 1889.

again differ slightly from your own calculation because of round-off error). The estimate *s* of the standard deviation is often called the *standard error*. When estimating a proportion—such as the frequency of an allele in a population—the standard error is calculated according to Equation 1.3 in Chapter 1.

In Chapter 1, the values 68%, 95%, and 99.7% quoted as the proportions of observations expected to fall within 1, 2, or 3 standard errors of the mean, respectively, emerge directly from Equation 2.1 for the normal distribution. In a normal distribution, the exact proportion of observations falling with any specified range of *x* equals the integral of Equation 2.1 across the specified range. For the normal distribution, the integral between the limits $\mu \pm \sigma$ equals 0.6827, that between $\mu \pm 2\sigma$ equals 0.9545, and that between $\mu \pm 3\sigma$ equals 0.9973. In data analysis, $\bar{x}$ and *s* are used in place of μ and σ. Incidentally, the integral of the normal distribution between the limits $\mu \pm 4\sigma$ equals 0.9999; this result says that fewer than one in 10,000 observations falls more than four standard deviations from the mean.

Central Limit Theorem

Galton was immensely impressed with the observation that many natural phenomena follow the normal distribution. He writes:

> I know of scarcely anything so apt to impress the imagination as the wonderful form of cosmic order expressed by the "law of frequency of error" [the normal distribution]. Whenever a large sample of chaotic elements is taken in hand and marshaled in the order of their magnitude, this unexpected and most beautiful form of regularity proves to have been latent all along. The law would have been personified by the Greeks if they had known of it. It reigns with serenity and complete self-effacement amidst the wildest confusion. The larger the mob and the greater the apparent anarchy, the more perfect is its sway. It is the supreme law of unreason.

It is, indeed, remarkable to consider that pure, blind chance is the reason for this "unexpected and most beautiful form of regularity."

The theoretical basis of the normal distribution is known in probability theory as the **central limit theorem**. Roughly speaking, the central limit theorem states that the sum of a large number of independent random quantities always converges to the normal distribution. For our purposes, "independent" in this context means that information about any one of the observations gives no improvement in the ability to predict any other of the observations. A large number of independent random quantities is apparently what Galton meant by "a large sample of chaotic elements." The central limit theorem explains in part why so many continuously distributed traits conform to the normal distribution. Most continuous traits are *multifactorial*, meaning that they are influenced by "many factors," typically several or many genes acting together with environmental factors. Among human beings, for example, the obvious differences between normal people in hair color, eye color, skin color, stature, weight, and other such traits are not usually traceable to single genes. They result from the combined effects of several or many genes as well as numerous environmental effects acting together as "a large sample of chaotic elements," which often produce, in the aggregate, a normal distribution of phenotypes.

It should be emphasized that the "large number" of random elements specified in the central limit theorem need not be excessive. As an example, Figure 2.2 is a bar graph of 100 observations in which each "observation" consists of the sum of nine consecutive random numbers chosen with equal probability from anywhere in the range (–1, +1). For the sum of nine random numbers in this range, the theoretical mean equals 0 and the theoretical standard deviation equals 1.73; the sample values were $\bar{x} = -0.12$ and $s = 1.70$. Expressed as a deviation from the mean in multiples of the standard error, the number of observations in each category is shown at the top of the bar in Figure 2.2. Because the expected numbers are 2.5, 13.5, 68, 13.5, and 2.5, the fit to a normal distribution is obviously very good. In this example, therefore, fewer than 10 "chaotic elements," when added together, yields "this unexpected and most beautiful form of regularity."

PROBLEM 2.1   At an International Health Exhibition in London in 1884, Galton set up an "anthropometric laboratory" that carried out tens of thousands of measurements covering a wide range of human traits. Among the traits was "strength of pull," expressed as the number of pounds that a person could pull with one arm against a resisting force in a sort of arm-wrestling contraption (Galton 1889). The data for 519 males aged 23–26 years fell into the following categories (the number in parentheses is the number of males in each category): 40–50 lbs (10), 50–60 (42), 60–70 (140), 70–80 (168), 80–90 (113), 90–100 (22), 100–110 (24). Using the midpoint of each category as the strength of pull for all males in that category, estimate the mean and standard deviation of strength of pull. Assuming that strength of pull has a normal distribution with parameters equal to these estimates, what is the expected proportion of males whose strength of pull exceeds 112 pounds?

ANSWER   The values of x_i are 45, 55, 65, and so forth. Then $\Sigma f_i = 519$, $\Sigma f_i x_i = 38{,}675$, and $\Sigma f_i x_i^2 = 2{,}963{,}375$. Hence, $\bar{x} = 74.5$ lbs, $s^2 = 156.8$ lbs^2,

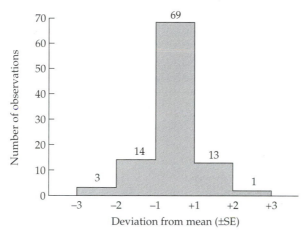

Figure 2.2 Distribution of 100 values of the sum of nine random numbers from the interval (−1, +1).

and so $s = 12.5$ lbs. (Answers may differ slightly because of round-off error.) A strength of pull of 112 lbs is three standard errors above the mean; hence a proportion of only $(1 - 0.997)/2 = 0.0015$ (about one in 667) males is expected to have a phenotype exceeding this value.

Discrete Mendelian Variation

Discrete Mendelian variation (also called simple Mendelian variation) refers to phenotypic differences resulting from segregation of the alleles of a single gene. Environmental effects on the trait are small enough, relative to hereditary differences, that the transmission of alleles determining the trait can be traced through pedigrees. An example of discrete Mendelian variation is the inheritance of red, pink, or white flower color in snapdragons (Chapter 1). This case is exceptionally convenient for genetic studies because of the intermediate phenotype of the heterozygote. However, most of the phenotypic variation in natural populations is multifactorial. In human beings, for example, although simple Mendelian variation accounts for many inherited disorders, each of the disorders is relatively rare.

Ironically, simple Mendelian variation is more easily detected by studying genes and their products than by studying phenotypes. Because the mechanisms of transcription, RNA processing, and translation are relatively free of the gene interactions and environmental effects that complicate the analysis of multifactorial traits at the phenotypic level, there is a direct connection between DNA sequences and alleles and a nearly direct connection between genes and their products. Indeed, the correspondence between DNA sequences and alleles is one-to-one: different alleles have different DNA sequences irrespective of whether the alleles affect phenotype. Likewise, alleles with nonsynonymous codon differences in a protein-coding region result in different amino acid sequences irrespective of what the polypeptide does in metabolism or how the difference in sequence affects the organism.

Hence, an efficient way to detect simple Mendelian variation is to study molecules—and therein lies a paradox. As evolutionary biologists, population geneticists are interested in observable phenotypes that are likely to be subject to natural selection: morphology, rate of development, mating behavior, age of reproduction, longevity, and so forth (in short, the types of traits that attracted Galton). On the other hand, genetic studies are most readily carried out with simple Mendelian variation detected as differences between molecules. The paradox is that differences in molecules among healthy organisms are not usually related in any obvious way to differences in phenotype. Thus, there is a gap in being unable to specify exactly which types of molecular differences underlie the evolutionary process. The irony of the situation is similar to that described by the physiologist Albert Szent-Gyorgyi:

> My own scientific life was a descent from higher to lower dimensions, led by the desire to understand life. I went from animals to cells, from cells to bacteria, from bacteria to molecules, from molecules to electrons. The story had its irony, for molecules and electrons have no life at all. On my way, life ran out between my fingers.

The gap between genotype and phenotype results from the complex interactions between genes and environment in the determination of physiology, development, and behavior. In evolutionary biology, the complexity is even greater because the key issue is the relative ability of organisms to survive and reproduce in their environments. Nevertheless, the disconnect between differences in molecules and evolutionary adaptations is by no means inevitable, permanent, or insurmountable. It is already clear that the study of the relation between genetic variation and evolutionary adaptation must be high on the agenda of evolutionary biology for the next century, and already there are many examples in which the relation is quite well established.

EXPERIMENTAL METHODS FOR DETECTING GENETIC VARIATION

For nearly 50 years, the workhorse method for revealing genetic variation has been **electrophoresis** because small differences in rate of migration in an electrophoretic field can be used to distinguish between nearly identical macromolecules. A typical laboratory setup for electrophoresis is illustrated

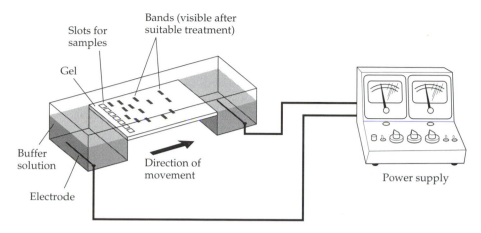

Figure 2.3   One type of laboratory apparatus for electrophoresis. The procedure is widely used to separate protein or DNA molecules. In conventional gels, DNA fragments smaller than about 20 kb migrate approximately in proportion to the logarithm of their molecular weights.

schematically in Figure 2.3. The tray contains a thick layer of a gel, typically starch, acrylamide, or agarose; it may be placed horizontally (as shown in the illustration) or vertically (with the gel sandwiched between two glass plates). Each sample of material is placed in a small slot near the edge of the gel. Connected to each edge of the gel is a chamber containing a buffered solution and electrodes. In electrophoresis, an electric current is applied across the gel for several hours. Molecules in the samples—usually proteins or nucleic acids are of greatest interest—move through the gel in response to the electric field. Molecules of different size and charge move at different rates. After the electrophoresis is finished, the positions of the molecule or molecules of interest are revealed by any of several procedures.

Protein Electrophoresis

In protein electrophoresis, used primarily to study enzyme molecules, the position to which a particular enzyme migrates is revealed by soaking the gel in a solution containing a substrate for the enzyme along with a dye that precipitates where the enzyme-catalyzed reaction takes place. A dark band thus appears in the gel at the position of the enzyme. If the enzyme present in a sample has an amino acid replacement that results in a difference in the overall ionic charge of the molecule, then the enzyme will have a somewhat altered **electrophoretic mobility** and move at a different rate. The electrophoretic mobility changes because enzymes of the same size and shape move at a rate determined largely by the ratio of the number of positively charged amino acids (primarily lysine, arginine, and histidine) to the number of negatively charged ones (principally aspartic acid and glutamic acid). Electrophoresis can therefore be used to detect a mutation that results in a difference in electrophoretic mobility of the enzyme it encodes.

One possible result of an electrophoresis experiment is shown in the hypothetical gel in Figure 2.4A, in which all samples manifest an enzyme with the same electrophoretic mobility. The result indicates a *monomorphic sample* because there is only one electrophoretic pattern observed. Another kind of result is shown in Figure 2.4B, in which *polymorphism* is observed in the types of electrophoretic patterns. When polymorphic enzyme bands are observed, genetic tests typically indicate that organisms with only a fast-migrating enzyme are homozygous for a *fast* allele (F/F) and those with only a slow-migrating enzyme are homozygous for a *slow* allele (S/S). Organisms with both enzyme bands are heterozygous for the alleles (F/S). Simple Mendelian inheritance of the polymorphism is indicated by, for example, the finding that matings of two heterozygotes produce, on the average, $\frac{1}{4}\,F/F$, $\frac{1}{2}\,F/S$, and $\frac{1}{4}\,S/S$ progeny. Two enzyme bands appear in heterozygotes whenever the active enzyme consists of a single polypeptide chain (rather than two or more polypeptide chains aggregated together) because heterozygotes produce a different polypeptide chain from each allele.

Enzymes that differ in electrophoretic mobility as a result of allelic differences in a single gene are called **allozymes**. Hence, allozyme variation in a population is an indication of simple Mendelian genetic variation. Allozyme variation is widespread in almost all natural populations studied by

(A) Monomorphic sample

(B) Polymorphic sample

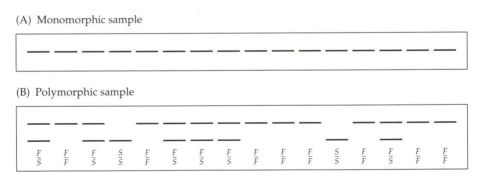

Figure 2.4   Monomorphism and polymorphism. (A) Hypothetical gel showing protein monomorphism. All samples have an enzyme with the same electrophoretic mobility. (B) Hypothetical gel showing allozyme polymorphism. Eight samples are homozygous for an allele (*F*) that codes for a rapidly migrating enzyme; two samples are homozygous for a different allele (*S*) that codes for a slowly migrating enzyme; and six samples are heterozygous (*F/S*) and therefore exhibit enzyme bands corresponding to both alleles.

electrophoresis, including organisms such as bacteria, plants, *Drosophila*, mice, and human beings.

PROBLEM 2.2   A sample of 35 organisms from a Texas population of the wild annual plant *Phlox drummondii* were examined for the electrophoretic mobility of the enzyme alcohol dehydrogenase (Levin 1978). Two alleles affecting electrophoretic mobility were found— Adh^a and Adh^b. The genotype frequencies observed in the sample were 0.04 Adh^a/Adh^a, 0.32 Adh^a/Adh^b, and 0.64 Adh^b/Adh^b. Estimate the allele frequency of Adh^a and its standard error.

ANSWER   Let p represent the allele frequency of Adh^a. Then $\hat{p} = 0.04 + 0.32/2 = 0.20$. The standard error equals $\sqrt{(0.20)(1-0.20)/(2 \times 35)} = 0.05$.

PROBLEM 2.3   From a natural population of *Drosophila melanogaster* in Raleigh, North Carolina, 660 fertilized females were trapped and used to found a large laboratory population (Mukai et al. 1974). After about five months (10 generations), 489 third chromosomes in the population were examined for allozymes coding for the enzymes esterase-6 (alleles $E6^F$ and $E6^S$), esterase-C (alleles EC^F and EC^S), and octanol dehydrogenase (alleles Odh^F and Odh^S). The order of the genes in the third chromosome is known to be $E6$–EC–Odh. The results were as follows:

$E6^F EC^F Odh^F$	152		$E6^S EC^F Odh^F$	264
$E6^F EC^F Odh^S$	7		$E6^S EC^F Odh^S$	13
$E6^F EC^S Odh^F$	15		$E6^S EC^S Odh^F$	29
$E6^F EC^S Odh^S$	1		$E6^S EC^S Odh^S$	8

Estimate the allele frequencies and their standard errors for $E6^F$ and $E6^S$, for EC^F and EC^S, and for Odh^F and Odh^S. What number of each of the chromosome types is expected assuming that the alleles are associated at random?

ANSWER   For esterase-6, there were 175 $E6^F$ and 314 $E6^S$ alleles, yielding $\hat{p} = 175/489 = 0.358$ for $E6^F$ and $\hat{q} = 314/489 = 0.642$ for $E6^S$; the standard error is the same for both estimates and equals $\sqrt{(0.358)(0.642)/489} = 0.022$. For the other alleles, the estimates and their standard errors are 0.892 ± 0.014 for EC^F and 0.108 ± 0.014 for EC^S; and 0.941 ± 0.011 for Odh^F and 0.059 ± 0.011 for Odh^S. Assuming random combinations, the expected number of each chromosome type equals the product of the allele frequencies times 489. For example, for $E6^F EC^F Odh^F$, the expected number is $0.358 \times 0.892 \times 0.941 \times 489 = 146.8$. The expected numbers (observed in parentheses) for all eight chromosome types are: 146.8 (152), 263.4 (264), 9.2 (7), 16.6 (13), 17.8 (15), 32.0 (29), 1.1 (1), 2.0 (8). The model of random combinations of alleles fits very well.

The Southern Blot Procedure

Like polypeptides, DNA fragments can be separated by electrophoresis. Unlike a polypeptide, which has a predetermined size according to the number of amino acids it contains, a molecule of chromosomal DNA is randomly sheared into fragments of various size during purification. Therefore, in any DNA preparation, the DNA fragments containing a particular sequence have a range of sizes depending on where on each side of the sequence the chromosomal DNA became sheared. Fortunately, there is a class of enzymes that cleaves DNA at particular sites along the molecule. Consequently, when chromosomal DNA is cleaved with such an enzyme, each DNA fragment containing a particular sequence is cut at the same sites on either side and so will have the same length.

The enzymes that cleave DNA at particular sites are called **restriction enzymes**. Each type of restriction enzyme cuts double-stranded DNA at all sites at which there is a particular nucleotide sequence called the **restriction site** of the enzyme. Examples of restriction enzymes and their restriction sites are shown in Figure 2.5; the cuts are made at the positions of the arrows. For example, the enzyme *Alu*I cuts at sites of the four-nucleotide sequence AGCT, and *Eco*RI cuts at the six-nucleotide sequence GAATTC. Most restriction enzymes used in population studies have either four-nucleotide or six-nucleotide restriction sites.

DNA is also unlike an enzyme in that it lacks any catalytic activity that can be used to determine the location of a band in a gel. On the other hand, any single strand of DNA is able to form a double-stranded molecule by

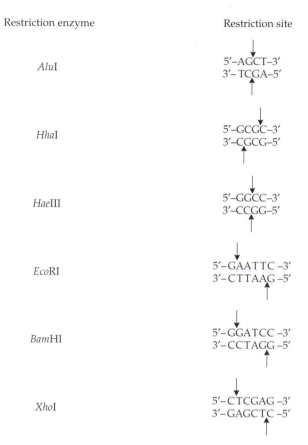

Restriction enzyme	Restriction site

*Alu*I

5′–AGCT–3′
3′–TCGA–5′

*Hha*I

5′–GCGC–3′
3′–CGCG–5′

*Hae*III

5′–GGCC–3′
3′–CCGG–5′

*Eco*RI

5′–GAATTC –3′
3′–CTTAAG –5′

*Bam*HI

5′–GGATCC –3′
3′–CCTAGG –5′

*Xho*I

5′–CTCGAG –3′
3′–GAGCTC –5′

Figure 2.5   Restriction enzymes cleave DNA molecules at sites of specific, short nucleotide sequences. More than 500 different restriction enzymes are commercially available. They are essential tools in DNA analysis and gene cloning. The cleavage site in each DNA strand is indicated by the arrow.

pairing with another strand having the complementary base sequence. This pairing of complementary DNA strands is the physical basis of the most widely used procedure for identifying DNA fragments in a gel; the procedure, illustrated in Figure 2.6, is a **Southern blot**. The reagent used for identification is a molecule of DNA called the **probe**, which contains the nucleotide sequence of interest. Probe DNA is usually obtained from a gene that has been cloned (for example, into a bacterial cell) or by amplification with the polymerase chain reaction (described in the next section). In the Southern procedure, DNA restriction fragments that have been separated by electrophoresis are rendered single-stranded by soaking in a solution of sodi-

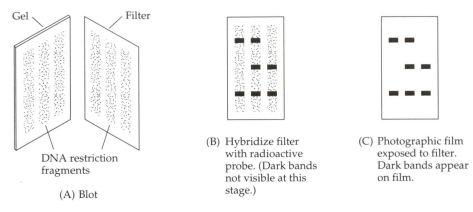

Gel   Filter

DNA restriction
fragments

(A) Blot

(B) Hybridize filter
with radioactive
probe. (Dark bands
not visible at this
stage.)

(C) Photographic film
exposed to filter.
Dark bands appear
on film.

Figure 2.6   Southern blot procedure. (A) DNA fragments separated by electrophoresis are transferred and chemically attached to a filter. (B) The filter is
mixed with radioactive probe DNA, which sticks to homologous DNA molecules in the filter. (C) After washing, the filter is exposed to photographic film,
which develops dark bands caused by radioactive emissions from the probe.

um hydroxide, then blotted onto a nitrocellulose or nylon filter where subsequent chemical treatment attaches them (Figure 2.6A). The filter is then
bathed in a solution containing probe DNA that has been rendered radioactive (part B). As the solution cools, the probe DNA strands form double-
stranded molecules with their complementary counterparts on the filter, and
careful washing removes all of the probe DNA that has remained unpaired.
The filter is sandwiched with photographic film, where radioactive disinte-
grations from the bound probe result in visible bands (part C). Alternatively,
the probe may be chemically modified and the bands visualized by fluores-
cence or staining.

Genetic differences resulting in the presence or absence of restriction sites
can be identified because they change the length of characteristic restriction
fragments. An example is illustrated in Figure 2.7. The upper part of each
panel shows the location of restriction sites in the DNA molecules in a diploid
genotype. The *a*-type molecule contains one additional restriction site not
present in the *A*-type molecule. The lower part of the figure demonstrates
that, with suitable probe DNA, all three genotypes can be distinguished by
their pattern of restriction fragments. A difference in the length of a restriction fragment found segregating in natural populations is called a **restriction
fragment length polymorphism** or **RFLP**. Because RFLPs are widely distributed throughout the genome of human beings and other organisms, they
have assumed major importance in population genetics.

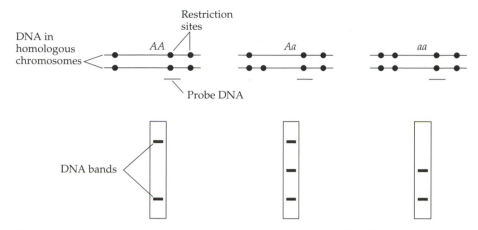

Figure 2.7   Restriction fragment length polymorphisms (RFLPs) result from the presence or absence of particular restriction sites in DNA. In this example, the DNA molecule designated A contains three restriction sites, and the one designated a contains four. Genotypes AA, Aa, and aa each yield a different pattern of bands in Southern blot using the indicated probe DNA.

The Polymerase Chain Reaction

The **polymerase chain reaction** (**PCR**) for the amplification of specific DNA sequences is of great utility in population genetics for the production of probe DNA or for the direct determination of the amount of nucleotide sequence variation present in natural populations. The method is outlined in Figure 2.8. The original DNA sequence to be amplified is shown in black and the newly synthesized DNA strands in gray. The small ovals represent synthetic oligonucleotides that are complementary in sequence to the ends of the region to be amplified. The oligonucleotides are called **primer** sequences because they anneal to the ends of the sequence to be amplified and are used as primers for chain elongation by DNA polymerase. Primer oligonucleotides are typically 18–22 nucleotides in length. DNA to be used as the template in a PCR reaction is first mixed with both primers along with a thermostable DNA polymerase in a buffer solution. The PCR amplification takes place in cycles. In the first cycle, the DNA is heated to separate the strands and then cooled in the presence of a vast excess of the primer oligonucleotides. Then elongation of the primers produces double-stranded molecules. The second cycle of PCR is similar to the first but, after the second cycle, there are four copies of each original molecule. The cycle is repeated from 20 to 30 times, each resulting in a doubling of the number of molecules. The theoretical result of n rounds of amplification is 2^n copies of each template molecule originally present.

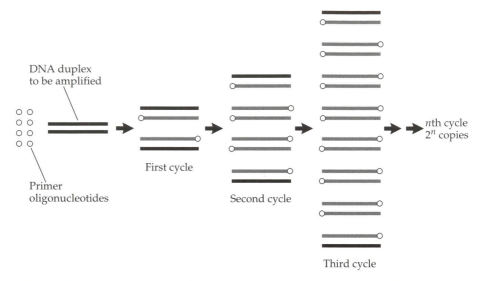

DNA duplex
to be amplified

Primer
oligonucleotides

First cycle

Second cycle

Third cycle

*n*th cycle
2^n copies

Figure 2.8   The polymerase chain reaction (PCR). Short primer oligonucleotides are used as primers to initiate DNA replication from opposite ends of a
DNA duplex to be amplified. After each round of replication, the DNA is heated
to separate the strands and then cooled to allow new primers to anneal. Repeated rounds of replication result in an exponential increase in the number of target molecules.

PCR amplification is very useful in generating large quantities of a specific DNA sequence without the need for cloning. The main limitation of the
technique is that the DNA sequences at the ends of the region to be amplified
must be known so that primer oligonucleotides can be synthesized. There are
many applications in which this requirement is met. In population genetics,
for example, PCR can be used to amplify different alleles present in natural
populations.

PROBLEM 2.4   PCR was used to amplify five alleles (designated *f–j*)
of the gene *Rh3* coding for a light-sensitive protein in the eye of
Drosophila simulans, a species of fruit fly closely related to *D.
melanogaster*. The resulting DNA fragments were sequenced (Ayala et
al. 1993). The data show the nucleotide present at each of 16 polymorphic nucleotide sites found in the first 500 nucleotide sites in the
amino acid coding region of the gene; the remaining 484 nucleotide
sites were monomorphic in this sample. Any nucleotide site that is an

exact multiple of three is at the third position of a codon. In this region of the gene:

(*a*) what proportion of polymorphic nucleotide sites are in third positions of codons? What can you infer from this observation?
(*b*) what proportion of nucleotide sites are polymorphic?
(*c*) why is the standard error formula not appropriate for the estimate in part (*b*)?

Nucleotide site in gene.

Allele	132	142	162	192	198	201	207	240	246	351	354	372	375	405	417	483
f	T	C	T	A	C	C	T	C	C	T	C	G	G	T	T	A
g	T	C	C	T	A	C	C	T	C	C	T	G	G	T	T	T
h	C	T	C	C	C	C	C	T	C	T	T	T	G	C	T	A
i	C	T	C	C	C	C	C	T	T	C	T	G	A	C	T	T
j	C	T	C	C	C	T	C	T	T	T	T	G	G	C	C	A

ANSWER   (*a*) Among the 16 polymorphic sites, only site 142 is not an exact multiple of three, hence 15/16 = 94% of the polymorphic sites are in the third codon position. The inference is that many of the nucleotide polymorphisms are silent (synonymous) in that they do not alter the amino acid sequence of the polypeptide. (In fact, all 16 are silent polymorphisms, including the C → T change in 142, which alters the codon from CUA → UUA, both of which code for leucine.) (*b*) A total of 16/500 = 3.2% of the nucleotide sites are polymorphic in this region of the gene. (*c*) The binomial standard error is not appropriate in this case because the nucleotides within a gene are not independent samples; they are genetically closely linked. (A suitable estimate of the standard error is given later in this chapter.)

POLYMORPHISM AND HETEROZYGOSITY

Monomorphism or polymorphism of a gene in a sample is usually of interest only insofar as it indicates monomorphism or polymorphism of the gene in the population as a whole. In a population, a **polymorphic gene** is one for which the most common allele has a frequency of less than 0.95 (some authors prefer a more stringent cutoff at 0.99). Conversely, a **monomorphic gene** is one that is not polymorphic. The cutoff at 0.95 (sometimes 0.99) in the definition of polymorphism is arbitrary, but it serves to focus attention on those genes in which allelic variation is common. In any large population, rare alleles are observed for virtually every gene. An allele is considered a rare allele if its frequency is less than 0.005; in human beings, between one and two people per thousand are heterozygous for rare alleles of any gene. Many rare alleles are deleterious and are presumably maintained in the population by recurrent mutation. The definition of polymorphism is an attempt to focus on genes that have alleles with frequencies too high to be explained solely by recurrent mutation to harmful alleles. With the 0.95 definition of polymorphism given above, and if alleles are combined at random into genotypes, then at least 9.5% of the population is heterozygous for the most common allele (because $2 \times 0.95 \times 0.05 = 0.095$).

Allozyme Polymorphisms

Polymorphism of alleles that determine allozymes is extremely widespread. Figure 2.9 summarizes the results of electrophoretic surveys of 14 to 71 (mostly around 20) genes in populations of 243 species. Each point in the figure (except that for human beings) gives the type of organism studied and the number of species examined. The axis labeled Polymorphism refers to the estimated proportion of genes that are polymorphic by the 0.95 criterion. The axis labeled Heterozygosity refers to the average heterozygosity in each group. The average heterozygosity is the estimated proportion of genes expected to be heterozygous in an average organism; it is estimated as the proportion of heterozygous genotypes for each gene averaged over all genes. For example, the data for Europeans include an English population in which 10 enzyme genes were examined (Harris 1966). Of the 10 genes, three were found to be polymorphic, from which the estimated proportion of polymorphic genes in the genome is $3/10 = 0.3$. The observed proportion of heterozygous genotypes for each of the three polymorphic genes was 0.509 (for red-cell acid phosphatase), 0.385 (for phosphoglucomutase), and 0.095 (for adenylate kinase); the average heterozygosity in this sample—taking into account the additional seven genes for which the observed heterozygosity was 0—is therefore $(0.509 + 0.385 + 0.095 + 7 \times 0)/10 = 0.099$.

The vertical and horizontal bars on the point corresponding to *Drosophila* indicate the size of the standard error of the estimate. Therefore, the bars indicate the limits of polymorphism and heterozygosity within which about 68% of the species are expected to fall. Among *Drosophila* species, approximately 68% have a proportion of polymorphic genes in the range 0.30–0.56 and an average heterozygosity in the range 0.09–0.19. Such bars could be attached to each point; their lengths would be comparable to those for *Drosophila*, indicating substantial variability in polymorphism and heterozygosity among species within groups.

Figure 2.9 has no simple summary because of the immense variability in polymorphism and heterozygosity found within each group of organisms (as

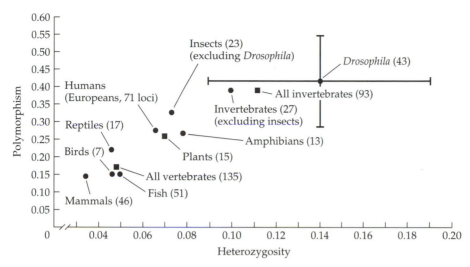

Figure 2.9 Estimated levels of heterozygosity and proportion of polymorphic genes derived from allozyme studies of various groups of plants and animals. The number of species studied is shown in parenthesis beside each point. Squares denote averages for plants, invertebrates, and vertebrates. The bars across the *Drosophila* point indicate the standard error within which about 68% of the species are expected to fall. Other groups have similarly large standard errors. (Data from Nevo 1978.)

indicated by the length of the variability bars corresponding to *Drosophila*). On the whole, there is a positive relationship between amount of polymorphism and degree of heterozygosity. This relationship is as expected because the greater the fraction of polymorphic genes in a population, the more genes that are expected to be heterozygous on the average. The overall mean polymorphism in Figure 2.9 is 0.26 ± 0.15, and the mean heterozygosity is 0.07 ± 0.05. Vertebrates have the lowest average amount of genetic variation among the groups in Figure 2.9, plants come next, and invertebrates have the highest. *Drosophila* is the most genetically variable group of higher organisms so far studied, and mammals the least variable. Human beings are fairly typical of large mammals: An extensive electrophoretic survey of 104 genes in a sample including all major human races gave estimates of polymorphism of 0.32 and heterozygosity of 0.06 (Harris et al. 1977). The one obvious conclusion that can be reached from Figure 2.9 is that allozyme polymorphisms are widespread among higher organisms. Genetic variation is even more prevalent among some prokaryotes. For example, natural isolates of the mammalian intestinal bacterium *Escherichia coli* exhibit levels of genetic polymorphism two or three times greater than vertebrates (Selander et al. 1987).

Although genetic polymorphisms are widespread, they are not universal. For example, both major subspecies of the cheetah *Acinonynx jubatus* are virtually monomorphic (O'Brien et al. 1987). A survey of 49 enzymes among 30 animals from the East African subspecies (*A. j. raineyi*) yielded only two polymorphic genes and estimates of polymorphism of 0.04 and heterozygosity of 0.01; among 98 animals from the South African species (*A. j. jubatus*), the estimate of polymorphism was 0.02 and that of heterozygosity 0.0004. Most unusual was the finding of skin-graft acceptance between unrelated cheetahs from the South African subspecies. Graft acceptance means that the cheetah population is monomorphic for the major histocompatibility locus, which is abundantly polymorphic in other mammals. Apparently, the cheetah, which was worldwide in its range at one time but presently numbers less than 20,000 animals, underwent at least two severe constrictions in population number resulting in the loss of most of its genetic variability.

How Representative Are Allozymes?

The generality of estimates of polymorphism based on electrophoresis is somewhat uncertain. The amount of polymorphism may be underestimated because conventional electrophoresis fails to detect many amino acid replacements. For example, in a study of 14 myoglobin proteins from various species including cetaceans (whales, dolphins and porpoises), no more than eight could be distinguished by conventional electrophoresis; however, 13 could be distinguished by varying the pH value of the electrophoresis buffer (McLellan and Inouye 1986). Some amino acid replacements can be detected because they render the enzyme sensitive to high temperatures; a test for temperature sensitivity increased the number of identified alleles of the gene coding for xanthine dehydrogenase in *Drosophila pseudoobscura* from 6 to 37 and increased the estimate of average heterozygosity from 0.44 to 0.73 (Singh et al. 1976). On the other hand, although more elaborate techniques reveal additional alleles of genes known to be polymorphic, thus increasing estimates of heterozygosity, genes classified as monomorphic by means of routine electrophoresis tend to remain monomorphic, and so estimates of polymorphism remain much the same as before.

Electrophoretic surveys might also overestimate the amount of polymorphism because the enzymes typically surveyed are those found in relatively high concentration in tissues or body fluids ("Group I enzymes") and often lack the high substrate specificity of enzymes implicated in central metabolic processes ("Group II enzymes"). For example, among 10 Group I and 11 Group II enzymes in *Drosophila*, estimates of polymorphism and heterozygosity were 0.70 and 0.24 in the former and 0.27 and 0.04 in the latter (Gillespie and Langley 1974). In summary, protein electrophoresis is a convenient method for detecting polymorphisms, but it is difficult to extrapolate from electrophoretic surveys of enzymes to the entire genome because the enzymes may not be representative.

Polymorphisms in DNA Sequences

One inevitable limitation of protein electrophoresis is the inability to detect variation in a nucleotide sequence that does not alter the amino acid sequence. A polymorphism is **silent** if it is present in the coding region but does not alter the amino acid sequence; many nucleotide differences in third-codon position are of this type. A polymorphism is **noncoding** if it affects nucleotides in noncoding regions such as the upstream region, the downstream region, or introns. Silent and noncoding polymorphisms may have subtle effects on the organism, and the alleles may be affected by natural selection; the polymorphic alleles are silent or noncoding only in the sense that they all code for the same amino acid sequence. An example of extensive silent polymorphism in *Drosophila* is illustrated in Figure 2.10 for alleles of the gene coding for alcohol dehydrogenase. This gene has an electrophoretic polymorphism that is widespread in natural populations with two predominant alleles, slow (*Adh-S*) and fast (*Adh-F*). The molecular difference is that, in the fourth and last exon of the gene, the codon for amino acid number 193 in *Adh-S* is AAG (lysine) and in *Adh-F* is ACG (threonine). The enzymes differ not only in electrophoretic mobility. The product of the fast allele has a greater enzymatic activity and is also synthesized in greater amount than that of the slow allele.

The data in Figure 2.10 are derived from studies of RFLPs in the *Adh* region of 1533 flies isolated from 25 populations throughout eastern North America (Berry and Kreitman 1993). A total of 113 haplotypes were identified. A **haplotype** is a unique combination of genetic markers present in a chromosome. In Figure 2.10, the haplotypes indicated with squares are *Adh-F* and those with circles are *Adh-S*. The number inside each symbol is the relative abundance of the haplotype (1 being the most frequent, 2 the next most frequent, and so forth). A straight line connecting two haplotypes indicates that they differ by a single change. Figure 2.10 includes 93 haplotypes related to at least one other by a singe change; the other 20 haplotypes observed in the study include additional changes. The main point of the *Adh* example is that natural populations contain a great abundance of different types of nucleotide-sequence variation that does not affect amino acid sequence.

Nucleotide Polymorphism and Nucleotide Diversity

Sequence data can be used quantitatively to estimate the level of genetic variation at the nucleotide level. The data in Problem 2.4 are typical and so will be used to exemplify the calculations. The level of **nucleotide polymorphism**, symbolized θ, is the proportion of nucleotide sites that are expected

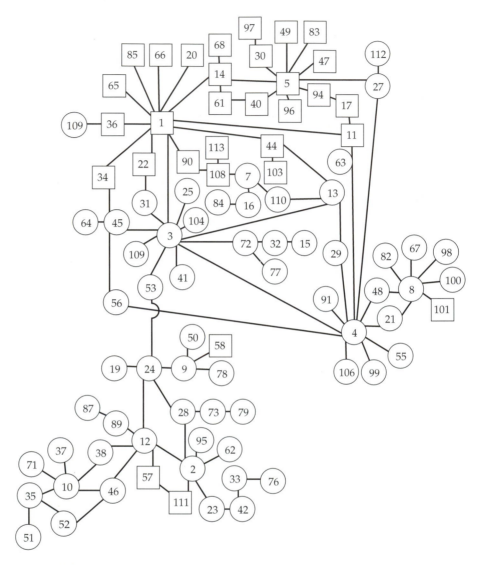

Figure 2.10   Haplotypes of alleles in the *Adh* region of *Drosophila melanogaster* from the East Coast of North America. Each line in the network connects two haplotypes differing by a single molecular difference. An additional 20 haplotypes, differing by more than one change from those in the network, are not shown. Squares indicate the *Adh-F* allele, circles the *Adh-S* allele. (From Berry and Kreitman 1993.)

to be polymorphic in any sample of size 5 from this region of the genome. The estimate θ equals the proportion of nucleotide polymorphism observed in the sample, often symbolized as S, divided by

$$a_1 = \sum_{i=1}^{n-1} \frac{1}{i} \qquad\qquad 2.4$$

where n is the size of the sample. In this case, $S = 16/500 = 0.032$ for a sample of size $n = 5$, so that $a_1 = 1/1 + 1/2 + 1/3 + 1/4 = 2.083$. The estimate of θ, per nucleotide site, is therefore

$$\hat{\theta} = \frac{S}{a_1} = \frac{0.032}{2.083} = 0.015 \qquad\qquad 2.5$$

As noted in Problem 2.4, the variance of $\hat{\theta}$ is not binomial because, owing to genetic linkage, successive nucleotides cannot be regarded as realizations of independent trials. An approximation to the variance can be derived under the assumption that the nucleotides at a site are functionally equivalent or invisible to natural selection; the mathematical details are beyond the scope of this book, but the result is quite simple. The variance of $\hat{\theta}$, per nucleotide site, is given by

$$V(\hat{\theta}) = \frac{\hat{\theta}}{ka_1} + \frac{a_2\hat{\theta}^2}{a_1^2} \qquad\qquad 2.6$$

where a_1 is as defined in Equation 2.4, k is the number of nucleotides in each sequence (in our example, $k = 500$), and a_2 is a function of the number of alleles n in the sample, namely

$$a_2 = \sum_{i=1}^{n-1} \frac{1}{i^2} \qquad\qquad 2.7$$

For $n = 2$ through 10, the values of a_2 are 1, 1.25, 1.36, 1.42, 1.46, 1.49, 1.51, 1.523, 1.54. In the case at hand, $n = 5$ and the estimated variance of $\hat{\theta} = 0.015/(500 \times 2.083) + 1.42 \times 0.015^2/2.083^2 = 9.2131 \times 10^{-5}$. The standard error of $\hat{\theta}$ is the square root of the variance or, in this case, 0.0096 per nucleotide site.

A second quantity used to assess polymorphisms at the DNA level is the **nucleotide diversity**, typically denoted π, which is the average proportion of nucleotide differences between all possible pairs of sequences in the sample. In a sample of n sequences, there are $n(n-1)/2$ pairwise comparisons. For the data in Problem 2.4, $n = 5$, and so there are 10 pairwise comparisons. The pairwise comparisons may be considered for each nucleotide in turn and the differences averaged later. For the polymorphic sites in Problem 2.4, the number of pairwise differences is 6 (= 2 × 3) for sites 132, 142, 246, 351, 405, and 483; it is 4 (= 1 × 4) for sites 162, 198, 201, 207, 240, 354, 372, 375, and 417; and it is 7 for site 192. Among the 484 monomorphic nucleotides in Problem 2.4, the number of pairwise differences is 0. The average proportion of pairwise differences between the sequences in the sample is the estimate— $\hat{\pi}$ —of the **nucleotide diversity**; hence,

$$\hat{\pi} = (6 \times 6 + 4 \times 9 + 1 \times 7 + 0 \times 484)/(10 \times 500) = 0.016$$

The variance of $\hat{\pi}$ is estimated as follows:

$$\mathrm{Var}(\hat{\pi}) = \frac{b_1}{k}\hat{\pi} + b_2\hat{\pi}^2 \qquad\qquad 2.8$$

where k is again the length of the sequences in nucleotides and where

$$b_1 = \frac{n+1}{3(n-1)} \qquad\qquad 2.9$$

$$b_2 = \frac{2(n^2 + n + 3)}{9n(n-1)} \qquad\qquad 2.10$$

For example, when $n = 5$, then $b_1 = 0.5$ and $b_2 = 0.37$, and so $\mathrm{Var}(\hat{\pi}) = (0.5/500) \times 0.016 + 0.37 \times 0.016^2 = 0.000107$; the standard error of $\hat{\pi}$ is the square root, or 0.010.

The estimates of θ and π based on nucleotide sequences are not readily convertible to levels of polymorphism and heterozygosity expected at the protein level. The main reason is that most observed nucleotide polymorphisms are either silent or noncoding and so do not change the amino acid sequence of the polypeptide. The level of protein polymorphism is determined to a large extent by the degree to which the amino acid sequence is constrained by natural selection against variant sequences (or, in some cases, by natural selection for variant sequences), and constraints at the protein level are not generally predictable from θ and π.

On the other hand, there is a theoretical relation between θ and π that is expected under the simplifying assumption that the alleles are invisible to natural selection. The theoretical basis of relation between θ and π is discussed in connection with the neutral theory of molecular evolution in Chapter 8, but the expected relation is that $\theta = \pi$. For the data in Problem 2.4, for example, $\hat{\theta} = 0.015$; this number is to be compared with $\hat{\pi} = 0.016$, and so the agreement with expectation is quite good. (On the other hand, the sample size is very small.)

Estimates of nucleotide polymorphism and diversity can also be carried out with restriction-site data in the form of restriction fragment length polymorphisms (RFLPs). The simplest way to proceed is to analyze the restriction sites in turn. Each monomorphic restriction site is regarded as identifying six adjacent monomorphic nucleotides (or four monomorphic nucleotides, if the enzyme has a four-base restriction site). Each polymorphic restriction site is regarded as identifying five monomorphic nucleotides and one polymorphic nucleotide (or three monomorphic and one polymorphic, if the enzyme has a four-base restriction site). In other words, each restriction site polymorphism is supposed to result from polymorphism of a single nucleotide in the restriction site. Pairwise comparisons to estimate π are carried out under this assumption. The reasoning is illustrated in the following problem.

PROBLEM 2.5 Restriction-site variation was studied around the gene for alcohol dehydrogenase *(Adh)* in a population of *D. melanogaster* descended from animals trapped at a Dutch fruit market in Groningen (Cross and Birley 1986). The region contained a total of 23 sites for five restriction enzymes, each having a six-base restriction site. A total of 16 sites were cut in all flies in the sample. The accompanying table documents the presence (+) or absence (−) of each of the seven polymorphic sites in a sample of 10 chromosomes. Estimate the proportion of polymorphic nucleotides $\hat{\theta}$, the nucleotide diversity $\hat{\pi}$, and the standard error of each. Does the relation $\theta = \pi$ seem to hold for these estimates?

*Bam*HI	*Hind*III	*Pst*I	*Xho*I	*Pst*I	*Eco*RI	*Eco*RI
+	−	−	+	+	−	−
+	−	−	−	−	+	+
−	−	+	−	−	+	−
−	+	−	+	−	+	+
−	+	−	+	−	+	+
−	+	−	+	−	+	+
−	+	−	+	−	+	+
−	+	−	+	−	+	+
−	+	−	+	−	+	+
−	−	−	+	−	+	−

ANSWER Consider first the nucleotide polymorphisms. The 16 monomorphic sites identify $16 \times 6 = 96$ monomorphic nucleotides; the 7 polymorphic sites identify $7 \times 5 = 35$ monomorphic sites and 7×1

polymorphic nucleotides (assuming only 1 nucleotide is altered for each restriction site that is lost). Altogether, there are 138 nucleotides of which 7 are polymorphic. Because $n = 10$, then $a_1 = 2.83$ and $a_2 = 1.54$. The estimate of θ is therefore $\hat{\theta} = (7/138)/2.83 = 0.0179$ per nucleotide site and $\mathrm{Var}(\hat{\theta}) = 0.0179/(138 \times 2.83) + (1.54 \times 0.0179^2/2.83^2) = 1.0778 \times 10^{-4}$. The standard error of $\hat{\theta}$ is therefore 0.0104 per nucleotide site. For estimating π, there are $10 \times 9/2 = 45$ pairwise comparisons, and a restriction site with i "plus" and $(10 - i)$ "minus" means that the polymorphic nucleotide site results in $i \times (10 - i)$ pairwise mismatches. Therefore, the total number of mismatches for each of the restriction sites, from left to right, equals 16, 24, 9, 16, 9, 9, and 21, respectively, totaling 104. In addition, there are 16×6 nucleotides (from the monomorphic sites) and 7×5 nucleotides (from the polymorphic sites) for which the number of pairwise mismatches equals 0. Therefore, $\hat{\pi} = 104/(45 \times 23 \times 6) = 0.017$. For $n = 10$, $b_1 = 0.407$ and $b_2 = 0.279$, and so $\mathrm{Var}(\hat{\pi}) = 0.0001277$; hence the standard error equals 0.011. In these data, $\hat{\theta} = 0.018$ and $\hat{\pi} = 0.017$, which are in very good agreement. However, the sample size is too small to generalize this conclusion.

Uses of Genetic Polymorphisms

Whether studied through allozymes or nucleotide sequences, natural genetic variation has many uses. Genetic variation provides a set of built-in markers for the genetic study of organisms in their native habitats, including organisms for which domestication or laboratory rearing is unfeasible or for which conventional genetic manipulation is impossible.

Genetic polymorphisms are useful in investigating the genetic relationships among subpopulations in a species. The principle is that alleles are shared among subpopulations because of migration, and therefore similarity in allele frequencies among subpopulations can be used to estimate the rate of migration (Chapter 4). Within subpopulations, alleles are shared because of common ancestry. For example, the Ainu people of Northern Japan have numerous Caucasoid-like features, including their facial features, light skin, and hairy bodies, yet their genetic polymorphisms clearly show them to be more closely related to other Mongoloid groups (Watanabe et al. 1975). Among the most informative alleles, the Ainu people possess the *D(Chi)* allele of transferrin protein and the *Di*[a] allele of the Diego blood group, both of which are virtually restricted to Mongoloid populations. Conversely, the Ainu people lack several alleles that are polymorphic in Caucasoids.

From a practical point of view, genetic polymorphisms are useful in human populations as genetic markers that may be genetically linked to harmful genes that cause disease. In kinships with a family history of the disease, the genetic markers can be used to determine which members of the kindred are likely to be carriers of the harmful gene. The markers can also be used in early diagnosis of persons likely to be affected. RFLPs and other types of DNA polymorphisms that are linked to disease genes have also demonstrated their utility as probes for identifying recombinant DNA clones containing the defective genes. The nearby genetic markers enable the defective gene and its function to be identified, thus serving as a first step in the search for effective treatments.

Particularly useful in population genetics are DNA markers with a large number of alleles of moderate frequency. In most organisms, many regions of the genome have multiple alleles consisting of a short sequence of bases repeated in tandem. Multiple alleles result because the number of copies of the repeated sequence may differ from one chromosome to the next. The genotypes are even more variable because each genotype carries two alleles. One of the practical applications of the use of such polymorphisms is in *DNA typing*, in which the alleles in the DNA from a suspect are matched with those from a crime-scene sample. The examination of a sufficient number of such highly variable regions provides a basis for distinguishing one person from another because no two people (with the exception of identical twins) have the same genotype. Genetic variability of this sort is used in determining paternity as well as in criminal investigations. The experimental methods of DNA typing, and certain relevant issues in population genetics, are discussed in Chapter 4.

DNA typing has also been applied to studies of the natural mating systems of plants and animals because, with the large number and high specificity of DNA types, close relatives can be detected in populations. In behavioral studies, DNA typing can determine whether organisms that perform mutually altruistic acts are genetically related. Polymorphisms of other types can also be informative about mating systems. For example, the observed frequencies of genotypes can be used to estimate the amount of self-fertilization in populations of monoecious plants or hermaphroditic animals.

From the standpoint of evolutionary biology, sequences of genes and patterns of polymorphism can be used to make inferences about evolutionary history and about the evolutionary process. The sequences of macromolecules contain within themselves a record of their evolutionary history. Organisms with a shared ancestry usually have similar gene sequences. Conversely, similarity in sequence can be regarded as a measure of shared ancestry. As an index of shared ancestry, sequence similarity provides a means of inferring the ancestral relationships among a group of organisms (*molecular phyloge-*

netics, discussed in Chapter 8). The rates and patterns of change in sequence within species and between closely related species also contain a record of evolutionary forces at work. Within the past 20 years, population genetics has gone from a data-poor field to a data-rich field, and numerous new methods of data analysis and hypothesis testing have been developed.

MULTIPLE-FACTOR INHERITANCE

We have seen that Galton and Mendel chose opposite types of traits for their studies of variation: Galton chose continuous traits, Mendel discrete traits. The choices reflected a deep difference of opinion in the manner in which inheritance should be studied. Galton's approach was empirical, based on the observed similarity between relatives such as parents and offspring. Mendel's approach was theoretical, based on unobserved segregating factors that determined the patterns of inheritance. Even after the rediscovery of Mendel's paper in 1900, the disciples of Galton (called "biometricians") dismissed its significance, claiming that the postulated Mendelian factors were not only irrelevant for continuous traits but also inadequate to explain the observed correlations between relatives. The Mendelians argued that segregation and independent assortment could explain continuous traits just as well as discrete traits. The acrimonious dispute between the biometricians and the Mendelians continued for nearly 20 years.

The dispute abated substantially with a 1918 paper by the statistician Ronald Aylmer Fisher (1890–1962) entitled "The correlation between relatives on the supposition of Mendelian inheritance." Fisher examined a mathematical model of multifactorial inheritance and deduced the expected correlations between relatives. He showed that the kinds of data available for continuous traits were not only compatible with Mendelian inheritance but were also predicted by it.

The spirit of Fisher's model is shown in Figure 2.11, which illustrates the genetic variation expected among the progeny of a cross between genotypes that are heterozygous for each of three unlinked genes. The alleles of the genes are represented A/a, B/b, and C/c, and the genetic variation resulting from segregation and independent assortment is evident in the various degrees of shading. If we assume a trait in which each uppercase allele adds one unit to the phenotype and in which each lowercase allele is without effect, then the *aa bb cc* genotype has a phenotype of 0 and the *AA BB CC* genotype has a phenotype of 6. Thus there are seven possible phenotypes (0–6) among the progeny. The distribution of phenotypes is shown in the bar graph in Figure 2.12. The smooth curve is the normal distribution approximating the data, which has a mean of 3 and a variance of 1.5. In Figure 2.10, we have assumed that all of the variation in phenotype results from differences in genotype. If there were also random environmental factors affecting the trait, as well as a greater number of genes, then the bars in Figure 2.12

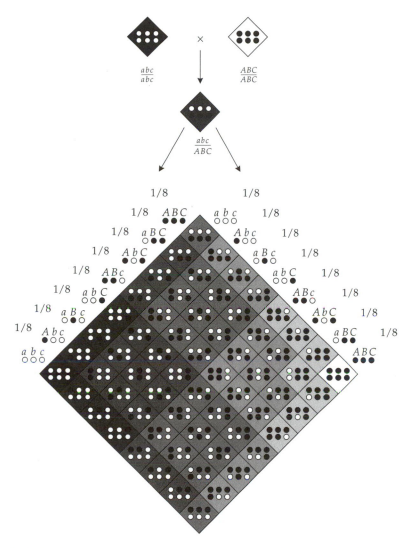

Figure 2.11   Result of segregation of three independent pairs of alleles affecting the same trait. Each allele that is indicated by an uppercase letter is assumed to contribute one unit to the phenotype. The phenotypes range from 0 to 6 and, in the cross between triple heterozygotes, are formed in the proportions 1 : 6 : 15 : 20 : 15 : 6 : 1.

would become less distinct and a normal distribution approximated even better. The result is the central limit theorem at work producing Galton's "supreme law of unreason."

Fisher's model was a good deal more complex than that in Figure 2.10, allowing for differences in the effects of alleles, differences in allele frequen-

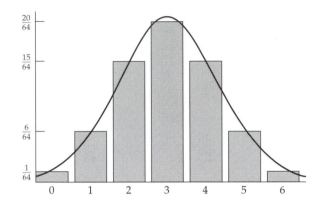

Figure 2.12   Distribution of phenotypes from the cross in Figure 2.11 and the approximating normal distribution. The normal curve has mean 3 and variance 1.5.

cy, various types of dominance relations, and the effects of random environmental factors. The work was pathbreaking in demonstrating that continuous variation could be explained by multiple interacting Mendelian factors. Fisher's model was complex for its time and the paper a difficult one. It is not clear even now what practical role Fisher's paper may have played in ending the controversy between the biometricians and the Mendelians. Not many people seem to have read it. On the other hand, it is the seminal paper that marked the reconciliation of the theories of Galton and Mendel.

SUMMARY

Galton examined the statistical relations between the distributions of phenotypic traits in successive generations. Most of the traits he studied were continuous traits, like height or weight, which are measured on a quantitative scale. Galton was very taken with the observation that the phenotypes of many continuous traits are distributed according to the bell-shaped curve known as the normal distribution. The peak of the normal distribution is determined by the mean and the spread is determined by the variance. Phenotypic variation in natural populations is usually in the form of differences in continuous traits. Most continuous traits are also multifactorial, that is, determined by the combined effects of multiple genetic and environmental factors. The normal distribution is often encountered in practice because of the central limit theorem, which states that the limiting distribution of the sum of a large number of independent random quantities is normal.

Mendel studied discrete variation, such as round versus wrinkled peas, resulting from segregation of the alleles of a single gene. Simple Mendelian variation is the rule for genes and their products. Genetic variation in protein molecules can be identified by such techniques as protein electrophoresis. Proteins differing in electrophoretic mobility that are coded by alternative alleles of the same gene are called allozymes. Allozyme variation is widespread in most organisms. Based on electrophoretic surveys of human populations, about 30% of all enzyme-coding genes are polymorphic (in the sense that the most common allele has a frequency less than 0.95), and about 7% of the loci are heterozygous in an average person. Plants and invertebrates have even higher levels of allozyme variation. Although there is wide variation among species, *Drosophila* averages about 40% polymorphic loci with an average heterozygosity of 14%.

Genetic variation at the DNA level can be detected with the Southern blot procedure, in which DNA fragments produced by a restriction enzyme are separated by electrophoresis and identified by hybridization with a homologous labeled probe sequence. Polymorphisms in the length of restriction fragments (restriction fragment length polymorphisms) are abundant throughout the genome and have applications in studies of genetic linkage in many organisms. DNA studies are also often carried out with the polymerase chain reaction (PCR), in which multiple cycles of primer annealing, DNA replication, and strand separation are used to exponentially amplify the DNA sequence flanked by the oligonucleotide primers. Amplified DNA may be sequenced, used as a probe, or manipulated in other ways.

Polymorphisms in nucleotide sequence are abundant in natural populations, particularly in noncoding regions and at silent sites in coding regions (especially at third codon positions, in which a nucleotide substitution need not result in an amino acid replacement). For a sample of DNA sequences, the amount of nucleotide polymorphism is the proportion of nucleotide sites occupied by two or more bases (A, T, G, C) in the sample. The nucleotide diversity is the average proportion of nucleotide differences between all sequences in the sample taken in pairwise comparison. The estimates of nucleotide polymorphism and nucleotide diversity are not readily compared with allozyme data because much of the observed sequence variation is either noncoding or silent.

There is often a disconnect between molecular variation and phenotypic variation because differences in phenotype among healthy organisms cannot usually be attributed to differences in specific molecules. Indeed, there is a sort of disconnect between simple Mendelian inheritance and continuous variation because the segregation of any pair of alleles affecting a continuous trait is obscured by the segregation of other pairs of alleles as well as by the effects of the environment. In the early years after the rediscovery of Mendel's paper, there was considerable controversy whether Mendelian factors could account for the patterns of variation and correlation among relatives noted by Galton and others. The issue was resolved theoretically by R. A. Fisher's 1918 paper on the correlation between relatives on the supposition of Mendelian inheritance. Closing the gap between the study of evolution at the level of phenotypes and at the level of molecular genotypes remains one of the major challenges in population genetics.

PROBLEMS

1. Shell widths of mussels are approximately normally distributed. If the mean is 70 mm and the standard deviation is 10 mm, what fraction of the population is smaller than 80 mm?
2. Following Problem 1, what fraction of the population is between 80 and 90 mm in width?
3. Calculate the mean, variance, standard deviation, and standard error of the mean for the following bristle counts: 13, 14, 13, 15, 14, 15, 12, 13, 14, 16, 12, 15, 13, 14.
4. Measurements of body weight of a very large sample from a species of mouse have a mean of 60 g and variance of 64 g^2. In one area it was suspected that environmental contamination had reduced the size of the mice. A sample of 100 mice from this area had a mean of 58 g and a sample variance of 64 g^2. Is this sample population significantly smaller in size than the population examined with the very large sample?
5. A standard means for using a computer to generate normally distributed random numbers is to take 12 uniform random numbers and add them up. After scaling the sum by a constant that depends on the mean and the variance, the result represents a sample from the normal distribution one wants. Why does this approach work?
6. One statement of the central limit theorem is that the sum of independent, identically distributed random variables has a limiting normal distribution. If the variables that are being added exhibit positive covariance in successive measures (as opposed to being independent), how would the sum deviate from the normal distribution predicted by the central limit theorem?
7. Allozyme gels reveal a sample with 64 *FF,* 32 *FS* and 4 *SS* females, but there seem to be 40 *FF* males and 10 *SS* males with no heterozygotes. How do you explain these data?
8. Many proteins exist in an active form only as dimers, with two molecules joined either by hydrogen bonding or even by covalent cysteine bridges. If an enzyme is only active as a dimer, and there is electrophoretic variation in a population with two alleles (*F* and *S*), what do you think a heterozygote would look like on a gel? What would a heterozygote look like if only tetramers were active?

9. How many copies of a fragment of DNA should be present after 30 rounds of PCR, assuming perfect efficiency?

10. *Taq* polymerase does not have perfect fidelity in copying DNA sequences, and the result is that PCR products have some variation in sequence. Why do you suppose it is still possible to sequence PCR-amplified DNA to obtain the true sequence? When might the errors caused by *Taq* polymerase cause errors in the final sequence?

11. Many new ways of scoring DNA variation at individual nucleotide sites are becoming available, including an oligonucleotide ligation assay, "Taqman," template-directed dye-terminator incorporation (TDI), and hybridization to dense oligonucleotide arrays known as DNA chips. An important criterion for the utility of any of these methods is that it must be very accurate. Why is accuracy so critical?

12. Four sequences of a 1200 bp gene gave the following counts of pairwise differences: 4, 7, 5, 3, 6, 5. What is the estimate of nucleotide diversity for this sample?

13. In forensic applications of genetics, if the DNA types from a crime scene and a suspect do not match, the confidence one has in the conclusion is much greater than if the types do match. Why?

CHAPTER 3

Organization of Genetic Variation

RANDOM MATING · *HARDY-WEINBERG PRINCIPLE* · *CHI-SQUARE TEST*

MULTIPLE ALLELES · *LINKAGE DISEQUILIBRIUM*

T HE WORD *POPULATION* HAS so far been used in an informal, intuitive sense to refer to a group of organisms belonging to the same species. Further discussion and clarification of the concept is necessary at this time. In population genetics, the word **population** does not usually refer to an entire species; it refers instead to a group of organisms of the same species living within a sufficiently restricted geographical area that any member can potentially mate with any other member (provided that they are of the opposite sex). Precise definition of such a unit is difficult and varies from species to species because of the almost universal presence of some sort of *geographical structure* in species—some typically nonrandom pattern in the spatial distribution of organisms. Members of a species are rarely distributed homogeneously in space: there is almost always some sort of clumping or aggregation, some schooling, flocking, herding, or colony formation. Population subdivision is often caused by environmental patchiness, areas of favorable habitat intermixed with unfavorable areas. Such environmental patchiness is obvious in the case of, for example, terrestrial organisms on islands in an archipelago, but patchiness is a common feature of most habitats—freshwater lakes have shallow and deep areas, meadows have marshy and dry areas, forests have sunny and shady areas. Population subdivision can also be caused by social behavior, as when wolves form packs. Even the human population is clumped or aggregated—into towns and cities, away from deserts and mountains.

The local interbreeding units of possibly large, geographically structured populations are of some interest because it is within such local units that adaptive evolution takes place through systematic changes in allele frequency. Such local interbreeding units—often called **local populations** or **demes**—are the fundamental units of population genetics. Local populations are the actual, evolving units of a species. Unless otherwise specified (or clear from context), the term *population* as used in this book means *local population*. Local populations are sometimes also referred to as *Mendelian populations* or *subpopulations*.

RANDOM MATING

In sexual organisms, genotypes are not transmitted from one generation to the next. Genotypes are broken up in gamete formation by the processes of segregation and recombination, and they are assembled anew in each generation in fertilization: genotypes → gametes → genotypes. The frequency of a specified genotype in a population is the **genotype frequency**. The formation of a genotype in newly fertilized eggs is determined by the opportunity for the relevant gametes to come together in fertilization, and the opportunity for gametes to come together in fertilization is determined by the matings that take place among organisms of reproductive age in the previous generation. To put the matter in a slightly different way, the genotypes of the mating pairs determine the genotypes of the progeny. Furthermore, there are mathematical relationships between the frequencies of mating pairs and the frequencies of progeny genotypes. Such mathematical relationships are usually inferred from models in which the types of matings in the population are specified. One of the important models in population genetics is that of **random mating**, in which mating pairs have the same frequencies as if they were formed by random collisions between genotypes. The chance that an organism mates with another having a prescribed genotype is therefore equal to the frequency of the prescribed genotype in the population. For example, suppose that in some population the genotype frequencies of *AA*, *Aa*, and *aa* are 0.16, 0.48, and 0.36, respectively; if mating is random, *AA* males mate with *AA*, *Aa*, and *aa* females in the proportions 0.16, 0.48, and 0.36, respectively; these same proportions apply to the mates of *Aa* and *aa* males.

Superficial appearances to the contrary, random mating is not a simple or trivial process. One complication is that random mating depends on the trait: mating can be random with respect to some traits but nonrandom with respect to other traits at the same time and in the same population. For example, it is perfectly consistent for a human population to undergo random mating with respect to blood groups, allozyme phenotypes, restriction fragment length polymorphisms, and many other characteristics, but at the same time to engage in nonrandom mating with respect to other traits such as skin color and height. A second complication is population substructure. Paradoxical as it may seem, random mating may be observed within each of the subpopulations constituting a larger population, but random mating may still fail to hold in the population as a whole. (The reason for this paradox is discussed in Chapter 4.) In spite of these and other complications, random mating plays an important role in models in population genetics because random mating often serves as a point of departure for considering more realistic situations.

Nonoverlapping Generations

One of the most important mathematical models in population genetics is the **nonoverlapping generation model**, in which the cycle of birth, maturation, and death includes the death of all organisms present in each generation before the members of the next generation mature. The nonoverlapping generation model is diagrammed in Figure 3.1. The model applies literally only to organisms with a very simple sort of life history, such as certain short-lived insects or annual plants that have a short growing season. In such plants, all members of any generation germinate at about the same time, mature together, shed their pollen, are fertilized almost simultaneously, and die immediately after producing the new generation. This sort of hypothetical population, with its simple life history, is used in population genetics as a first approximation to populations that have more complex life histories. Although at first glance the model seems hopelessly oversimplified, calcula-

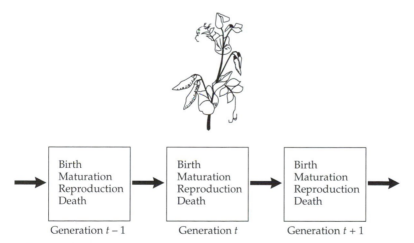

Figure 3.1   The nonoverlapping generation model. The life history of the organism is assumed to be like that of an annual plant (or any short-lived organism), and the generations are assumed to be separated in time (discrete generations). Although the model is simple, it provides a convenient first approximation to populations with more complex life histories.

tions of expected genotype frequencies based on the model are adequate for many purposes. In some applications, the nonoverlapping generation model turns out to be a useful approximation even for populations with a long and complex life history, such as human beings.

The Hardy-Weinberg Principle

Genotype frequencies are determined in part by the pattern of mating. In this section, we consider the consequences of random mating in the model with nonoverlapping generations. To deduce the genotype frequencies under random mating, additional assumptions are needed. First, the allele frequencies should not change from one generation to the next because of systematic evolutionary forces, the most important of which are mutation, migration, and natural selection. For the moment, these evolutionary forces are assumed to be absent or negligibly small in magnitude. (Their effects are discussed in Chapters 5 and 6.) Second, the population must be large enough in size that the allele frequencies are not subject to change merely because of sampling error. Variation in allele frequency owing to sampling error in small populations is called *random genetic drift* and is the subject of Chapter 7. Although random genetic drift is present unless the population is infinite in size, the magnitude of the effect on allele frequency over a small number of generations is usually sufficiently small that the process can be ignored if population size is 500 or more. The qualifier "over a small number of generations" is important because the effects of random genetic drift are cumulative. Considered over a sufficiently large number of generations, random genetic drift can be important even in populations of size 10^6 or more.

Before proceeding further, it may be helpful to summarize the assumptions that we are making:

• The organism is diploid.
• Reproduction is sexual.
• Generations are nonoverlapping.
• The gene under consideration has two alleles.
• The allele frequencies are identical in males and females.
• Mating is random.
• Population size is very large (in theory, infinite).
• Migration is negligible.
• Mutation can be ignored.
• Natural selection does not affect the alleles under consideration.

Collectively, these assumptions summarize the *Hardy-Weinberg model*, named after the English mathematician G. H. Hardy (1877–1947) and the German physiologist Wilhelm Weinberg (1862–1937), who, in 1908, independently formulated the model and deduced its theoretical predictions of genotype frequency.

In the Hardy-Weinberg model, the mathematical relation between the allele frequencies and the genotype frequencies is given by

$$AA: p^2 \qquad Aa: 2pq \qquad aa: q^2 \qquad\qquad 3.1$$

in which p^2, $2pq$, and q^2 are the frequencies of the genotypes AA, Aa, and aa in zygotes of any generation, p and q are the allele frequencies of A and a in gametes of the previous generation, and $p + q = 1$. The frequencies displayed in Equation 3.1 constitute the **Hardy-Weinberg principle** or the **Hardy-Weinberg equilibrium** (**HWE**).

One rationale for the Hardy-Weinberg principle displayed in Equation 3.1 is based on the outcome of repeated and independent trials. With random mating, the choices of male gamete and female gamete are independent trials, and so pairs of gametes carrying the alleles AA, Aa, or aa are expected in proportions given by $(p\,A + q\,a)^2 = p^2\,AA + 2pq\,Aa + q^2\,aa$. A graphical illustration of the rationale of independent trials is shown in Figure 3.2. The chance of two A-bearing gametes coming together is $p \times p = p^2$ and that of two a-bearing gametes coming together is $q \times q = q^2$; for the heterozygote, the chance is $p \times q + q \times p = 2pq$ because the female gamete could carry A and the male gamete carry a, or the other way around.

Male gametes

	Allele	A	a
	Frequency	p	q

Allele	Frequency		
A	p	AA $\;\;\;$ p^2	Aa $\;\;\;$ pq
a	q	aA $\;\;\;$ qp	aa $\;\;\;$ q^2

Female gametes

Summed frequencies in zygotes:

AA: $P' = p^2$
Aa: $Q' = pq + qp = 2pq$
aa: $R' = q^2$

Figure 3.2 Cross-multiplication square showing Hardy-Weinberg frequencies resulting from random mating with two alleles.

TABLE 3.1   **DEMONSTRATION OF THE HARDY-WEINBERG PRINCIPLE**

Mating	Frequency of mating (parents)	Frequency of zygotes (progeny)		
		AA	Aa	aa
$AA \times AA$	P^2	1	0	0
$AA \times Aa$	$2PQ$	$\frac{1}{2}$	$\frac{1}{2}$	0
$AA \times aa$	$2PR$	0	1	0
$Aa \times Aa$	Q^2	$\frac{1}{4}$	$\frac{1}{2}$	$\frac{1}{4}$
$Aa \times aa$	$2QR$	0	$\frac{1}{2}$	$\frac{1}{2}$
$aa \times aa$	R^2	0	0	1
	Totals (next generation)	P'	Q'	R'

therefore

$$P' = P^2 + 2PQ/2 + Q^2/4 = (P + Q/2)^2 = p^2$$
$$Q' = 2PQ/2 + 2PR + Q^2/2 + 2QR/2 = 2(P + Q/2)(R + Q/2) = 2pq$$
$$R' = Q^2/4 + 2QR/2 + R^2 = (R + Q/2)^2 = q^2$$

Random Mating of Genotypes versus Random Union of Gametes

Figure 3.2 implicitly assumes an important premise: that random mating of genotypes is equivalent to random union of gametes. A demonstration of this premise in the case of two alleles is outlined in Table 3.1, in which pairs of genotypes are chosen at random to form matings. The genotype frequencies of AA, Aa, and aa in the parental generation are written as P, Q, and R, respectively, where $P + Q + R = 1$. In terms of the genotype frequencies, the allele frequencies p of A and q of a are as follows:

$$p = (2 \times P + Q)/2 = P + Q/2$$

$$q = (2 \times R + Q)/2 = R + Q/2$$

3.2

Note that $p + q = P + Q + R = 1.0$; this result is a consequence of the fact that the gene has only two alleles.

With two alleles of a gene, there are six possible types of matings. When mating is random, these mating types take place in proportion to the genotypic frequencies in the population, and the types of mating pairs are given by successive terms in the expansion of $(P\ AA + Q\ Aa + R\ aa)^2$. For example, the proportion of $AA \times AA$ matings is $P \times P = P^2$. Similarly, the proportion of $AA \times Aa$ matings is $2 \times P \times Q$ because the mating can include either an AA male with an Aa female (proportion $P \times Q$) or an Aa male with an AA female (proportion $Q \times P$). The frequencies of these and the other types of matings are given in the second column of Table 3.1.

The genotypes of the zygotes produced by the matings are given in the last three columns of Table 3.1. The offspring frequencies follow from Mendel's law of segregation, which states that an Aa heterozygote produces an equal number of A-bearing and a-bearing gametes. The AA and aa homozygotes produce only A-bearing and only a-bearing gametes, respectively. Thus, the mating $AA \times aa$ produces all Aa zygotes, the mating $AA \times Aa$ produces $\frac{1}{2}$ AA and $\frac{1}{2}$ Aa zygotes, the mating $Aa \times Aa$ produces $\frac{1}{4}$ AA, $\frac{1}{2}$ Aa, and $\frac{1}{4}$ aa zygotes, and so forth.

The genotype frequencies of AA, Aa, and aa zygotes after one generation of random mating are denoted in Table 3.1 as P', Q', and R', respectively. These values are calculated as the sum of the cross-products shown at the bottom of the table. The genotype frequencies simplify to $P' = p^2$, $Q' = 2pq$, and $R' = q^2$, where p and q are the allele frequencies given in Equation 3.2. Note that the parental genotype frequencies—P, Q, and R—were completely arbitrary except for the requirement that $P + Q + R = 1$. Therefore, the Hardy-Weinberg frequencies are attained after one generation of random mating irrespective of the genotype frequencies in the parental generation.

PROBLEM 3.1   A four-base cleavage site for the restriction enzyme BanI is located within a large intron of the larval transcript of the gene coding for alcohol dehydrogenase in $D.\ melanogaster$. Cleavage at this site was found in 29 of 60 chromosomes isolated from a population sampled at a farmer's market in Raleigh, North Carolina (Kreitman and Aguadé 1986). Letting B and b represent the presence or absence, respectively, of the BanI site in a chromosome, and assuming Hardy-Weinberg genotype frequencies, calculate the expected frequencies of the genotypes BB, Bb, and bb.

ANSWER   The estimated allele frequencies are $p = 29/60 = 0.48$ of B and $q = 1 - p = 0.52$ of b, and so the expected genotype frequencies with HWE are $p^2 = 0.23$ BB, $2pq = 0.50$ Bb, and $q^2 = 0.27$ bb.

PROBLEM 3.2   In an experimental population of $D.\ melanogaster$, the genotype frequencies for two alleles, $E6^F$ and $E6^S$, of the gene coding for esterase-6 were found to be consistent with Hardy-Weinberg proportions with allele frequencies of 0.3579 for $E6^F$ and 0.6421 for $E6^S$

(Mukai et al. 1974). Assuming that all of the assumptions of the Hardy-Weinberg model hold, particularly those pertaining to random mating in a large population with no mutation, selection, or migration, make a table of mating frequencies similar to Table 3.1 for the esterase-6 alleles. Then calculate the genotype frequencies expected in the next generation along with the corresponding allele frequencies.

ANSWER   The Hardy-Weinberg frequencies among parents are *FF*: 0.1281; *FS*: 0.4596, and *SS*: 0.4123. Therefore, the expected frequencies of the matings are: *FF* × *FF* (0.0164); *FF* × *FS* (0.1177); *FF* × *SS* (0.1056); *FS* × *FS* (0.2112); *FS* × *SS* (0.3790); and *SS* × *SS* (0.1700). The expected genotype frequencies among the zygotes are, for *FF*, 0.0164 + 0.1177/2 + 0.2112/4 = 0.1281; for *FS*, 0.1177/2 + 0.1056 + 0.2112/2 + 0.3790/2 = 0.4596; for *SS*, 0.2112/4 + 0.3790/2 + 0.1700 = 0.4123; note that these are the same as in the parental generation. The allele frequencies of *F* and *S* are again 0.3579 and 0.6421, respectively.

PROBLEM 3.3   Use a cross-multiplication square like that in Figure 3.2 to show that, when the allele frequencies differ in male and female parents, the Hardy-Weinberg frequencies are not attained after one generation of random mating. Use the symbols p_m and q_m for the frequencies of *A* and *a* in male gametes and the symbols p_f and q_f for the frequencies of *A* and *a* in female gametes. After the first generation of random mating, what are the genotype frequencies in male and female zygotes? What are the allele frequencies in male and female zygotes? What are the genotype frequencies in zygotes after the second generation of random mating? Are these in Hardy-Weinberg proportions?

ANSWER   This problem demonstrates the principle that, with random mating, the frequency of an allele in zygotes equals the average of the allele frequencies in the parents. If the allele frequencies in parents differ, then random mating results in Hardy-Weinberg proportions only after two generations. The first generation equalizes the allele frequencies in males and females, and the second generation yields the Hardy-Weinberg proportions. Using the suggested symbols, after one generation of random mating, the genotype frequencies are AA: $p_m \times p_f$, Aa: $p_m \times q_f + q_m \times p_f$, and aa: $q_m \times q_f$. These are not in the form x^2, $2x(1-x)$, and $(1-x)^2$ unless $p_m = p_f$ and $q_m = q_f$. However, the allele frequencies have become equal in the sexes at $\bar{p} = p_m\,p_f + (p_m q_f + q_m p_f)/2 = (p_m + p_f)/2$ and $\bar{q} = (q_m + q_f)/2$. The HWE is reached in one additional generation of random mating, in which the genotype frequencies in zygotes are $\bar{p}^2$, $2\bar{p}\bar{q}$, and $\bar{q}^2$.

Implications of the Hardy-Weinberg Principle

The Hardy-Weinberg principle has provided the foundation for many theoretical and experimental investigations in population genetics. However, the theory is far from profound, and the applicability is far from universal. Hardy especially seems to have regarded the Hardy-Weinberg principle as virtually self-evident. He writes, "I should have expected the very simple point which I wish to make to have been familiar to biologists." In fact, it was familiar to some biologists—the basic principle had been noted as early as 1903 by the Harvard geneticist William E. Castle (1867–1962). Castle's work was little known, however, and Hardy was writing to counter an argument put forth against Mendelism that phenotypic ratios of 3 dominant to 1 recessive should be encountered frequently in natural populations if the mechanism of Mendelian heredity were generally applicable. The immediate implication of the Hardy-Weinberg principle was to refute the 3 : 1 argument by showing that the genotypic ratio of $A-$: aa is determined by the allele frequencies and has no special tendency to attain one particular ratio as any other.

Beyond the virtue of simplicity, why would anyone want to consider a model based on so many restrictive and seemingly incorrect assumptions? And in what sense can such a simple model be considered fundamental? Among several reasons, two stand out. First, the Hardy-Weinberg model is a reference model in which there are no evolutionary forces at work other than those imposed by the process of reproduction itself. In this sense, the model is similar to models in mechanical physics where objects fall through the sky without wind resistance or roll down inclined planes without friction. The model affords a baseline for comparison with more realistic models in which evolutionary forces can change allele frequencies. Perhaps more importantly, the Hardy-Weinberg model separates life history into two intervals: gametes → zygotes and zygotes → adults. In constructing more complex and realistic models, one can often introduce the complications into the zygotes → adults part of the life cycle—for example, in considering the effects of migration into the population or of differential survival among the genotypes. With all sources of change in allele frequency accounted for in the zygotes → adults component, the gametes → zygotes component follows from the principle that random union of gametes and results in the Hardy-Weinberg proportions among zygotes. In other words, the Hardy-Weinberg model is fundamental in the sense that the approach of tracking allele and genotype frequencies through time can be generalized to more realistic situations.

One of the most important implications of the Hardy-Weinberg principle emerges when we calculate the allele frequencies of A and a in the next generation from the formulas for P', Q', and R' in Table 3.1. Using the result in Equation 3.2, the allele frequency of A among the zygotes equals $P' + Q'/2 = p^2 + 2pq/2 = p(p + q) = p$. Likewise, the allele frequency of a among zygotes equals $R' + Q'/2 = q^2 + 2pq/2 = q(q + p) = q$. Thus, the allele frequencies in the next generation are exactly the same as they were the generation before. With random mating, the allele frequencies remain the same generation after generation. In any generation, therefore, the genotype frequencies are p^2, $2pq$, and q^2 for AA, Aa, and aa, respectively, as given in Equation 3.1. The constancy of allele frequency—and therefore of the genotypic composition of the population—is the single most important implication of the Hardy-Weinberg principle. The constancy of allele frequencies implies that, in the absence of specific evolutionary forces to change allele frequency, the mechanism of Mendelian inheritance, by itself, keeps the allele frequencies constant and thus preserves genetic variation. A second item of interest is that the Hardy-Weinberg frequencies are attained in just one generation of random mating if the allele frequencies are the same in males and females. This, however, is true only with nonoverlapping generations; in populations with more complex life histories, the Hardy-Weinberg frequencies are attained gradually over a period of several generations.

It is important to note here that conventional statistical tests for Hardy-Weinberg proportions (such as the χ^2 test discussed below) are not very sensitive to deviations from the expected genotype frequencies. Consequently, the mere fact that observed genotype frequencies may happen to fit the Hardy-Weinberg proportions cannot be taken as evidence that all of the assumptions underlying the model are valid. The most that can be concluded is that, whatever departures from the assumptions there may be, they are not sufficiently large to result in deviations from HWE that are detectable with conventional statistical tests.

The Hardy-Weinberg Principle in Operation

Application of the Hardy-Weinberg principle can be illustrated with data on the MN blood groups in a British population. In a sample of 1000 people (Race and Sanger 1975), the observed phenotypes were 298 blood group M (indicating genotype *MM*), 489 blood group MN (indicating genotype *MN*), and 213 blood group N (indicating genotype *NN*). To determine whether these genotype frequencies are in accord with HWE, the allele frequencies of *M* and *N* must first be estimated. The estimated allele frequency $\hat{p}$ of *M* is $1085/2000 = 0.5425$ and that $\hat{q}$ of *N* is $915/2000 = 0.4575$. (For the details, see Problem 1.4 in Chapter 1.) Were the population in HWE, we would expect the genotype frequencies of *MM*, *MN*, and *NN* to be p^2, $2pq$, and q^2, respectively, where *p* and *q* are the allele frequencies in the underlying population from which the sample was drawn. Because *p* and *q* are parameters, their true values are unknown. However, in testing for HWE we can substitute the estimated values to obtain the expected proportions *MM*: $(0.5425)^2 = 0.2943$, *MN*: $2(0.5425)(0.4575) = 0.4964$, and *NN*: $(0.4575)^2 = 0.2093$, respectively. Because the sample size is 1000, the expected numbers of the *MM*, *MN*, and *NN* genotypes are $0.2943 \times 1000 = 294.3$, $0.4964 \times 1000 = 496.4$, and $0.2093 \times 1000 = 209.3$, respectively.

At this point, it is convenient to tabulate the data into three columns, the first giving the genotypes, the second giving the observed numbers, and the third giving the expected numbers:

MM	298	294.3
MN	489	496.4
NN	213	209.3

With the data so arrayed, it is evident that the fit between the observed numbers and the expected numbers, though not perfect because of chance statistical fluctuations in the number of each genotype that may be included in any given sample, is nevertheless very close. To verify this conclusion, we will apply a conventional statistical test to the data in order to assess quantitatively the closeness of fit. A test commonly employed in population genetics is called the **chi-square test**, which is based on the value of a number, called χ^2, calculated from the data as

$$\chi^2 = \sum \frac{(obs - exp)^2}{exp} \qquad\qquad 3.3$$

where *obs* refers to the observed number in any genotypic class, *exp* refers to the expected number in the same genotypic class, and the Σ sign denotes that the values are to be summed over all genotypic classes. In the case at hand,

$$\chi^2 = (298 - 294.3)^2/294.3$$

$$+ (489 - 496.4)^2/496.4$$

$$+ (213 - 209.3)^2/209.3$$

$$= 0.222$$

To be completely unambiguous, some statisticians prefer use of the symbol X^2 for the realized value of the test statistic defined Equation 3.3, in order to distinguish between the test statistic and the true χ^2 distribution itself. The distinction should certainly be kept in mind, but we will not recognize it formally with different symbols.

Associated with any χ^2 value is a second number called the **degrees of freedom** for that χ^2. In general, the number of degrees of freedom (*df*) associated with a χ^2 value equals

df = Number of classes of data

– Number of parameters estimated from the data

– 1

In the MN example, there are three classes of data and one parameter (p) estimated from the data, and so $df = 3 - 1 - 1 = 1$. Note that a degree of freedom is not subtracted for estimating q because of the relation $q = 1 - p$; that is, once p has been estimated, the estimate of q is automatically fixed, and so we deduct just the one degree of freedom corresponding to p.

Calculation of χ^2 and its associated degrees of freedom is carried out in order to obtain a number for assessing goodness of fit; the number is determined from Figure 3.3. To use the chart, find the value of χ^2 along the horizontal axis, then move vertically from this value until the line for the number of degree of freedom is intersected, then move horizontally from the point of intersection to the vertical axis and read the corresponding probability value P. In our case, with $\chi^2 = 0.222$ and one degree of freedom, the corresponding probability value is about $P = 0.67$. The probability associated with a particular χ^2 test has the following interpretation: it is the probability that chance alone could produce a deviation between the observed and expected values at least as great as the deviation actually realized. Thus, if the probability is large, it means that chance alone could account for the deviation, and it strengthens our confidence in the validity of the model used to obtain the expectations—in this case, the Hardy-Weinberg model. On the other hand, if the probability associated with the χ^2 is small, it means that chance alone is not likely to lead to a deviation as large as actually realized, and it undermines our confidence in the validity of the model. Where exactly the cutoff should be between a "large" probability and a "small" one is, of course, not obvious, but there is an established guideline to follow. If the probability is less than 0.05, then the goodness of fit is considered sufficiently poor that the model is judged invalid for the data; alternatively, if the probability is greater than 0.05, the fit is considered sufficiently close that the model is not rejected. Because the probability in the MN example is 0.67, which is greater than 0.05, we have no reason to reject the hypothesis that the genotype frequencies are in Hardy-Weinberg proportions for this gene.

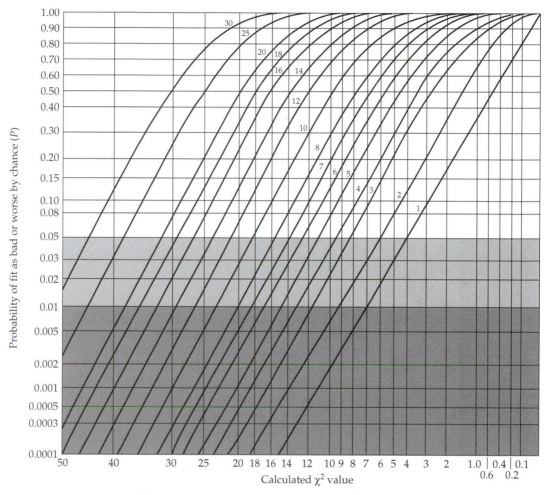

Figure 3.3   Graph of χ^2. To use the graph, find the value of χ^2 along the horizontal axis, then read the probability value for the appropriate number of degrees of freedom from the vertical axis. (From Hartl 1994.)

PROBLEM 3.4   In the Ss blood group, related to the MN system, three phenotypes corresponding to the genotypes *SS*, *Ss*, and *ss* can be identified by appropriate reagents. Among the same 1000 British people who gave the MN data above, the observed number of each genotype for the Ss blood groups were 99 *SS*, 418 *Ss*, and 483 *ss*.

Estimate the allele frequency of S (p) and s (q) and carry out a χ^2 test of goodness of fit between the observed genotype frequencies and their Hardy-Weinberg expectations. Is there any reason to reject the hypothesis of Hardy-Weinberg proportions for this gene?

ANSWER $\hat{p} = 0.308$ and $\hat{q} = 0.692$. The expected numbers of SS, Ss, and ss are 94.86, 426.27, and 478.86, respectively. The $\chi^2 = 0.377$ with one degree of freedom. The associated probability from Figure 3.3 is about 0.55, so there is no reason to reject the hypothesis of HWE.

Complications of Dominance

Dominance obscures the one-to-one relation between phenotype and genotype, but the allele frequencies can still be estimated if one is willing to assume HWE. For a polymorphic gene with two alleles in which one of the alleles is dominant, only two phenotypic classes can be distinguished—the dominant phenotype and the recessive phenotype. An example is the D allele in the human Rh blood groups, which codes for an Rh^+ antigen present on the surface of red blood cells. An alternative allele designated d, fails to code for the antigen. The allele D is dominant over d because both DD and Dd genotypes produce the Rh^+ antigen. The genotypes DD and Dd therefore have the Rh^+ phenotype and are said to be *Rh positive*; the dd genotype has the phenotype Rh^- and is said to be *Rh negative*. At the molecular level, the Dd genotype might be expected to produce only half as much antigen as DD because it contains only one D allele, but the phenotype is nevertheless Rh positive.

Among American Caucasians, the frequency of Rh^+ is about 85.8% and the frequency of Rh^- is about 14.2% (Mourant et al. 1976). Given only the phenotype frequencies, the data cannot be used to calculate the genotype frequencies because we have no way of knowing what proportion of Rh^+ phenotypes are DD and what proportion are Dd. However, if we are willing to assume random mating, then the relative proportions DD and Dd genotypes are given by the Hardy-Weinberg principle. Assuming random mating and HWE, the genotype frequencies are given by p^2, $2pq$, and q^2, where p is the allele frequency of D. An estimate of q can therefore be obtained by setting $q^2 = 0.142$ (the frequency of the homozygous recessive phenotype), and so $\hat{q} = \sqrt{0.142} = 0.3768$. More generally, if R is the frequency of homozygous recessive genotypes found in sample of n organisms, then $\hat{q}$ and its standard error are estimated as

$$\hat{q} = \sqrt{R}$$

$$SE(\hat{q}) = \sqrt{\frac{1-R}{4n}}$$

3.4

With $\hat{q}$ estimated from Equation 3.4 as 0.3768, then $\hat{p} = 1 - 0.3768 = 0.6232$, and the frequencies of DD, Dd, and dd are expected to be $p^2 = (0.6232)^2 = 0.3884$, $2pq = 2(0.6232)(0.3768) = 0.4696$, and $q^2 = (0.3768)^2 = 0.1420$, respectively. The proportion of Rh^+ people that are actually heterozygous is therefore $0.4696/(0.4696 + 0.3884) = 54.7\%$. However, when there is dominance, there is no possibility for a χ^2 test of goodness of fit to HWE because there are 0 degrees of freedom. The lack of degrees of freedom is the reason why the calculated frequencies of Rh^+ and Rh^- $(0.3884 + 0.4696 = 0.858$ and 0.142, respectively) fit the observed frequencies exactly.

PROBLEM 3.5 The Basque people, who live in the Pyrenees mountains between France and Spain, have one of the highest frequencies of the d allele in the Rh system so far reported. In one study of 400 Basques, 230 were found to be Rh^+ and 170 Rh^- (Mourant et al. 1976). Estimate the frequencies of the D and d alleles, the genotype frequencies, and the proportion of Rh^+ people who are heterozygous Dd. What is the standard error of the estimate $\hat{q}$?

ANSWER: $\hat{q} = \sqrt{(170/400)} = 0.65$, $\hat{p} = 0.35$, and the estimated genotype frequencies of DD, Dd, and dd are 0.121, 0.454, and 0.425, respectively. The proportion of Dd among Rh^+ phenotypes in the Basque population is $0.454/(0.121 + 0.454) = 79\%$. The standard error of $\hat{q}$ equals $\sqrt{[(1 - 0.425)/1600]} = 0.02$.

The Hardy-Weinberg principle also finds application in studies of industrial melanism, one of the most famous and best-studied cases of evolution in action (Kettlewell 1973). **Industrial melanism** refers to the evolution of black (melanic) color patterns in several species of moths that accompanied progressive pollution of the environment by coal soot during the industrial revolution.

(The various color forms of the moths are known as **morphs**.) The evolution of melanism has been observed in Great Britain, West Germany, Eastern Europe, the United States, and in other heavily industrialized areas. The species that evolve melanism are typically large moths that fly by night and rest in a sort of cataleptic state by day, often on the trunks of trees, using their cryptic black-and-white mottled color pattern for concealment from visually cued predators such as hedge sparrows, redstarts, and robins (Figure 3.4). Of nearly 800 species of large moths in the British Isles, where industrial melanism has been most intensively studied, about 100 species are industrial melanics (Bishop and Cook 1975). The best known of these are the peppered moth (*Biston betularia*) and the scalloped hazel moth (*Gonodontis bidentata*). In most instances, the melanic color pattern has been found to be due to a single dominant allele.

PROBLEM 3.6   In one study of a heavily polluted area near Birmingham, England, Kettlewell (1956) observed a frequency of 87% melanic *Biston betularia*. Estimate the frequency of the dominant allele leading to melanism in this population and the frequency of melanics that are heterozygous.

Figure 3.4   Melanic and nonmelanic moths, showing camouflage of light moths on light background and dark moths on dark. (Photograph by H. B. D. Kettlewell.)

ANSWER   The observed frequency of homozygous recessives is $R = 0.13$, and so the frequency of recessive allele is estimated as $\hat{q} = \sqrt{(0.13)} = 0.36$. Assuming random mating, the expected frequencies of dominant homozygotes, heterozygotes, and recessive homozygotes are 0.41, 0.46, and 0.13, respectively. The proportion of melanics that are heterozygous is $0.46/0.87 = 52.9\%$.

Frequency of Heterozygotes

The Hardy-Weinberg principle also has important implications for the frequency of heterozygotes carrying rare recessive alleles. The graphs in Figure 3.5 depict the frequencies of *AA*, *Aa*, and *aa* in a population in HWE. The heterozygotes are most frequent when the allele frequencies are 0.5. Suppose that the allele *a* is a recessive, and consider the curves as the allele frequency of *a* goes toward 0. As *a* becomes rare, the frequencies of recessive homozygotes and heterozygotes both decrease, but the frequency of the recessive homozygote is much lower. As the frequency of *a* goes to 0, the frequency of recessive homozygotes goes to 0 at a rate of q^2, whereas the frequency of heterozygotes goes to 0 at a rate of $2pq$. The result is that the ratio of heterozy-

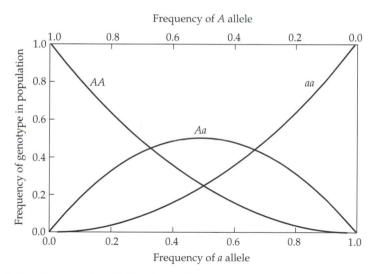

Figure 3.5   Frequencies of *AA*, *Aa*, and *aa* genotypes with HWE. Note that, as either allele becomes more rare, the frequency of homozygotes for that allele is much lower than the frequency of heterozygotes.

gotes to recessive homozygotes increases without limit as the recessive allele becomes rare.

To illustrate the principle, suppose $q = 0.10$; then $2pq/q^2 = 18$, meaning that there are 18 times as many heterozygotes as recessive homozygotes. For $q = 0.01$, to take a more extreme example, the ratio is 198; and for $q = 0.001$, the ratio is 1998. These examples demonstrate that when a recessive allele is rare, most genotypes containing the rare allele are heterozygous.

Quantitatively, the ratio of heterozygotes to homozygotes equals $2pq/q^2 = 2/q - 2$ which, for small q, is approximately $2/q$. Consequently, the excess of heterozygotes over homozygotes becomes progressively greater as the recessive allele becomes more rare. To take a real example, consider cystic fibrosis, an autosomal-recessive defect in chloride transport characterized by abnormal glandular secretions, impaired digestion, frequent respiratory infections, and other serious symptoms. The frequency of the homozygous recessive genotype in newborn Caucasians is approximately 1 in 1700. For this allele, $\hat{q} = \sqrt{(1/1700)} = 0.024$. Assuming random mating, the frequency of heterozygotes is estimated as $2(0.024)(1 - 0.024) = 0.047$, or about 1 in 21. In other words, although only 1 person in 1700 is actually affected with cystic fibrosis, 1 person in 21 is a heterozygous carrier of the harmful allele.

PROBLEM 3.7   Phenylketonuria is a defect in phenylalanine metabolism caused by lack of a functioning allele. Over 200 defective alleles have been identified and most affected individuals are actually heterozygous for two different defective alleles. The condition affects about 1 in 10,000 newborn Caucasians. Estimate the frequency of heterozygotes for the normal and a defective allele under the assumption of random mating.

ANSWER   About 1 person in 50 carries a defective allele.

SPECIAL CASES OF RANDOM MATING

In this section we extend the Hardy-Weinberg principle to multiple alleles and to genes located on the X chromosome.

Three or More Alleles

Genotype frequencies under random mating for genes with three alleles are shown in Figure 3.6. Here it is convenient to label the alleles as A_1, A_2, and A_3

Male gametes

Allele	A_1	A_2	A_3
Frequency	p_1	p_2	p_3

Female gametes

Allele	Frequency			
A_1	p_1	A_1A_1 p_1^2	A_1A_2 p_1p_2	A_1A_3 p_1p_3
A_2	p_2	A_2A_1 p_2p_1	A_2A_2 p_2^2	A_2A_3 p_2p_3
A_3	p_3	A_3A_1 p_3p_1	A_3A_2 p_3p_2	A_3A_3 p_3^2

Figure 3.6 Cross-multiplication square showing Hardy-Weinberg frequencies for three autosomal alleles.

and the corresponding allele frequencies as p_1, p_2, and p_3. Because there are only three alleles, $p_1 + p_2 + p_3 = 1$. With three alleles there are six diploid genotypes, and under random mating their expected frequencies are as follows:

$$A_1A_1 \quad p_1^2$$
$$A_1A_2 \quad 2p_1p_2$$
$$A_2A_2 \quad p_2^2$$
$$A_1A_3 \quad 2p_1p_3$$
$$A_2A_3 \quad 2p_2p_3$$
$$A_3A_3 \quad p_3^2$$

These frequencies can be obtained by expanding $(p_1 A_1 + p_2 A_2 + p_3 A_3)^2$, which the cross-multiplication square in Figure 3.6 does automatically.

Application of Figure 3.6 can be illustrated with the familiar ABO blood groups in humans. The ABO blood groups are controlled by three alleles designated I^O, I^A, and I^B. Genotypes I^AI^A and I^AI^O have blood type A; genotypes

$I^B I^B$ and $I^B I^O$ have blood type B, genotype $I^O I^O$ has blood type O, and genotype $I^A I^B$ has blood type AB. In one test of 6313 Caucasians in Iowa City, the number of people with blood types A, B, O, and AB was found to be 2625, 570, 2892, and 226, respectively (Mourant et al. 1976). The best estimates of allele frequency in this case are $\hat{p}_1 = 0.2593$ (for I^A), $\hat{p}_2 = 0.0625$ (for I^B), and $\hat{p}_3 = 0.6755$ (for I^O). (Estimation of allele frequencies for the ABO blood groups is complicated because of dominance; for methods see Cavalli-Sforza and Bodmer 1971 and Vogel and Motulsky 1986.) The expected (and observed) numbers of the four blood-type phenotypes are therefore:

A: $(0.2593^2 + 2 \times 0.2593 \times 0.6755) \times 6313 = 2636.0$ (observed 2625)

B: $(0.0652^2 + 2 \times 0.0652 \times 0.6755) \times 6313 = 582.9$ (observed 570)

O: $0.6755^2 \times 6313 = 2880.6$ (observed 2892)

AB: $(2 \times 0.2593 \times 0.0652) \times 6313 = 213.5$ (observed 226)

The χ^2 for goodness of fit to Hardy-Weinberg proportions is 1.11. There is one degree of freedom for this test: 4 (to start with) − 1 (for fixing the total at 6313) − 1 (for estimating $\hat{p}_1$ from the data) − 1 (for estimating $\hat{p}_2$ from the data); a degree of freedom is not deducted for estimating $\hat{p}_3$ because $\hat{p}_3 = 1 - \hat{p}_1 - \hat{p}_2$. For a χ^2 of 1.11 with one degree of freedom, the associated probability from Figure 3.3 is about 0.30, and so the Iowa City population gives no evidence against Hardy-Weinberg proportions for this gene.

PROBLEM 3.8   In a sample of 1617 Spanish Basques, the numbers of A, B, O, and AB blood types observed were 724, 110, 763, and 20, respectively (Mourant et al. 1976). The best estimates of allele frequency are $\hat{p}_1 = 0.2661$ (for I^A), $\hat{p}_2 = 0.0411$ (for I^B), and $\hat{p}_3 = 0.6928$ (for I^O). Calculate the expected numbers of the four phenotypes and carry out a χ^2 test for goodness of fit to the Hardy-Weinberg expectations.

ANSWER   The expected numbers of A, B, O, and AB are 710.7, 94.8, 776.1, and 35.4, respectively. The χ^2 equals 9.61 with one degree of freedom, for which the corresponding probability is 0.0025. Because a deviation as large or larger than that observed would be expected by chance in only 0.0025 samples (that is, about 1 in 400), there is very good reason to reject the hypothesis that the genotypes are in Hardy-Weinberg proportions in this population. The reason for the discrepancy is not known. One likely possibility is migration into the population by people with allele frequencies that are significantly different from those among the Basques themselves.

PROBLEM 3.9   Among many aboriginal American Indian tribes, the allele frequency of I^B is extremely low. For example, a sample of 600 Papago Indians from Arizona included 37 A and 563 O blood types (Mourant et al. 1976). What are the best estimates of the allele frequencies of I^A, I^B, and I^O in this population, and what are the expected genotype frequencies assuming random mating?

ANSWER   There are no I^B alleles in the sample, so the best estimate of p_2 is 0. Thus, there are only two alleles I^A and I^O with I^A dominant. The best estimate of p_3 is thus obtained from Equation 3.4 as $\sqrt{(563/600)} = 0.9687$ and that of p_1 as $1 - p_3 = 0.0313$. The expected genotype frequencies are $0.0313^2 = 0.0010$ for $I^A I^A$, $2(0.0313)(0.9687) = 0.0606$ for $I^A I^O$, and $0.9687^2 = 0.9384$ for $I^O I^O$.

In general, if there are n alleles

$$A_1, A_2, \ldots, A_n$$

with respective frequencies

$$p_1, p_2, \ldots, p_n$$

(and $p_1 + p_2 + \cdots + p_n = 1$), then the genotype frequencies expected under random mating are

$$\begin{aligned} p_i^2 \quad & \text{for } A_i A_i \text{ homozygotes} \\ 2p_i p_j \quad & \text{for } A_i A_j \text{ heterozygotes} \end{aligned}$$

3.5

Equation 3.5 may be applied to data on allozyme polymorphisms in *Drosophila persimilis* in California. One sample of 108 adult flies from the Fish Creek population included four alleles of the gene *Xdh*, which codes for xanthine dehydrogenase. We may call the alleles *Xdh-1*, *Xdh-2*, *Xdh-3*, and *Xdh-4*; their respective frequencies were estimated as $\hat{p}_1 = 0.08$, $\hat{p}_2 = 0.21$, $\hat{p}_3 = 0.62$, and $\hat{p}_4 = 0.09$ (Prakash 1977). With four alleles, there are four possible homozygotes (for example, *Xdh-1/Xdh-1*) and six possible heterozygotes (for example, *Xdh-1/Xdh-2*). In a random-mating population, the frequency of any homozygous genotype is expected to be the square of the corresponding allele frequency. For example, the frequency of *Xdh-1/Xdh-1* is expected to be p_1^2, and the frequency of any heterozygous genotype is expected to be two times the product of the corresponding allele frequencies. For example, the frequency of *Xdh-1/Xdh-2* is expected to be $2p_1p_2$. The Hardy-Weinberg frequencies for all 10 possible genotypes can be obtained by expanding the expression $(0.08 \; Xdh\text{-}1 + 0.21 \; Xdh\text{-}2 + 0.62 \; Xdh\text{-}3 + 0.09 \; Xdh\text{-}4)^2$.

PROBLEM 3.10   Four alleles of the gene *Adh* coding for alcohol dehydrogenase were found in a Texas population of *Phlox cuspidata* (Levin 1978). The alleles may be designated *Adh-1*, *Adh-2*, *Adh-3*, and *Adh-4*. Their frequencies were estimated as 0.11, 0.84, 0.01, and 0.04, respectively. What are the expected Hardy-Weinberg proportions of the 10 genotypes?

ANSWER   *Adh-1/Adh-1*: $0.11^2 = 0.0121$; *Adh-1/Adh-2*: $2(0.11)(0.84) = 0.1848$; *Adh-2/Adh-2* $= 0.84^2 = 0.7056$; *Adh-1/Adh-3* $= 2(0.11)(0.01) = 0.0022$; *Adh-2/Adh-3* $= 2(0.84)(0.01) = 0.0168$; *Adh-3/Adh-3* $= 0.01^2 = 0.0001$; *Adh-1/Adh-4* $= 2(0.11)(0.04) = 0.0088$; *Adh-2/Adh-4* $= 2(0.84)(0.04) = 0.0672$; *Adh-3/Adh-4* $= 2(0.01)(0.04) = 0.0008$; *Adh-4/Adh-4* $= 0.04^2 = 0.0016$. It should be pointed out that the observed genotype frequencies were nowhere near the Hardy-Weinberg expectations because *Phlox cuspidata* undergoes a substantial frequency of self-fertilization (about 78%), which violates the assumption of random mating. How to deal with such departures from random mating is discussed in Chapter 4.

X-Linked Genes

An important exception to the rule that diploid organisms contain two alleles of every gene applies to genes on the X and Y chromosomes. In mammals and many insects, females have two copies of the X chromosome whereas males have one X chromosome and one Y chromosome. The X and Y chromosomes segregate, and so half the sperm from a male carry the X chromosome and half carry the Y chromosome. Although the Y chromosome carries very few genes other than those involved in the determination of sex and male fertility, the X chromosome carries as full a complement of genes as any other chromosome. Genes on the X chromosome are called **X-linked genes**, and the important consequence of X linkage is that a recessive allele on the X chromosome in a male is expressed phenotypically because the Y chromosome lacks any compensating allele. For X-linked genes with two alleles, therefore, there are three female genotypes (AA, Aa, and aa) but only two male genotypes (A and a).

The consequences of random mating with two X-linked alleles are shown in Figure 3.7, where the alleles are denoted X^A and X^a. Note that in females, which have two X chromosomes, the genotype frequencies are as given by the Hardy-Weinberg principle in Equation 3.1; in males, which have only one

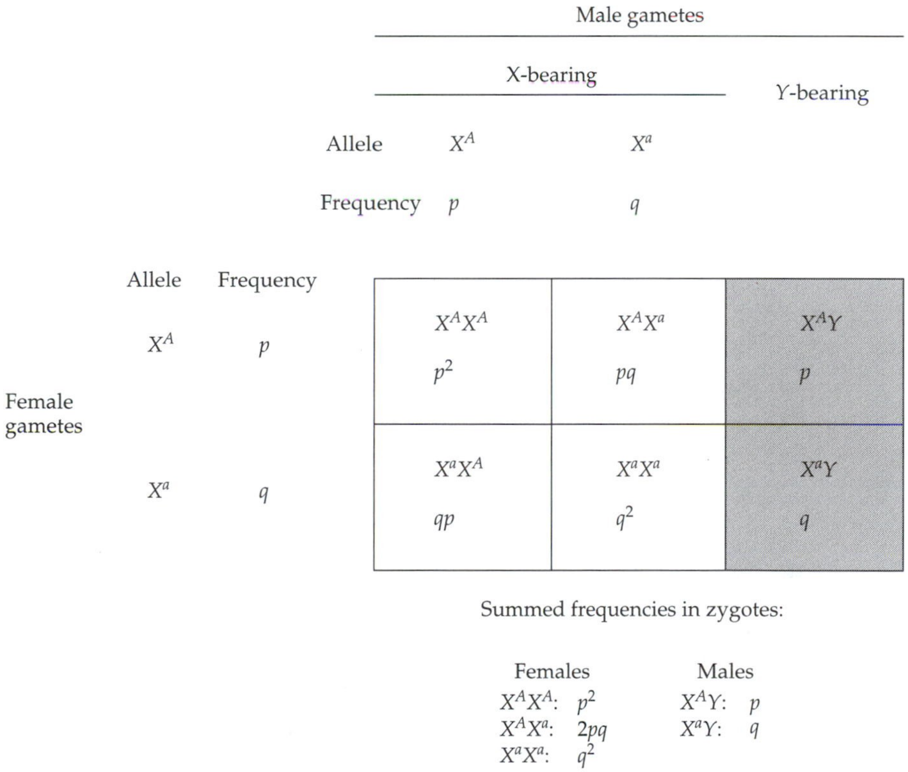

Summed frequencies in zygotes:

	Females		Males	
	$X^A X^A$:	p^2	$X^A Y$:	p
	$X^A X^a$:	$2pq$	$X^a Y$:	q
	$X^a X^a$:	q^2		

Figure 3.7   Consequences of random mating with X-linked genes. Genotype frequencies in females equal the Hardy-Weinberg frequencies, and genotype frequencies in males equal the allele frequencies.

X chromosome, the genotype frequencies are equal to the allele frequencies. The calculations in Figure 3.7 are valid only if the allele frequencies are identical in eggs and sperm. When they differ, approximate equality of allele frequencies in the sexes is usually attained for X-linked genes in a period of 10 or so generations of random mating because, in each generation, any allele frequency in female zygotes is the average of the frequency of the allele in male and female parents in the previous generation.

PROBLEM 3.11   The human Xg blood group is controlled by an X-linked gene with two alleles, designated Xg^a and Xg. Two phenotypes can be distinguished by means of the appropriate antisera, Xg(a+) and Xg(a–). Xg^a is dominant to Xg, and so females of genotype Xg^a/Xg^a and Xg^a/Xg have blood type Xg(a+), whereas females of genotype Xg/Xg are phenotypically Xg(a–). Males of genotype Xg^a have blood type Xg(a+); those of genotype Xg have blood type Xg(a–). In a sample of 2082 British people, there were 967 Xg(a+) females, 667 Xg(a+) males, 102 Xg(a–) females, and 346 Xg(a–) males (Race and Sanger 1975). The best estimates of allele frequency are $\hat{p} = 0.675$ (for Xg^a) and $\hat{q} = 0.325$ (for Xg). Calculate the expected numbers in the four phenotypic classes, assuming random-mating proportions, and carry out a χ^2 test for goodness of fit. (The number of degrees of freedom in this case is 1: there are four degrees of freedom to start with; one must be deducted for using the observed number of males in calculating the expectations for males; one must be deducted for using the observed number of females in calculating their expectations; and one more must be deducted for estimating p from the data.)

ANSWER   The expected numbers of Xg(a+) and Xg(a–) males are $0.675 \times 1013 = 683.8$ and $0.325 \times 1013 = 329.2$, respectively. The expected numbers of Xg(a+) and Xg(a–) females are $[0.675^2 + 2(0.675)(0.325)] \times 1069 = 956.1$ and $0.325^2 \times 1069 = 112.9$, respectively. The χ^2 equals 2.45 which, as noted above, has one degree of freedom. The associated probability is about 0.12 (Figure 3.3), and so there is no reason to reject the hypothesis of random-mating proportions.

One of the important features of random mating for X-linked genes is that phenotypes resulting from a recessive allele will be more common in males than in females. In Problem 3.11, for example, the proportion of Xg(a–) males is $346/1013 = 34\%$, whereas the proportion of Xg(a–) females is only

102/1069 = 10%. There is always an excess of affected males because q (which equals the proportion of males with the recessive phenotype) will always be greater than q^2 (which is the proportion of females with the recessive phenotype). Indeed, the discrepancy grows larger as the recessive allele becomes more rare. For example, with the X-linked "green" type of color blindness, q = 0.05 in Western Europeans, and so the ratio of affected males to affected females is $q/q^2 = 1/q = 1/0.05 = 20$. In contrast, for the X-linked "red" type of color blindness, q = 0.01 and so, in this case, the ratio of affected males to affected females is $1/0.01 = 100$.

PROBLEM 3.12   *California populations of Drosophila persimilis have two alleles of an X-linked gene coding for allozymes of phosphoglu-comutase-1 (Policansky and Zouros 1977). The alleles may be designated Pgm-1^A and Pgm-1^B; their estimates frequencies were 0.25 and 0.75, respectively. Assuming random-mating proportions, what are the expected genotype frequencies in males and females?*

ANSWER   In males, $Pgm\text{-}1^A$ at 0.25 and $Pgm\text{-}1^B$ at 0.75. In females, $Pgm\text{-}1^A/Pgm\text{-}1^A$ at $0.25^2 = 0.0625$; $Pgm\text{-}1^A/Pgm\text{-}1^B$ at $2(0.25)(0.75) = 0.3750$; $Pgm\text{-}1^B/Pgm\text{-}1^B$ at $0.75^2 = 0.5625$.

Before leaving the subject of X-linkage, it is necessary to point out that certain species—among them, birds, moths, and butterflies—have the sex-chromosome situation backwards. In these species, females are XY and males XX. The consequences of random mating are the same as otherwise, except that the sexes are reversed.

LINKAGE AND LINKAGE DISEQUILIBRIUM

With random mating, the alleles of any gene are combined at random into genotypes according to frequencies given by the Hardy-Weinberg proportions. To be specific, imagine a gene with two alleles, call them A_1 and A_2, at frequencies p_1 and p_2, respectively, where $p_1 + p_2 = 1$. Then the Hardy-Weinberg principle tells us that genotypes A_1A_1, A_1A_2, and A_2A_2 are expected in the proportions p_1^2, $2p_1p_2$, and p_2^2, respectively, provided that mating is random.

Similarly, we may consider a different gene with alleles B_1 and B_2 at frequencies q_1 and q_2, respectively, where $q_1 + q_2 = 1$. Then the Hardy-Weinberg principle tells us again that the genotype frequencies of B_1B_1, B_1B_2, and B_2B_2

are expected in the proportions q_1^2, $2q_1q_2$, and q_2^2, respectively, provided that mating is random. Thus, the A_1 allele is in random association with the A_2 allele, and the B_1 allele is in random association with the B_2 allele. Strange as it may seem, the alleles of the A gene may nevertheless fail to be in random association with the alleles of the B gene. The precise meaning of "random association" is illustrated in Figure 3.8. In this figure the squares refer to the alleles present in gametes, not to genotypes as in earlier diagrams. When the alleles of the genes are in random association, the frequency of a gamete carrying any particular combination of alleles equals the product of the frequencies of those alleles. Genes that are in random association are said to be in a state of **linkage equilibrium**, and genes not in random association are said to be in **linkage disequilibrium**. With linkage equilibrium, therefore, the gametic frequencies are:

$$
\begin{array}{lll}
A_1B_1: & p_1 \times q_1 \\[4pt]
A_1B_2: & p_1 \times q_2 \\[4pt]
A_2B_1: & p_2 \times q_1 \\[4pt]
A_2B_2: & p_2 \times q_2
\end{array}
$$

3.6

With random mating and the other simplifying assumptions listed earlier (including a large population with no mutation, migration, or selection), linkage equilibrium between genes is eventually attained. However, linkage equilibrium is attained gradually, and the rate of approach can be very slow. The slow approach to linkage equilibrium stands in contrast to the attainment of HWE with alleles of a single gene, which typically requires just one generation (when generations are nonoverlapping) or a relatively small number of generations (when generations are overlapping).

The rate of approach to linkage equilibrium depends on the rate of recombination in genotypes heterozygous for both genes. There are two types of double heterozygotes:

$$A_1B_1/A_2B_2$$

$$A_1B_2/A_2B_1$$

In the first case, the genotype was formed by the union of an A_1B_1 gamete with an A_2B_2 gamete. In the second case, the genotype was formed by the union of an A_1B_2 gamete with an A_2B_1 gamete. For the moment, consider the genotype A_1B_1/A_2B_2. The gametes produced by this genotype are of four types: (1) A_1B_1, (2) A_2B_2, (3) A_1B_2, and (4) A_2B_1. Gametic types 1 and 2 are known as **nonrecombinant gametes** because the alleles are associated in the same manner as in the previous generation (specifically, A_1 with B_1 and A_2 with B_2). Gametic types 3 and 4 are known as **recombinant** gametes because

Alleles of A gene

Allele	A_1	A_2
Frequency	p_1	p_2

	Allele	Frequency		
			A_1B_1 $\quad$ p_1q_1	A_2B_1 $\quad$ p_2q_1
	B_1	q_1		
Alleles of B gene				
	B_2	q_2	A_1B_2 $\quad$ p_1q_2	A_2B_2 $\quad$ p_2q_2

Figure 3.8 Random association between two alleles of each of two genes, showing expected gametic frequencies when the alleles are in linkage equilibrium.

the alleles are associated differently than in the previous generation (specifically, A_1 with B_2 and A_2 with B_1).

Because of Mendelian segregation, the frequency of gametic type 1 equals that of type 2, and the frequency of gametic type 3 equals that of type 4. That is, the two nonrecombinant gametes are formed in equal frequencies, and the two recombinant gametes are formed in equal frequencies. However, the overall frequency of recombinant gametes (type 3 + type 4) does not necessarily equal the overall frequency of nonrecombinant gametes (type 1 + type 2) except in special cases. The term **recombination fraction**, usually symbolized r, refers to the proportion of recombinant gametes produced by a double heterozygote. Suppose, for example, that the genotype A_1B_1/A_2B_2 produces gametes A_1B_1, A_2B_2, A_1B_2, and A_2B_1 in the proportions 0.38, 0.38, 0.12, and 0.12, respectively. Then the recombination fraction between the genes is $r = 0.12 + 0.12 = 0.24$.

The recombination fraction between genes depends on whether they are present on the same chromosome and, if so, on the physical distance between them. For genes on different chromosomes, the recombination fraction is $r = 0.5$ because the four possible gametic types are produced in equal frequency. For genes on the same chromosome, the recombination fraction depends on their distance apart, because each chromosome aligns side-by-side with its partner chromosome in meiosis and can undergo a sort of breakage and reunion resulting in an exchange of parts between the partner chromosomes. The closer two genes are, the less likely that a breakage and reunion takes place in the region between the genes; the farther apart two genes are, the more likely such an event becomes. The smallest possible recombination fraction is $r = 0$, which would imply that the two genes are so close together that a break never takes place between them. The largest possible recombination fraction is $r = 0.5$, which is found when genes are very far apart on the same chromosome or, as noted above, when they are on different chromosomes. Genes for which the recombination fraction is less than 0.5 must necessarily be on the same chromosome, and such genes are said to be **linked**.

To sum up, if the recombination fraction between the A and B genes is denoted r, then the genotype A_1B_1/A_2B_2 produces the following types of gametes:

$$A_1B_1 \text{ with frequency } (1 - r)/2$$

$$A_2B_2 \text{ with frequency } (1 - r)/2$$

$$A_1B_2 \text{ with frequency } r/2$$

$$A_2B_1 \text{ with frequency } r/2$$

The situation in A_1B_2/A_1B_2 genotype is much the same, but there is one important difference. In this case, the A_1B_1 and A_2B_2 gametes are the *recombinant types*, and the A_1B_2 and A_2B_1 gametes are the *nonrecombinant types.* Thus, the genotype A_1B_2/A_1B_2 produces the following types of gametes:

$$A_1B_1 \text{ with frequency } r/2$$

$$A_2B_2 \text{ with frequency } r/2$$

$$A_1B_2 \text{ with frequency } (1 - r)/2$$

$$A_2B_1 \text{ with frequency } (1 - r)/2$$

PROBLEM 3.13   The genes for the human MN and Ss blood groups discussed in Problem 3.4 are close together on the same chromosome. Suppose that the recombination fraction between the genes is $r = 0.01$. What types and frequencies of gametes would be produced by a person of genotype MS/Ns? By a person of genotype Ms/NS?

ANSWER   The MS/Ns genotype produces gametic types MS, Ns, Ms, and NS in proportions $(1 - 0.01)/2 = 0.495$, $(1 - 0.01)/2 = 0.495$,

0.01/2 = 0.005, and 0.01/2 = 0.005, respectively. The *Ms*/*NS* genotype produces exactly the same gametic types, but their frequencies are 0.005, 0.005, 0.495, and 0.495, respectively.

The recombination fraction between genes is important in population genetics because it governs the rate of approach to linkage equilibrium. To be precise, consider a population in which the actual frequencies of the chromosome types among gametes are as follows:

$$A_1B_1: \quad P_{11}$$

$$A_1B_2: \quad P_{12}$$

$$A_2B_1: \quad P_{21}$$

$$A_2B_2: \quad P_{22}$$

where $P_{11} + P_{12} + P_{21} + P_{22} = 1$. In terms of the gametic frequencies, linkage equilibrium is defined as the state in which $P_{11} = p_1q_1$, $P_{12} = p_1q_2$, $P_{21} = p_2q_1$, and $P_{22} = p_2q_2$ (see Figure 3.8).

Suppose that the genes are not in linkage equilibrium. To determine how rapidly linkage equilibrium is approached, we need to deduce the gametic frequencies in the next generation. Consider first the A_1B_1 gamete. In any one generation, a chromosome carrying A_1B_1 either could have undergone recombination between the genes (an event with probability r, where r is the recombination fraction), or could have escaped recombination between the genes (an event with probability $1 - r$). Among the A_1B_1 chromosomes that did not undergo recombination, the frequency of A_1B_1 is the same as it was in the previous generation; among the chromosomes that did undergo recombination, the frequency of A_1B_1 chromosomes is simply the frequency of $-B_1/A_1-$ genotypes in the previous generation, where the dash in place of the A and B allele means that the identity of that particular allele is irrelevant. Because mating is random, the overall frequency of $-B_1/A_1-$ genotypes is p_1q_1. Putting all the steps in the argument together, the frequency of A_1B_1 in any generation, call it P_{11}', is related to the frequency P_{11} in the previous generation by the equation

$$P_{11}' = (1 - r) \times P_{11} \qquad \text{[for the nonrecombinants]}$$

$$+ r \times p_1q_1 \qquad \text{[for the recombinants]}$$

Subtraction of p_1q_1 from both sides leads to

$$P_{11}' - p_1q_1 = (1-r)(P_{11} - p_1q_1) \qquad\qquad 3.7$$

Equation 3.7 becomes simplified somewhat by defining D as the difference $P_{11} - p_1q_1$. Then D_n is the value of D in the nth generation, and Equation 3.7 implies that $D_n = (1 - r)D_{n-1}$. The solution of this equation is found by successive substitution as

$$D_n = (1-r)D_{n-1} = (1-r)^2 D_{n-2} = \cdots = (1-r)^n D_0 \qquad 3.8$$

where D_0 is the value of D in the founding population. Because $1 - r < 1$, $(1 - r)^n$ goes to zero as n becomes large, but how rapidly $(1 - r)^n$ goes to zero depends on r; the closer r is to zero, the slower the rate. This principle is illustrated in Figure 3.9. Recall here that $r = 0.5$ corresponds either to genes far apart in the same chromosome or to genes in different chromosomes. Because $(1 - r)^n$ goes to zero, D goes to zero, and therefore P_{11} goes to p_1q_1 unless there are other offsetting processes. Analogous arguments hold for gametes containing A_1B_2, A_2B_1, or A_2B_2, and so P_{12}, P_{21}, and P_{22} go to p_1q_2, p_2q_1, and p_2q_2, respectively. Thus, linkage equilibrium is attained at a rate determined by the value of r.

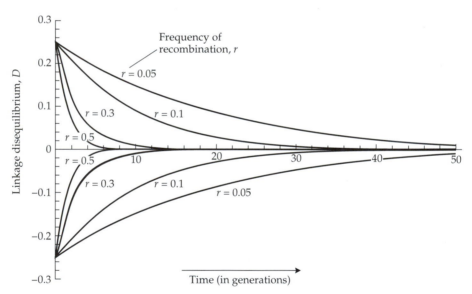

Figure 3.9   Linkage disequilibrium between genes gradually disappears when mating is random, provided there is no countervailing force building it up. The rate of approach to linkage equilibrium depends on the recombination frequency between the genes. The disappearance of linkage disequilibrium is gradual even with free recombination ($r = \frac{1}{2}$). In these examples, the frequencies of both alleles at both loci equal $\frac{1}{2}$, and the initial linkage disequilibrium is either at its maximum ($D = 0.25$) or minimum ($D = -0.25$) value, given these allele frequencies.

The value of D that holds for $P_{11} - p_1q_1$ also holds for the other possible gametes, as follows

$$P_{11} = p_1q_1 + D$$
$$P_{12} = p_1q_2 - D$$
$$P_{21} = p_2q_1 - D$$
$$P_{22} = p_2q_2 + D$$

The quantity D is often called the **linkage disequilibrium parameter**. In terms of the gametic frequencies, D can be shown to satisfy

$$D = P_{11}P_{22} - P_{12}P_{21} \qquad 3.9$$

With random mating and no countervailing forces, the value of D changes according to Equation 3.8, and $D = 0$ corresponds to linkage equilibrium. Furthermore, P_{11}, P_{12}, P_{21}, and P_{22} must all be nonnegative and so, for any prescribed allele frequencies p_1, p_2, q_1, and q_2, the smallest possible (D_{min}) and largest possible (D_{max}) values of D are as follows

$$D_{min} = \text{the larger of } -p_1q_1 \text{ and } -p_2q_2$$
$$\qquad 3.10$$
$$D_{max} = \text{the smaller of } p_1q_2 \text{ and } p_2q_1$$

In studies of linkage disequilibrium, estimation of the gametic frequencies P_{11}, P_{12}, P_{21}, and P_{22} usually requires complex statistical procedures rather than straightforward chromosome-counting methods because there are 10 genotypes but usually no more than nine phenotypes. (There are 10 genotypes because A_1B_1/A_2B_2 and A_1B_2/A_2B_1 must be distinguished.)

An example of linkage disequilibrium is found in the genes controlling the MN and Ss blood groups in human populations. Earlier in this chapter, we cited data from 1000 Britishers with respect to the MN blood groups and showed that the genotypes *MM*, *MN*, and *NN* are in Hardy-Weinberg proportions. In Problem 3.4, data from the same 1000 people were analyzed with respect to the Ss blood groups, and genotypes *SS*, *Ss*, and *ss* were also found to satisfy the Hardy-Weinberg proportions. In order to discuss linkage disequilibrium between the genes, it will be convenient to use the symbols p_1 and p_2 for the allele frequencies of *M* and *N*, respectively, and the symbols q_1 and q_2 for the allele frequencies of *S* and *s*, respectively. The earlier analyses yielded estimates of $\hat{p}_1 = 0.5425$ and $\hat{p}_2 = 0.4575$ for *M* and *N* and $\hat{q}_1 = 0.3080$ and $\hat{q}_2 = 0.6920$ for *S* and *s*. Were the loci in linkage equilibrium, the gametic frequencies would be p_1q_1 for MS, p_1q_2 for Ms, p_2q_1 for NS, and p_2q_2 for Ns. Therefore, among the 1000 genotypes (a total of 2000 chromosomes), the expected numbers are as shown in the third column below (the second column gives the observed numbers):

$$MS \quad 474 \quad 0.5425 \times 0.3080 \times 2000 = 334.2$$

$$Ms \quad 611 \quad 0.5425 \times 0.6920 \times 2000 = 750.8$$

$$NS \quad 142 \quad 0.4575 \times 0.3080 \times 2000 = 281.8$$

$$Ns \quad 773 \quad 0.4575 \times 0.6920 \times 2000 = 633.2$$

The χ^2 for goodness of fit is 184.7 with one degree of freedom: 4 (to start with) $-1 -1$ (for estimating p_1 from the data) -1 (for estimating q_1 from the data) $= 1$. The associated probability is so small as to be off the chart in Figure 3.3, and consequently it is very much less than 0.0001. This result means that chance alone would produce a fit as poor or poorer substantially less than one time in 10,000, and so the hypothesis that the loci are in linkage equilibrium can confidently be rejected.

To quantify the amount of linkage disequilibrium, we must estimate the gametic frequencies P_{11}, P_{12}, P_{21}, and P_{22}:

$$MS: \quad \hat{P}_{11} = 474/2000 = 0.2370$$

$$Ms: \quad \hat{P}_{12} = 611/2000 = 0.3055$$

$$NS: \quad \hat{P}_{21} = 142/2000 = 0.0710$$

$$Ns: \quad \hat{P}_{22} = 773/2000 = 0.3865$$

Thus, D can be estimated as $\hat{D} = \hat{P}_{11}\hat{P}_{22} - \hat{P}_{12}\hat{P}_{21} = 0.07$. From Equation 3.10, D_{max} is given by $p_1 q_2$ or $p_2 q_1$, whichever is smaller; in this case, $p_1 q_2 = 0.38$ and $p_2 q_1 = 0.14$, hence $D_{max} = 0.14$. Therefore, $\hat{D}/D_{max} = 0.07/0.14 = 50\%$, and so we conclude that the amount of disequilibrium between the genes controlling the MN and Ss blood groups is about 50% of its theoretical maximum. In most local populations of sexual organisms that regularly avoid extreme inbreeding (mating between relatives) values of D are typically zero or close to zero (indicating linkage equilibrium) unless the genes are very closely linked. This overall conclusion is exemplified in the following problems.

PROBLEM 3.14 In *Drosophila melanogaster*, the genes *E6–EC–Odh* are linked in chromosome 3. The *E6* and *EC* genes are rather loosely linked ($r = 0.122$), whereas *EC* and *Odh* are tightly linked ($r = 0.002$). The recombination fractions are those in females, as recombination does not take place in males of this species. Using the data from the experimental population given in Problem 2.3 (page 47), carry out an analysis to determine whether there is linkage disequilibrium between $E6$ and EC. If there is linkage disequilibrium, what is its magnitude relative to the theoretical maximum (or minimum) value?

ANSWER   For the data given in Problem 2.3, the observed numbers of the four chromosomal types $E6^F EC^F$, $E6^F EC^S$, $E6^S EC^F$, and $E6^S EC^S$ were 159, 16, 277, and 37, respectively. The estimated allele frequencies of $E6^F$, $E6^S$, EC^F, and EC^S are 0.3579, 0.6421, 0.8916, and 0.1084, respectively. Assuming linkage equilibrium, the expected numbers of the four chromosomal types are 156.0, 19.0, 280.0, and 34.0, respectively. The χ^2 value with one degree of freedom is 0.828, for which the associated probability is about 0.4. Thus, there is no reason to reject the hypothesis that $E6$ and EC are in linkage equilibrium in this experimental population.

PROBLEM 3.15   Carry out an analysis of linkage disequilibrium for the genes EC and Odh, using the data in Problem 2.3 (page 47). A convenient shortcut to obtaining the χ^2 value is first to calculate

$$\rho = D/(p_1 p_2 q_1 q_2)^{1/2}$$

by substituting, on the right-hand side, the estimated values for each of the parameters. The value of χ^2 is numerically equal to $\rho^2 N$, where N is the total number of chromosomes examined. The biological meaning of ρ is that it is the correlation between alleles present in the same chromosome.

ANSWER   For the data given in Problem 2.3, the observed numbers of the chromosomal types $EC^F Odh^F$, $EC^F Odh^S$, $EC^S Odh^F$, and $EC^S Odh^S$ were 416, 20, 44, and 9, respectively. The estimated allele frequencies of EC^F, EC^S, Odh^F, and Odh^S are 0.8916, 0.1084, 0.9407, and 0.0593, respectively, and $\hat{D} = (416 \times 9 - 20 \times 44)/489^2 = 0.0120$. Thus, $\rho = 0.0120/(0.8916 \times 0.1084 \times 0.9407 \times 0.0593)^{1/2} = 0.1631$. Consequently, $\chi^2 = 0.1631^2 \times 489 = 13.0$ with one degree of freedom, for which the associated probability is 0.0004. Thus, there is significant linkage disequilibrium between these genes. The value of D_{max} is the smaller of 0.053 and 0.102, and so $D_{max} = 0.053$. The magnitude of the linkage disequilibrium, relative to its theoretical maximum, is $0.012/0.053 = 22.6\%$. The χ^2 can also be calculated from the expected numbers of the four gametic types, which are 410.1, 25.9, 49.9, and 3.1, respectively.

PROBLEM 3.16   Use the formula for χ^2 in Problem 3.15 to evaluate the statistical significance of the linkage disequilibrium between alleles of the gene for alcohol dehydrogenase in *Drosophila melanogaster* and the presence or absence of an *Eco*RI restriction site located 3500 nucleotides downstream. The data are from a population descended from animals trapped at a Dutch fruit market in Groningen (Cross and Birley 1986).

Adh^F *Eco*RI site present: 22

Adh^F *Eco*RI site absent: 3

Adh^S *Eco*RI site present: 4

Adh^S *Eco*RI site absent: 5

ANSWER $\hat{D} = 0.085$ and $\chi^2 = \rho^2 N = 0.453^2 \times 34 = 7.0$ with one degree of freedom; the associated probability value is approximately 0.01. The linkage disequilibrium is statistically significant and has a value of 49% of its maximum possible value.

Linkage disequilibrium in local populations, such as seen in the preceding examples, can be caused by linkage disequilibrium in the founding population that has not yet had time to dissipate due to the small value of r. Another possible cause of linkage disequilibrium is admixture of populations with differing gametic frequencies. A third possibility is natural selection differentially favoring some genotypes over others to such an extent that it overcomes the natural tendency for D to go to zero.

Several examples in which linkage disequilibrium typically is present in natural populations should be mentioned here. One case concerns plants that ordinarily undergo self-fertilization, and examples are discussed in Chapter 4 in connection with the discussion of inbreeding. Another case involves certain inversions that are polymorphic in populations of certain species of *Drosophila*, most notably *D. pseudoobscura* and *D. subobscura* and their relatives. A chromosome with an **inversion**, as the name implies, has a certain segment of its genes in reverse of the normal order. Because of the inverted segment, the process of chromosome breakage and reunion in meiosis cannot be completed in the normal manner, with the result that the alleles in the inverted segment are usually unaffected by recombination and so they remain linked together. Because inversions prevent recombination, each inversion represents a sort of "supergene," and natural selection accumulates beneficially interacting alleles within each inversion. The beneficially interacting alleles are said to show **genetic coadaptation**.

Linkage disequilibrium can also arise as an artifact of admixture of subpopulations that differ in allele frequencies. Organisms that are subdivided into local populations are said to have **population substructure**. An example of linkage disequilibrium arising from subpopulation admixture is illustrated in Table 3.2. In this example, subpopulation 1 and subpopulation 2 are both in linkage equilibrium for the alleles of the A and B genes. Subpopulation 1 has an allele frequency of 0.05 for both A_1 and B_1, and subpopulation 2 has an allele frequency of 0.95 for both A_1 and B_1. An equal mixture of organisms from both subpopulations has the gametic frequencies shown in the last column of Table 3.2. The allele frequencies of A_1 and B_1 are both 0.50 in the

TABLE 3.2   LINKAGE DISEQUILIBRIUM FROM ADMIXTURE OF SUBPOPULATIONS

Chromosome	Frequency	Subpopulation 1	Subpopulation 2	Equal mixture
$A_1 B_1$	P_{11}	0.0025	0.9025	0.4525
$A_1 B_2$	P_{12}	0.0475	0.0475	0.0475
$A_2 B_1$	P_{21}	0.0475	0.0475	0.0475
$A_2 B_2$	P_{22}	0.9025	0.0025	0.4525
$D = P_{11}P_{22} - P_{12}P_{21}$		0	0	0.2025
	D_{min}	−0.0025	−0.0025	−0.2500
	D_{max}	0.0475	0.0475	0.2500

mixture, but there is substantial linkage disequilibrium between the alleles, as shown at the bottom of the table. In the mixed population, D equals 81% of its theoretical maximum value. The sole cause of the disequilibrium is the differing allele frequencies in the subpopulations. Furthermore, the considerations in Table 3.2 make no assumption that A and B are on the same chromosome, hence linkage disequilibrium may result from population admixture even for genes on different chromosomes. If subpopulations become permanently mixed and undergo random mating, then Equation 3.8 implies that the induced linkage disequilibrium is expected to decrease at the rate r per generation, where r is the recombination fraction between the A and B genes. For unlinked genes, $r = \frac{1}{2}$.

SUMMARY

In any population, the genotype frequencies among zygotes are determined in large part by the patterns in which genotypes of the previous generation come together to form mating pairs. In random mating, genotypes form mating pairs in the proportions expected from random collisions. For a gene with two alleles A and a in a random-mating population, the expected genotype frequencies of AA, Aa, and aa are given by p^2, $2pq$, and q^2, respectively, where p and q are the allele frequencies of A and a, respectively, with $p + q = 1$. The expected genotype frequencies with random mating constitute the Hardy-Weinberg equilibrium (HWE). The rate at which the HWE frequencies are attained depends on the life history of the organism. In an organism with nonoverlapping generations, such as an annual plant, each generation is separated in time from the preceding and the following generation; in this case, the Hardy-Weinberg frequencies are attained in one generation of random mating provided that the allele frequencies are equal in the sexes. In an organism with nonoverlapping generations, the approach to HWE is gradual. Statistical tests of HWE are often based on the χ^2 test, but this test is relatively weak in detecting departures from the expected frequencies, especially those caused by admixture of subpopulations differing in allele frequency.

One of the principal implications of the HWE is that the allele frequencies and the genotype frequencies remain constant from generation to generation, hence genetic variation is maintained. Another major implication is that, when an allele is rare, the population contains many more heterozygotes for the allele than it contains homozygotes for the allele.

Extensions of the HWE include multiple alleles and X-linked genes. With multiple alleles, the expected frequency of a homozygous genotype A_iA_i equals p_i^2, and the expected frequency of a heterozygous genotype A_iA_j equals $2p_ip_j$, where p_i and p_j are the allele frequencies of A_i and A_j. With X-linked alleles, the genotype frequencies in females (XX) are given by the HWE but those in males (XY) are given by the allele frequencies. Consequently, for a recessive X-linked mutation with allele frequency q, the proportion of affected males (q) always exceeds the proportion of affected females (q^2); the rarer the recessive allele, the greater is the excess of affected males.

Nonrandom association between the alleles of different genes is measured by the linkage disequilibrium parameter D. Random association between alleles of different genes is called linkage equilibrium, and it is indicated by $D = 0$. When $D \neq 0$, the alleles are said to be in linkage disequilibrium. Ordinarily, unless there is some countervailing process that maintains linkage disequilibrium between two genes, D is expected to go to zero at a rate determined by the recombination fraction between the genes. For unlinked genes, D decreases by one-half in each generation; for genes that recombine with a frequency r, D decreases by the fraction r in each generation. Significant linkage disequilibrium is usually found in natural populations for genes that are tightly linked, for genes that are within or near an inverted segment of chromosome, or for genes in plant species that regularly undergo self-fertilization. Significant linkage disequilibrium can also result from admixture of two or more subpopulations differing in allele frequencies.

PROBLEMS

1. Phenylketonuria is an autosomal recessive form of severe mental retardation. About one in 10,000 newborn Caucasians are affected. Assuming random mating, what is the frequency of heterozygous carriers?
2. Mourant et al. (1976) cite data on 400 Basques from Spain, of which 230 were Rh^+ and 170 were Rh^-. Estimate the allele frequencies of D and d. How many of the Rh^+ individuals are expected to be heterozygous?
3. Kelus (cited in Mourant et al. 1976) reports a study of 3100 Poles, of whom 1101 were MM, 1496 were MN, and 503 were NN. Calculate the allele frequencies and the expected numbers of the three genotypes and carry out a χ^2 test for goodness of fit to random-mating proportions.
4. Consider an autosomal gene with four alleles A_1, A_2, A_3, and A_4 with respective frequencies 0.1, 0.2, 0.3, and 0.4. Calculate the expected genotype frequencies under random mating.
5. Show that the proportion of heterozygous offspring from a heterozygous parent is $\frac{1}{2}$ in a population undergoing random mating for a single gene with two alleles.
6. If random mating with two alleles gives frequencies D, H, and R for homozygous dominant, heterozygote, and homozygous recessive, show that $DR = H^2/4$.
7. When mating is random for a gene with two alleles A and a at frequencies p and q, show that the genotype frequencies of AA, Aa, and aa are

approximately $1 - 2q$, $2q$, and 0 when q is so small that q^2 is approximately 0.

8. In a population undergoing random mating for a single gene with a dominant and recessive allele, show that the allele frequency of the recessive allele among individuals with the dominant phenotype is $q/(1 + q)$, where q is the allele frequency of the recessive in the whole population.

9. The frequency of one form of recessive X-linked color blindness is 5% among European males. What is the expected frequency of this form of color blindness among females? What fraction of females would be heterozygous carriers?

10. For a trait due to a rare X-linked recessive gene, show that the frequency of carrier females is approximately equal to two times the frequency of affected males.

11. What is the analogue of the Hardy-Weinberg principle for a gene with two alleles in a tetraploid?

12. Given the following table of allele frequencies:

	Gene				
	1	*2*	*3*	*4*	*5*
Allele 1	0.63	0.94	0.995	1.0	0.78
Allele 2	0.37	0.06	0.005	–	0.12
Allele 3	–	–	–	–	0.06
Allele 4	–	–	–	–	0.04

What is the proportion (P) of polymorphic genes (using the definition in the text)? Assuming random mating and linkage equilibrium, what is the average heterozygosity (H) for the set of genes?

13. Charles Darwin could have discovered segregation had he known what to look for, as Mendelian segregation occurred in at least one of his own experiments. Darwin (cited in Iltis 1932) studied flower shape in the snapdragon *Antirrhinum*. In a cross between a true-breeding strain with regular (peloric) flowers and a true-breeding strain with irregular (normal) flowers, all of the F_1's were normal. Crosses of $F_1 \times F_1$ yielded 88 normal and 37 peloric plants. Perform a χ^2 test assuming a $3 : 1$ ratio in the F_2. Is the peloric or normal allele dominant?

14. For a mating between triple dominant/recessive heterozygotes of three unlinked genes, there are eight phenotypic classes among the offspring. What are the expected phenotypic ratios? Mendel carried out such an experiment and obtained the phenotypic ratio $269 : 98 : 86 : 88 : 30 : 34 : 27 : 7$ among a total of 639 progeny. (He complained that this experiment required the most time and effort of any of his crosses.) Calculate the χ^2 and associated probability.

15. If one gene has alleles A_1 and A_2 at frequencies p_1 and p_1, and another gene has alleles B_1, B_2, and B_3 at frequencies q_1, q_2, and q_3, what are the expected frequencies of gametes with linkage equilibrium assuming that $p_1 = 0.3$, $q_1 = 0.2$, and $q_2 = 0.3$?

16. For two genes with alleles A_1 and A_2 and B_1 and B_2, respectively, with p_1 and p_2 the allele frequencies of A_1 and A_2, and q_1 and q_2 those of B_1 and B_2, let $p_1 = 0.7$ and $q_1 = 0.3$.

   a. What are the frequencies of all possible gametes assuming linkage equilibrium?

   b. What are the frequencies of all possible gametes if there is linkage disequilibrium with D equal to 50% of its theoretical maximum?

17. Use the result in Problem 8 to show that the frequency of homozygous recessive genotypes from dominant × dominant matings is $[q/(1 + q)]^2$ and from dominant × recessive matings is $q/(1 + q)$. Note that the latter is equal to the square root of the former. (These proportions are called Snyder's ratios and were once used to test traits for simple recessive inheritance.)

CHAPTER 4

Population Substructure

HIERARCHICAL STRUCTURE ▪ **F** *STATISTICS* ▪ *WAHLUND EFFECT*
DNA TYPING ▪ *ASSORTATIVE MATING* ▪ *INBREEDING*
INBREEDING COEFFICIENT

P OPULATION SUBSTRUCTURE is almost universal among organisms. Many organisms naturally form subpopulations in the form of herds, flocks, schools, colonies, or other types of aggregations. In addition, natural habitats are typically patchy, with favorable areas inter-mixed with unfavorable areas. Through time, even uniformly favorable areas can be disrupted by floods, fires, or other perils. When there is population subdivision, there is almost inevitably some genetic differentiation among the subpopulations. By **genetic differentiation** we mean the acquisition of allele frequencies that differ among the subpopulations. Genetic differentiation may result from natural selection favoring different genotypes in different subpopulations, but it may also result from random processes in the transmission of alleles from one generation to the next or from chance differences in allele frequency among the initial founders of the subpopulations. This chapter considers some of the consequences of population subdivision as well as other types of nonrandom mating.

HIERARCHICAL POPULATION STRUCTURE

A population is said to have a **hierarchical population structure** if the subpopulations can be grouped into progressively inclusive levels in which, at each grouping, the next lower levels are included ("nested") within the next higher ones. To consider a concrete example, imagine we were interested in the population structure of a widespread species of freshwater fish. The lowest population level consists of a local interbreeding population of animals within a stream. A stream may contain more than one such local population. The next-higher level in the hierarchy may be the organization of streams into groups feeding the same river. Another higher level may be rivers within watersheds. An even higher level of organization may be watersheds within continents. The aggregation of subpopulations into progressively more inclusive groups may continue for as many levels as is convenient and informative. It is inevitably somewhat arbitrary how the groups at each level are combined to form the next higher level in the hierarchy. The objective of the classification is informativeness: one tries to group the subpopulations in such a way as to highlight the genetic similarities and differences among them. If there were so much migration of fish among subpopulations that all members of the species constituted essentially a single, random-mating population, then there would be no need to define a hierarchical population structure because it would be uninformative. However, most organisms do have significant population substructure.

Reduction in Heterozygosity

One of the important consequences of population substructure is a reduction in the average proportion of heterozygous genotypes relative to that expected under random mating. The reason for the reduction in heterozygosity may be understood by considering the hypothetical example in Figure 4.1. The outline is the floor plan of a large barn. The organisms of interest are the mice concentrated primarily into two subpopulations of equal size at the west and east ends of the barn. The movement of mice between the subpopulations is prevented by a large population of hungry and vigilant cats in the central area. The occasional mouse that comes out of its refuge is quickly eaten. (These hypothetical mice have not been endowed with the ingenuity to find alternative routes between the west and east ends of the barn, like sneaking along the rafters.) Because of chance effects in the founding of the subpopulations, the west and east subpopulations are completely homozygous for alternative alleles of a gene. All the mice in the west subpopulation are AA, and all those in the east subpopulation are aa. In technical terms, the west subpopulation is **fixed** for the A allele (its allele frequency equals 1), and the east subpopulation is fixed for the a allele. The genotype frequencies of AA, Aa, and aa in the west subpopulation are 1, 0, and 0, respectively, and those in the east subpopulation are 0, 0, and 1, respectively. Within each subpopulation there is random mating, and the genotype frequencies, though extreme, still satisfy the Hardy-Weinberg principle. In particular, the frequencies of AA, Aa, and aa within each subpopulation are given by p^2, $2pq$, and q^2, where $p = 0$ in the east subpopulation, and $p = 1$ in the west

West   East

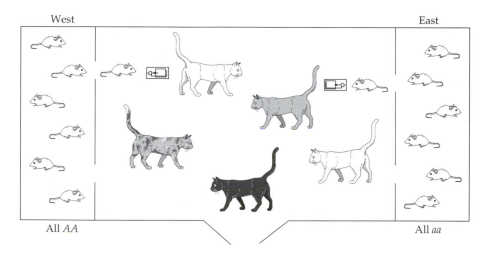

All *AA*   All *aa*

Figure 4.1   An extreme example of the general principle that a difference in allele frequency among subpopulations results in a deficiency of heterozygotes. The floor plan is that of a hypothetical barn. The mouse subpopulations in the east and west enclaves are completely isolated owing to the cats in the middle. The west subpopulation is fixed for the *A* allele and the east subpopulation for the *a* allele. Trapping mice at random in the area patrolled by the cats would yield an overall allele frequency of ½ but no heterozygotes.

subpopulation. Therefore, within any one of the subpopulations in Figure 4.1, the frequency of heterozygotes equals the frequency expected with HWE.

The situation regarding the total population in Figure 4.1 is very different, however, as there is an overall deficiency of heterozygotes. By "total population" in this context, we mean the aggregate of all mice without regard to the population substructure. Suppose we were unaware of the population substructure in the barn. We might then suppose that the barn contained a single randomly mating population. To study the total population of the barn, we trap mice at random in the center area, catching the occasional escapee from the cats. Because the subpopulations are fixed for either *A* or *a*, half the time we would trap an *AA* homozygote and half the time an *aa* homozygote. Consequently, we estimate the allele frequency of *A* as $\hat{p} = \frac{1}{2}$. Assuming random mating and Hardy-Weinberg genotype frequencies in the total population, the expected genotype frequencies of *AA*, *Aa*, and *aa* are given by the HWE as $\hat{p}^2$, $2\,\hat{p}\hat{q}$, and $\hat{q}^2$. Because the overall allele frequency of *A* among the trapped animals is ½, we would naively expect a fraction $2 \times \frac{1}{2} \times \frac{1}{2} = \frac{1}{2}$ of the animals to be heterozygous. In fact, we would have caught no heterozygotes at all!

This rather paradoxical result—that there is a deficiency of heterozygotes in the total population even though random mating takes place within each subpopulation—is a consequence of the difference in allele frequency among the subpopulations. Were the allele frequencies in both subpopulations the same, it would not matter whether we sampled from the west subpopulation, the east subpopulation, or from the area in between. We would recover genotypes in Hardy-Weinberg proportions because both subpopulations are genotypically identical and in HWE. In an organism with hierarchically structured subpopulations, there is an analogous deficiency of heterozygotes at each level in the hierarchy. The following section examines the heterozygosities in more detail.

Average Heterozygosity

In the Mohave desert, local populations of the annual plant *Linanthus parryae* are polymorphic for white versus blue flowers. The plant is diminutive, averaging just 1 cm in height, and when the plant is in bloom, the ground cover of white flowers justifies the popular name "desert snow." Blue flowers result from homozygosity for a recessive allele. The geographical distribution of the frequency q of the recessive allele across a region of the Mohave desert is illustrated in Figure 4.2. Each allele frequency is based on an examination of approximately 4000 plants over an area of about 30 square miles (Epling and Dobzhansky 1942).

Judging from the allele-frequency map in Figure 4.2, the highest frequencies of the blue-flower allele are largely concentrated at the west and east ends of the region in question. The unequal allele frequencies across the range imply a decrease in average heterozygosity, relative to HWE, analo-

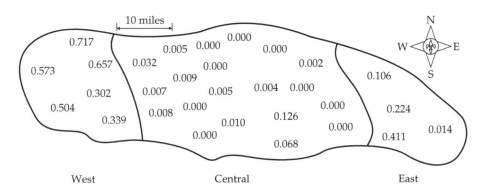

Figure 4.2   Estimated frequency of a recessive allele for blue flower color in populations of *Linanthus parryae* in an area of approximately 900 square miles in the Mohave desert. Each allele frequency is based on an examination of approximately 4000 plants over an area of about 30 square miles. (After Wright 1943a.)

gous to the mouse example in Figure 4.1, though not as extreme. Figure 4.2 shows the estimated allele frequency in each of 30 subpopulations. Suppose each of the subpopulations is regarded as a random-mating unit in HWE for the flower-color alleles. The average heterozygosity among the subpopulations can be denoted as H_S, where the subscript indicates subpopulation. The calculations are shown in the third column in Table 4.1; the heterozygosity in each subpopulation is calculated as $2pq$, where p and q are the estimated frequencies of the alleles for white versus blue flower color, respectively, in each subpopulation. The H_S tabulated at the bottom is the average of all the

TABLE 4.1   HIERARCHICAL STRUCTURE OF *LINANTHUS PARRYAE*

Region	Subpopulations		Regions		Total	
	Allele frequency	Heterozygosity	Average allele frequency	Heterozygosity	Average allele frequency	Heterozygosity
W	0.573	0.4893				
	0.717	0.4058				
	0.504	0.5000				
	0.657	0.4507				
	0.302	0.4216				
	0.339	0.4482	0.5153	0.4995		
C	9×0.000	0.0000				
	0.032	0.0620				
	0.007	0.0139				
	0.008	0.0159				
	0.005	0.0100				
	0.009	0.0178				
	0.005	0.0100				
	0.010	0.0198				
	0.068	0.1268				
	0.002	0.0040				
	0.004	0.0080				
	0.126	0.2202	0.0138	0.0272		
E	0.106	0.1895				
	0.224	0.3476				
	0.411	0.4842				
	0.014	0.0276	0.1888	0.3062	0.1374	0.2371
Average heterozygosity	$H_S = 0.1424$		$H_R = 0.1589$		$H_T = 0.2371$	

Source: Data from Wright 1943a.

subpopulation heterozygosities (counting the value 0.000 a total of nine times because of the nine different subpopulations in which $q = 0.000$).

A second hierarchical level of population substructure is that of region—west (W), central (C), or east (E). To calculate the heterozygosity expected from HWE in each region, we first estimate the average allele frequency in the region by taking the mean allele frequency across all subpopulations in the region. For example, the average allele frequency q in region E is (0.106 + 0.224 + 0.411 + 0.014)/4 = 0.1888. In each region, the heterozygosity expected from HWE is calculated as $2pq$, where p and q are the average allele frequencies in the region. In region E, therefore, the regional heterozygosity equals $2 \times (1 - 0.1888) \times 0.1888 = 0.3062$. The average heterozygosity within regions at the bottom of column 5 is denoted H_R; it is the weighted average of the regional heterozygosities, where each regional heterozygosity is weighted by the number of subpopulations in the region. In this example, $H_R = (6 \times 0.4995 + 20 \times 0.0272 + 4 \times 0.3062)/30 = 0.1589$.

Yet another hierarchical level of population substructure in Figure 4.2 is the total population—the aggregate population obtained by conceptually uniting all subpopulations to form a single random mating unit. The average allele frequency is the mean allele frequency across all subpopulations, and $q = 0.1374$. Then H_T is calculated as $2pq = 2 \times 0.8626 \times 0.1374 = 0.2371$.

To sum up:

- H_S is the average HWE heterozygosity among organisms within random-mating subpopulations.
- H_R is the average HWE heterozygosity among organisms within regions.
- H_T is the average HWE heterozygosity among organisms within the total area.

The concepts of hierarchical population structure and the various levels of heterozygosity were originally developed by Sewall Wright (1889–1988) to quantify genetic differences among subgroups at the various levels; he called his theory **isolation by distance** (Wright 1943a, 1943b). The motivation for developing such a method was summarized in the following passage. The term **panmixia** is a synonym for random mating.

> Study of statistical differences among local populations is an important line of attack on the evolutionary problem. While such differences can only rarely represent first steps toward speciation in the sense of the splitting of the species, they are important for the evolution of the species as a whole. They provide a possible basis for intergroup selection of genetic systems, a process that provides a more effective mechanism for adaptive advance of the species as a whole than does the mass selection which is all that can occur under panmixia.

Furthermore, the reduction in heterozygosity resulting from population substructure is intimately related to the reduction in heterozygosity caused by inbreeding—mating between relatives—as we shall see later in this chapter. Indeed, the relation of population substructure to inbreeding can be understood by interpreting each subpopulation as a sort of "extended family" or set of interconnected pedigrees. Organisms in the same subpopulation will often share one or more recent or remote common ancestors, and so a mating between organisms in the same subpopulation will often be a mating between relatives. The larger the subpopulation, and the more recently it has been isolated, the smaller this inbreeding effect; nevertheless the analogy to inbreeding is valid.

Wright's F Statistics

To quantify the inbreeding effect of population substructure, Wright (1921) defined what has come to be called the **fixation index**. This index equals the reduction in heterozygosity expected with random mating at any one level of a population hierarchy relative to another, more inclusive level of the hierarchy. The fixation index is a useful index of genetic differentiation because it allows an objective comparison of the overall effect of population substructure among different organisms without getting into details of allele frequencies, observed levels of heterozygosity, and so forth. The genetic symbol for a fixation index is F embellished with subscripts denoting the levels of the hierarchy being compared. For example, F_{SR} is the fixation index of the subpopulations relative to the regional aggregates:

$$F_{SR} = \frac{H_R - H_S}{H_R} \qquad 4.1$$

In words, Equation 4.1 defines F_{SR} as the decrease of heterozygosity among subpopulations within regions ($H_R - H_S$), relative to the heterozygosity among regions (H_R). For the *Linanthus* example in Table 4.1, $F_{SR} = (0.1589 - 0.1424)/0.1589 = 0.1036$.

At the next level of the hierarchy, we may define the fixation index F_{RT} as the proportionate reduction in heterozygosity of the regional aggregates relative to the total combined population:

$$F_{RT} = \frac{H_T - H_R}{H_T} \qquad 4.2$$

The data in Table 4.1 shows that $F_{RT} = (0.2371 - 0.1589)/0.2371 = 0.3299$. Comparison of this value with F_{SR} above already makes it clear that there is substantially more variation among regions (as measured by F_{RT}) as there is among subpopulations within regions (as measured by F_{SR}). The comparison of the fixation indices at the two levels gives quantitative expression to the regional differences apparent in Figure 4.2.

The fixation index F_{ST} compares the least inclusive to the most inclusive levels of the population hierarchy and measures all effects of population substructure combined:

$$F_{ST} = \frac{H_T - H_S}{H_T} \qquad\qquad 4.3$$

From Table 4.1, $F_{ST} = (0.2371 - 0.1424)/0.2371 = 0.3993$. The overall reduction in average heterozygosity is therefore close to 40% of the total heterozygosity—a very substantial effect.

The **hierarchical F-statistics** defined in Equations 4.1 through 4.3 are all types of fixation indices, but they differ in the reference populations: F_{SR} is concerned with subpopulations (S) relative to the regional aggregates (R), F_{RT} is concerned with the regional groupings relative to the total population (T), and F_{ST} is concerned with the subpopulations relative to the total population. The index F_{ST} is the most inclusive measure of population substructure. The mathematical relation between the three types of F statistics is demonstrated in the following problem.

PROBLEM 4.1   Show that F_{SR}, F_{RT}, and F_{ST} are related by the equation

$$(1 - F_{SR}) \times (1 - F_{RT}) = 1 - F_{ST}$$

ANSWER   From Equation 4.1, $F_{SR} = 1 - (H_S/H_R)$, or $1 - F_{SR} = H_S/H_R$. Equation 4.2 implies that $F_{RT} = 1 - (H_R/H_T)$, or $1 - F_{RT} = H_R/H_T$. Finally, Equation 4.3 implies that $F_{ST} = 1 - (H_S/H_T)$, or $1 - F_{ST} = H_S/H_T$. Now multiply the expressions for $1 - F_{SR}$ and $1 - F_{RT}$ together to obtain $(1 - F_{SR}) \times (1 - F_{RT}) = (H_S/H_R) \times (H_R/H_T) = H_S/H_T = (1 - F_{ST})$.

For examining the overall level of genetic divergence among subpopulations, F_{ST} is the informative statistic. Although F_{ST} has a theoretical minimum of 0 (indicating no genetic divergence) and a theoretical maximum of 1 (indicating fixation for alternative alleles in different subpopulations), the observed maximum is usually much less than 1. Wright (1978) has suggested the following qualitative guidelines for the interpretation of F_{ST}:

• The range 0 to 0.05 may be considered as indicating *little* genetic differentiation.

- The range 0.05 to 0.15 indicates *moderate* genetic differentiation.
- The range 0.15 to 0.25 indicates *great* genetic differentiation.
- Values of F_{ST} above 0.25 indicate *very great* genetic differentiation.

On the other hand, Wright also notes that, among subpopulations, "differentiation is by no means negligible if F_{ST} is as small as 0.05 or even less."

PROBLEM 4.2   Some subpopulations of *Drosophila melanogaster* show an altitudinal gradient in the allozymes of alcohol dehydrogenase in which the frequency of the *Adh-F* allele increases with altitude. The data in the accompanying table are estimates of the allele frequency of *Adh-F* in seven samples of adult flies captured either in the mountains, in the foothills, or on the plains of the Caucasus Mountains of the former Soviet Union. Each allele frequency is based on electrophoresis of approximately 300 adult flies (Grossman et al. 1970). Calculate the *F* statistics F_{SE} (subpopulations within elevations), F_{ET} (elevations within the total), and F_{ST} (subpopulations relative to the total). What do the magnitudes of the *F* statistics suggest regarding genetic differentiation among subpopulations in the frequency of *Adh-F* with respect to altitude?

Elevation	Allele frequency	Elevation	Allele frequency	Elevation	Allele frequency
Mountain	0.321	Foothill	0.131	Plain	0.082
Mountain	0.226	Foothill	0.109	Plain	0.088
				Plain	0.035

ANSWER   Let p represent the allele frequency of *Adh-F*. For each subpopulation, the HWE heterozygosity equals $2p(1-p)$, which for the seven samples are 0.4359 and 0.3498 (mountain), 0.2277 and 0.1942 (foothill), and 0.1506, 0.1605, and 0.0676 (plain). The average of these values is H_S, which equals 0.2266. At each of the elevations, the average allele frequency is the mean across the subpopulations sampled at that elevation. For mountain, foothill, and plain, these means equal 0.274, 0.120, and 0.068, respectively, yielding the elevation HWE heterozygosities 0.3974, 0.2112, and 0.1273, respectively. (Your results may differ slightly according to the number of significant digits you carry along.) The average of the elevation heterozygosities equals the mean elevation heterozygosity (H_E), and it is the weighted average $(2 \times 0.3974 + 2 \times 0.2112 + 3 \times 0.1273)/7 = 0.2285$. Finally, the allele frequency for the total heterozygosity is equal to the mean allele frequency across subpopulations, which is 0.142, yielding a total HWE heterozygosity (H_T) of 0.2433. The F statistics are $F_{SE} = (H_E - H_S)/H_E = 0.0081$, $F_{ET} = (H_T - H_E)/H_T = 0.0609$, and $F_{ST} = (H_T - H_S)/H_T = 0.0684$. [As a check, note that $(1 - F_{SE}) \times (1 - F_{ET}) = 1 - F_{ST}$.] Judging from the magnitudes of the F statistics, it is clear that most of the differentiation among subpopulations is correlated with altitude; there is very little genetic differentiation among subpopulations at each elevation.

The method of estimating the F statistics by replacing the parameters in Equations 4.1 through 4.3 with their observed or estimated values is not necessarily the best, particularly with small samples. Ideally, estimates of the F statistics should correct for the effects of sampling a limited number of subpopulations, as well as for the effects of sampling a limited number of organisms in each subpopulation. Methods for making these corrections have been suggested but are quite complex and raise additional issues. For an excellent discussion, see Weir and Cockerham (1984). Important issues are also addressed in Wright (1978, pp. 86–89), Curie-Cohen (1982), Nei and Chesser (1983), and Nei (1986). We will use the uncorrected estimation procedure, which is adequate for purposes of illustration.

Genetic Divergence among Subpopulations

The fixation index F_{ST} defined in Equation 4.3 serves as a convenient and widely used measure of genetic differences among subpopulations. The identification of the causes underlying a particular value of F_{ST} observed in a natural population is often difficult. Allele frequencies among subpopulations can become different because of random processes (random genetic drift) as well as by natural selection with complications from migration among the subpopulations. Difficulties in the assignment of cause do not, however, invalidate the usefulness of F_{ST} as an index of genetic differentiation.

The levels of genetic divergence among human subpopulations and among subpopulations of several other species are presented in Table 4.2. The values of F_{ST} imply that genetic divergence between human subpopulations is quite small. Of the total genetic variation found in three major races (Caucasoid, Negroid, and Mongoloid), only 7% (0.07) is ascribable to genetic

TABLE 4.2   TOTAL HETEROZYGOSITY (H_T),AVERAGE HETEROZYGOSITY AMONG SUBPOPULATIONS (H_S), AND FIXATION INDEX (F_{ST}) FOR VARIOUS ORGANISMS

Organism	Number of populations	Number of loci	H_T	H_S	F_{ST}
Human (major races)	3	35	0.130	0.121	0.069
Human, Yanomama Indian villages	37	15	0.039	0.036	0.077
House mouse (Mus musculus)	4	40	0.097	0.086	0.113
Jumping rodent (Dipodomys ordii)	9	18	0.037	0.012	0.676
Drosophila equinoxialis	5	27	0.201	0.179	0.109
Horseshoe crab (Limulus)	4	25	0.066	0.061	0.076
Lycopod plant (Lycopodium lucidulum)	4	13	0.071	0.051	0.282

Source: Protein electrophoretic data from Nei 1975.

differences among races. About 93% of the total genetic variation is found within races. Similarly, of the total genetic variation found in the native Yanomama Indians of Venezuela and Brazil, only 7.7% (0.077) is due to differences in allele frequency among villages. This result implies that 92.3% of the total genetic variation is found within any single village. Values of F_{ST} for other organisms are quite variable, presumably because F_{ST} is influenced by the size of the subpopulations—which is a major determinant of the magnitude of random changes in allele frequency—by the amount and pattern of migration between subpopulations, and by other factors, including natural selection.

Table 4.2 provokes a brief discussion of the sensitive term *race* because the term is prone to misunderstanding or misuse. In population genetics, a **race** is a group of organisms in a species that are genetically more similar to each other than they are to the members of other such groups. Populations that have undergone some degree of genetic divergence as measured by, for example, F_{ST}, therefore qualify as races. Using this definition, the human population contains many races. Each Yanomama village represents, in a certain sense, a separate "race," and the Yanomama as a whole also form a distinct "race." Such fine distinctions are rarely useful, however. It is usually more convenient to group populations into larger units that still qualify as races in the definition given. These larger units often coincide with races based on physical characteristics such as skin color, hair color, hair texture, facial features, and body conformation. Contemporary anthropologists tend to avoid "race" as a descriptive term for human groups because cultural and linguistic differences, which are also important, are often discordant with genetic differences and sometimes discordant with each other.

Here it must be pointed out that the data in Table 4.2, which indicate much more genetic variation within than among human races, may be misleading. The conclusion is based primarily on genes determining allozymes, and it certainly is not true for genes influencing skin color, hair color, hair texture, and other traits that most people think of in connection with the word "race." However, skin color and other prominent racial characteristics are used to delineate races precisely because racial differences for these traits are rather large, so the genes involved cannot be representative of the entire genome. On the other hand, allozyme loci may not be very representative of the genome either. See Nei and Roychoudhury (1982) for a review of the genetic relationship and evolution of human races.

ISOLATE BREAKING: THE WAHLUND PRINCIPLE

The flip side of the coin of heterozygosity is homozygosity because a diploid organism that is not heterozygous must be homozygous. Mathematically, *homozygosity = 1 − heterozygosity*. Therefore, a corollary of the deficit in average heterozygosity, relative to HWE, that results from population substructure is that there is an equal excess in average homozygosity. If the population substructure is eliminated and the former subpopulations undergo random mating, the average homozygosity decreases, and the average heterozygosity increases by an equal amount. The phenomenon that the average homozygosity decreases when subpopulations join together is called **isolate breaking** or the **Wahlund principle**, after the Swedish statistician and human geneticist Sten Gösta William Wahlund (1901–1976) who first described the effect (Wahlund 1928).

The subpopulations of hypothetical mice in Figure 4.1 afford an illustration of the Wahlund principle. As long as the cats keep the subpopulations separate, the homozygosity equals 1 because the west subpopulation is genotypically *AA* and the east subpopulation is genotypically *aa*. If the cats were to disappear and the subpopulations of mice came together and practiced random mating, the genotype frequencies would be ¼ *AA*, ½ *Aa*, and ¼ *aa*. The homozygosity in the fused population is ¼ + ¼ = ½, which is a substantial decrease over the average in the subpopulation prior to fusion and random mating. Not only is the total homozygosity reduced by population fusion, so is the average frequency of each homozygous genotype. Consider *aa*, for example. Prior to fusion, the average frequency of *aa* across both subpopulation equals ½; after fusion and random mating, the frequency of *aa* equals ¼.

   In human population genetics, the Wahlund principle is usually cited for its implication that fusion of subpopulations results in a decrease in the average frequency of children born with a genetic disease resulting from homozygosity for a rare recessive allele, particularly an allele with a relatively high frequency in one of the subpopulations. Examples of harmful recessive alleles at high frequency in some human subpopulations include, in Caucasians, the alleles for α_1-antitrypsin deficiency ($q \cong 0.024$) and cystic fibrosis ($q \cong 0.022$); in blacks, sickle-cell anemia ($q \cong 0.05$ in American blacks, up to $q \cong 0.1$ in some African populations); in the Hopi and some other Southwest American Indian tribes, albinism ($q \cong 0.07$); and, in Ashkenazi Jews, Tay-Sachs disease ($q \cong 0.013$).

   The Wahlund principle for a recessive allele in two subpopulations is illustrated in Figure 4.3A. The west subpopulation has allele frequency q_1 and genotype frequency q_1^2; the east subpopulation has allele frequency q_2 and genotype frequency q_2^2. The average frequency of the homozygous recessive

(A) Separate subpopulations

$$P(a) = q_1$$

$$P(aa) = q_1^2$$

$$P(a) = q_2$$

$$P(aa) = q_2^2$$

$$\text{Average: } R_{separate} = \frac{q_1^2 + q_2^2}{2}$$

(B) Fused subpopulations

$$P(a) = \frac{q_1 + q_2}{2}$$

$$P(aa) = \left(\frac{q_1 + q_2}{2}\right)^2$$

$$R_{fused} = \left(\frac{q_1 + q_2}{2}\right)^2$$

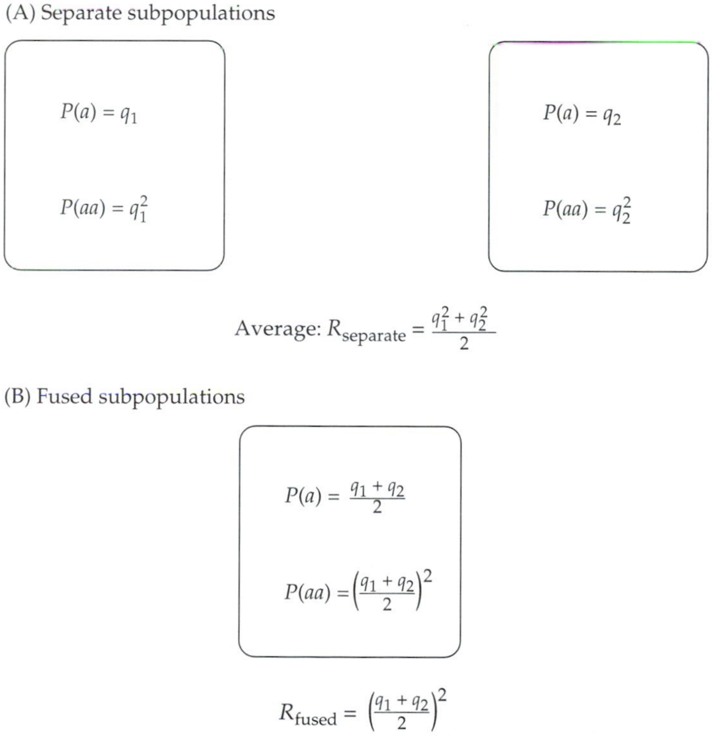

Figure 4.3   Illustration of the Wahlund principle. The frequency of homozygous recessives after population fusion and random mating is less than the average frequency before fusion. The difference in frequency of the homozygous recessives equals the variance in allele frequency among the subpopulations.

across both subpopulations equals $(q_1^2 + q_2^2)/2$. The result of fusion of the subpopulations is shown in part B. Assuming that the subpopulations are equal in size, the allele frequency in the combined population is $\bar{q} = (q_1 + q_2)/2$, and the genotype frequency with HWE equals $\bar{q}^2$. Therefore, were the subpopulations in part A to fuse and come into HWE, the average frequency of homozygous recessives would be reduced by an amount given by:

$$R_{\text{separate}} - R_{\text{fused}} = \frac{q_1^2 + q_2^2}{2} - \bar{q}^2$$

$$= \frac{1}{2}(q_1 - \bar{q})^2 + \frac{1}{2}(q_2 - \bar{q})^2 \qquad 4.4$$

$$= \sigma_q^2$$

In Equation 4.4, we leave it as an exercise to verify that the expressions in q_1 and q_2 on the first and second lines are equal. The symbol σ_q^2 is the variance in allele frequency among the original subpopulations. Because the variance is always nonnegative, isolate breaking always decreases the average frequency of homozygous recessives unless the allele frequencies are equal to begin with. Furthermore, the result in Equation 4.4 is true for any number of subpopulations of equal or unequal size; in words:

> Fusion of subpopulations with random mating and HWE decreases the average frequency of homozygous recessives by an amount equal to the variance in allele frequency among the original subpopulations.

To illustrate the effect of isolate breaking, imagine a subpopulation of gray squirrels that has a high frequency of albinism equal to 16%. (Albinism is an inherited absence of pigment resulting from a homozygous recessive gene.) In a nearby forest there is another subpopulation of equal size in which the albino mutation is absent, so that the allele frequency in this subpopulation is 0. Overall, the average frequency of albinos in the two populations is $(0.16 + 0)/2 = 8\%$. If the two subpopulations fused with random mating and HWE, the allele frequency of the albino mutation in the fused population would be $(0.4 + 0)/2 = 0.2$, and the frequency of the homozygous recessive would equal $0.2^2 = 4\%$. The frequency of albinos in the fused population is substantially smaller than the average frequency in the original subpopulations.

PROBLEM 4.3   Tay-Sachs disease is an autosomal-recessive degenerative disorder of the brain that usually leads to death in infancy or early childhood. Among Ashkenazi Jews, the incidence of the condition is about 1 in 6000 births but, in other groups, the incidence is about 1 in 500,000 births (Myrianthopoulos and Aronson 1966). What incidence of the disease would be expected among the offspring of matings of Ashkenazi Jews with members of other groups? If these offspring were to mate randomly among themselves, what incidence of the disease would be expected in future generations?

ANSWER   The allele frequency of the Tay-Sachs mutation among Ashkenazi Jews is estimated as $\hat{q}_1 = \sqrt{(1/6{,}000)} = 1.291 \times 10^{-2}$; in other groups, $\hat{q}_2 = \sqrt{(1/500{,}000)} = 1.414 \times 10^{-3}$. In matings between members of the two groups, the expected frequency of homozygous recessives is $q_1 q_2 = 1.826 \times 10^{-5}$, or about 1 in 55,000 births. There is actually a greater reduction in the first generation than in subsequent generations because each mating in the first generation combines a high-risk gamete with a low-risk gamete. The allele frequency in the first-generation offspring is $(q_1 + q_2)/2 = 7.162 \times 10^{-3}$ and, with HWE in subsequent generations, the homozygous recessive frequency stabilizes at $(7.162 \times 10^{-3})^2 = 5.130 \times 10^{-5}$, or about 1 in 19,000 births. (The fact that homozygous recessives do not reproduce has been ignored because the effect is negligible.)

Wahlund's Principle and the Fixation Index

Equation 4.4 applies equally well to *AA* homozygotes as to *aa* homozygotes. Therefore, letting P represent the frequency of homozygous *AA* genotypes, we can write

$$P_{\text{separate}} - P_{\text{fused}} = \sigma_p^2 \qquad\qquad 4.5$$

When there are only two alleles, the total reduction in homozygosity must be the summation of Equations 4.4 and 4.5, which equals $\sigma_p^2 + \sigma_q^2$. Because there are only two alleles, it is also true that $\sigma_p^2 = \sigma_q^2$, which we will write as σ^2. Hence, the total reduction in homozygosity from the Wahlund effect upon population fusion and HWE can be expressed as follows:

$$\text{Reduction in total homozygosity} = 2\sigma^2$$

On the other hand, the reduction in total homozygosity with population fusion must also equal the increase in heterozygosity—the term $H_T - H_S$ in Equation 4.3—which is the numerator of F_{ST}. Hence, $F_{ST} = (H_T - H_S)/H_T = 2\sigma^2/H_T$. However, H_T is the heterozygosity with HWE using the average allele frequencies—$\bar{p}$ and $\bar{q}$—across subpopulations. Therefore, the connection between the fixation index F_{ST} and the variance in allele frequency is

$$F_{ST} = \frac{\sigma^2}{\overline{p}\,\overline{q}}$$

4.6

Consequently, the F statistics at the various levels of a hierarchical population are related to the variances in allele frequencies among the subpopulations grouped together at the various levels. Equation 4.6 affords a convenient method of estimating F_{ST} from allele-frequency data. For example, among the subpopulations of *Linanthus* in Figure 4.2, the variance in allele frequency is 0.0473. Earlier we calculated the average allele frequencies as $\overline{p} = 0.8626$ and $\overline{q} = 0.1374$. Hence, $\sigma^2/(\overline{p} \times \overline{q}) = 0.3993$, which confirms the previous calculation that $F_{ST} = 0.3993$. (The values as stated may differ slightly from yours because they were calculated with more than four significant digits.)

PROBLEM 4.4   The data in the accompanying table are the allele frequencies of several genes in three human subpopulations: (A) blacks from West Africa; (B) blacks from Claxton, Georgia; and (C) whites from Claxton, Georgia (Adams and Ward 1973). Each gene has two predominant alleles and may, for purposes of this problem, be considered to have only two alleles. The genes control the MN blood group (alleles M and N), the Ss blood group (alleles S and s), the Duffy blood group (alleles Fy^a and Fy^b), the Kidd blood group (alleles Jk^a and Jk^b), the Kell blood group (alleles Js^a and Js^b), the enzyme glucose-6-phosphate dehydrogenase (alleles $G6PD^-$ and $G6PD^+$, and β-hemoglobin (alleles β^S and β^+). For each gene, use Equation 4.6 to estimate F_{ST} for the comparison A versus B and for the comparison A versus C. Classify each F_{ST} as indicating *little, moderate, great,* or *very great* genetic differentiation according to Wright's qualitative guidelines. *Note:* In comparing two subpopulations with two alleles in each, the variance in allele frequency is $\sigma^2 = (p_1 - p_2)^2/4$.

	Subpopulation		
Gene	A Blacks (West Africa)	B Blacks (Georgia)	C Whites (Georgia)
M	0.474	0.484	0.507
S	0.172	0.157	0.279
Fy^a	0	0.045	0.422
Jk^a	0.693	0.743	0.536
Js^a	0.117	0.123	0.002
$G6PD^-$	0.176	0.118	0
β^S	0.090	0.043	0

ANSWER The estimates and their qualitative interpretations are as shown in the table. It is clear that the degree of genetic divergence between West African blacks and Georgia blacks, as assessed by the average F_{ST} value, is relatively small. However, some of the genes show substantial genetic divergence between blacks and whites. Note, however, that the fixation index can differ substantially from one gene to another. This compilation of genes includes gradations of genetic divergence ranging from *little* to *very great*.

Gene	A versus B	A versus C
M	0.0001 *(little)*	0.0011 *(little)*
S	0.0004 *(little)*	0.0164 *(little)*
Fy^a	0.0230 *(little)*	0.2676 *(very great)*
Jk^a	0.0031 *(little)*	0.0260 *(little)*
Js^a	0.0001 *(little)*	0.0591 *(moderate)*
$G6PD^-$	0.0067 *(little)*	0.0965 *(moderate)*
β^s	0.0089 *(little)*	0.0471 *(little)*
Average	0.0060 *(little)*	0.0734 *(moderate)*

Genotype Frequencies in Subdivided Populations

In many organisms in which the population structure is hierarchical, it is useful to be able to calculate directly the average genotype frequencies across all subpopulations. Equations 4.4 through 4.6 make it possible to deduce the average genotype frequencies. Consider first Equation 4.4, which pertains to the genotype frequency of AA. The quantity called D_{separate} is what we wish to calculate: it is the average frequency of AA across subpopulations. The quantity D_{fused} equals $\bar{p}^2$—the genotype frequency of AA with population fusion and HWE. The value of σ_p^2 is also known from Equation 4.6: it equals $F_{ST} \times \bar{p} \times \bar{q}$. Putting all this together, the average genotype frequency of AA across subpopulations must equal $\bar{p}^2 + F_{ST}\bar{p}\bar{q}$. Likewise, interpreting Equation 4.4 in the same manner as Equation 4.5 yields the average genotype frequency of aa across subpopulations as $\bar{q}^2 + F_{ST}\bar{p}\bar{q}$.

Because every genotype that is not homozygous must be heterozygous, the average genotype frequency of heterozygotes across subpopulations is given by $1 - (\bar{p}^2 + F_{ST}\bar{p}\bar{q}) - (\bar{q}^2 + F_{ST}\bar{p}\bar{q})$. Note that $1 - \bar{p}^2 - \bar{q}^2 = 2\bar{p}\bar{q}$ and so the average frequency of heterozygotes simplifies to $2\bar{p}\bar{q} - 2\bar{p}\bar{q}F_{ST}$.

The genotype frequencies in a subdivided population are important enough to be displayed:

$$AA: \quad \bar{p}^2 + \bar{p}\,\bar{q}F_{ST}$$
$$Aa: \quad 2\bar{p}\,\bar{q} - 2\bar{p}\,\bar{q}F_{ST} \qquad\qquad 4.7$$
$$aa: \quad \bar{q}^2 + \bar{p}\,\bar{q}F_{ST}$$

These genotype frequencies are the average genotype frequencies across all subpopulations. They do not obey the Hardy-Weinberg principle because there is an excess of homozygotes and a deficiency of heterozygotes relative to HWE. The result is somewhat paradoxical because, within any particular subpopulation, the genotype frequencies do obey the Hardy-Weinberg principle with whatever allele frequencies are found in that subpopulation. The reason for the validity of HWE within each subpopulation is the assumption of random mating within each subpopulation. The reason for the departure from HWE in the population as a whole is that the subpopulations differ in allele frequency. Because the allele frequencies differ, random mating within each subpopulation is not equivalent to random mating among all the organisms in the entire population.

From the expressions in Equation 4.7, it is clear that the value of F_{ST} determines the degree of departure from HWE. If $F_{ST} = 0$, the second term in each expression vanishes, and the genotype frequencies reduce to the HWE; on the other hand, $F_{ST} = 0$ means that there is no variation in allele frequency among the subpopulations for the gene in question. Because F_{ST} may vary from one gene to the next, other genes in the same subpopulations may have nonzero values of F_{ST}. The extreme case is $F_{ST} = 1$, which happens when two subpopulations are fixed for alternative alleles. In this case, the average allele frequencies are ½ for each allele and the average genotype frequencies of AA, Aa, and aa across subpopulations are ½, 0, and ½, respectively. This case is illustrated in Figure 4.1.

POPULATION GENETICS IN DNA TYPING

The term **DNA typing** means the application of molecular genetics to highly polymorphic genetic markers for the purpose of matching DNA samples from unknown people with those of known suspects. Applications include paternity testing, in which DNA from a child is matched against that of an accused father, and criminal investigation, in which a crime-scene sample of DNA from blood, semen, or other sources is matched against that of one or more suspects. DNA typing undoubtedly ranks with the use of fingerprints as a major innovation in personal identification.

In theory, DNA typing is not as powerful as ordinary fingerprinting. Fingerprints result from the pattern of raised skin ridges that carry sweat glands. The ridge pattern on each finger may form an arch, loop, whorl, or other design. The ridges vary in pattern from one person to the next so greatly that each person has unique fingerprints suitable for personal identification. When the fingers are formed in the embryo, the fingertips develop as fluid-filled pads. The fluid is later resorbed and the expanded skin collapses, forming the ridges. There is a strong random component to the manner in which the skin collapses, and so the details of the fingerprint pattern differ in each finger and in each person. Even identical twins have different fingerprints. However, certain general features of the fingerprints are strongly inherited—for example, the total number of ridges on all the fingers, without regard to pattern. The Dionne quintuplets—five Canadian girls born in 1934, all formed from the splitting of a single fertilized egg—had total ridge counts ranging between 99 and 102; by comparison, their older siblings had total ridge counts of 69, 78, and 139.

It is the random component in fingerprint ridge pattern that makes fingerprints so powerful for personal identification. DNA types are inherited and so are not necessarily unique in each person. Even for a highly polymorphic marker in which both parents are heterozygous—for example, in the mating $A_iA_j \times A_kA_l$—any particular genotype in an offspring has a $\frac{1}{4}$ chance of being matched in a sibling owing to Mendelian segregation. Thus, strong evidence that an unknown DNA sample comes from a particular suspect can come only from the matching of a combination of genotypes across a number of polymorphic loci. The strength of the evidence increases with the number of loci that are examined and number of alleles present in the population. The greater the number of loci, and the more highly polymorphic the loci, the stronger the evidence linking the suspect to the unknown sample. Although matching DNA types may provide strong evidence that a suspect is the source of an unknown sample, a DNA mismatch is usually conclusive. When the DNA of a suspect contains alleles that are clearly not present in the unknown sample, then the sample must have originated from a different person.

Polymorphisms Based on a Variable Number of Tandem Repeats (VNTR)

The type of polymorphism usually used in DNA typing in the United States is illustrated in Figure 4.4. Each allele of a locus is defined by the size of a restriction fragment that hybridizes with a locus-specific probe in a Southern blot (Chapter 2). The restriction fragments differ in size according to the number of copies they contain of a short sequence of nucleotides repeated in tandem. When there are more copies of the repeating unit, the restriction fragment is of greater size. A polymorphic gene of this type is called a **VNTR** polymorphism, which means that the restriction fragments contain a **variable number of tandem repeats**. VNTRs are employed in DNA typing because many alleles are possible because of the variable number of repeating units. Although many alleles may be present in the population as a

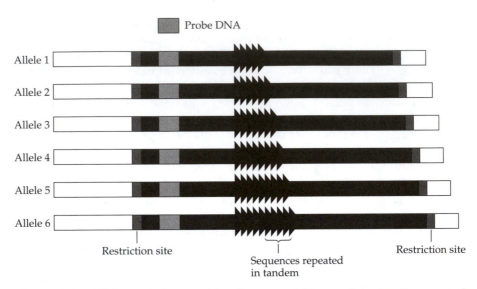

Figure 4.4   Allelic variation resulting from a variable number of units repeated in tandem in a nonessential region of a gene. The probe DNA detects a restriction fragment for each allele. The length of the fragment depends on the number of repeating units present. (From Hartl 1994.)

whole, any one person can have no more than two alleles of each VNTR locus. An example of a VNTR used in DNA typing is shown in Figure 4.5. The lanes in the gel labeled M contain multiple DNA fragments of known size to serve as molecular-weight markers. Each numbered lane contains DNA from a different person. Two typical features of VNTRs are to be noted:

- Most people are heterozygous for two VNTR alleles with restriction fragments of different size. Heterozygosity is indicated by the presence of two distinct bands. In Figure 4.5, only the persons numbered 2 and 5 appear to be homozygous for a particular allele.
- The restriction fragments from different people cover a wide range of sizes. The variability in size indicates that the population as a whole contains many VNTR alleles.

Figure 4.5 also makes it clear why VNTR polymorphisms are useful in DNA typing: each of the 13 people has a different DNA type (pattern of bands) for this VNTR and therefore could be distinguished from any other person. On the other hand, the uniqueness of each DNA type in Figure 4.5 results in part from the small sample size. If more people were examined, then DNA types that matched by chance might well be found among unre-

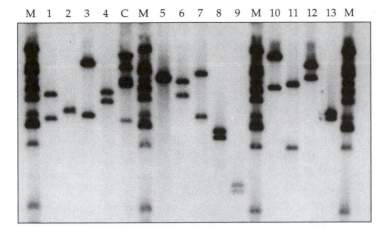

Figure 4.5 Genetic variation in a VNTR used in DNA typing. Each numbered lane contains DNA from a single person. After digestion of the DNA with a restriction enzyme, the fragments are separated by electrophoresis and hybridized with a radioactive probe DNA. The lanes labeled M contain molecular-weight markers; lane C is another type of internal control. (Courtesy of R. W. Allen.)

lated people. For example, in one study of five VNTR loci, the chance of a match between unrelated people ranged from 1/20 to 1/200, depending on the locus (Herrin 1993). Although less common, chance matches for two VNTR loci can also be found among unrelated people. The same study found two-locus matches at frequencies of 1/2,500 to 1/50,000. Even chance matches for three VNTR loci are far from impossible. In one study of Italians from Milan, three-locus matches were found at a frequency of approximately 1/1,200 (Krane et al. 1992). Because of the possibility of chance matches between VNTR types, applications of DNA typing are usually based on at least three loci and preferably more. Matches at 7 to 9 VNTR loci are virtually definitive of identity—barring technical errors in the DNA typing itself (such as mislabeling of blood samples) and except for identical twins.

DNA typing can be exclusionary as well as incriminating. For example, if the DNA type of a suspected rapist does not match the DNA type of semen taken from the victim, then the suspect could not be the perpetrator—unless there is some reason to suspect that the test itself was faulty. For example, Figure 4.6 shows the DNA profiles of nine VNTR loci among three suspects and from evidence recovered in seven serial rape cases. The label M denotes molecular-weight markers (present in four lanes in each panel), S1–S3 denotes three suspects in the cases, and U1–U7 denotes DNA from semen samples recovered from the seven victims. Suspects S1 and S3 are excluded by the DNA typing, but S2 matches at all nine loci. Based on this and other evidence, a jury convicted suspect S2 of 81 criminal counts related to these and other cases. He was sentenced to 139 years in prison and will not become eligible for parole until the year 2087.

Match Probabilities with Hardy-Weinberg Equilibrium and Linkage Equilibrium

If a person is found whose DNA type matches that of a sample found at the scene of a crime, how is the significance of the match to be evaluated? The significance of the match depends on the likelihood of it happening by chance, and hence matches of rare DNA types are more telling than matches of common DNA types. Initially, the method for estimating the frequency of a DNA type in the population was to use a cross-multiplication square like that in Figure 3.6, extended to multiple alleles, to calculate the expected frequency of the particular genotype for each VNTR locus; this calculation assumes Hardy-Weiberg equilibrium (HWE). The locus-by-locus frequencies were then multiplied together to obtain the expected frequency of the multilocus match; this calculation assumes linkage equilibrium. With HWE and linkage equilibrium, the expected frequency of a DNA type in the population as a whole is calculated as

$$\underset{\substack{\text{homozygous} \\ \text{loci}}}{\prod p_i^2} \quad \times \quad \underset{\substack{\text{heterozygous} \\ \text{loci}}}{\prod 2 p_i p_j} \qquad\qquad 4.8$$

where capital Π means chain multiplication. The first multiplication is across all loci presumed to be homozygous owing to the presence of a single band in the gel; for each locus, p_i is the frequency of the allele that is homozygous. The second multiplication is across all heterozygous loci and, for each locus, the factor is two times the product of the frequencies of the alleles that are heterozygous. Because human subpopulations can differ in their allele frequencies, the calculation would be carried out using allele frequencies among Caucasians for white suspects, using those among blacks for black suspects, and using those among Hispanics for Hispanic suspects.

Effects of Population Substructure

The multiplication in Equation 4.8 makes a number of assumptions about human populations: (1) that the Hardy-Weinberg principle holds for each locus, (2) that each locus is statistically independent of the others so that the multiplication across loci is justified, and (3) that the only level of population substructure that is important for DNA typing is that of race. Critics of the multiplication rule argued that genetically important subpopulations

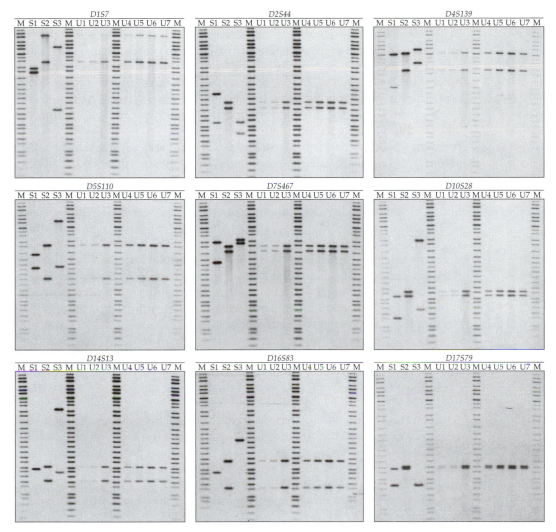

Figure 4.6 An example of DNA typing. Suspect S2 matches evidence samples in seven rape cases (U1–U7) for each of nine VNTR loci (*D1S7, D2S44, D4S139*, and so forth). Suspects S1 and S3 do not match and are excluded. The lanes labeled M contain molecular-weight markers. (Courtesy of Steven L. Redding, Office of the Hennepin County District Attorney, Minneapolis, and Lowell C. Van Berkom and Carla J. Finis, Minnesota Bureau of Criminal Apprehension.)

need not coincide with racial designations. For example, the term "Hispanic" includes a mixture of different subpopulations with variable amounts of Spanish, native American Indian, and African ancestry. Similarly, there are potentially important differences in allele frequency among Caucasian populations (for example, Finnish people versus Italians) and among black populations (for example, blacks from Africa versus blacks from Trinidad). Furthermore, if the allele frequencies of different VNTRs differ among subpopulations, then the loci are not statistically independent—even if they are genetically unlinked—and so the multiplication across loci is unjustified. Because of population substructure, DNA matches across multiple VNTRs could be more common among people within a particular ethnic group than among people drawn at random from the population as a whole, and so calculations of genotype frequency should be based on the ethnic group of the accused person and not on the race as a whole. On the other side, defenders of the multiplication rule argued that population substructure would have a relatively minor effect on the final outcome of the calculation and that what matters most is not a high degree of accuracy but rather a general sense of whether a particular multilocus genotype is rare or common. After much acrimony in the scientific community and in courts of law, a panel of the National Research Council (NRC 1992) recommended a compromise called the *ceiling principle* in which a modified multiplication procedure was adopted using, for each allele frequency, a "ceiling" equal to the larger of either 0.10 or the upper 95% confidence limit of the highest frequency of the allele observed among at least three racial databases.

Even this recommendation proved controversial because some population geneticists regarded the compromise formula as too conservative. Continuing controversy prompted the formation of a second panel of the National Research Council (NRC 1996), which recommended the use of a modified product rule that takes moderate population substructure into account. According to this recommendation, in most cases the match probability may be calculated according to the left-hand side of the following:

$$\prod_{\substack{\text{homozygous} \\ \text{loci}}} 2p_i \times \prod_{\substack{\text{heterozygous} \\ \text{loci}}} 2p_i p_j > \prod_{\substack{\text{homozygous} \\ \text{loci}}} [p_i^2 + p_i(1-p_i)F_{ST}] \times \prod_{\substack{\text{heterozygous} \\ \text{loci}}} [2p_i p_j - 2p_i p_j F_{ST}]$$

In this expression, p_i and p_j have the same meaning as in Equation 4.8, and F_{ST} is the fixation index among the subpopulations in the larger whole (typically a major racial group). The use of the calculation is justified by the inequality. Each factor on the right-hand side of this inequality is the per-locus genotype frequency calculated from Equation 4.7, which takes F_{ST} into account. The left-hand side is greater than the right-hand side because, for each homozygous locus, it can be shown that $2p_i > p_i^2 + p_i(1 - p_i)F_{ST}$; and, for each heterozygous locus, it is clear that $2p_ip_j > 2p_ip_j - 2p_ip_j F_{ST}$ because $F_{ST} > 0$. Equally as important as the calculation itself, the committee emphasized, was the principle that no probability value should be cited unless accompanied by an appropriate 95% confidence interval to indicate its degree of reliability. The 1996 report also enumerated a number of special situations in which alternative formulas are required because of population substructure or inbreeding.

INBREEDING

When matings take place between relatives, the pattern of mating is called **inbreeding**. In human beings, the closest degree of inbreeding usually encountered in most societies is first-cousin mating. Many plants regularly undergo self-fertilization, and some insects regularly practice brother-sister mating. Inbreeding need not unite close relatives, however. As we shall see, a certain level of inbreeding is inescapable in small subpopulations because the members of a subpopulation typically share recent or remote common ancestors. The common ancestry between mating pairs constitutes inbreeding. Hence, the genetic differentiation among subpopulations described by the hierarchical F statistics can be interpreted as a sort of inbreeding effect resulting from population substructure. The relationship between population substructure and inbreeding is a subtle one, but it has profound consequences in population genetics.

Genotype Frequencies with Inbreeding

The main effect of population substructure is a decrease in average heterozygosity among subpopulations, relative to the heterozygosity expected with random mating in a hypothetical total population. Likewise, the main effect of inbreeding is to produce organisms with a decrease in heterozygosity, relative to the heterozygosity expected with random mating in the same subpopulation. The decrease in heterozygosity due to inbreeding can be illustrated with the example of repeated self-fertilization. Consider a self-fertilizing population of plants that consists of $1/4$ AA, $1/2$ Aa, and $1/4$ aa genotypes, which are in Hardy-Weinberg proportions. Because each plant undergoes self-fertilization, the AA and aa genotypes produce only AA and aa offspring, respectively, and the Aa genotypes produce $1/4$ AA, $1/2$ Aa, and $1/4$ aa offspring. After one generation of self-fertilization, therefore, the genotype frequencies of AA, Aa, and aa are:

$$AA: \qquad 1/4 \times 1 + 1/2 \times 1/4 = 3/8$$

$$Aa: \qquad \tfrac{1}{2} \times \tfrac{1}{2} = \tfrac{2}{8}$$

$$aa: \qquad \tfrac{1}{4} \times 1 + \tfrac{1}{2} \times \tfrac{1}{4} = \tfrac{3}{8}$$

These genotype frequencies are no longer in Hardy-Weinberg proportions. There is a deficiency of heterozygous genotypes and an excess of homozygous genotypes. After a second generation of self-fertilization, the genotype frequencies are $\tfrac{7}{16}$ AA, $\tfrac{2}{16}$ Aa, and $\tfrac{7}{16}$ aa, which have an even greater deficiency of heterozygotes. Note, however, that the allele frequency of A remains constant. Denoting the allele frequency of A as p, then:

In the initial population: $\qquad p = \tfrac{1}{4} + \tfrac{1}{2} \times \tfrac{1}{2} = \tfrac{1}{2}$

After one generation of selfing: $\qquad p = \tfrac{3}{8} + \tfrac{1}{2} \times \tfrac{2}{8} = \tfrac{1}{2}$

After two generations of selfing: $\qquad p = \tfrac{7}{16} + \tfrac{1}{2} \times \tfrac{2}{16} = \tfrac{1}{2}$

The example of self-fertilization illustrates the general principle that inbreeding, by itself, does not change the allele frequency. One assumption required for constant allele frequencies under inbreeding is that all genotypes must have an equal likelihood of survival and reproduction, which is to say that no natural selection takes place. If there is selection, then the allele frequencies can change with inbreeding (or, for that matter, with any mating system).

The effects of inbreeding can be made quantitative by comparing the proportion of heterozygous genotypes among inbred organisms with the proportion of heterozygous genotypes expected with random mating. To be precise, consider a gene with two alleles, A and a, at respective frequencies p and q (with $p + q = 1$). Suppose that the frequency of heterozygous genotypes in a subpopulation of inbred organisms is some quantity H_I. Were the subpopulation undergoing random mating, the HWE frequency of heterozygous genotypes would be $2pq$. However, for the sake of generality, we will denote the random-mating heterozygosity by the symbol H_0. The effects of inbreeding can be defined as the proportionate reduction in heterozygosity relative to random mating. This value is expressed mathematically as $(H_0 - H_I)/H_0$; this ratio is usually denoted by the symbol F, which is called the **inbreeding coefficient**. At this point, the use of F for the inbreeding coefficient may seem a poor choice in view of the use of F_{ST} and related symbols for measuring the effects of population substructure, but we will see in a few moments that inbreeding and population substructure are intimately related.

Thus we define

$$F = \frac{(H_0 - H_I)}{H_0} \qquad\qquad 4.9$$

In biological terms, F measures the fractional reduction in heterozygosity of an inbred subpopulation relative to a random-mating subpopulation with the same allele frequencies. Because $H_0 = 2pq$, the frequency of heterozygous genotypes in the inbred subpopulation can be written in terms of F as $H_I = H_0 - H_0F = H_0(1 - F) = 2pq(1 - F)$.

The frequency of AA homozygous genotypes in an inbred subpopulation can also be expressed in terms of F. Suppose that the proportion of AA genotypes is denoted P. Because the allele frequency of A is p, we must have, by Equation 4.9 that $P + H_I/2 = p$. But $H_I = 2pq(1 - F)$, and so $P = p - 2pq(1 - F)/2$.

PROBLEM 4.5   Use the relation $P = p - 2pq(1 - F)/2$ and the fact that $p + q = 1$ to show that $P = p^2 + pqF$. Show also that P can be written as $P = p^2(1 - F) + pF$.

ANSWER $P = p - 2pq(1 - F)/2 = p - pq(1 - F) = p - pq + pqF = p(1 - q) + pqF = p^2 + pqF$. This establishes the first identity. Then, substituting for q in the second term, $P = p^2 + p(1 - p)F = p^2 + pF - p^2F = p^2(1 - F) + pF$.

Problem 4.5 shows that the frequency of AA genotypes in an inbred subpopulation equals $p^2(1 - F) + pF$. In a similar manner, it can be shown that the frequency of aa genotypes is $q^2(1 - F) + qF$.

In summary, in a subpopulation of organisms with inbreeding coefficient F, the genotype frequencies are expected in the proportions:

$$AA : p^2(1 - F) + pF = p^2 + pqF$$
$$Aa : 2pq(1 - F) = 2pq - 2pqF \qquad 4.10$$
$$aa : q^2(1 - F) + qF = q^2 + pqF$$

The expressions at the far right in Equation 4.10 facilitate comparison of the genotype frequencies expected with inbreeding relative to those expected with HWE. With inbreeding, there is a deficiency of heterozygotes equal to $2pqF$ and an excess of each homozygous class equal to half the deficiency of heterozygotes. The biological reason that the missing heterozygotes are allocated equally to the two homozygous classes is that each heterozygous genotype contains one A and one a allele. Notice that when there is no inbreeding ($F = 0$), the genotype frequencies are in the familiar Hardy-Weinberg proportions; with complete inbreeding ($F = 1$), the inbred subpopulation consists entirely of AA and aa homozygotes in the frequencies p and q, respectively.

If a gene has multiple alleles $A_1, A_2, \ldots, A_n$ at respective frequencies $p_1, p_2, \ldots, p_n$ (with $p_1 + p_2 + \cdots + p_n = 1$), then in a population with inbreeding coefficient F, the frequencies of A_iA_i homozygotes and A_iA_j heterozygotes are as follows:

$$p_i^2(1 - F) + p_iF$$
$$2p_ip_j(1 - F)$$

4.11

We are now in a position to apply the Equations 4.10 and 4.11 to real data.

PROBLEM 4.6   Plants able to undergo self-fertilization are said to be self-compatible. In a population of self-compatible plants, if each plant undergoes self-fertilization a fraction s of the time and otherwise mates randomly, then it can be shown (Crow and Kimura 1970; Hedrick and Cockerham 1986) that F very quickly attains the value $F = s/(2 - s)$. *Phlox cuspidata* is self-compatible, and for this species the amount of self-fertilization is estimated at approximately $s = 0.78$ (Levin 1978). From s we can predict the inbreeding coefficient as $F = 0.78/(2 - 0.78) = 0.64$. In a Texas population of *P. cuspidata*, Levin (1978) found two electrophoretic alleles of the phosphoglucomutase-2 gene, designated Pgm-2^a and Pgm-2^b. In a sample of 35 plants, there were 15 Pgm-$2^a/Pgm$-2^a, 6 Pgm-$2^a/Pgm$-2^b, and 14 Pgm-$2^b/Pgm$-2^b genotypes. Are these numbers consistent with the estimate $F = 0.64$? (*Note:* The χ^2 in this case has one degree of freedom because only the allele frequency is estimated from the data; if F also were estimated from the data, rather than being calculated independently from the degree of self-fertilization, then there would be zero degrees of freedom and no goodness-of-fit test would be possible.)

ANSWER   The allele frequencies of Pgm-2^a and Pgm-2^b are estimated as $(30 + 6)/70 = 0.514$ and $1 - 0.514 = 0.486$, respectively. The hypothesis is that $F = 0.64$, and so $1 - F = 0.36$. The expected numbers of the genotypes aa, ab, and bb are, respectively, $[(0.514)^2(0.36) + (0.514)(0.64)](35) = 14.8$, $[2(0.514)(0.486)(0.36)](35) = 6.3$, and $[(0.486)^2(0.36) + (0.486)(0.64)](35) = 13.9$. With these expectations, the $\chi^2 = 0.02$ with one degree of freedom, and the associated probability is about 0.96. The fit to the inbreeding model is excellent.

PROBLEM 4.7   Assuming that $F = 0.64$ in Texas populations of *Phlox cuspidata*, calculate the genotype frequencies expected from the four alleles of the gene *Adh* coding for alcohol dehydrogenase by using the allele frequencies 0.11 (*Adh-1*), 0.84 (*Adh-2*), 0.01 (*Adh-3*), and 0.04 (*Adh-4*) from Problem 3.10 in Chapter 3.

ANSWER   Using the expressions in Equation 4.11, the expected genotype frequencies are: *Adh-1/Adh-1* = 0.0748, *Adh-1/Adh-2* = 0.0665, *Adh-2/Adh-2* = 0.7916, *Adh-1/Adh-3* = 0.0008, *Adh-2/Adh-3* = 0.0060, *Adh-3/Adh-3* = 0.0064, *Adh-1/Adh-4* = 0.0032, *Adh-2/Adh-4* = 0.0242, *Adh-3/Adh-4* = 0.0003, *Adh-4/Adh-4* = 0.0262.

Relation Between the Inbreeding Coefficient and the F Statistics

There is an intimate relation between the inbreeding coefficient F and the hierarchical F statistics examined in the first section of this chapter. Each of the hierarchical F statistics is also a type of inbreeding coefficient that measures the reduction in heterozygosity at any level of a population hierarchy, relative to a higher level. The connection between the inbreeding coefficient and the F statistics is indicated by the formal similarity between Equation 4.7 and the right-hand side of Equation 4.10. To incorporate the inbreeding coefficient F from mating between relatives into the hierarchical framework, we will embellish it with the subscript IS. In words, F_{IS} is the inbreeding coefficient of a group of inbred organisms relative to the subpopulation to which they belong. The value of F_{IS} is the reduction in heterozygosity of the inbred organisms, and the genotype frequencies among the inbred organisms are given by Equation 4.10 with p and q equal to the allele frequencies in the relevant subpopulation. Within each subpopulation there is random mating, and so the genotype frequencies are given by the HWE. Among the subpopulations, however, there is a reduction in average heterozygosity, relative to the total population, because mates within subpopulations often share remote common ancestors. The sharing of remote common ancestors explains the apparent paradox that inbreeding accumulates even when there is random mating within a subpopulation. The reduction in heterozygosity attributable to this type of inbreeding, relative to the total population, is measured by F_{ST}, and the appropriate formulas for the genotype frequencies, averaged across the subpopulations, are given in Equation 4.7, in which $\bar{p}$ and $\bar{q}$ are the average allele frequencies among the subpopulations.

A population geneticist is often interested not only in F_{IS} but also in F_{IT}. The former is the heterozygosity of a group of organisms relative to the subpopulation to which they belong; the latter is the heterozygosity of the inbred organisms relative to the total population. Hence, F_{IT} is the most inclusive measure of all inbreeding. It embraces not only the effects of mating between close relatives within a subpopulation but also the accumulated inbreeding resulting from mating between remote relatives at all levels of the population hierarchy. An expression for F_{IT} is implicit in the definitions. For consistency, we will use the symbol H_S to denote the heterozygosity in a particular subpopulation. Hence, Equation 4.9 defining F_{IS} may be rewritten as:

$$F_{IS} = \frac{H_S - H_I}{H_S} \qquad\qquad 4.12$$

Similarly, if we use H_T to denote the heterozygosity in the total population, the analogous equation defining F_{IT} is:

$$F_{IT} = \frac{H_T - H_I}{H_T} \qquad\qquad 4.13$$

Consequently, $1 - F_{IS} = H_I/H_S$ and $1 - F_{IT} = H_I/H_T$. However, the remarks in Problem 4.1 also indicate that $1 - F_{ST} = H_S/H_T$, and so by multiplication,

$$(1 - F_{IS})(1 - F_{ST}) = 1 - F_{IT} \qquad\qquad 4.14$$

Hence, if we know both F_{IS} and F_{ST}, then we can obtain F_{IT} from Equation 4.14. The value of F_{ST} that results from mating between remote relatives in a subpopulation of limited size is taken up in Chapter 7. The value of F_{IS} resulting from mating between close relatives within a subpopulation can be calculated from the pedigree of the inbred organisms by using an alternative probability interpretation of F_{IS} defined in the next section.

The Inbreeding Coefficient as a Probability

The inbreeding coefficient F_{IS}—which we will again call simply F unless the subscripts are needed for clarity—has an interpretation in terms of probability in addition to its interpretation in terms or heterozygosity spelled out in Equation 4.12. The probability interpretation is important in the calculation of F from pedigrees. To express the inbreeding coefficient in terms of probability, imagine the two alleles of a gene present in a single inbred organism. Because the organism is inbred, the parents share one or more common ancestors. The two alleles present in the inbred organism could have been derived from the same ancestral allele by DNA replication in one of the common ancestors. In this case, the alleles are said to be **identical by descent (IBD)**, and the genotype of the inbred organism is said to be **autozygous**. Conversely, the alleles may not be replicas of a single ancestral allele, in which case the alleles are not identical by descent, and the genotype is said to be **allozygous**. The probability interpretation of the inbreeding coefficient is that F is the probability that the two alleles of a gene in an inbred organism are IBD (autozygous). Note that the concepts of autozygosity and allozygosity have nothing to do with the state of an allele—whether the allele is A or a, for example. The concepts are concerned only with common ancestry. If the alleles are replicas of a single allele in a common ancestor, they are autozygous; otherwise, they are allozygous.

Interpreted as the probability of autozygosity, the inbreeding coefficient is clearly a relative concept. F measures the probability of autozygosity relative to some ancestral subpopulation. In defining the ancestral subpopulation, we arbitrarily assume that all alleles present in the ancestral population are not identical by descent. The inbreeding coefficient of an organism in the present population is then the probability that the two alleles of a gene in the inbred organism arose by replication of a single allele more recently than the time at which the ancestral population existed. The ancestral population need not be remote in time from the present one. Indeed, the ancestral population, usually presumed to be noninbred ($F_{IS} = 0$), typically refers to the population existing just a few generations previous to the present one, and F_{IS} in the present population then measures inbreeding that has accumulated in the span of these few generations. (Technically, any prior inbreeding is allocated to F_{ST}.) Because the span of time is usually short, the possibility of mutation can safely be ignored. Autozygous genotypes must therefore be homozygous for some allele of the gene under consideration. On the other hand, allozygous genotypes can be either homozygous or heterozygous.

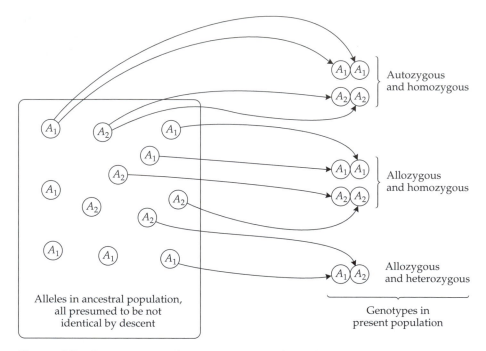

Figure 4.7   In a genotype that is autozygous, homologous alleles are derived from a single DNA sequence in an ancestor, and they are therefore identical by descent. In an allozygous genotype, homologous alleles are not identical by descent. As shown here, allozygous genotypes may be heterozygous or homozygous, but autozygous genotypes must be homozygous (except in the unlikely event that one allele has mutated).

Figure 4.7 illustrates how the concepts of autozygosity and allozygosity are related to those of homozygosity and heterozygosity. The essential point is that two alleles can be **identical by state** (**IBS**), which means that they have the same sequence of nucleotides along the DNA, without being identical by descent. The concept of identity by descent pertains to the ancestral origin of an allele and not to its chemical makeup. Although, as shown in Figure 4.7, two distinct alleles that are identical by state (for example, two A_1 alleles or two A_2 alleles) may come together in fertilization and thereby make the inbred organism homozygous, the alleles in the ancestral population are, by definition, not identical by descent, and so the genotype is allozygous. Similarly, although a heterozygous genotype must be allozygous (ignoring mutation), a homozygous genotype may be either autozygous or allozygous (see Figure 4.7).

The probability interpretation of the inbreeding coefficient results in the same expected genotype frequencies as the heterozygosity interpretation set out in Equation 4.10. To verify the equivalence, we need only consider the implications of the probability definition for a subpopulation of inbred organisms. For this purpose, imagine a subpopulation in which the organisms have average inbreeding coefficient F. Consider the alleles of a gene present in any one of the inbred organisms. Either of two things must be true: the alleles must either be allozygous (probability $1 - F$) or be autozygous (probability F). If the alleles are allozygous, then the probability that the chosen organism has any particular genotype is simply the probability of that genotype in a random-mating population, because, by chance, the inbreeding has not affected this particular gene. On the other hand, if the alleles are autozygous, then the chosen organism must be homozygous, and the probability of homozygosity for any particular allele is simply the frequency of the allele in the subpopulation as a whole. (Because the alleles in question are autozygous, knowing which allele is present in one chromosome immediately tells you that an identical allele is in the homologous chromosome.) These considerations hold regardless of the number of alleles but, to simplify matters, suppose there are only two alleles A and a at frequencies p and q (with $p + q = 1$). The probability that an organism has genotype AA is therefore $p^2(1 - F) + pF$. In this expression, the first term refers to cases in which the alleles are allozygous and the second to cases in which the alleles are autozygous. Similarly, the probability that an organism has genotype aa is $q^2(1 - F) + qF$. Heterozygous Aa genotypes then have the frequency $2pq(1 - F)$ since alleles that are heterozygous must be allozygous.

The genotype frequencies with inbreeding are summarized graphically in Figure 4.8. The box is divided vertically into two parts, corresponding to genes whose alleles remain allozygous in spite of the inbreeding and those whose alleles are autozygous because of the inbreeding. The division is in the proportion $1 - F : F$. Within the allozygous part of the box, the horizontal panels correspond to the allozygous genotypes AA, Aa, and aa, which are the Hardy-Weinberg frequencies. Within the autozygous part of the box, the horizontal panels correspond to the autozygous genotypes AA and aa, which are in the proportions $p : q$. The formulas for the genotype frequencies with inbreeding are given in Table 4.3. Note that the genotype frequencies are exactly the same as those given in the Equations 4.10. This result shows that the autozygosity definition of F and the heterozygosity definition of F, though superficially quite different, are actually equivalent.

Corresponding to the probability interpretation of F_{IS}, there is also a probability interpretation of F_{ST}. However, the comparison is not between homologous alleles in the same organism but between homologous alleles drawn at random from the same subpopulation. Specifically, F_{ST} is the probability of IBD between two alleles drawn at random from the same subpopulation. However, the inbreeding at this level is not realized as a departure from HWE but rather as differences in allele frequency among the subpopulations (Equation 4.6). The variance in allele frequency, in turn, results in a departure

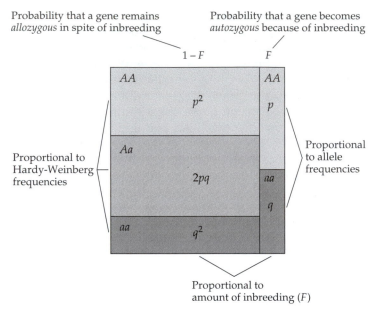

Figure 4.8   Graphical representation of the effects of inbreeding on genotype frequencies. Some genes remain allozygous in spite of the inbreeding, and among these the genotype frequencies of *AA*, *Aa*, and *aa* are given by the Hardy-Weinberg principle. Other genes are autozygous because of the inbreeding, and among these the genotype frequencies of *AA* and *aa* are given by the allele frequencies. There are no heterozygotes in the autozygous case because the two alleles present at an autozygous locus are, by definition, identical by descent.

TABLE 4.3   **GENOTYPE FREQUENCIES WITH INBREEDING**

	Frequency in Population				
Genotype	With inbreeding coefficient F			With F = 0 (random mating)	With F = 1 (complete inbreeding)
AA	$p^2(1-F)$	+	pF	p^2	p
Aa	$2pq(1-F)$			$2pq$	0
aa	$q^2(1-F)$	+	qF	q^2	q
	Allozygous genes	Autozygous genes			

from HWE in the genotype frequencies when averaged across subpopulations (Equation 4.7). The probability interpretation of F_{ST} makes the meaning of Equation 4.14 transparent. It says that, in the total population, a pair of alleles will escape being IBD $(1 - F_{IT})$ only if they escape the effects of mating between close relatives $(1 - F_{IS})$ and, independently, if they escape the cumulative inbreeding effects of mating between remote relatives due to population substructure $(1 - F_{ST})$.

Genetic Effects of Inbreeding

In outcrossing species, which means species that regularly avoid inbreeding, close inbreeding is generally harmful. The effects are seen most dramatically when inbreeding is complete or nearly complete. Although nearly complete autozygosity can be approached in most species by many generations of brother-sister mating, autozygosity of entire chromosomes can easily be accomplished in *Drosophila* by the sort of mating scheme shown in Figure 4.9. In this diagram, *Cy* (*Curly* wings) and *Pm* (*Plum-colored* eyes) are dominant mutations present in certain laboratory second chromosomes that carry several long inversions to prevent recombination. In step A, a wildtype fly is mated with *Cy/Pm*; four genotypes of offspring are produced because the wildtype fly is heterozygous for two different wildtype chromosomes. From each cross in A, a single *Cy* son is chosen and mated with *Cy/Pm*. This step is shown in part B. Three classes of progeny are produced (because *Cy/Cy* is lethal); moreover, from each mating the *Cy/+* progeny all carry wildtype second chromosomes that are IBD because they originated by replication of a single chromosome in the previous generation. In the cross in part C, the *Cy/+* progeny from part B are mated among themselves; the expected progeny are +/+ and *Cy/+* in the ratio ⅓ : ⅔, and the wildtype homozygotes have second chromosomes that are IBD. For chromosome 2, these flies are completely inbred. In the mating D, *Cy/+* flies carrying two different wildtype chromosomes are crossed; again the expected progeny are +/+ and *Cy/+* in the ratio ⅓ : ⅔, but in this case the wildtype flies are heterozygous for different copies of chromosome 2 and are not completely inbred.

For the matings in part C and part D, an estimate $\hat{v}$ of the viability (ability to survive) of the +/+ genotype, relative to that of the *Cy/+* genotype, is given by

$$\hat{v} = \frac{2 \times \text{Number } (+/+)}{1 + \text{Number } (Cy/+)} \qquad 4.15$$

where Number (+/+) and Number (*Cy/+*) are the counts of wildtype and *Curly* offspring, respectively (Haldane 1956). The addition of 1 to the denominator makes the estimate of v almost unbiased. When the total number of offspring is large, $\hat{v}$ is essentially equal to two times the number of wildtype offspring divided by the number of *Curly* offspring.

Results of an experiment using the procedure in Figure 4.9 are shown in Figure 4.10. It is evident that the homozygous genotypes (shaded histogram) are relatively poor in viability. In fact, about 37% of the homozygotes are lethal. Moreover, among the homozygotes that have viabilities within the normal range of heterozygotes (open histogram), virtually all can be shown to have reduced fertility (Sved 1975; Simmons and Crow 1977). Inbreeding so close as to make entire chromosomes homozygous is rare in outcrossing species, except in the kind of experiment in Figure 4.9, but the effects are clearly very harmful and provide a new dimension of genetic diversity. In the case of allozymes, genetic diversity results from common alleles that do not perceptibly impair viability or fertility when homozygous. In the case of inbreeding, the effects are mainly due to rare alleles that are severely detrimental when homozygous. (The fact that the alleles are rare is shown by the small proportion of lethal or near-lethal heterozygotes.) Figure 4.10 shows that natural populations of *Drosophila* contain considerable hidden genetic variation in the form of rare deleterious recessive alleles.

Detrimental effects of inbreeding, called **inbreeding depression**, are found in virtually all outcrossing species, and the more intense the inbreeding, the more harmful the effects. Inbreeding in human beings is also generally harmful, but the effect is difficult to measure because the degree of inbreeding is less than that in experimental organisms; the effects may also vary from population to population. Nevertheless, children of first-cousin matings are, on the average, less capable than noninbred children in any number of ways (for example, higher rate of mortality, lower IQ scores)—although it should be emphasized that many such children are within the normal range of abilities and some are quite gifted. As in most organisms, inbreeding depression is largely due to the

Figure 4.9   Mating scheme to extract wildtype chromosomes (in this case, the second chromosome) from populations of *Drosophila melanogaster*. *Cy* (*Curly* wings) and *Pm* (*Plum* eye color) are dominant mutations contained in certain special laboratory chromosomes that have multiple inversions to prevent recombination. From each mating of the type in part A, a single *Cy* son (containing one wildtype second chromosome) is selected. This son is backcrossed (part B) in order to reproduce many replicas of the second chromosome; the *Cy* progeny are selected for further mating, and the other progeny are discarded. Brother-sister mating as in part C is expected to produce ¼ *Cy/Cy*, ½ *Cy/+*, and ¼ *+/+* zygotes (where + denotes the wildtype second chromosome); the *Cy/Cy* zygotes do not survive, and so the surviving offspring are ⅔ *Cy/+* (*Curly* wings) and ⅓ *+/+* (wildtype straight wings). Mating as in part D, between a female containing one wildtype second chromosome and a male carrying a different one, are also expected to produce ⅔ *Curly*-winged and ⅓ straight-winged progeny. However, in mating C, the straight-winged flies are homozygous for a single wildtype second chromosome; whereas; in mating D, the straight-winged flies are heterozygous for two different wildtype second chromosomes.

(A) Mate and select single *Curly*-winged son.

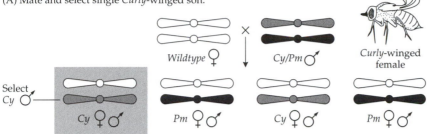

(B) Backcross a single *Cy* male from (A) and select *Curly* sons and daughters, which are heterozygous.

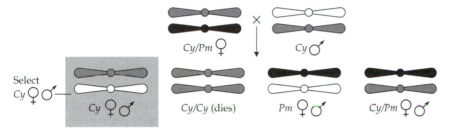

(C) Mate heterozygotes for same wildtype chromosome and count proportion of non-*Curly* offspring.

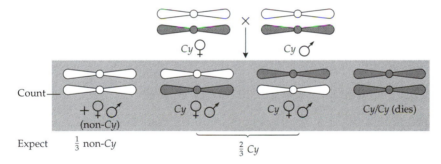

(D) Mate heterozygotes for different chromosomes and count proportion of non-*Curly* offspring.

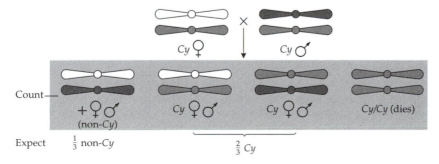

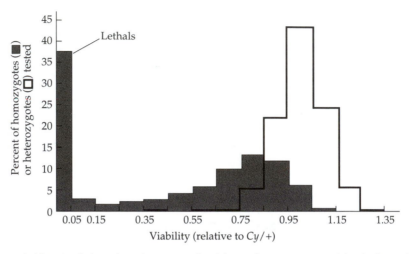

Figure 4.10   Viability distributions of wildtype homozygotes (shaded area) and wildtype heterozygotes (black outline) of second chromosomes extracted from *Drosophila melanogaster* according to the mating scheme in Figure 4.9. The histograms depict results of testing 691 homozygous combinations and 688 heterozygous combinations. Note that, in this sample, nearly 37% of the wildtype chromosomes are lethal when homozygous, and many more have viabilities substantially below normal. (Data from Mukai et al. 1974.)

increased homozygosity of rare recessive alleles, and so inbreeding effects in human beings are seen most dramatically in the increased frequency of genetic abnormalities due to harmful recessive alleles among the children of first-cousin matings. The increased frequency of such conditions results from the genotype frequencies given in Table 4.3. If a denotes a rare deleterious recessive allele, $\frac{1}{16}$ then, among the children of first-cousin matings, the frequency of aa is $q^2(1 - \frac{1}{16}) + q\,(\frac{1}{16})$ because, for these children, $F = \frac{1}{16}$, as will be shown in the next section. On the other hand, with random mating, the frequency of recessive homozygotes is q^2. Thus, the risk of an affected offspring from a first-cousin mating relative to that from a mating of nonrelatives is given by

$$\frac{q^2\left(1 - \frac{1}{16}\right) + q\left(\frac{1}{16}\right)}{q^2} = 0.9375 + \frac{0.0625}{q} \qquad 4.16$$

For example, when $q = 0.01$, the increased risk is approximately 7; that is, a first-cousin mating has seven times the chance of producing a homozygous recessive child as compared to a mating between nonrelatives when the frequency of the harmful recessive allele is 0.01. There is clearly a dramatic inbreeding effect—and the rarer the frequency of the deleterious recessive allele, the greater the effect.

PROBLEM 4.8   Relative to the risk with random mating, calculate the risk of a homozygous recessive offspring from a mating of second cousins ($F = \frac{1}{64}$) when the recessive allele frequency is $q = 0.01$.

ANSWER   In general, the relative risk is given by $[q^2(1 - F) + qF]/q^2 = (1 - F) + F/q$. For $F = \frac{1}{64}$, this becomes $0.9844 + 0.0156/q$, and the value for $q = 0.01$ is approximately 2.5.

Calculation of the Inbreeding Coefficient from Pedigrees

Computation of F from a pedigree is simplified by drawing the pedigree in the form shown in Figure 4.11A, where the lines represent gametes contributed by parents to their offspring. The same pedigree is shown in conventional form in Figure 4.11B. The organisms in gray in part B are not represented in part A because they have no ancestors in common and therefore do not contribute to the inbreeding of the organism denoted I. The inbreeding coefficient F_I of I is the probability that I is autozygous for the alleles of

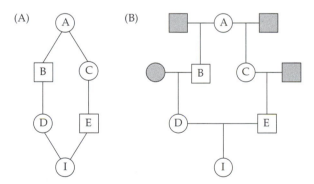

Figure 4.11   (A) Convenient way to represent pedigrees for calculation of the inbreeding coefficient. In this case, the pedigree shows a mating between half-first cousins. (B) Conventional representation of the same pedigree as in part A. Squares represent males, circles represent females, and the shaded organisms in part B are not depicted in part A because they do not contribute to the inbreeding of the inbred organism designated I.

an autosomal gene under consideration. The first step in calculating F_I is to locate all the common ancestors in the pedigree, because an allele could become autozygous in I only if it were inherited through both of I's parents from a common ancestor; in this case, there is only one common ancestor, namely, A. The next step in calculating F_I, which is carried out for each common ancestor in turn, is to trace all the paths of gametes that lead from one of I's parents back to the common ancestor and then down again to the other parent of I. These paths are the paths along which an allele in a common ancestor could become autozygous in I. In Figure 4.11A, there is only one such path: DBACE, in which the common ancestor is underlined for book-keeping purposes, an especially useful procedure in complex pedigrees.

The third step in calculating F_I is to calculate the probability of autozygosity in I due to each of the paths in turn. For the path DBACE, the reasoning is illustrated in Figure 4.12. Here the black dots represent alleles transmitted along the gametic paths, and the number associated with each step is the probability of identity by descent of the alleles indicated. For all steps except that around the common ancestor, the probability is $\frac{1}{2}$ because, with Mendelian segregation, the probability that a particular allele present in a parent is transmitted to a specified offspring is $\frac{1}{2}$. To understand why $\frac{1}{2}(1 + F_A)$ is the probability associated with the loop around the common ancestor, denote the alleles in the common ancestor as α_1 and α_2. These symbols are used to avoid confusion with conventional allele symbols designating functional types of alleles, such as A for dominant and a for recessive. The pair of gametes contributed by A could contain $\alpha_1\alpha_1$, $\alpha_2\alpha_2$, $\alpha_1\alpha_2$, or $\alpha_2\alpha_1$, each with a probability of $\frac{1}{4}$ because of Mendelian segregation. In the first two cases, the alleles are clearly identical by descent; in the second two cases,

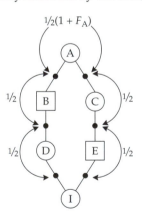

Figure 4.12   Loops for the pedigree in Figure 4.11A, showing probabilities that designated alleles (solid dots) are identical by descent. Each loop is independent of the others, so their probabilities multiply. Thus, the inbreeding coefficient of organism I is $F_I = (\frac{1}{2})^5(1 + F_A)$, where F_A represents the inbreeding coefficient of the common ancestor.

the alleles are identical by descent only if α_1 and α_2 are already identical by descent, which means that A is autozygous. The probability that A is autozygous is, by definition, the inbreeding coefficient of A, F_A. Hence, the probability for the step around the common ancestor A is $\frac{1}{4} + \frac{1}{4} + \frac{1}{4}F_A + \frac{1}{4}F_A = \frac{1}{2} + \frac{1}{2}F_A = \frac{1}{2}(1 + F_A)$. Because each of the steps in Figure 4.12 is independent of the others, the total probability of autozygosity in I due to the path through A is $\frac{1}{2} \times \frac{1}{2} \times \frac{1}{2}(1 + F_A) \times \frac{1}{2} \times \frac{1}{2}$, or $(\frac{1}{2})^5(1 + F_A)$. Note that the exponent on the $\frac{1}{2}$ is simply the total number of ancestors in the path. In general, if a path through a common ancestor A contains i individuals, the probability of autozygosity due to that path is

$$(\tfrac{1}{2})^i(1 + F_A)$$

Thus, the inbreeding coefficient of I in Figure 4.11A is $(\frac{1}{2})^5(1 + F_A)$. Assuming that A is not inbred ($F_A = 0$), the inbreeding coefficient of I reduces to $(\frac{1}{2})^5 = \frac{1}{32}$.

In pedigrees of greater complexity, there is more than one common ancestor and there may be more than one path through any of the common ancestors. The paths are mutually exclusive because autozygosity due to an allele inherited along one path excludes autozygosity due to an allele inherited along a different path. Thus, the total inbreeding coefficient is the sum of the probabilities of autozygosity due to each path considered separately. The whole procedure for calculating F is summarized in an example of a first-cousin mating in Figure 4.13. In a first-cousin mating, there are two common

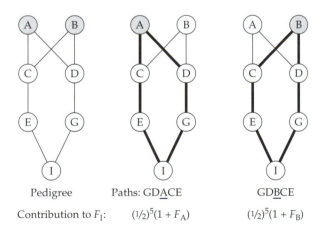

Pedigree	Paths: GD<u>A</u>CE	GD<u>B</u>CE
Contribution to F_I:	$(\frac{1}{2})^5(1 + F_A)$	$(\frac{1}{2})^5(1 + F_B)$

Figure 4.13 On the left is a pedigree of individual I, the offspring of a first-cousin mating. On the right are the two paths through common ancestors (heavy lines) used in calculating the inbreeding coefficient of I. Below each path is the contribution to F_I due to that path, calculated as in Figure 4.12. Each path is mutually exclusive of the others, and so their probabilities add. Thus, the total inbreeding coefficient of I is the sum of the two separate contributions. If $F_A = F_B = 0$, then $F_I = \frac{1}{16}$.

ancestors (A and B) and two paths (one each through A and B). The total inbreeding coefficient of I is the sum of the two separate contributions shown in Figure 4.13. If A and B are both noninbred, then $F_A = F_B = 0$, and so $F_I = (\frac{1}{2})^5 + (\frac{1}{2})^5 = \frac{1}{16}$; this result is the probability that I is autozygous at the specified locus. Alternatively, F_I can be interpreted as the average proportion of all genes in I in which the alleles present are autozygous.

In general, for any autosomal gene, the formula for calculating the inbreeding coefficient F_I of an inbred organism I is

$$F_I = \sum_A \left(\frac{1}{2}\right)^i (1 + F_A) \qquad\qquad 4.17$$

in which the summation Σ over A means summation over all possible paths through all common ancestors, i is the number of organisms in each path, and A is the common ancestor in each path.

PROBLEM 4.9   The accompanying pedigree depicts two generations of brother-sister mating. Calculate the inbreeding coefficient of I, assuming that none of the common ancestors is inbred. (Altogether, there are four common ancestors and six paths.)

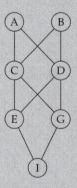

ANSWER $F_I = (\frac{1}{2})^3(1 + F_C) + (\frac{1}{2})^3(1 + F_D) + (\frac{1}{2})^5(1 + F_A) + (\frac{1}{2})^5(1 + F_A) + (\frac{1}{2})^5(1 + F_B) + (\frac{1}{2})^5(1 + F_B)$. When the common ancestors are assumed to be noninbred, then $F_A = F_B = F_C = F_D = 0$, and so $F_I = \frac{3}{8}$.

Regular Systems of Mating

In plant and animal breeding, it is often important to know how rapidly the inbreeding coefficient increases when a strain is propagated by a regular system of mating, such as repeated self-fertilization, sib mating, or backcrossing to a standard strain. The reasoning involved in calculating the inbreeding coefficient for any generation is illustrated in Figure 4.14 for repeated self-fertilization. In this figure, the labels $t-1$ and t refer to the inbred organisms after $t-1$ and t generations of self-fertilization. The loop around the ancestor in generation $t-1$ designates the probability that the two indicated alleles are identical by descent. Here the formula in Equation 4.17 applies with only one path and only one ancestor in the path, and so $F_t = (\frac{1}{2})^1(1 + F_{t-1})$, where F_t is the inbreeding coefficient in generation t. This equation is easy to solve in terms of the quantity $1 - F_t$, which is often called the **panmictic index**, panmixia being a synonym for random mating. Multiplying both sides of the equation for F_t by -1 and then adding $+1$ to each side leads to $1 - F_t = 1 - \frac{1}{2}(1 + F_{t-1}) = 1 - \frac{1}{2} - \frac{1}{2}F_{t-1} = \frac{1}{2}(1 - F_{t-1})$, or

$$1 - F_t = \left(\frac{1}{2}\right)^t (1 - F_0) \qquad\qquad 4.18$$

where F_0 is the inbreeding coefficient in the initial generation when the repeated self-fertilization begins. Self-fertilization therefore leads to an extremely rapid increase in the inbreeding coefficient. When $F_0 = 0$, then $F_1 = \frac{1}{2}$, $F_2 = \frac{3}{4}$, $F_3 = \frac{7}{8}$, $F_4 = \frac{15}{16}$, and so on. The increase in F under self-fertilization and several other regular systems of mating is shown in Figure 4.15.

Many plants reproduce predominantly by self-fertilization, including crop plants such as soybeans, sorghum, barley, and wheat. As expected of

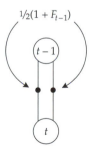

Figure 4.14   Increase in F resulting from continued self-fertilization. The organism in generation t is the offspring of self-fertilization of the organism in generation $t-1$. The loop shows that $F_t = \frac{1}{2}(1 + F_{t-1})$.

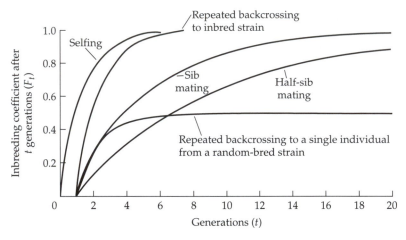

Figure 4.15 Theoretical increase in the inbreeding coefficient F for regular systems of mating: selfing, sib mating, half-sib mating, and repeated backcrossing to a single organism from a random-bred strain. In each case, the initial value of F is assumed to be $F_0 = 0$.

highly self-fertilizing species, each plant is highly homozygous for alleles such as those determining allozymes. Yet the proportion of polymorphic genes is comparable to that found in outcrossing species. Polymorphisms are found because self-fertilization does not eliminate genetic variation; it simply reorganizes genetic variation into homozygous genotypes. On the other hand, self-fertilizing species do contain fewer deleterious recessives than do outcrossing species, presumably because the increased homozygosity permits harmful recessives to be eliminated from the population by natural selection. One other important point about naturally self-fertilizing species: The high homozygosity of all genes implies that recombination rarely results in new gametic types not already present in the parent. Therefore, predominance of selfing has the effect of retarding the approach to linkage equilibrium because the approach to linkage equilibrium is through recombination in double heterozygotes (*AB/ab* and *Ab/aB* in the case of two alleles at each locus); with extreme inbreeding, such double heterozygotes are rare. Indeed, the most extreme examples of linkage disequilibrium have been found in predominantly self-fertilizing species such as barley (*Hordeum vulgare*) and wild oats (*Avena barbata*).

Barley, which regularly undergoes more than 99% self-fertilization, provides an extreme example of linkage disequilibrium between two unlinked esterase genes (Clegg et al. 1972). A population that had originated as a complex cross was maintained for 26 generations under normal agricultural conditions without conscious selection. The population was polymorphic for two alleles B_1 and B_2 of an *Esterase-B* gene and also polymorphic for two alleles D_1 and D_2 of an *Esterase-D* gene. The gametic types were found in the following proportions. For all practical purposes, these numbers also refer to homozygous genotypes because there is such close inbreeding.

$$B_1D_1 \quad 1501 \quad (1642.6)$$

$$B_1D_2 \quad 754 \quad (613.7)$$

$$B_2D_1 \quad 720 \quad (577.1)$$

$$B_2D_2 \quad 74 \quad (215.6)$$

(The numbers in parentheses are the expected numbers based on the assumption of linkage equilibrium, calculated as in Chapter 3.) The χ^2 value in this case is 172.7 with one degree of freedom. The associated probability is much less than 0.0001, and so there is undoubtedly linkage disequilibrium. For the above data, the linkage disequilibrium parameter (Equation 3.9) is $D = -0.046$, which is about 66% of its theoretical minimum.

One of the dramatic successes of plant breeding has come from the crossing of inbred lines to produce high-yielding hybrid corn. Yield of a genetically heterogeneous, outcrossing variety of corn can be improved by selecting the plants with the highest yields in each generation to be the progenitors of the next generation; such artificial selection results in only gradual improvement, however (see Chapter 9). If a large number of self-fertilized lines are established from a heterogeneous population, each line declines in yield as inbreeding proceeds, owing to the forced homozygosity of deleterious recessives. Many lines become so inferior that they have to be discontinued. Self-fertilized lines are not likely to become homozygous for exactly the same set of deleterious recessives, however, and when different lines are crossed to produce a hybrid, the hybrid becomes heterozygous for these genes. Alleles favoring high yield in corn are generally dominant, and there may also be genes in which the heterozygous genotypes have a more favorable effect on yield than do the homozygous genotypes; in any case, the hybrid has a much higher yield than either inbred parent. The phenomenon of enhanced hybrid performance is called **hybrid vigor** or **heterosis**. In practice, inbred lines are crossed in many combinations to identify those that produce the best hybrids. Yields of hybrid corn are typically 15 to 35% greater than yields of outcrossing varieties, and the successful introduction of hybrid corn has been remarkable. Virtually all corn acreage in the United States today is planted with hybrids, as compared to 0.4% of the acreage in 1933 (Sprague 1978).

ASSORTATIVE MATING

When choice of mates is based on phenotypes, mating is said to be **assortative**. Most assortative mating is *positive assortative mating*; this term means that mating pairs have, on the average, more similar phenotypes than expected with random mating. The qualifier "on the average" is important. Even when mating is random, some mating pairs are phenotypically similar, and so positive assortative mating refers only to those situations in which mating partners are phenotypically more similar than would be expected by chance encounters.

There are also examples of *negative assortative mating*—sometimes called *disassortative mating*—in which mating pairs are more dissimilar than expected by chance. One case of negative assortative mating is a polymorphism known as **heterostyly** found in most species of primroses (*Primula*) and their relatives. The heterostyly polymorphism refers to the relative lengths of the styles and stamens in the flowers (Figure 4.16). (In botanical terminology, the style is a stalk bearing the stigma, which is the female organ that receives pollen; the stamen is the male organ bearing anthers, in which the pollen is produced.) Most populations of primroses contain approximately equal proportions of two types of flowers, one known as *pin*, which has a tall style and short stamens, and the other known as *thrum*, which has a short style and tall stamens. In heterostyly, insect pollinators that work high on the flowers pick up mostly thrum pollen and deposit it on pin stigmas, whereas pollinators that work low in the flowers pick up mostly pin pollen and deposit it on

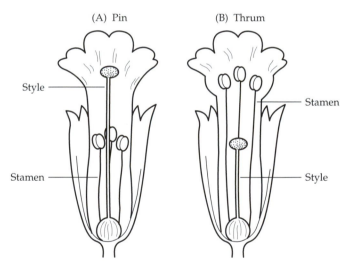

Figure 4.16   Diagrams of cross sections of (A) pin and (B) thrum flowers of the primrose, *Primula*. The pin flowers have a long style and short stamens; the thrum flowers have a short style and long stamens. The differences in flower morphology assist in the maintenance of negative assortative mating mediated by insect pollinators.

thrum stigmas. Negative assortative mating therefore takes place because pins mate preferentially with thrums. Additional floral adaptations facilitate the negative assortative mating. For example, pollen grains from pin flowers fit the receptor cells of thrum stigmas better than they do their own, and pollen grains from thrum flowers germinate better on pin stigmas than they do on their own.

The pollination biology of flowering plants also provides examples of positive assortative mating. For example, when the length of time in which any plant flowers is short relative to the total duration of the flowering season, then plants that flower early in the season are preferentially pollinated by other early flowering plants, and those that flower late are preferentially pollinated by other late flowering ones. Thus, there is positive assortative mating for flowering time.

In human beings, positive assortative mating is observed for height, IQ score, and certain other traits, although assortative mating varies in degree in different populations and is absent in some. As might be expected, positive assortative mating is found for certain socioeconomic variables. In one study in the United States, the highest correlation found between married couples was in the number of rooms in their parents' homes. Negative assortative mating is apparently quite rare in human populations.

In certain species of *Drosophila*, a curious type of nonrandom mating is a phenomenon called *minority male mating advantage*, in which females mate preferentially with males with rare phenotypes. For example, in a study of experimental populations of *D. pseudoobscura* containing flies homozygous for either a recessive *orange* eye-color mutation or a recessive *purple* eye-color mutation, Ehrman (1970) found that, when 20% of the males were orange, the orange-eyed males participated in 30% of the observed matings; conversely, when 20% of the males were purple, the purple-eyed males participated in 40% of the observed matings.

The consequences of positive assortative mating are complex. They depend on the number of genes that influence the trait in question, on the number of different possible alleles of the genes, on the number of different phenotypes, on the sex performing the mate selection, and on the criteria for mate selection. Traits for which mating is assortative are rarely determined by the alleles of a single gene, however. Most such traits are polygenic, so reasonably realistic models of assortative mating tend to be rather complex. Here we should note one obvious, qualitative consequence of positive assortative mating: since like phenotypes tend to mate, assortative mating generally increases the frequency of homozygous genotypes in the population at the expense of heterozygous genotypes, and thus the phenotypic variance in the population increases. (Negative assortative mating generally has the opposite effect.)

SUMMARY

Species that are spread over a large geographical area are usually divided into subpopulations. Matings between organisms within the same subpopulation are more likely than matings between organisms in different subpopulations. Geographical subdivision of a population is called population substructure. The genetic consequences of population substructure result from the fact that the frequencies of alleles may differ from one subpopulation to the next. When the allele frequencies differ, the average heterozygosity among the subpopulations is smaller than that expected with random mating in the total population. Many populations are subdivided into groups within larger groups, a kind of structure called a hierarchical population structure. The F statistics are a quantitative measure of the reduction in heterozygosity at various levels in a population hierarchy. For example, F_{SR} is the proportionate reduction in average heterozygosity among subpopulations (S) as compared to that expected with HWE within regions (R): $F_{SR} = (H_R - H_S)/H_R$. Similarly, F_{RT} is the proportionate reduction in average heterozygosity among regions (R) as compared to that expected with HWE in the total population (T): $F_{RT} = (H_T - H_R)/H_T$. The fixation index F_{ST} combines the effects due to subdivision into subpopulations within regions and regions within the total population: $F_{ST} = (H_T - H_S)/H_T$. Generally speaking, an F statistic with a value smaller than 0.05 indicates little genetic differentiation; a value from 0.05 to 0.15 indicates moderate genetic differentiation, from 0.15 to 0.25 indicates great genetic differentiation, and above 0.25 indicates very great genetic differentiation among subpopulations.

When subpopulations undergo fusion and random mating, the deficiency of heterozygotes is eliminated. Said another way around, the excess of homozygous genotypes in a subdivided population is eliminated by population fusion and random mating. This effect of population fusion is called the Wahlund principle. Quantitatively, the Wahlund principle implies that population fusion and random mating will cause a reduction in the frequency of any homozygous genotype by an amount equal to the variance in allele frequency among the original subpopulations. For two alleles, the Wahlund effect is related to the fixation index by the relation $F_{ST} = \sigma^2/(\bar{p} \times \bar{q})$. In terms of the fixation index, the average genotype frequencies across subpopulations are: AA with average frequency $\bar{p}^2(1 - F_{ST}) + \bar{p}F_{ST}$, Aa with average frequency $2\,\bar{p}\bar{q}\,(1 - F_{ST})$, and aa with average frequency $\bar{q}^2(1 - F_{ST}) + \bar{q}F_{ST}$. Despite the departure from HWE when genotype frequencies are averaged across subpopulations, within each subpopulation mating is random and the genotype frequencies are in HWE for the allele frequencies in the subpopulation.

Inbreeding means mating between relatives. The most important effect of inbreeding is that replicas of a single allele in a common ancestor may be transmitted down both sides of the pedigree and come together in fertilization to produce the inbred organism. In such a case, the inbred organism is said to be autozygous, and the alleles are identical by descent (IBD). Otherwise the inbred organism is allozygous. The inbreeding coefficient F is the probability that the two homologous genes in an inbred organism are IBD. With close inbreeding among parents with relatively recent common ancestors, the value of F can be calculated from elementary probability considerations using the formula $F = \Sigma \, (\frac{1}{2})^i(1 + F_A)$, where the summation is over all paths from one parent to the other through each common ancestor, i is the number of organisms in the path, and F_A is the inbreeding coefficient of the common ancestor in the path. Among organisms in which the inbreeding coefficient is F, the genotype frequencies of a gene with two alleles are, for AA, $p^2(1 - F) + pF$; for Aa, $2pq(1 - F)$; and for aa, $q^2(1 - F) + qF$. Hence, one of the most important consequences of close inbreeding is an increased risk of homozygosity of rare recessive alleles—$q^2(1 - F) + qF$ for inbred organisms versus q^2 for noninbred organisms. In human populations, a substantial proportion of children affected with rare, homozygous recessive genetic diseases have first-cousin parents, although first-cousin mating is infrequent.

Population substructure results in an accumulation of inbreeding because mating pairs within subpopulations will often have remote relatives in common, even when mates are chosen at random. Thus, the inbreeding coefficient F resulting from nonrandom mating within a subpopulation should be designated F_{IS}. The total inbreeding resulting from nonrandom mating combined with all levels of population substructure is given by the expression $(1 - F_{IT}) = (1 - F_{IS}) \times (1 - F_{ST})$.

PROBLEMS

1. Two diploid random mating populations have allele frequencies $q + \varepsilon$ and $q - \varepsilon$ for a recessive allele of a gene. What are the frequencies of homozygotes before and after population fusion?

2. Show that $F_{IT} = F_{IS} + F_{IT} - F_{IS}F_{IT}$ and interpret the expression.

3. Calculate F_{ST} among the three random-mating populations below based on the specified allele frequencies. What is the maximum value of F_{ST} in this situation?

Population	Population 1	Population 2	Population 3
Allele 1	0.1	0.2	0.3
Allele 2	0.3	0.3	0.3
Allele 3	0.6	0.5	0.4

4. Calculate F_{IS}, F_{ST}, and F_{IT} for the populations with the genotype frequencies shown in the following table:

	Population 1	Population 2
Genotype AA	0.056	0.072
Aa	0.288	0.256
aa	0.656	0.672

5. Suppose two subpopulations with equal allele frequencies of two linked genes have an amount of linkage disequilibrium that is equal but opposite in sign. What is the amount of linkage disequilibrium in a population formed by mixing equal numbers of individuals from the two populations?
6. Show that $p^2(1 - F) + pF = p^2 + pqF = p - (1 - F)pq$, when $q = 1 - p$.
7. With two alleles and $p = \frac{1}{2}$, what are the expected genotype frequencies in a random mating population and among the offspring of first cousins? How great is the decrease in heterozygosity in the inbred population relative to the random mating population?
8. If the frequency of an autosomal recessive disorder is $1/1600$ among unrelated parents, what is the expected frequency among the offspring of first cousins?
9. For a recessive allele at frequency q in a population in which one percent of the matings are between first cousins, but otherwise occur at random, the proportion of affected individuals having first-cousin parents is $(1 + 15q)/(1 + 1599q)$. Calculate for $q = 0.1, 0.05, 0.1, 0.005$, and 0.001. Interpret the result of the equation when $q = 1$.
10. In a population of monoecious plants in Hardy-Weinberg proportions for two alleles with allele frequency p, what is the variance in allele frequency among plants? What is the variance if the population were completely inbred? If a random mating population were to undergo self-fertilization, what would the variance be when the inbreeding coefficient equals F?
11. The measure of genetic divergence G_{ST} is very useful for multiple alleles in multiple subpopulations. G_{ST} can be defined as $(J_S - J_T)/(1 - J_T)$, where p_i is the frequency of the ith allele, $J_S = \Sigma Avg(p_i)$ and $J_T = \Sigma[Avg(p_i)]^2$ (Nei 1987). The summation means summation over all alleles, and Avg means the average over all subpopulations. For the random mating population below, calculate F_{ST} and G_{ST}.

	Population 1	Population 2
Allele 1	0.2	0.6
Allele 2	0.3	0.0
Allele 3	0.5	0.4

12. G_{ST} for multiple alleles is actually a weighted average of F_{ST} values $G_{ST} = \Sigma p_i(1 - p_i)F_{ST(i)}/\Sigma p_i(1 - p_i)$, where the summation is over all alleles, p_i is the average frequency of the ith allele among the subpopulations, and $F_{ST(i)}$ is the F_{ST} value for the ith allele calculated as if the gene ha

only two alleles with frequencies p_i and $1 - p_i$ in each subpopulation. Calculate $F_{ST(i)}$ for each allele in the preceding problem and confirm numerically that the weighted average equals G_{ST}.

13. In calculating F from pedigrees for X-linked genes, why are paths with two or more consecutive males not counted?

14. What is the coefficient of relationship between I and J in the accompanying pedigree, where I and J are the offspring of a pair of first cousins (A, B) mated with another pair of first cousins (C, D)?

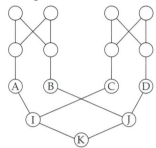

15. Assuming $F_A = F_B = 0$, calculate the inbreeding coefficient for each of the individuals $C - I$ in the accompanying pedigree.

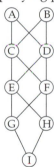

16. If a population is maintained by self-fertilization in even-numbered generations and by random mating in odd-numbered generations, what happens to the inbreeding coefficient?

17. For a gene with two alleles and $p = 0.3$, what are the expected genotype frequencies after five generations of sib mating? What are the expected genotype frequencies after one additional generation of random mating?

18. What is the inbreeding coefficient in a population of size 50 that undergoes
    a. 47 generations of random mating followed by three generations of sib mating?
    b. 50 generations of random mating?

19. In gametophytic self-incompatibile plants, the pollen can only fertilize ovules whose genotype has neither allele borne by the haploid pollen. In

a plant population at equilibrium with three gametophytic self-incompatibility alleles, what is the probability that a pollen grain will land on a compatible style?

20. Two-way hybrid corn is produced by crossing two different inbred lines; three-way hybrids are produced by crossing a two-way hybrid with an unrelated inbred; and four-way hybrids are produced by crossing two different two-way hybrids. What is the inbreeding coefficient of the offspring of randomly mated two-way, three-way, or four-way hybrids? (*Hint:* Consider the allele frequencies in gametes.)

21. Derive a recursion equation for F_t for repeated parent-offspring mating (see pedigree), and calculate F_t for $t = 0$ to 5.

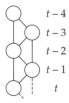

22. Derive a recursion equation for F_t for repeated backcrossing to a single noninbred individual A (see pedigree). Calculate F_t for $t = 0$ to 5 and the equilibrium value.

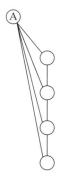

CHAPTER 5

Sources of Variation

MUTATION   •   INFINITE ALLELES MODEL   •   NEUTRAL MUTATIONS
RECOMBINATION   •   MIGRATION   •   TRANSPOSABLE ELEMENTS

G ENETICS INCLUDES several processes that create new types of genetic variation in populations or that allow for the reorganization of previously existing variation either within genomes or among subpopulations. The ultimate source of genetic variation is **mutation**, by which we mean any heritable change in the genetic material. Mutation therefore includes a change in the nucleotide sequence of a single gene as well the formation of a chromosome rearrangement, such as an inversion or a translocation. Recombination brings mutations of different genes together into the same chromosome. Migration enables mutations to spread among subpopulations. A **transposable element** is a DNA sequence able to replicate and insert into any of a large number of sites in the genome. By insertion in or near a gene, a transposable element can alter the level or pattern of gene expression; recombination between transposable elements can result in a chromosome rearrangement, for example, an inversion. In this chapter, we consider the processes by which genetic variation is created.

MUTATION

Mutation is the ultimate source of genetic variation for evolutionary change. However, most wildtype genes mutate at a very low rate, typically in the range from 10^{-4} to 10^{-6} new mutations per gene per generation. Even a low mutation rate can create many new mutant alleles because, in a large population, each of a large number of genes is at risk of mutating. In a population of size N diploid organisms, there are $2N$ copies of each gene, each of which can mutate in any generation. Mutations are rare, but in a large population there are many alleles at risk. For example, if the mutation rate (probability of mutation) is 10^{-9} per nucleotide pair per generation, then in each human gamete, the DNA of which contains 10^9 nucleotide pairs, there would be an average of three new mutations in each generation; each newly fertilized egg would carry, on the average, six new mutations. The present-day human population of approximately 6 billion people would therefore be expected to carry approximately 36 billion new mutations that were not present even one generation earlier.

Irreversible Mutation

Although mutation may create a new allele, the initial frequency of the mutant allele must be very small if the population size is large. A single new mutant allele in a diploid population of size N has an initial frequency of $1/2N$. New mutations in subsequent generations may augment the number of mutant alleles, but recurrent mutation alone increases the allele frequency of the mutant very slowly. Consider an example in which A is the wild-type allele and a the mutant form. If there is exactly one new mutation per generation, then the allele frequency of a increases according to the series $1/2N, 2/2N, 3/2N, \ldots$ and, if N is large (for example, $N = 10^6$), then the increase is very slow indeed. Hence, the tendency for allele frequency to change as a result of recurrent mutation (**mutation pressure**) is very small. On the other hand, the cumulative effects of mutation over long periods of time can become appreciable.

A useful model for thinking about mutation is the Hardy-Weinberg model of Chapter 3, but with mutation permitted. For the moment, we focus on mutations that have so little effect on the ability of the organism to survive and reproduce that natural selection does not appreciably influence their frequency. We will also assume that mutation is *irreversible*, which means that a cannot reverse-mutate to A. To avoid complications resulting from change in allele frequency due to chance, we will assume a population that is infinite in size.

Consider a gene with two alleles, A and a, and suppose that A mutates to a at a rate of μ mutations per A allele per generation. In other words, each A allele has a probability of μ of mutating to a in any generation. We will symbolize the allele frequency of A as p and that of a as q and keep track of generations with subscripts. Hence, p_t and q_t are the allele frequencies of A and a, respectively, in the tth generation, where $t = 0, 1, 2, \ldots$. In any generation, $p_t + q_t = 1$ because A and a are the only alleles considered.

Next we will deduce a formula for the allele frequency p_t in terms of the allele frequency p_{t-1} in the previous generation. In generation t, p_t includes all the A alleles in generation t that did not mutate in that generation, and so

$$p_t = p_{t-1} \times (1 - \mu)$$

However, by the same reasoning, p_{t-1} includes all A alleles in generation $t-1$ that did not mutate in that generation, and so $p_{t-1} = p_{t-2} \times (1-\mu)$. Substituting this equation into the one above yields

$$p_t = p_{t-2} \times (1-\mu)^2$$

Continuing in the same manner leads eventually to

$$p_t = p_0 (1-\mu)^t \qquad 5.1$$

The effect of mutation pressure on allele frequency is illustrated in Figure 5.1 for the case $\mu = 10^{-4}$. The allele frequency of A decreases very slowly, almost linearly at first because the governing term in Equation 5.1, $(1-\mu)^t$, is approximated by $1 - \mu t$ when t is sufficiently small. After 1000 generations, the allele frequency of A is still 0.90; however, at $t = 10,000$ generations, $p_t = 0.37$; and at $t = 20,000$ generations, $p_t = 0.14$.

One instructive way to analyze Equation 5.1 is to consider the time required to reduce the allele frequency of A by half. To find the "half-life" of the process, set $p_t = 0.5 \times p_0$; this relationship implies that $0.5 = (1-\mu)^t$. Taking logarithms of both sides, we obtain

$$t_{1/2} = \ln(0.5)/\ln(1-\mu) \cong 0.6931/\mu$$

In the example in Figure 5.1, $t_{1/2} = 6931$ generations. A decrease in μ by a factor of 10 increases $t_{1/2}$ accordingly, to approximately 69,310 generations for $\mu = 10^{-5}$ and to approximately 693,100 generations for $\mu = 10^{-6}$. The fact

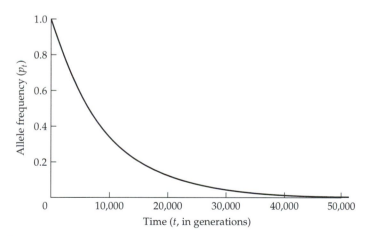

Figure 5.1 Change in frequency under mutation pressure. In this example, an allele A mutates to a at a rate of $\mu = 1 \times 10^{-4}$ per generation; p_t is the allele frequency of A in generation t. We assume that $p_0 = 1$. With the given value of μ, the allele frequency decreases by half every 6931 generations.

that mutation pressure is a weak force for changing allele frequency is illustrated by the long half-lives calculated for realistic values of the mutation rate.

As noted with reference to Equation 5.1, the approximation $p_t = p_0(1 - \mu t)$ is quite accurate for small values of t. With respect to the allele frequency of the mutant allele a, the approximation can also be written as $q_t = q_0 + \mu t$, provided that q_0 is small. This approximation implies that the allele frequency of the a allele increases linearly with time with a slope equal to μ. Because μ is small, however, the linear increase in q_t is difficult to detect experimentally except in very large populations. A large population size can be attained in a bacterial **chemostat**, which is a device for maintaining a population of bacteria in a continuous state of growth and cell division (Figure 5.2). The linear increase in q_t from mutation pressure observed in a

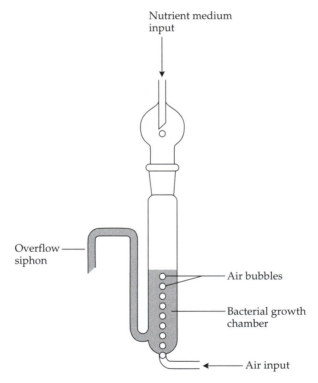

Figure 5.2   Diagram of a bacterial chemostat. Nutrient medium drips in at the top, but a constant volume is maintained by means of an overflow siphon. The air coming in at the bottom provides oxygen. At the steady state, the rate of inflow of nutrient equals the rate of outflow. Cells within the chemostat are in a continuous state of division, but the population does not increase in size because, in any interval of time, the number of new cells produced by division is balanced by the number washed out through the siphon.

chemostat is shown in Figure 5.3. Note the abrupt increase in mutation rate (indicated by the increase in slope) shortly after the addition of caffeine, a bacterial mutagen.

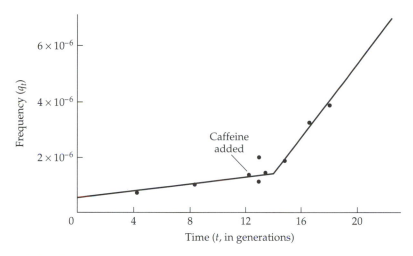

Figure 5.3   Estimation of mutation rate in a bacterial chemostat. This example concerns the rate of mutation of a gene in *Escherichia coli* that confers resistance to infection by the bacteriophage T5. The frequency q_t is the frequency of T5-resistant cells after t generations of growth. The mutation rate is estimated as the slope of the straight-line segments. Prior to the addition of caffeine, the slope was $\mu = 7.2 \times 10^{-8}$ per generation. After addition of caffeine at a concentration of 150 mg/l, the slope increased about tenfold to $\mu = 66 \times 10^{-8}$ per generation. In this experiment, the generation time was 5.5 hours. (From Novick 1955.)

PROBLEM 5.1   A genetic factor has been described in *Drosophila mauritiana* that results in the spontaneous deletion of the transposable genetic element *mariner* at a frequency of approximately one percent per generation for each copy (Bryan et al. 1987). In a population containing an autosomal site at which a *mariner* insertion is fixed (homozygous), how many generations would be required for the frequency of flies that are homozygous for a deletion of the element to exceed five percent? Assume that the population is large, that mating is random, that the excision factor is fixed, and that deletion of the element does not affect survival or reproduction.

ANSWER   Let p_t be the frequency of chromosomes in which the *mariner* element remains undeleted in generation t, and let $\mu = 0.01$ be the probability of deletion of the element per generation. For this situation, Equation 5.1 applies with $\mu = 0.01$ and $p_0 = 1$. The frequency of deletion homozygotes is greater than five percent when $(1 - p_t)^2 > 0.05$, or $p_t < 1 - (.05)^{1/2} = 0.776$. Thus, t should be greater than $\ln(0.776)/\ln(0.99) = 25.2$ generations.

Reversible Mutation

In this section, in addition to forward mutation of A to a, we also allow reverse mutation from a to A. In this case, the mutation pressure on the allele frequency p is in both directions: forward mutation tends to decrease p, reverse mutation tends to increase p. Eventually, an equilibrium is reached in which the frequency p remains constant from generation to generation. At this point, the loss of A alleles from forward mutation is exactly offset by the gain of A alleles from reverse mutation.

To deduce the point of equilibrium, suppose that the rate of forward mutation from A to a is μ per generation and that the rate of reverse mutation from a to A is v per generation. Let p_t and q_t denote the allele frequencies of A and a in generation t, so that $p_t + q_t = 1$. An A allele in generation t can originate in either of two ways. It could have been an A allele in generation $t - 1$ that escaped mutation to a (which happens with probability $1 - \mu$), or it could have been an a allele in generation $t - 1$ that mutated to A (which happens with probability v). In symbols,

$$P_t = p_{t-1}(1 - \mu) + (1 - p_{t-1})v \qquad 5.2$$

To solve equations of this type, a useful trick is to determine whether the relation can be expressed in the form $p_t - A = (p_{t-1} - A)B$, where A and B are constants dependent only on μ and v. Simplifying, we obtain $p_t = p_{t-1}B + A(1 - B)$. Putting Equation 5.2 into the same form yields $p_t = p_{t-1}(1 - \mu - v) + v$. Equating like terms, we deduce that $B = 1 - \mu - v$ and $A(1 - B) = v$. Consequently, $A = v/(\mu + v)$. Hence, we can rewrite Equation 5.2 in the form

$$p_t - \frac{v}{\mu + v} = \left(p_{t-1} - \frac{v}{\mu + v} \right)(1 - \mu - v) \qquad 5.3$$

Because the relation between p_{t-1} and p_{t-2} is the same as that between p_t and p_{t-1}, the solution to Equation 5.3 is

$$p_t - \frac{v}{\mu + v} = \left(p_0 - \frac{v}{\mu + v} \right)(1 - \mu - v)^t \qquad 5.4$$

To understand what happens to the allele frequency in the long run, consider Equation 5.4 in the case when t is very large, for example 10^5 or 10^6 generations. Even though $1 - \mu - v$ is ordinarily close to 1, the value of t eventually becomes so large that $(1 - \mu - v)^t$ becomes approximately 0. Thus, the whole right-hand term in Equation 5.4 goes to 0, and so p_t eventually attains a value that remains the same generation after generation. Such a value of p is called an **equilibrium** value, which we will denote by $\hat{p}$. In case of reversible mutation, the equilibrium is found by equating the left-hand side of Equation 5.4 to 0; hence

$$\hat{p} = \frac{v}{\mu + v} \qquad 5.5$$

The manner in which p_t converges to its equilibrium value is shown in Figure 5.4 for the case $\mu = 10^{-4}$ and $v = 10^{-5}$. Note that, whatever the initial frequency of A, the allele frequency of A eventually goes to $\hat{p}$, which in this example equals $0.00001/(0.0001 + 0.00001) = 0.091$. Figure 5.4 also indicates that mutation pressure is usually very weak in changing allele frequency, inasmuch as the population requires thousands or tens of thousands of generations to reach equilibrium.

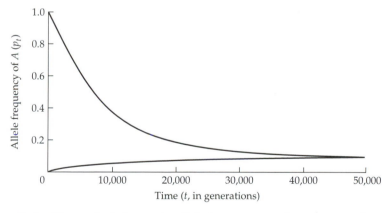

Figure 5.4 Theoretical change in allele frequency under pressure of reversible mutation. The attainment of near-equilibrium values requires tens of thousands of generations for realistic mutation rates. In this example, the forward mutation rate ($A \rightarrow a$) is $\mu = 10^{-4}$ and the reverse mutation rate ($a \rightarrow A$) is $v = 10^{-5}$. The equilibrium allele frequency of A, calculated from Equation 5.5, is 0.091.

PROBLEM 5.2   The bacterium *Salmonella typhimurium* has a genetic switching mechanism that regulates the production of alternative forms of a protein component of the cellular flagella. There are two alleles, which we will call *A* (for the "specific-phase" flagellar property) and *a* (for the "group-phase" flagellar property). Switching back and forth between *A* and *a* takes place rapidly enough that Equation 5.4 can be applied. The transition from *A* to *a* has a rate of $\mu = 8.6 \times 10^{-4}$ per generation, and that of *a* to *A* has a rate of $v = 4.7 \times 10^{-3}$ per generation. These rates are orders of magnitude larger than mutation rates typically observed for other genes. The reason is that the change from *A* to *a* and back again does not result from mutation in the conventional sense but from intrachromosomal recombination (Simon et al. 1980). Formally, however, we can treat the system as one with reversible mutation. In cultures initially established with the frequency of *A* at $p_0 = 0$, Stocker (1949) found that the frequency increased to $p = 0.16$ after 30 generations and to $p = 0.85$ after 700 generations. In cultures initiated with $p_0 = 1$, the frequency decreased to 0.88 after 388 generations and to 0.86 after 700 generations. How do these values agree with those calculated from Equation 5.4 using the estimated mutation rates? What is the predicted equilibrium frequency of *A*?

ANSWER   Note that $v/(\mu + v) = 0.845$. This is the predicted equilibrium frequency (Equation 5.5). Also, $1 - \mu - v = 0.99444$, and this quantity determines the rate of approach to equilibrium. For the cultures with $p_0 = 0$, the predicted values are $p_{30} = 0.845 - (0.845)(0.99444)^{30} = 0.13$ and $p_{700} = 0.845 - (0.845)(0.99444)^{700} = 0.83$. For the cultures with $p_0 = 1$, the predicted values are $p_{388} = 0.845 + (0.155)(0.99444)^{388} = 0.86$ and $p_{700} = 0.845 + (0.155)(0.99444)^{700} = 0.85$. The predicted values are in very good agreement with the observations.

Probability of Fixation of a New Neutral Mutation

The assumption of an infinite population size is not very realistic. In an improved model in which the population is finite, the change in frequency of a mutant allele depends not only on the mutation pressure but also on random sampling from generation to generation. The sampling process, called

random genetic drift, results in chance changes in allele frequency. The process is illustrated in Figure 5.5. The squares represent the $2N$ alleles in the adult population in generation t. Each allele is assigned a unique label—α_1, α_2, α_3, ..., α_{2N}—to temporarily mask its identity as either A or a. The circles represent the essentially infinite pool of gametes in generation t. In the gamete pool, each labeled allele has a frequency of $1/2N$. The squares at the bottom represent two diploid genotypes in generation $t + 1$ formed by random sampling from the pool of gametes. By chance, the two alleles forming a

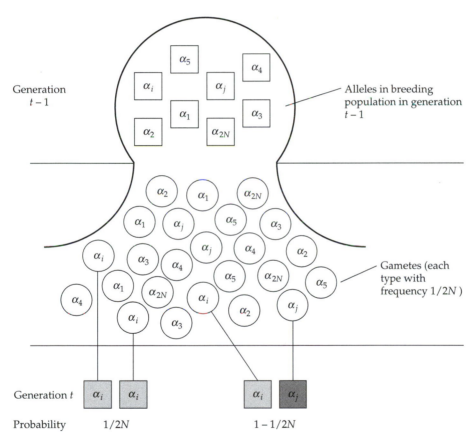

Figure 5.5   Random sampling of alleles in a finite population increases the probability of identity by descent (IBD). Two randomly chosen alleles, illustrated in the squares at the bottom, may be IBD either because they are replicas of the same allele in the immediately preceding generation ($\alpha_i\alpha_i$) or because they are replicas of the same allele in a more remote generation ($\alpha_i\alpha_j$).

genotype may be replicas of the same allele in the previous generation, for example, $\alpha_i\alpha_i$. Alternatively, the two alleles forming a genotype may come from different alleles in the previous generation, for example, $\alpha_i\alpha_j$.

The random sampling from the gamete pool means that some alleles may be overrepresented in generation $t + 1$, relative to their frequency in generation t, and some alleles may be underrepresented. Indeed, any particular allele has a good chance of being unrepresented in generation $t + 1$, and hence the lineage of that allele is terminated. To be precise, each allele in generation t has a chance of approximately $1/e = 0.368$ of not being represented in generation $t + 1$. To understand why, consider the allele designated α_1. The frequency of α_1 in the gamete pool is $1/2N$, and the frequency of all other alleles together is therefore $1 - 1/2N$. Because the genotypes in generation $t + 1$ are formed by the random selection of $2N$ alleles from the pool of gametes, the distribution of the number of α_1 and non-α_1 alleles present in generation $t + 1$ is given by successive terms in the binomial expansion (Chapter 1):

$$\left[\frac{1}{2N}\alpha_1 + \left(1 - \frac{1}{2N}\right)\alpha\right]^{2N} \qquad 5.6$$

in which α represents the collection of all alleles other than α_1. Hence, the probability that α_1 is not represented in generation $t + 1$ is

$$\left(1 - \frac{1}{2N}\right)^{2N} \approx 1/e = 0.368 \qquad 5.7$$

The approximation is very good even when N is quite small. For example, when $N = 10$, the left-hand side of Equation 5.7 equals 0.358, and, when $N = 20$, the left-hand side equals 0.363.

The important implication of Equation 5.7 is that, owing to random genetic drift, the ancestral lineage of each allele faces a substantial risk of extinction in each generation. As time goes on, the lineages progressively disappear, one or a few at a time. Eventually, a time is reached at which all lineages except one have become extinct. At that time, every allele in the population is identical by descent with a particular allele present in an ancestral population.

The ultimate extinction of all but one lineage implies the answer to the question: What is the probability that a single new mutation eventually becomes fixed in a population of size $2N$? The reasoning is illustrated in Figure 5.6. Parts A and B show all the alleles present in the current generation, immediately after a new mutation (shaded circle) has been created. After a sufficient number of generations have passed, each of the alleles in the descendant population will descend from a single allele, chosen at random, in the current population. In part A, the descendant alleles all derive

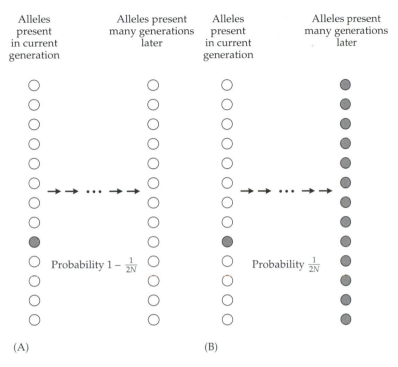

| Alleles present in current generation | Alleles present many generations later | Alleles present in current generation | Alleles present many generations later |

Probability $1 - \frac{1}{2N}$ (A)

Probability $\frac{1}{2N}$ (B)

Figure 5.6   In a finite population, the lineages of all alleles must trace back to a single allele in some ancestral population. Here, a particular allele of interest in a diploid population of size N is indicated by the shaded circle. (A) The probability the designated allele is not destined to be the common ancestor of all alleles many generations in the future is $1 - 1/2N$. (B) The probability the designated allele is destined to be the common ancestor of all alleles many generations in the future is $1/2N$. Hence, the probability of ultimate fixation of a newly arising neutral allele is $1/2N$.

from one of the nonmutants in the current population; the nonmutant alleles have frequency $1 - 1/2N$, and so this is the probability of ultimate fixation of a nonmutant. In part B, the descendant alleles all derive from the mutant, and so $1/2N$ is the probability of ultimate fixation of a new mutant allele. More generally, for neutral alleles, which do not affect the survival or reproduction of the organism, the probability of ultimate fixation of a selectively neutral allele in a finite population is equal to the frequency of the neutral allele in the initial population.

For the lucky few neutral alleles that are eventually fixed, the process takes a long time: on the average, $4N$ generations. The method by which this result can be deduced is considered in Chapter 7.

The Infinite-Alleles Model

Recall from Chapter 2 that many genes have more than two alleles represented among the organisms in a natural population. It is therefore of some importance to determine the expected level of genetic variation under mutation pressure. A convenient measure of genetic variation is the heterozygosity (the proportion of heterozygous genotypes). If a gene has a greater heterozygosity than expected from mutation pressure alone, then other forces that operate in nature must tend to preserve genetic variation. On the other hand, if a gene has a smaller heterozygosity than expected, then other forces must tend to eliminate genetic variation.

The heterozygosity of a gene is a function of the number of alleles and their relative frequencies. In principle, the number of alleles of any gene could be very large. For example, a gene coding for a protein of 300 amino acids has a coding sequence 900 nucleotides in length. Because each nucleotide site could be occupied by either an A, T, G, or C, the total number of possible alleles is 4^{900}, which equals about 10^{542}. Hence, we can suppose that every new mutation creates an allele that does not already exist in the population. This is called the **infinite-alleles model** of mutation. The infinite-alleles model is but one way to specify the characteristics of new mutations. Although it represents a somewhat simplified view of mutation, it nevertheless provides a useful standard of comparison for other models or for observed allele frequencies.

In the infinite-alleles model, two alleles that are identical by state must also be identical by descent because of the assumption that each mutation creates a unique allele. Hence, in this model, homozygous genotypes must be autozygous. To measure the homozygosity, therefore, we need to calculate the autozygosity. This can be done with reference to the finite-population model Figure 5.5. As in Chapter 4, we let F_t be the probability that, in generation t, two alleles randomly chosen from a population are identical by descent. In the context of Figure 5.5, the randomly chosen alleles are combined in pairs to make genotypes, and so F_t is also the probability of autozygosity in generation t. We will use the $\alpha_i\alpha_i$ and $\alpha_i\alpha_j$ genotypes in generation t in Figure 5.5 to derive an expression for F_t in terms of F_{t-1}, N, and the mutation rate μ. First, consider the genotype $\alpha_i\alpha_i$. What is the probability that this genotype has alleles that are identical by descent? The alleles must be identical by descent provided that neither allele has mutated in the course of one generation, and so the probability of identity by descent in this case is $(1 - \mu)^2$. Now consider the genotype $\alpha_i\alpha_j$. These alleles are identical by descent only if two randomly chosen alleles in generation $t - 1$ are identical by descent, and if neither allele mutated in the course of one generation, and so the probability of identity by descent in this case is $F_{t-1}(1 - \mu)^2$. Because each of the labeled α's in Figure 5.5 has the same frequency in the gamete pool (namely, $1/2N$), the probability of a combination like $\alpha_i\alpha_i$ is $1/2N$ and the probability of a combination like $\alpha_i\alpha_j$ is $1 - 1/2N$. Putting all this together, the recurrence equation for F_t is

$$F_t = \left(\frac{1}{2N}\right)(1-\mu)^2 + \left(1 - \frac{1}{2N}\right)(1-\mu)^2 F_{t-1} \qquad 5.8$$

Eventually an equilibrium value of F, call it $\hat{F}$, is attained in which the increase in autozygosity from random genetic drift in any generation is exactly offset by the decrease in autozygosity from new mutations. The equilibrium can be found by equating $F_t = F_{t-1} = \hat{F}$ in Equation 5.8 and solving. Ignoring terms in μ^2 and those in μ/N because they are expected to be negligibly small, the solution is

$$\hat{F} = \frac{1}{1 + 4N\mu} \qquad 5.9$$

to an excellent approximation. Therefore, the number of selectively neutral alleles increases under mutation pressure until $\hat{F}$ satisfies Equation 5.9. Being the equilibrium value of the probability of identity by descent, $\hat{F}$ is also the equilibrium value of the autozygosity. Because of the assumption in the infinite-alleles model that each allele in the population arises only once, all genotypes that are homozygotes must also be autozygous. Therefore, $\hat{F}$ can also be interpreted as the equilibrium value of the proportion of homozygous genotypes.

It is an odd feature of Equation 5.9 that it gives the equilibrium homozygosity of a population without explicit reference to allele frequencies. The natural way to write the homozygosity expected with random mating for n alleles with frequencies $p_1, p_2, p_3, \ldots, p_n$, is

$$\sum_{i=1}^{n} p_i^2 = p_1^2 + p_2^2 + \ldots + p_n^2 \qquad 5.10$$

We thus have two expressions for the equilibrium homozygosity in the forms of Equatons 5.9 and 5.10. Because the two equations refer to the same thing, they must equal each other, and so $\sum p_i^2 = \hat{F} = 1/(4N\mu + 1)$. Alternative approaches leading to essentially the same result are discussed in Sved and Latter (1977).

The homozygosity is the proportion of homozygous genotypes in a population; the heterozygosity is the proportion of heterozygous genotypes. Hence, homozygosity and heterozygosity are opposite sides of the same coin. Therefore, if the homozygosity in a population is given by $\hat{F} = 1/(4N\mu + 1)$, then the heterozygosity is given by $1 - \hat{F} = 4N\mu/(4N\mu + 1)$. These functions for the equilibrium homozygosity and heterozygosity are plotted against $4N\mu$ in Figure 5.7. The illustration shows that there is a rather narrow range

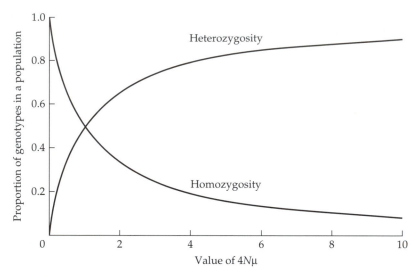

Figure 5.7 Plot of average homozygosity and average heterozygosity for the infinite-alleles model. Intermediate values of heterozygosity are maintained over only a small range of $4N\mu$.

of $4N\mu$ over which an intermediate level of genetic variation (heterozygosity) is maintained. For example, the equilibrium heterozygosity is in the range 0.2 to 0.8 only when $4N\mu$ is in the range 0.25 to 4.

A complication in the interpretation of Equation 5.10 is that any number of distributions of allele frequency can result in the same homozygosity. For example, a population in HWE with the four alleles at frequencies $p_1 = 0.7$, $p_2 = 0.1$, $p_3 = 0.1$, and $p_4 = 0.1$ has a homozygosity of $\Sigma p_i^2 = 0.52$; likewise, a population in HWE with two alleles at frequencies $p_1 = 0.6$ and $p_2 = 0.4$ also has a homozyogosity of 0.52. The problem that many distributions of allele frequency can result in the same homozygosity can be sidestepped by assuming that all alleles are equally frequent. If the population contains n equally frequent alleles, then $p_1 = p_2 = p_3 = \ldots = p_n = 1/n$; the homozygosity is calculated from Equation 5.10 as $\Sigma p_i^2 = n(1/n)^2 = 1/n$. At equilibrium, therefore, $1/n = \hat{F} = 1/(4N\mu + 1)$, or $n = 4N\mu + 1$. The number n of equally frequent alleles is called the **effective number of alleles**, often symbolized as n_e. Diverse distributions of allele frequency can be compared in terms of their effective number of alleles. Biologically speaking, n_e is the number of equally frequent alleles that would be required to produce the same homozygosity as observed in an actual population. In the examples given at the beginning of this paragraph, the four-allele population and the two-allele population with identical homozygosities of 0.52 also have the same effective number of alleles, namely $n_e = 1/0.52 = 1.92$.

PROBLEM 5.3   An allozyme study of a Caribbean population of *Drosophila willistoni* (Ayala and Tracy 1974) yielded the following estimated allele frequencies for the loci *Adk-1* (adenylate kinase-1), *Lap-5* (leucine amino peptidase-5), and *Xdh* (xanthine dehydrogenase).

	Adk-1	Lap-5	Xdh
Allele 1	0.574	0.801	0.446
Allele 2	0.309	0.177	0.406
Allele 3	0.114	0.014	0.092
Allele 4	0.003	0.004	0.034
Allele 5	—	0.004	0.014
Allele 6	—	—	0.004
Allele 7	—	—	0.002
Allele 8	—	—	0.002

Estimate the effective number of alleles of each gene.

ANSWER   The effective number of alleles is estimated as the reciprocal of Σp_i^2. For *Adk-1*, $n_e = 2.28$; for *Lap-5*, $n_e = 1.49$; and for *Xdh*, $n_e = 2.68$. Note that the effective number of alleles is determined more by the uniformity of allele frequencies than by the actual number of alleles. For example, *Lap-5* has more actual alleles than *Adk-1* but a smaller effective number of alleles.

Neutral Mutations

The hypothesis that many genetic polymorphisms result from selectively neutral alleles maintained by a balance between the effects of mutation and random genetic drift is known as the **neutral theory** or the theory of **selective neutrality** (Kimura 1968; King and Jukes 1969). Mutation introduces new alleles into a population, and random genetic drift determines whether a neutral allele will ultimately be fixed or lost. (Loss is the usual outcome.) At equilibrium, there is a balance between mutation and random genetic drift, so that, on the average, each new allele gained by mutation is balanced against an existing allele that is lost (or, more rarely, fixed). The balance point for the homozygosity in the infinite-alleles model is given in Equation 5.9.

In essence, the neutrality hypothesis states that many mutations have so little effect on the organism that their influence on survival and reproduction is negligible. The frequencies of neutral alleles are not, therefore, determined by natural selection. Consequently, if the neutrality hypothesis is true, then many polymorphisms may have no particular significance in the adaptation of a species to its environment. From the perspective of adaptation, selectively neutral polymorphisms are mere evolutionary "noise" and, regardless of how much their study may reveal about population structure and random genetic drift, they tell us little or nothing about adaptive genetic changes in evolution. Kimura (1968) gave the irony a positive spin by noting that "if my chief conclusion [about the prevalence of neutral alleles] is correct, then we must recognize the great importance of random genetic drift . . . in forming the genetic structure of biological populations." Quite so. Indeed, while neutral alleles are unsuitable for the study of genetic adaptation, the very fact that they are invisible to natural selection makes them ideal for mapping the geographical structure of populations and for tracing the ancestral lineages of DNA sequences to make inferences about the phylogenetic relationships between species.

Because the neutrality hypothesis is of fundamental importance in population genetics and evolution, it has been a subject of considerable discussion. The neutrality hypothesis was put forward in the late 1960s at a time when most of the genome was supposed to have a protein-coding function. Introns and other noncoding sequences were unknown. Today it is clear that only about 4 percent of the mammalian genome codes for proteins. The low coding density affords ample scope for mutations that have little or no effect on fitness, including some (but by no means all) mutations in introns, pseudogenes, spacers between genes, noncoding DNA in the centromeric region of chromosomes, and so forth.

There is still considerable controversy whether amino acid polymorphisms are selectively neutral or nearly neutral. To assess the plausibility of the neutrality hypothesis, many aspects of the model must be compared with the situation in actual populations. One aspect of the hypothesis developed in the preceding section concerns the homozygosity to be expected with the infinite-alleles model. Using an observed allozyme homozygosity, we can estimate the effective number of alleles n_e and, from the expression $n_e = 4N\mu + 1$, estimate the corresponding value of $N\mu$. If the resulting values are grossly unreasonable, we can safely reject the infinite-alleles version of the neutrality hypothesis (or at least argue that actual populations cannot be in equilibrium).

Recall from Chapter 2 that observed values of heterozygosity of allozyme genes range from 0.04 to 0.14 in most organisms (see Figure 2.9). Observed homozygosities therefore range from $1 - 0.04 = 0.96$ to $1 - 0.14 = 0.86$, which corresponds to estimated n_e in the range $1/0.96 = 1.04$ to $1/0.86 = 1.16$. Estimates of $N\mu$, calculated as $(n_e - 1)/4$, therefore range from 0.01 to 0.04. The fact that the maximum estimated value of $N\mu$ differs from the minimum by a factor of only about four is surprising, inasmuch as the population number in different species ranges over a factor of 10^4 or more. The apparently too uniform distribution of allozyme homozygosities among diverse organisms has been interpreted as implying that the neutrality hypothesis is wrong for amino acid polymorphisms. On the other hand, estimates of the population number in natural populations are generally imprecise because the studies are very difficult, and estimates of μ, which in this case is the mutation rate to *neutral* alleles, are even more uncertain.

Figure 5.8A shows a second type of test of the adequacy of the neutrality hypothesis in explaining observed levels of genetic variation of allozyme genes. The shaded histogram is the observed distribution of heterozygosity of 74 genes in Caucasians. The histogram outlined in solid lines is a computer-generated theoretical distribution expected with the infinite-alleles model. The observed average heterozygosity is 0.099, and the theoretical heterozygosity is 0.091. The correspondence between the histograms is fairly good,

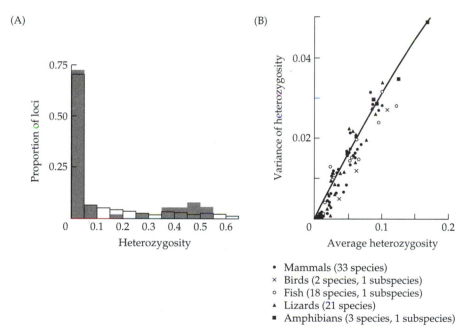

Figure 5.8   (A) Observed distribution of allozyme heterozygosity among genes in Caucasians (shaded) along with theoretical distribution for selective neutrality (solid lines). (B) Mean and variance of heterozygosity among allozyme genes in vertebrates. The solid line is the theoretical curve for the infinite-alleles model when the mutation rate to neutral alleles varies among genes in such a manner that the variance in mutation rate equals the square of the mean mutation rate. (After Nei et al. 1976.)

but the observed distribution seems to include too many genes with heterozygosities in the range of 0.35 to 0.55. (For a possible explanation, see Fuerst et al. 1977.)

A third type of test of the neutrality hypothesis is shown in Figure 5.8B, which presents data on the mean and variance of heterozygosity in 77 vertebrate species. The curve is the theoretical expectation from the infinite-alleles model when the rate of selectively neutral mutation varies among genes (Nei et al. 1976). At first glance, the fit in Figure 5.8B is impressive. On the other hand, the observed points are sufficiently scattered that any number of other curves might fit at least as well. Evidently, statistical comparisons of this sort are too lacking in power to distinguish between the hypotheses.

A brief consideration of the phrase *lacking in power* may be in order. The neutral theory is useful in being a sort of starting point, or null hypothesis, which provides predictions about the relationships among observed quantities that can be confirmed or rejected. Statistical tests of the neutral theory are similar to other types of statistical tests in that two distinct types of possible errors must be balanced. If the tests are too demanding (for example, in failing to allow for the effects of random sampling error), then data may often result in rejection of the hypothesis even when it is true. False rejection is called **Type I error**. On the other hand, if the statistical test allows too much latitude in the data, then data will seldom result in rejection of the hypothesis even when it is false. False acceptance is called **Type II error**. The tradeoff between Type I error and Type II error is that the probability of Type I error cannot be decreased without increasing the probability of Type II error, and vice versa. By convention, statisticians usually adopt a 5 percent criterion for rejection of the null hypothesis even when it is true. This is the familiar 5% level of statistical significance, and it means that there is a 5% chance of rejecting a true hypothesis (Type I error). With this convention, the probability of a Type II error (failing to reject a false hypothesis) falls where it may, and a test with a relatively high probability of Type II error is said to be *lacking in power*.

Although the comparisons in Figure 5.8 are lacking in power and hence are inconclusive in their support of the neutrality hypothesis, many other observations and types of data have been brought to bear in assessing the hypothesis. These data often rely on comparison of nucleotide sequences of DNA in different genes or in different species. These types of comparisons and the conclusions from them are discussed further in Chapter 7.

LINKAGE AND RECOMBINATION

In the context of genetic variation, the importance of recombination is that it allows linked alleles to become associated in many different combinations. In a random mating diploid population, as discussed in Chapter 3, linked alleles come into random association (*linkage equilibrium*) at a rate determined by the frequency of recombination r (Equation 3.8). If r is small, it may require many generations for linkage equilibrium to be attained. For example, the average rate of recombination between adjacent nucleotides in *Drosophila* is 2.7×10^{-8}, with wide variation in different parts of the genome, and so nucleotide polymorphisms in the same region of the genome are often in linkage disequilibrium. Consequently, the ultimate fate of a new mutation may depend to a considerable extent on the effects of other polymorphisms with which it is very closely linked. The effect of recombination on the fate of genetic variation is the subject of this section.

Presumed Evolutionary Benefit of Recombination

Evolutionary biologists have long taken it for granted that recombination is important in evolution because it accelerates the rate of formation of beneficial gene combinations. A graphical representation of the process is illustrated in Figure 5.9. In part A are two large populations, one with no recombination (an asexual species) and one with recombination (a sexual species). Each has three favorable mutations, *a*, *b*, and *c*, which ultimately become incorporated into the genome. In the asexual species, the mutations are incorporated sequentially because each favorable mutation must take place in the genetic background of the one before. The process is slow because each favorable mutation must be nearly fixed before there is a high chance that the next favorable mutation takes place in the proper genetic background. In contrast, in the sexual population, there is no such problem. Recombination between the genes allows that triple mutant *abc* to be formed almost immediately.

The evolutionary advantage of recombination outlined in Figure 5.9A does not apply as strongly to the small populations in Figure 5.9B. In a small population, three favorable mutations are unlikely to be present simultaneously, and so the fixation of the favorable alleles proceeds sequentially in a sexual as well as in an asexual species.

Recombination and Polymorphism

Because recombination between adjacent nucleotides is infrequent, nearby nucleotide sites tend to evolve together. Owing to genetic linkage, forces that tend to maintain genetic diversity or that tend to reduce genetic diversity will act regionally. Therefore, the level of polymorphism found in any region of the genome is expected to be correlated with the level of polymorphism in a closely linked region. Evolutionary forces thus leave their mark on the level and type of genetic variation found within closely linked regions of the genome.

In *D. melanogaster*, an important pattern of genetic polymorphism associated with degree of linkage is illustrated in Figure 5.10. A region of the

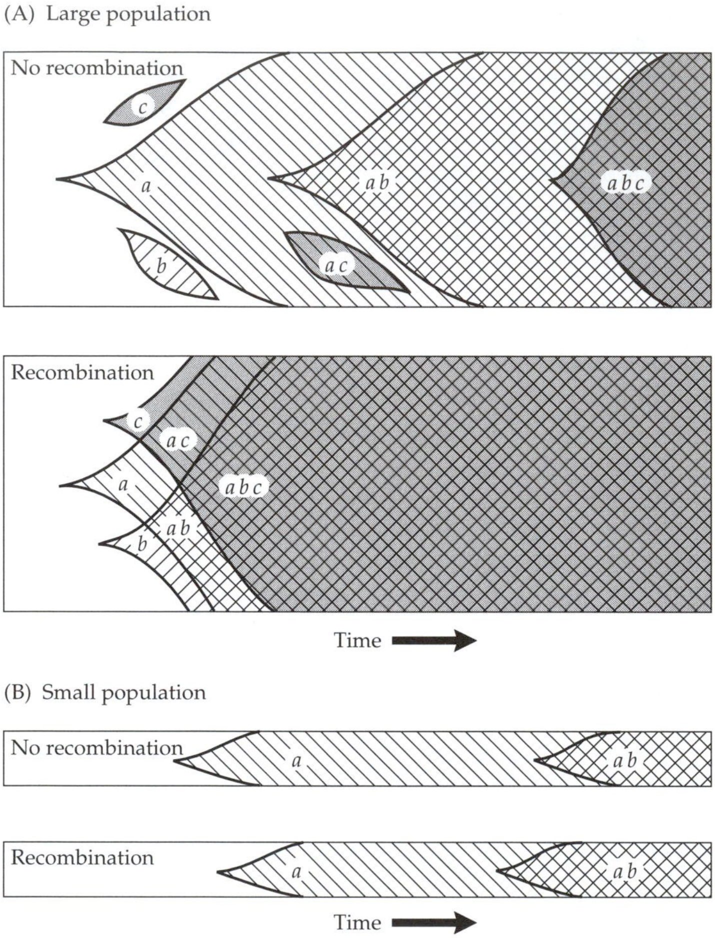

(A) Large population

(B) Small population

Figure 5.9   Evolutionary effect of recombination. (A) In a large population of an asexual species with no recombination (top panel), the favorable mutations *a*, *b*, and *c* must be incorporated into the genome sequentially because there is no mechanism to bring the favorable mutations together; each favored mutation must reach a high frequency to have a reasonable chance that the next favorable mutation will take place in the proper genetic background. With recombination (bottom panel), recombination between the favorable genes enables the triple mutant *a b c* to be formed very rapidly. (B) The beneficial effect of recombination is diminished in a very small population because, in a small population, multiple favorable mutations are unlikely to be present simultaneously. (From Crow and Kimura 1970.)

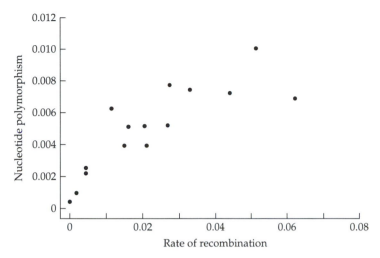

Figure 5.10   Observed relation between the level of nucleotide polymorphism and the rate of recombination in *Drosophila*. (From Aquadro et al. 1994.)

genome in which the rate of recombination per nucleotide is reduced, such as near the tip or near the base of each chromosome arm, also tends to have a reduced level of genetic polymorphism even though the rates of mutation are uniform across the chromosome (Aquadro et al. 1994). In Figure 5.10, the level of polymorphism is expressed as the proportion of nucleotide sites that are polymorphic (called θ in Chapter 2). For the regions plotted, θ ranges over more than a factor of 10, so there is clearly an important effect of close linkage in reducing the level of polymorphism.

In theory, the reduction in the level of polymorphism in regions of tight linkage could be explained by either of two diametrically opposed mechanisms. In one mechanism, the reduction results from the fixation of favorable mutations. In the other mechanism, the reduction results from the elimination of harmful mutations. These explanations have somewhat different implications for the pattern of polymorphism in regions of tight linkage, and so they can be distinguished experimentally.

Consider first the consequences of fixation of a favorable mutation. On its way to fixation, any new favorable mutation may carry along a small surrounding region of the genome and render the region monomorphic. The monomorphism will not usually be complete. Some degree of polymorphism may remain in the region, either because new mutations happen in the process of fixation or because of rare recombination events that take place.

The process in which a favorable mutation becomes fixed in a population is called a **selective sweep**. During a selective sweep of a favorable allele, any neutral alleles sufficiently tightly linked go along for the ride and are said to be **hitchhiking**. The main effect of hitchhiking is that a small region around the favored allele will be overrepresented in the population. In other words, there will be an apparent excess of rare genetic variants owing to the over-representation of the region that profited from the hitchhiking.

Consider next the consequences of a harmful mutation. For concreteness, consider the genetic map diagrammed in Figure 5.11A, in which the short vertical lines indicate adjacent nucleotide sites. One site that can undergo neutral mutation is embedded in the middle surrounded by sites that can

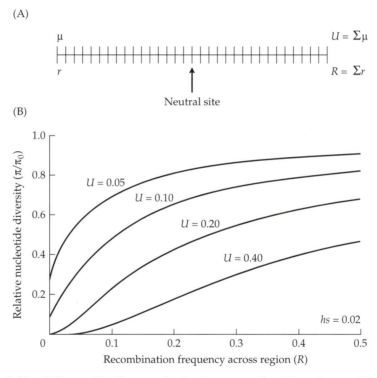

Figure 5.11 Effects of background selection on nucleotide polymorphism. (A) A region of a chromosome containing a set of genes (tick marks) that can mutate to detrimental alleles; within this set of genes is a single neutral site. The mutation rate per locus is μ and the rate of recombination between adjacent loci is r. (B) Relative nucleotide diversity as a function of U, the total mutation rate, and R, the total recombination rate, across the chromosomal region. Note the positive correlation between level of nucleotide polymorphism and rate of recombination.

undergo harmful mutations only. The rate of harmful mutation per site per generation is denoted μ, and the rate of recombination between adjacent sites is denoted r.

Suppose further that each mutation, even when heterozygous, is sufficiently harmful that any chromosome in which a mutation is present is ultimately doomed. In the absence of recombination, the fate of a chromosome depends on whether it is free of harmful mutations because, under our assumptions, no chromosome can persist for long unless it is free of mutations. The effect of harmful mutation, which in this context is called **background selection**, is to reduce the number of chromosomes that can contribute to the ancestry of remote generations. Indeed, the effect of background selection is identical to that of a reduction in population size except that the reduction applies, not to the genome as a whole, but to a tightly linked region (Charlesworth et al. 1993). Background selection therefore reduces the level of genetic polymorphism. Looser linkage means that a linked neutral mutation can escape the fate of a harmful neighboring mutation by recombination with a mutation-free chromosome. Hence, the tighter the linkage, the greater the reduction in polymorphism due to background selection. Although there is a reduction in the level of polymorphism, background selection does not skew the distribution of rare polymorphisms because, for all practical purposes, the harmful allele merely causes one chromosome to drop out of the population, much as if it were to go extinct by chance (Braverman et al. 1995).

Although the evidence is not yet conclusive, the model of background selection appears to provide a better explanation of the *Drosophila* data than does the model of selective sweeps (Hudson and Kaplan 1995; Charlesworth et al. 1995). The evidence is that rare nucleotide polymorphisms are found at a frequency that would be expected given the overall level of polymorphism (Braverman et al. 1995). There is no evidence for a skewed distribution toward rare variants that the model of selective sweeps would predict.

The effect of background selection on the level of genetic variation is shown graphically in Figure 5.11B for the genetic map diagrammed in part A. The curves are plotted from the formula

$$\pi = \pi_0 e^{-U/(2hs+R)} \qquad\qquad 5.11$$

(Hudson and Kaplan 1995). The symbol π is the nucleotide diversity, defined as the average proportion of nucleotide differences between all possible pairs of sequences (Chapter 2); π_0 is the value of π in the absence of background selection. U and R refer to the diagram in part A. U is the total mutation rate per diploid genome, summed across all genes in the region; and R is the total rate of recombination across the region, summed over each of the intervals between genes. The quantity hs measures the degree of harmfulness of each deleterious mutation in a heterozygous genotype; the extremes are $hs = 0$, when there is no effect in the heterozygote, and $hs = 1$, when the heterozygote is lethal. The model on which Equation 5.11 is based includes the assumption that hs is small but not 0.

The curves in Figure 5.11B are for the specific value $hs = 0.02$, which means that a genotype that is heterozygous for one deleterious mutation has a 2% reduction in survival compared with a homozygous nonmutant. For each curve, the relative nucleotide diversity (π/π_0) decreases as the total recombination rate R decreases. This result means that, with tighter linkage, each detrimental mutation that is eliminated takes with it a larger surrounding region of chromosome. The relative nucleotide diversity also decreases as the total mutation rate increases; that is, greater background selection eliminates a greater number of chromosomes. Together, tight linkage and a moderate or high total mutation rate can result in a very substantial decrease in relative nucleotide diversity, reducing it to a level of 20% or less of that expected in the absence of background selection. In view of the reduction in genetic variation in regions of reduced recombination observed in *Drosophila* (Figure 5.10), the implication of Equation 5.11, along with the absence of a skewed distribution toward rare variants, suggests that much of the effect results from background selection.

Piecewise Recombination in Bacteria

Many prokaryotic organisms make use of mechanisms of recombination in which a piece of DNA that is small, relative to the size of the entire genome, is transferred from a donor cell into a recipient cell. These mechanisms include *transformation*, in which free DNA is taken up by the recipient from the surrounding medium; *transduction*, in which a DNA fragment is carried from the donor to the recipient by means of a virus particle; and *conjugation*, in which a replica of the chromosome from a donor cell is transferred into a recipient cell by a gradual process requiring cell-to-cell contact, but the chromosome usually breaks before the transfer is complete. Because relatively short patches of the genome participate in recombination, these processes differ in their evolutionary implications from meiotic recombination in eukaryotes.

The main effect of short-patch recombination is that long-range linkage disequilibrium tends to be maintained. For example, in enteric bacteria, such as *Escherichia coli*, which are part of the normal intestinal flora, linkage disequilibrium between allozyme loci is very strong (Whittam et al. 1983). At the level of DNA sequence, however, many genes have an obviously mosaic structure in which different segments have different phylogenetic histories (DuBose et al. 1988). An example from the *phoA* gene, coding for alkaline phosphatase in *E. coli*, is illustrated in Figure 5.12. Among the polymorphic nucleotide sites indicated, the unique nucleotide at each site is inscribed in a box. At the extreme ends of the gene, the alleles from strains RM217T and

Nucleotide site in *phoA* gene

Allele			

```
                 1 1 1 1 1 1 1 1 1 1 1 1 1 1 1 1 1 1 1 1 1
            6 8 0 0 0 0 0 0 4 4 4 4 5 5 5 5 6 7 7 7 8 8
            2 3 5 6 6 7 7 8 2 2 7 7 9 0 2 5 6 8 1 6 8 2 5
            7 1 9 1 8 4 7 1 5 8 4 9 7 9 4 1 0 3 2 9 2 6 0
                *               * *   *
RM217T   C A [G A] C G A G [G T G T G T T T T T] T G A A T
RM45E    [T] A C G C G A G T C A C T C C C C T T G A A T
RM224H   C [G] C G [T A G T] T C A C T C C C C [C C A T T C]
```

Figure 5.12   Evidence for recombination in the *phoA* gene in natural isolates of *E. coli*. The pair of strains at the top are more similar at the beginning and end of the gene, the pair of strains at the bottom are more similar in the central region. There is significant clustering of the nucleotide sites inscribed in boxes, as expected from recombination. (Data from DuBose et al. 1988.)

RM45E are the most closely related; in the middle of the gene, from nucleotide sites 1425 to 1560, there is a run of polymorphic nucleotides in which the similarity between RM217T and RM45E is lost, as if this part of the gene had been introduced by recombination with a more distantly related allele. Although short runs of similar or dissimilar nucleotides can also be the result of chance, chance effects can be ruled out by appropriate statistical tests for recombination (Stephens 1985; Sawyer 1989).

The finding that many genes have a mosaic ancestry through recombination seems at first to contradict the finding of significant linkage disequilibrium between more widely separated genes. The paradox is resolved by the fact that each recombination event is local; it replaces a relatively short stretch of the recipient chromosome, and the linkage phase between more distant alleles is maintained. The *E. coli* chromosome, therefore, consists of clonal segments from a common ancestor, which is called the **clonal frame** (Milkman and Bridges 1990, 1993), interrupted by short segments derived from recombination with diverse other clones. Even though the clonal frames are interrupted by relatively short recombinant segments, their integrity would ultimately be lost unless there were occasional selective events favoring particular genotypes.

Absence of Recombination in Animal Mitochondrial DNA

Studies in animal population genetics often focus on the DNA of mitochondria. The mitochondrial genome is informative about parentage because, in most species of animals, it is maternally inherited and does not undergo recombination. It is also a small molecule present in abundant quantities in most cells. In animals, mitochondrial DNA (mtDNA) is a circular molecule typically in the range from 15 to 20 thousand base pairs in length. It codes for fewer than 40 genes; approximately half code for ribosomal RNA or for transfer RNA used in mitochondrial protein synthesis, and the remaining genes code for proteins used in electron transport or oxidative phosphorylation. In many species, including mammals, parts of the mtDNA sequence evolve very rapidly in comparison with nuclear genes, and hence mtDNA can often be used to make inferences about population structure and recent population history.

An example of the utility of mtDNA in population studies is illustrated in Figure 5.13, which summarizes the result of examining the mtDNA of 87 pocket gophers, *Geomys pinetis,* collected across the geographic range of the species in Alabama, Georgia, and Florida (Avise et al. 1979). The mtDNA

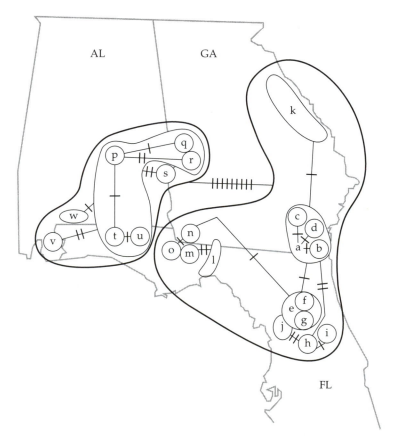

Figure 5.13 Lineage relationships between mtDNA types in pocket gophers. The lowercase letters are different mtDNA types grouped according to similarity and superimposed on a geographical map of the collection sites. The tick marks across the connecting lines are the numbers of inferred mutational steps. (From Avise 1994.)

from each gopher was digested in turn with each of six restriction enzymes, each cleaving the DNA at a different six-base recognition site. The resulting restriction fragments were separated by electrophoresis and compared among the animals to estimate the number of nucleotide differences affecting the restriction sites.

Among the 87 gophers, there were 23 distinct types of mtDNA, represented by the lowercase letters in Figure 5.13. Each of these types represents a maternal mtDNA lineage, distinct from other lineages. Animals that share an mtDNA type must have a female ancestor in common. The branching network in Figure 5.13 estimates the matriarchal phylogeny of the mtDNA. The straight lines connect related types of mtDNA, and the number of slashes across each line indicates the estimated number of nucleotide differences in the restriction sites between the mtDNA types. Groups of related mtDNA types are enclosed in thin black lines; the thickest lines delineate a western and an eastern subpopulation of gophers whose overall mtDNA sequence differs by an estimated 3%. Between the eastern and western subpopulations, there are 9 nucleotide differences among the sites cleaved by the restriction enzymes.

The mtDNA network in Figure 5.13 also resolves population subdivision within the western and eastern subpopulations. This subdivision is indicated by the mtDNA types circumscribed by the thin black lines. Some of the mtDNA types such as "k" and "p" are widespread, whereas others such as "b" and "q" are more local in their distribution. The local clones usually differ from the most widespread mtDNA type in the region by only one or two nucleotides among the sites cleaved by the restriction enzymes. The example in Figure 5.13 shows that, because of matrilineal inheritance and the absence of recombination in mtDNA, the network of mtDNA types can reveal a great deal about population substructure in natural populations.

MIGRATION

In a subdivided population, random genetic drift results in genetic divergence among subpopulations. **Migration**, which refers to the movement of organisms among subpopulations, is a sort of genetic glue that holds subpopulations together genetically and that sets a limit to how much genetic divergence can take place. To understand the homogenizing effects of migration, it is useful to study migration in several simple models of population structure.

One-Way Migration

When migration takes place predominantly from one population into another, without an equal amount of migration in the reverse direction, then there is said to be **one-way migration**. An illustration of one way migration

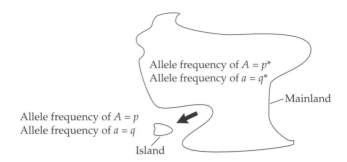

Allele frequency of $A = p^*$
Allele frequency of $a = q^*$

Mainland

Allele frequency of $A = p$
Allele frequency of $a = q$

Island

Figure 5.14   Model of one-way migration from a large land mass onto an island. The allele frequencies in the source population, p^* and q^*, are assumed to remain constant, whereas those in the recipient population, p_t and q_t, change with time.

between a large mainland population and a small island subpopulation is shown in Figure 5.14. For simplicity, we consider a gene with two alleles, A and a, with respective frequencies p^* and q^* on the mainland and p and q on the island. Suppose that, in any generation, a proportion m of zygotes in the island subpopulation originates as a random sample of organisms from the mainland. Then, if p and p' are the frequencies of A in the island subpopulation in two successive generations, it follows that

$$p' = (1-m)p + mp^*$$ 5.12

In Equation 5.12, m is called the **migration rate** between the mainland and the island. Subtracting p^* from both sides of Equation 5.12 and simplifying leads to the expression $p' - p^* = (1-m)(p-p^*)$; from this expression it follows immediately that $p_t - p^* = (1-m)^t(p_0-p^*)$, where p_t is the frequency of A in the island subpopulation in generation t. Hence,

$$p_t = p^* + (1-m)^t(p_0 - p^*)$$ 5.13

Equation 5.13 expresses mathematically what should be clear intuitively: With one-way migration, the allele frequency of A in the island subpopulation gradually approaches that of the mainland population, and the rate of approach is m per generation. As a check on Equation 5.13, note that, when $t = 0$, then $p_t = p_0$, as must be the case, and as t becomes large, $p_t \to p^*$.

As an evolutionary process that brings potentially new alleles into a population, migration is qualitatively similar to mutation. The major difference is quantitative: Generally speaking, the rate of migration among subpopulations of a species is vastly greater than the rate of mutation of a gene. The contrast is illustrated in Figure 5.15 for the unrealistic case in which the A

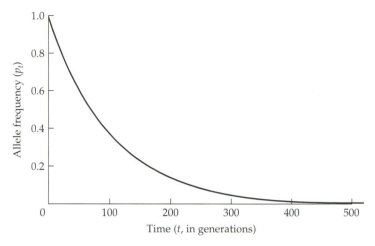

Figure 5.15   Change of allele frequency with one-way migration assuming that an allele A is initially fixed in the recipient population and absent in the source population. The migration rate is $m = 0.01$. Note that this is the same curve as in Figure 5.1 except that the horizontal axis is compressed to 500 generations. The time scale is different because, generally speaking, the migration rate m is much larger than the mutation rate μ.

allele present in an island subpopulation is absent on the mainland. In this case, Equation 5.13 becomes $p_t = p_0(1 - m)^t$, which has the same form as Equation 5.1 for one-way mutation except that m replaces μ. The identity in the shape of the curves is apparent, but the time axis in Figure 5.15 is compressed because, when $m = 0.01$, as in this example, compared with the value of $\mu = 0.0001$ in Figure 5.1, it requires only one generation of migration to change the allele frequency to the same extent as 100 generations of mutation.

Equation 5.13 holds more generally for one-way migration by letting p be the frequency of any allele in the population that receives the migrants and p^* be the frequency of the same allele in the population that supplies the migrants. Application of this equation to estimating the amount of genetic migration in certain human populations makes use of the allele-frequency data given in Problem 4.4 (page 126). The data pertain to blacks and whites in Claxton, Georgia, and blacks in West Africa. The case of the MN blood groups serves as an example. In West Africa, which for the purpose of this problem may be regarded as the ancestral black population, $p_0 = 0.474$ for the allele frequency of M. In present-day Claxton blacks, $p_t = 0.484$. The Claxton white population may reasonably be regarded as representative of the source of the migrants, and for Claxton whites, $p^* = 0.507$. Blacks came into the United States on a large scale from West Africa about 300 years ago, hence t is about 10 generations. Substituting these estimates into Equation 5.13, we obtain $0.484 = 0.507 + (1 - m)^{10}(0.474 - 0.507)$, from which we infer that $m = 0.035$ per generation. This estimate can be interpreted as implying that, in the genetic history of the population of Claxton blacks, about 3.5% of the alleles of the *MN* gene in any generation were newly introduced by genetic migration from whites. The apparent amount of migration estimated by this method differs from one locus to the next. It also differs according to the geographical region in which the white and black populations reside.

PROBLEM 5.4   Estimate the amount of migration from whites to blacks using allele frequencies for each of the other genes in Problem 4.4 in (page 126).

ANSWER   Ss blood group: $m = -0.013$ per generation; Duffy: $m = 0.011$; Kidd: $m = -0.028$; Kell: $m = -0.005$: G6PD, $m = 0.039$: hemoglobin β: $m = 0.071$.

Problem 5.4 illustrates some of the difficulties in estimating racial admixture from allele frequencies. The positive values of m vary widely, and the negative values are not consistent with the proposed model of migration. Cavalli-Sforza and Bodmer (1971) remark that "The weakness of the analysis is mostly due to the uncertainty of the origin of black Americans … and the variability of gene frequencies in the probable area of the slave markets in West Africa. In addition, it is unavoidable that gene frequencies have changed somewhat from their original values, due to drift or, in some cases, selection. The opportunities for admixture, and the time available for it, must also have varied widely." The most reliable gene among those in Problem 5.4 is probably that for the Duffy blood groups because the Fy^a allele is virtually nonexistent in all of West Africa. For this gene, the estimate of m is about one percent per generation, a result that is consistent with the average value for a large number of other genes (Cavalli-Sforza and Bodmer 1971).

The Island Model of Migration

In the **island model** of migration, a large population is split into many subpopulations dispersed geographically like islands in an archipelago. Examples of island population structure might include fish in freshwater lakes or slugs in dispersed garden plots. Each subpopulation is assumed to be so large that random genetic drift can be neglected. Consider an allele A with an average allele frequency among the subpopulations equal to $\bar{p}$. Migration is assumed to happen in such a way that the allele frequency among the migrants equals the average allele frequency among the subpopulations, namely, $\bar{p}$. The amount of migration is again measured by the parameter m, which equals the probability that a randomly chosen allele in any subpopulation comes from a migrant. Let us consider a particular subpopulation with an A allele frequency of p_t in generation t. For a randomly chosen allele in this subpopulation in generation t, the allele could have come from the same subpopulation in generation $t - 1$ with probability $1 - m$, in which case it is an A allele with probability p_{t-1}. Alternatively, the allele could have come from a migrant in generation $t - 1$ with probability m, in which case it is an A allele with probability $\bar{p}$. Because all evolutionary processes other than migration are ignored, $\bar{p}$ stays the same in all generations. Altogether,

$$p_t = p_{t-1}(1-m) + \bar{p}m \qquad\qquad 5.14$$

Equation 5.14 is similar to Equation 5.2 for mutation, and its solution in terms of p_0 is

$$p_t = \bar{p} + (1-m)^t(p_0 - \bar{p}) \qquad\qquad 5.15$$

The similarity with Equation 5.13 is apparent: in fact, the equations are identical except that the role of p^* in one-way migration is replaced with $\bar{p}$ in the island model. Perhaps less obvious is the similarity with Equation 5.4 for reversible mutation, in which case $v/(\mu + v)$ plays the role of $\bar{p}$ and $\mu + v$ plays the role of m. The correspondence between the equations again emphasizes the similarity between the effects of migration and those of mutation. The processes result in similar mathematical expressions because both mutation and migration act linearly on allele frequency, which means that p_t is a linear function of p_{t-1}. Although Equation 5.15 for migration is mathematically similar to Equation 5.4 for mutation, the biological implications are quite different. Because rates of migration are typically much greater than rates of mutation, changes in allele frequency are generally much faster with migration.

As an example of the use of Equation 5.15, suppose there are only two populations with initial allele frequencies of A of 0.2 and 0.8, respectively, with $m = 0.10$. Thus 10 percent of the organisms in either subpopulation in any generation are migrants having an allele frequency of A of $\bar{p} = (0.2 + 0.8)/2 = 0.5$. What is the allele frequency of A in the two populations after 10 generations? For the population with initial allele frequency 0.2, we substitute $p_0 = 0.2$, $\bar{p} = 0.5$, and $m = 0.10$ into Equation 5.15 to obtain $p_{10} = 0.5 + (1 - 0.10)^{10}(0.2 - 0.5) = 0.395$; for the other population, we substitute $p_0 = 0.8$, $\bar{p} = 0.5$, and $m = 0.10$, and so $p_{10} = 0.5 + (1 - 0.10)^{10}(0.8 - 0.5) = 0.605$.

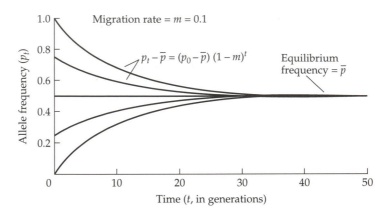

Figure 5.16   Change of allele frequency with time in five subpopulations exchanging migrants at the rate $m = 0.1$ per generation. Note the rapid convergence to a common equilibrium frequency.

Another example using Equation 5.15 is shown in Figure 5.16, where there are five subpopulations (initial frequencies 1, 0.75, 0.50, 0.25, and 0), again with $m = 0.10$. Note how rapidly the allele frequencies converge to the same value, in this case, 0.5.

How Migration Limits Genetic Divergence

It is remarkable how little migration is required to prevent significant genetic divergence among subpopulations as measured by, for example, the fixation index F_{ST}. To understand the homogenizing effect of migration, consider the model in Figure 5.5 (page 171), in which two alleles drawn at random from a subpopulation in generation t are replicas of the same allele in generation $t-1$ with probability $1/2N$ and replicas of different alleles in generation $t-1$ with probability $1 - 1/2N$. In the first case, the alleles are necessarily identical by descent; in the second case, they are identical by descent with probability F_{t-1}, where F is shorthand for F_{ST}. In either case, the identity by descent is unbroken only if neither allele is replaced by an allele from a migrant, and so

$$F_t = \left(\frac{1}{2N}\right)(1-m)^2 + \left(1 - \frac{1}{2N}\right)(1-m)^2 F_{t-1} \qquad 5.16$$

Illustrating again the analogy between migration and mutation, Equation 5.16 is identical to Equation 5.8 measuring the effect of mutation on the probability of identity by descent, except that m replaces μ. The equilibrium value $\hat{F}$ of F can be found by setting $\hat{F} = F_t = F_{t-1}$; after expanding the squared terms on the right hand side, and assuming that m is small enough, and N large enough, that terms in m^2 and m/N can be ignored, some rearrangement leads to

$$\hat{F} = \frac{1}{1 + 4Nm}$$ 5.17

As might be expected, Equation 5.17 is identical in form to Equation 5.9 for mutation but the biological implications are very different owing to the fact that the rate of migration is typically much greater than the rate of mutation.

The product Nm in Equation 5.17 has a straightforward biological interpretation. The total number of alleles in a subpopulation of size N diploid organisms is $2N$. In any generation, the proportion of alleles that are replaced by alleles from migrant organisms is m; hence the number of migrant alleles in any generation equals $2Nm$. However, $2Nm$ is also the total number of alleles in Nm diploid organisms, and so Nm can be interpreted as the absolute number of migrant organisms that come into each subpopulation in each generation.

Because the absolute number of migrants per generation equals Nm, Equation 5.17 implies that $\hat{F}$ decreases as the number of migrants increases. Indeed, the decrease in $\hat{F}$ with increasing Nm is extremely rapid, as shown in Figure 5.17. In the extreme case of complete genetic isolation between the subpopulations, $Nm = 0$ and $\hat{F} = 1$. The decrease is then so rapid that for:

- $Nm = 0.25$ (one migrant every fourth generation), $\hat{F} = 0.50$
- $Nm = 0.5$ (one migrant every second generation), $\hat{F} = 0.33$
- $Nm = 1$ (one migrant every generation), $\hat{F} = 0.20$
- $Nm = 2$ (two migrants every generation), $\hat{F} = 0.11$

The implication of Figure 5.17 is that migration is a potent force acting against genetic divergence among subpopulations. On the other hand, the homogenizing effect of migration should not be overestimated. The measure of genetic divergence in Figure 5.17 is F_{ST}, the value of which is determined by the variance in allele frequency among subpopulations (Equation 4.6) and so is affected primarily by polymorphic alleles that are at intermediate frequencies. Rare alleles present in one subpopulation but absent in others have hardly any effect on F_{ST}. Because rare alleles are rare, they are unlikely to be included among migrant organisms unless the migration rate is very great, and so rare alleles will tend to remain present in only one or a few subpopulations in a local area until such time as their frequency may become great enough to be dispersed by migration. An allele found in only one subpopulation is called a **private allele**. Next we shall see that the rate of migration can be estimated by an examination of the frequency of private alleles.

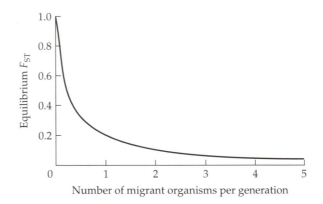

Figure 5.17   Decrease in the fixation index F_{ST} among subpopulations at equilibrium in the island model of migration. The curve is that in Equation 5.17 giving $\hat{F}$ as a function of Nm. In the island model, Nm is the number of migrant organisms that come into each subpopulation in each generation.

Estimates of Migration Rates

One method of estimating genetic migration in natural populations relies on the finding that, in theoretical models, the logarithm of Nm decreases approximately as a linear function of the average frequency of private alleles in samples from the subpopulations (Slatkin 1985). Data on the average frequency of private alleles has been compiled and analyzed by Slatkin (1985), and the resulting estimates of Nm and equilibrium values of F_{ST} are summarized in Table 5.1. There is obviously considerable variation in Nm among organisms. However, many of the values of Nm are smaller than about 2, which means that there is still considerable opportunity for genetic divergence among subpopulations.

A second kind of approach to estimating Nm in natural populations is illustrated in Figure 5.18, which gives the distribution of estimated values of F_{ST} among 61 genes in natural populations of *Drosophila melanogaster* (Singh and Rhomberg 1987). The average of the estimated values is $F_{ST} = 0.16$, which, assuming equilibrium, is an estimate of $1 + 4Nm$ (Equation 5.17). The estimate is therefore $Nm = [(1/0.16) - 1]/4 = 1.3$. This estimate is within the range for other *Drosophila* species in Table 5.1. However, there are many genes in Figure 5.18 that have F_{ST} values greater than 0.30. An analogous method of estimating Nm from the F_{ST} values of polymorphic nucleotides within a gene is discussed in Hudson et al. (1994a). In Chapter 7 we will consider how Nm can be estimated from the genealogies of genes.

Patterns of Migration

Migration in actual populations is more complex than is assumed in the island model of migration. In nature, migrants come primarily from nearby

TABLE 5.1   ESTIMATES OF *Nm* AND $\hat{F}_{ST}$

Species	Type of organism	Estimated Nm	Estimated $\hat{F}_{ST}$
Stephanomeria exigua	Annual plant	1.4	0.152
Mytilus edulis	Mullusc	42.0	0.006
Drosophila willistoni	Insect	9.9	0.025
Drosophila pseudoobscura	Insect	1.0	0.200
Chanos chanos	Fish	4.2	0.056
Hyla regilla	Frog	1.4	0.152
Plethodon ouachitae	Salamander	2.1	0.106
Plethodon cinereus	Salamander	0.22	0.532
Plethodon dorsalis	Salamander	0.10	0.714
Batrachoseps pacifica ssp. 1	Salamander	0.64	0.281
Batrachoseps pacifica ssp. 2	Salamander	0.20	0.556
Batrachoseps campi	Salamander	0.16	0.610
Lacerta melisellensis	Lizard	1.9	0.116
Peromyscus californicus	Mouse	2.2	0.102
Peromyscus polionotus	Mouse	0.31	0.446
Thomomys bottae	Gopher	0.86	0.225

Source: Data from Slatkin 1985.

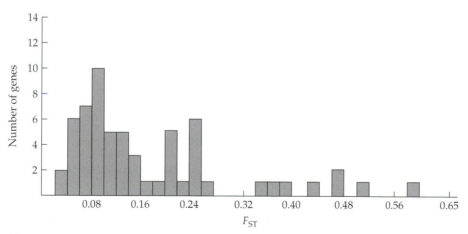

Figure 5.18   Distribution of estimated values of F_{ST} for 61 genes among natural populations of *Drosophila melanogaster*. Although the average value of F_{ST} suggests migration at a level of *Nm* between 1 and 2, about one-third of the genes have F_{ST} values greater than 0.20. (From Singh and Rhomberg 1987.)

populations. To the extent that nearby populations have similar allele frequencies, the effects of migration are smaller, and sometimes much smaller, than predicted by the island model. Populations in nature may be strung out along one dimension, such as a river bank. Populations may also be distributed regularly in two dimensions, or there may be one large population with an internal genetic structure caused by the tendency for mating to take place between organisms born in the same region. Analysis of the effects of migration in such complex population structures is usually very difficult. Among humans, migration rates depend on age, sex, marital status, socioeconomic status, population density, and many other factors. Migration rates also can change rapidly, and so a full-blown theory of migration has to be extremely complex.

The effects of migration on genetic differentiation of populations are seen dramatically in Figure 5.19. Part A pertains to the moth *Biston betularia*, part B to the moth *Gonodontis bidentata*. Both species have evolved melanic (blackened) forms in response to heavy air pollution, and the graphs give the frequency of the melanic forms in the two species. The geographical area in A includes Liverpool and Manchester, as viewed from rural Wales. Note the fall-off in frequency of melanics in the nonindustrial areas toward the front of the graph. *Biston betularia* exists in low population densities and must fly relatively long distances to find a mate. The resulting high rate of migration hinders differentiation of populations, hence the smooth surface. In contrast, *Gonodontis bidentata* exists in high population densities and the migration rate is low; hence there is substantial genetic differentiation among populations, as evidenced by the bumpy surface of the graph in part B.

TRANSPOSABLE ELEMENTS

A DNA sequence that can change its location within the genome is called a **transposable element**. In being able to create novel genome rearrangements, transposable elements are agents of genetic variation. A transposable element may insert into a coding region and inactivate a gene or insert into a regulatory region and change the pattern of expression of the gene. Also, pairs of transposable elements may undergo recombination and create novel chromosome rearrangements.

The process of transposition requires a protein, called **transposase**, which is usually encoded within the sequence of the transposable element itself. Most transposable elements undergo transposition through a replicative process with DNA or RNA intermediates. In most cases, transposition to a new location also leaves one copy of the transposable element behind in its original location, so transposable elements can increase in copy number in the genome. Some transposable elements are also able to regulate their own rate of transposition. Several major classes of transposable elements can be distinguished by their nucleotide sequence organization or by the details of their mechanisms of transposition or regulation.

(A)

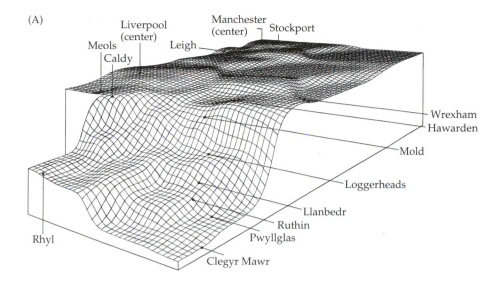

(B)

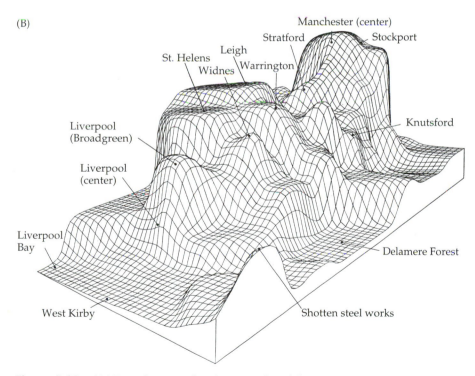

Figure 5.19   (A) Distribution of melanic moths of the species *Biston betularia* over an area including Liverpool and Manchester, as viewed from rural Wales. (B) Distribution of melanic moths of the species *Gonodontis bidentata* over a smaller area than in (A) but viewed from the same perspective. (From Bishop and Cook 1975.)

Factors Controlling the Population Dynamics of Transposable Elements

Transposable elements were originally discovered in maize as the cause of certain genetically unstable mutations. They are now known to be ubiquitous among prokaryotes and eukaryotes (Berg and Howe 1989). The ability of transposable elements to increase in copy number and create novel chromosomal rearrangements reveals a dynamic aspect of genome structure and evolution not previously recognized. Some transposable elements have become widely disseminated among organisms because of their ability to undergo horizontal transmission between reproductively isolated genomes. Often referred to as **selfish DNA** because transposition alone may be sufficient for persistence in the genome of a species, transposable elements also may occasionally create favorable mutations and thus become agents of adaptive evolution.

Models for the population dynamics of transposable elements usually incorporate several features:

- A rate of infection, in which genomes previously lacking the transposable element become infected with it.
- A rate of transposition, which determines how rapidly the copy number increases; the effects of regulation are taken into account by assuming that the rate of transposition is a decreasing function of copy number.
- A mechanism, or combination of mechanisms, for eliminating elements from the population; otherwise, the copy number would increase indefinitely. The usual assumption is that the presence of transposable elements in the genome decreases the ability of an organism to survive and reproduce, resulting in the elimination of some elements by means of natural selection, or that elements can be eliminated from the genome by means of genetic deletion.

Through the study of such models, the diversity and novel attributes of transposable elements have been incorporated into the concepts of population genetics; see, for example, Langley et al. (1983), Montgomery and Langley (1983), Kaplan and Brookfield (1983), Sawyer et al. (1987), Hartl and Sawyer (1988), Ajioka and Hartl (1989), Charlesworth et al. (1994).

Insertion Sequences and Composite Transposons in Bacteria

Bacteria contain several types of transposable elements. Among the simplest are **insertion sequences**, which are typically about 1000–2000 nucleotides in length and contain at least one long translational open reading-frame coding for the transposase protein. The transposase recognizes a short nucleotide sequence, inverted in orientation, present at each end of the insertion sequence, and so the element moves as an intact unit. The bacterium *Escherichia coli* contains several types of insertion sequences, each different but all sharing the same sequence organization with inverted repeats and at least one open reading frame. The factors controlling the population dynamics of insertion sequences can be deduced from the distribution of numbers of each element present among a sample of bacterial strains isolated from natural sources (Sawyer et al. 1987).

Population models of transposable elements in E. coli are greatly simplified because the organism has asexual reproduction, a low rate of recombination among strains, and a low rate of deletion of insertion sequences. The "state" of a bacterial strain with respect to a particular insertion sequence may be defined as the number of copies n of the element that are present. Among the factors that control the population dynamics are:

- The rate u at which uninfected cells become infected; u is the probability, per generation, that a cell initially in state $n = 0$ ends up in the state $n = 1$.
- The rate T of transposition in infected strains; T is the probability, per generation, that a cell in state $n > 0$ goes to state $n + 1$.
- The rate S at which reproduction of infected cells is less than that of uninfected cells. In terms of the exponential growth model in Chapter 1, if r_0 is the intrinsic rate of increase of uninfected cells (see Equation 1.7 on page 30) and r_0' is that of infected cells, then $S = r_0 - r_0'$.

The most general models of this type allow for T and S to be functions of n, but here we will assume that they are constant. Note, however, that the assumption that T is a constant implicitly defines a type of regulation because, if the probability of transition from state n to state $n + 1$ is independent of n, then the probability of transposition *per element* present in a strain must equal T/n and this fraction is a decreasing function of n.

Given constant values of u, T, and S, then it can be shown that a population of bacterial cells attains an equilibrium distribution of numbers of transposable elements in which the probability p_i that a cell contains exactly i copies of the transposable element is equal to

$$p_0 = \alpha \qquad \qquad 5.18a$$

and

$$p_i = (1 - \alpha)(1 - \phi)\phi^{i-1} \quad (i > 1) \qquad \qquad 5.18b$$

where $\alpha = 1 - (u/S)$ and $\phi = T/(T + S - u)$ (Sawyer and Hartl 1986, Sawyer et al. 1987).

Equation 5.18 can be applied to the concrete case of insertion sequence IS30 in E. coli, in which the distribution of numbers among 71 strains fits a model with $\alpha = \frac{1}{2}$ and $\phi = \frac{1}{2}$. With these parameters, the distribution simplifies to the remarkably simple formula $p_i = (\frac{1}{2})^i$ for $i \geq 0$. Among 71 strains, therefore, the observed and expected numbers of strains containing i elements are as indicated in Table 5.2. The strains with five or more elements have been grouped in order to carry out a χ^2 test of goodness of fit. This χ^2 test has three degrees of freedom because α and ϕ were estimated from the

---

TABLE 5.2   **NUMBER OF IS*30* ELEMENTS PRESENT IN 71 NATURAL ISOLATES OF *E. coli***

Number of copies of IS*30* element	Expected number of strains	Observed number of strains
0	35.5	36
1	17.8	16
2	8.9	13
3	4.4	2
4	2.2	2
≥ 5	2.2	2

Source: Data from Sawyer et al. 1987.

data. The value of χ^2 equals 3.48, which has an associated probability level of about 0.35. Thus, the simple model for IS*30* fits the observed data very well. Although the χ^2 test cannot be completely trusted in this case because of the small expected numbers in some of the categories, the conclusion is supported by a more exact statistical test (Sawyer et al. 1987). The following problem deals with the distribution of three other insertion sequences in *E. coli*.

PROBLEM 5.5   The distribution of IS*1* fits Equation 5.18 with $\alpha = 1/5$ and $\phi = 5/6$; IS2 fits the equation with $\alpha = 2/5$ and $\phi = 2/3$; and IS4 fits with $\alpha = 2/3$ and $\phi = 3/4$. Calculate the expected numbers for 71 strains and carry out a χ^2 test. (The observed numbers are from Sawyer et al. 1987.)

	No. copies					
	0	1	2	3	4	≥ 5
IS*1*	11	14	8	6	7	25
IS2	28	8	12	5	5	13
IS4	43	5	5	3	5	10

ANSWER   For IS*1*, the expected distribution is given by $p_0 = 1/5$, $p_i = (4/25)(5/6)^i$ for $1 \le i \le 4$, and $p_{\ge 5} = 1 - (p_0 + p_1 + p_2 + p_3 + p_4)$. For IS2, the expected distribution is $p_0 = 2/5$ and $p_i = (3/10)(2/3)^i$ $(1 \le i \le 4)$. For IS4, the expected distribution is $p_0 = 2/3$ and $p_i = (1/9)(3/4)^i$

$(1 \le i \le 4)$. Expected numbers, χ^2 values, and associated probabilities are:

| | No. copies | | | | | | | |
	0	1	2	3	4	≥ 5	χ^2	P value
IS1	14.2	9.5	7.9	6.6	5.5	27.4	3.58	0.35
IS2	28.4	14.2	9.5	6.3	4.2	8.4	6.31	0.10
IS4	47.3	5.9	4.4	3.3	2.5	7.5	4.00	0.28

As in the case of IS30, more exact statistical tests confirm the conclusion that the model fits. However, the distribution of IS1 has a very long tail, with nine strains containing from 15 to 20 copies and six strains containing from 21 to 30 copies; this distribution is approximated even more closely by a model in which the regulation of transposition decreases more gradually than T/n (Sawyer et al. 1987).

Apart from their own evolutionary dynamics, insertion sequences are important because they can mobilize other sequences in the genome. When two copies of an insertion sequence are on flanking sides of an unrelated sequence, the inverted repeats used in transposition are preferentially those at the extreme ends. This kind of insertion-sequence sandwich constitutes a composite transposable element or **transposon**, which transposes as a single unit. In a composite transposon, the central sequence can include one or more genes that confer a selective advantage on the host cell, such as a gene for resistance to an antibiotic; hence, the possession of the transposon would be favored in an environment containing the antibiotic.

Mobilization of genes for antibiotic resistance, heavy-metal resistance, and other functions is one of the principal evolutionary implications of transposable elements in bacteria. Transposable elements enable the piecewise assembly of specialized, infectious molecules called **plasmids**. Plasmids are autonomously replicating, circular molecules of DNA that exist within bacterial cells. Many plasmids contain genes that promote their transfer between different organisms. They may also contain genes, such as those for antibiotic resistance, that are highly advantageous to their hosts in certain environments. These genes are often contained in transposons, and they undoubtedly entered the plasmid through transposition from a different plasmid or from the genome of a previous host. Infectious plasmids containing multiple antibiotic-resistance genes are called **resistance transfer factors**, and they are a major source of multiple drug resistance in pathogenic bacteria.

Transposable Elements in Eukaryotes

Transposable elements can have important genetic consequences as mutagenic agents by the creation of novel genes, by alteration of the expression of genes in their vicinity, and in the genesis of major genomic rearrangements. Transposable elements also have important implications in population genetics and evolution. Several major classes of transposable elements have been identified that differ in the molecular mechanisms of transposition. Within each class, the members can also differ in DNA sequence. Based on similarity in DNA sequence, transposable elements typically can be grouped hierarchically into "subfamilies," in which the elements resemble each other quite closely; "families," in which they differ from one another somewhat more; and "superfamilies," in which the differences are relatively great. Transposable elements are widespread in both animals and plants. For example, *Drosophila melanogaster* contains multiple copies of each of 50 to 100 different families of transposable elements (Rubin 1983). Although few of these elements have been studied in detail from the standpoint of population genetics, indirect evidence suggests that most of the elements, like insertion sequences in bacteria, are mildly harmful to the host (Golding et al. 1986, Lohe et al. 1995).

Horizontal Transmission of Transposable Elements

Among the most widespread families of transposable elements is that of the *mariner*-like elements (MLEs), typified by the transposable element *mariner*. The molecular organization of the *mariner* element is illustrated in Figure 5.20A. The element is flanked by short (28 base pair) inverted repeats (IR) and includes a long open reading frame coding for the transposase protein (Hartl 1989). Insertion of the element is invariably adjacent to a 5'–TA–3' dinucleotide in the host genome and is accompanied by a duplication of the dinucleotide, so that the inserted *mariner* is flanked by 5'–TA–3'. The target sequence and dinucleotide, as well as features of the amino acid sequence of the transposase protein, identify a transposable element as an MLE.

MLEs are widely distributed among insects and other invertebrates (Robertson 1993; Robertson and MacLeod 1993). Figure 5.20B shows the distribution among species in the major insect orders (Coleoptera, Diptera, and so forth). The number of copies of an MLE per genome varies widely among species, ranging from a few copies to many thousands. The MLEs in Figure 5.20B have been grouped according to similarity in nucleotide sequence and arranged in the form of a tree with the root to the left and the tips of the branches to the right. There are several subfamilies of insect MLEs, denoted mauritiana, cecropia, honeybee, and so forth. MLEs in different subfamilies are typically 40 to 50% identical in nucleotide sequence, and those within the same subfamily are usually 60% or more identical. All of the insect MLEs are more closely related to each other than they are to an MLE found in the soil nematode *Caenorhabditis elegans*.

(A)

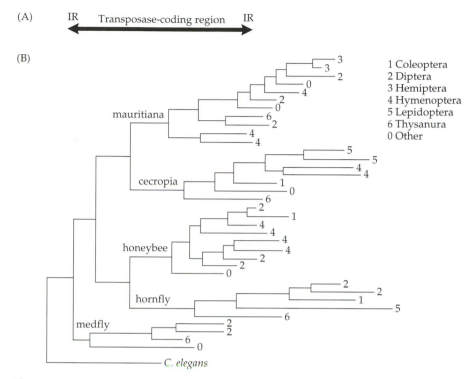

Figure 5.20   (A) The molecular organization of the transposable element *mariner* showing the inverted repeats flanking the transposase-coding region. (B) Distribution of MLEs among species representing major insect orders (numbered). Note that the MLEs can be grouped into subfamilies of elements (mauritiana, cecropia, and so forth) based on their similarity in sequence. *C. elegans* is the soil nematode *Caenorhabditis elegans*. (Data for B from Robertson 1993.)

Although MLEs are widespread, their distribution is "spotty," which means that, among closely related species, a particular type of MLE may be found in some species but not in others. Furthermore:

• Any species may contain MLEs from two or more different subfamilies.
• Closely related MLEs are often found in distantly related species.

An example of the second principle is an MLE found in *Drosophila erecta*, a close relative of *D. melanogaster*, which is 97% identical in nucleotide sequence with an MLE found in the cat flea *Ctenocephalides felis* (Lohe et al. 1995). For comparison, a gene coding for a subunit of the cellular sodium pump sequenced in both species shows only 39% nucleotide identity at third codon positions.

What process can account for the virtual identity between MLEs in species as distantly related as a *Drosophila* and a cat flea? One possibility is that the MLE was present in the common ancestor of the species a few hundred million years ago and then virtually stopped evolving, so that the sequences remain almost identical today. Unless the nucleotide sequence is very highly constrained, including third codon positions, this is a very unlikely possibility. Furthermore, if MLE sequences are so constrained, then why is there so much sequence variability within and among subfamilies? More likely than evolution stopping dead in its tracks for several hundred million years is the hypothesis of **horizontal transmission,** or the ability of an MLE to be transferred from a host species into the germline of a different, reproductively isolated species. To account for the *D. erecta–C. felis* case by horizontal transmission, an MLE would have to have been transmitted from a *D. erecta* ancestor to a *C. felis* ancestor (or the other way around) approximately 3 to 10 million years ago. Many additional examples of horizontal transmission of MLEs and other eukaryotic transposable elements have been discovered. Although the process of horizontal transmission certainly takes place, the rate at which it happens and the vectors and mechanisms are as yet unknown.

Once introduced into a genome, MLEs can persist through multiple speciation events (Maruyama and Hartl 1991). A lineage can, however, lose an MLE, as evidenced by *D. melanogaster*, which has lost an MLE (the *mariner* element itself) present in all its closest relatives. Two processes appear to contribute to loss of an MLE: (1) mutational inactivation, which may destroy the protein-coding function of an MLE or impair its ability to transpose; and (2) stochastic loss, by which we mean the elimination of an MLE from the genome as a result of random genetic drift. There might possibly also be a contribution from natural selection, depending on the extent to which presence of the MLE itself is deleterious. From the standpoint of the host species, an inactivating mutation in an MLE may be selectively neutral, or perhaps even favorable, inasmuch as natural selection may act to minimize the harmful mutagenic effects of transposition. Subsequent mutations in an already inactivated MLE are presumably selectively neutral and ultimately lost by chance. The role of mutational inactivation and stochastic loss in the evolutionary dynamics of MLEs is supported by the spotty distribution of MLEs among closely related species.

SUMMARY

Mutation provides the raw material for evolutionary change but, by itself, mutation pressure is a very weak force for changing allele frequency. If allele *A* mutates to allele *a* at a rate μ per generation, and *a* undergoes reverse mutation at a rate v per generation, then the equilibrium frequency of A is $v/(\mu + v)$, but the population may require tens of thousands or hundreds of thousands of generations to reach equilibrium. In the infinite-alleles model, the equilibrium value of F_{ST} for neutral alleles is given by $1/(4N\mu + 1)$, where μ is the mutation rate to selectively neutral alleles; $4N\mu + 1$ is called the effective number of alleles. For a neutral allele, the probability of ultimate fixation equals the frequency of the allele in the population. Statistical tests of the neutrality hypothesis based on the effective number of allozyme alleles or on the allozyme heterozygosity are inconclusive owing to lack of statistical power.

Recombination allows the formation of beneficial combinations of genes. In *Drosophila*, there is a positive correlation between the rate of recombination and the level of nucleotide polymorphism: regions of reduced recombination are less polymorphic. The reduced polymorphism could result from selective sweeps of favorable mutations or from background selection against detrimental mutations. In prokaryotes, there is extensive linkage disequilibrium over long genetic distances in spite of the fact that each gene may have a mosaic ancestry owing to intragenic recombination. The apparent paradox results because recombination in prokaryotes usually involves a short stretch of DNA and the process is infrequent. In animal mitochondrial DNA, the absence of recombination enables the identification of mitochondrial lineages.

Migration hinders genetic divergence among subpopulations. In finite populations, the equilibrium value of F_{ST} with migration is given by $1/(4Nm + 1)$, and only a few migrants per generation are sufficient to keep F_{ST} smaller than about 10%. On the other hand, a small amount of migration is usually not sufficient to disperse rare alleles among subpopulations, and so rare alleles are often unique to one or a few subpopulations.

Transposable elements are ubiquitous in the genomes of all organisms. Their tendency to increase in copy number through their ability to replicate and transpose is usually offset by the harmful effects of the insertions themselves; hence, there is an equilibrium distribution of copy number among organisms. Some transposable elements have direct or indirect beneficial effects; bacterial transposons that carry genes for antibiotic resistance provide an example. Bacterial transposons are disseminated among organisms and among species by transmission of infectious plasmids in which the transposons may reside. In eukaryotes, horizontal transmission can take place between species in spite of absolute reproductive isolation. Many transposable elements can be grouped into subfamilies, families, and superfamilies based on their degree of nucleotide sequence similarity. The *mariner*-like elements (MLEs) are exceptionally widespread among insects and other invertebrates. The innate tendency of an MLE to increase in copy number in a genome is offset by mutational inactivation and ultimately stochastic loss. These offsetting processes may explain the spotty distribution of MLEs observed among closely related species.

PROBLEMS

1. Most protein-coding genes have a forward mutation rate (normal to mutant) that is at least an order of magnitude greater than the reverse mutation rate (mutant back to normal). Why should this be the case?

2. A classical bacterial experiment demonstrated that mutations occur at random and not in response to specific selection pressures for them. The experiment used sterilized velvet to imprint the geometrical pattern of bacterial colonies on an agar surface in a petri dish (a "plate"), which was used to replicate the pattern by impressing the velvet on sterile nutrient agar in a selective plate containing an antibiotic. Colonies on the original plate giving resistant cells on the selective plate were dispersed into single cells, spread onto a nutrient agar plate without antibiotic, and allowed to multiply into colonies. This procedure was repeated until one or more colonies on the unselective media consisted exclusively of antibiotic resistant cells. How does this experiment prove the point?

3. Estimation of mutation rates from bacterial cultures can be tricky because, if a mutation occurs early in the life of a culture, the final frequency will be very high; but if it occurs late, the final frequency will be low. The fluctuation test is a method for getting around this problem by growing many smaller cultures and estimating the mutation rate from the proportion of cultures that contain no mutations using the zero term of the Poisson distribution $P_0 = \exp(-\mu N)$, where P_0 is the proportion of cultures with no mutations, μ is the mutation rate, and N is the average number of cells per culture. In one experiment for bacteriophage T1 resistance, $^{11}/_{20}$ cultures contained no mutations and the average number of cells per culture was 5.6×10^8. Estimate μ.

4. If recessive lethals occur independently in *Drosophila* autosomes, and the probability that an autosome contains one or more recessive lethals is 0.35 (a typical figure for chromosomes isolated from natural populations), what is the average number of recessive lethals per chromosome? Assume that the distribution of lethals is Poisson so that the probability of a chromosome containing exactly i lethals is $P_i = (m^i/i!)\exp(-m)$, where m is the mean.

5. The doubling dose of radiation is the quantity of radiation that induces as many mutations as occur spontaneously, so the total mutation rate of organisms exposed to the doubling dose equals two times the spontaneous mutation rate. Below are the induction rates per rad of x-rays (a

standard measure of dose) for various genetic end points in irradiated male mice, along with the spontaneous rates. What are the corresponding doubling doses?

	Induction rate/rad	Spontaneous rate
Dominant lethals	5×10^{-4}/gamete	2 to 10×10^{-2}/gamete
Recessive visibles	7×10^{-8}/locus	8×10^{-6}/locus
Reciprocal translocations	1 to 2×10^{-5}/cell	2 to 5×10^{-4}/cell

6. For irreversible mutation with a forward mutation rate $\mu = 5 \times 10^{-6}$, calculate the allele frequency p after 10, 100, 1000, and 10000 generations, assuming $p_0 = 1.0$.

7. If a transposable genetic element becomes fixed at a particular site but undergoes deletion at the rate of one percent per generation, how many generations are required to decrease the frequency of the element at the site to 90%?

8. The following data give the frequency q of bacteria resistant to a bacteriophage after t generations of chemostat growth. At $t = 12$ hours a novel metabolite was added to the medium.
   a. What is the basal rate of mutation to resistance?
   b. What is the effect of the novel metabolite on the mutation rate?

t	q	t	q
0	1×10^{-6}	16	7.04×10^{-6}
4	3×10^{-6}	20	7.08×10^{-6}
8	5×10^{-6}	24	7.12×10^{-6}
12	7×10^{-6}		

9. In the forward and reverse mutation model, what is the equilibrium frequency p of A if
   a. $\mu = 10^{-5}$ and $v = 10^{-6}$?
   b. μ is increased tenfold?
   c. v is increased tenfold?
   d. both are increased tenfold?

10. In the forward and reverse mutation model, show that the time required for the allele frequency to go halfway to equilibrium is approximately $t = 0.7/(\mu + v)$ generations. Use the approximation that $\ln(1 - x) \cong -x$ when x is small. What time is required to go halfway to equilibrium when $\mu = 10^{-5}$ and $v = 10^{-6}$?

11. In the irreversible mutation model, what is the frequency q_t of allele a in generation t if the mutation rate changes from generation to generation? If the equation $q_t = q_0 + \mu t$ is applied to this situation, what value corresponds to μ?

12. Suppose a gene has eight alleles at frequencies 0.55, 0.20, 0.09, 0.06, 0.04, 0.03, 0.02, and 0.01. What is the effective number of alleles? What would the effective number be if each allele had a frequency of 0.125?

13. Why is the effective number of alleles essentially independent of the number of rare alleles?

14. What is the equilibrium heterozygosity in a population of effective size 50 if new neutral mutations are introduced at a rate 10^{-5} by mutation and at a rate 10^{-3} by migration?

15. If the average number of alleles of a gene is $1 + x$ per diploid individual, where $0 < x < 1$, then what is the heterozygosity? (Note that one diploid individual is a random sample of two alleles.)

16. Calculate the autozygosity F after 200 generations in a random mating population of effective size $N = 50$.

17. In an isolated random mating population of effective size N, how many generations of random genetic drift are required to produce the same average inbreeding coefficient F as obtained in one generation of brother-sister mating (for which $F = 1/4$)? Use the approximation $[1 - 1/(2N)]^t \cong \exp(-t/2N)$.

18. If a mainland population of snails has an allele frequency of 0.8 and an island population has a frequency of 0.2, how many generations are required for the island population to achieve an allele frequency of 0.5, given a migration rate of 0.01?

19. If four populations with allele frequencies 0.2, 0.4, 0.6, and 0.8 undergo migration according to the island model with $m = 0.05$, what are the expected allele frequencies after 10 generations?

20. In the island model of migration, how does the variance in allele frequency among populations change as a function of m and t?

21. When random genetic drift is offset by migration among populations in the island model, what value of m is necessary to keep the equilibrium value of F smaller than 0.05?

CHAPTER 6

Darwinian Selection

NATURAL SELECTION • FITNESS • HAPLOID MODELS • DIPLOID MODELS
MUTATION-SELECTION BALANCE • COMPLEX MODES OF SELECTION
KIN SELECTION • INTERDEME SELECTION

 HUS FAR IN THIS BOOK, the term *natural selection* has been used in the informal, intuitive sense used by Darwin in *The Origin of Species* (1859):

> Owing to this struggle for life, variations, however slight and from whatever cause proceeding, if they be in any degree profitable to the individuals of a species, in their infinitely complex relations to other organic beings and to their physical conditions of life, will tend to the preservation of such individuals, and will generally be inherited by the offspring. The offspring, also, will thus have a better chance of surviving, for, of the many individuals of any species which are periodically born, but a small number can survive. I have called this principle, by which each slight variation, if useful, is preserved, by the term Natural Selection.

Modern formulations of natural selection are less literary and usually compacted into a form resembling a logical syllogism:

- In all species, more offspring are produced than can possibly survive and reproduce.
- Organisms differ in their ability to survive and reproduce—in part owing to differences in genotype.
- In every generation, genotypes that promote survival in the current environment are present in excess at the reproductive age and thus contribute disproportionately to the offspring of the next generation.

Through natural selection, therefore, alleles that enhance survival and reproduction increase gradually in frequency from generation to generation, and the population becomes progressively better able to survive and reproduce in the environment. The progressive genetic improvement in populations resulting from natural selection constitutes the process of evolutionary **adaptation.**

In the brief description of natural selection quoted above, Darwin uses the term *individual* three times. The unit of selection is the individual organism—not the species, not the subpopulation, not the sibship. It is the performance of the individual organism that matters. Each individual organism competes in the struggle for existence and survives or perishes on its own. Darwin also used the terms "struggle for existence" and "survival of the fittest" as synonyms for natural selection, but he emphasized that he employed the terms in their widest metaphorical sense to include not only the life of the organism but also the success of the organism in leaving progeny: fecundity is as important as survival. In this chapter, we shall see how Darwin's concept of "survival of the fittest" of individual organisms has been made more formal and quantitative and incorporated into models describing the change in allele frequency under natural selection. These models show that natural selection acts simultaneously on different components of fitness and can operate at different levels of population structure.

SELECTION IN HAPLOID ORGANISMS

Selection acts on the phenotype, not on the genotype, and the total phenotype is determined by many genes that interact with each other as well as with numerous environmental factors. However, in exploring the consequences of selection, it is convenient to focus on changes in the frequency of the alleles of a single gene. We shall begin by examining selection in its simplest form operating in a haploid, asexual organism, such as a species of bacteria. In haploids, selection is realized as differential population growth; hence we shall make reference to the discrete and continuous models of population growth examined in Chapter 1. The overall process of selection is identical whether population growth is in discrete or continuous generations, but the models have a somewhat different parameterization and it is necessary to relate the models to avoid confusion later.

Discrete Generations

Consider two bacterial genotypes, A and B, that reproduce asexually. For simplicity, we will assume the discrete model of population growth discussed in Chapter 1 and we set a and b equal to the rates of population growth of A and B, respectively. Equation 1.5 implies that $A_t = (1 + a)^t A_0$ and $B_t = (1 + b)^t B_0$, where A_t and B_t are the number of cells of genotype A and genotype B, respectively, at time t. Selection takes place when $a \neq b$. Figure 6.1A is an example in which the growth rates of A and B are $a = 0.04$ and $b = 0.05$, respectively. Both populations increase in size exponentially, but that of B increases faster than that of A. In most cases, we are not interested in the actual number of A cells or B cells but in the proportion of all cells that are of type A. Equivalently, we can examine the ratio of the number of A cells to that of B cells at time t, which is given by

$$\frac{A_t}{B_t} = \left(\frac{1+a}{1+b}\right)^t \frac{A_0}{B_0} = w^t \left(\frac{A_0}{B_0}\right) \tag{6.1}$$

The outcome of selection is determined by the ratio of a to b because, if $a < b$, then the ratio of A cells to B cells decreases until, ultimately, A is lost;

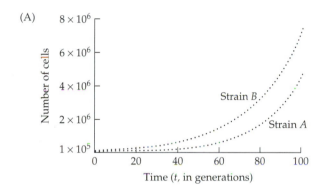

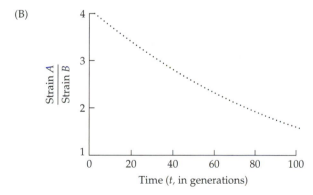

Figure 6.1   (A) Discrete population growth of two hypothetical bacterial strains, A and B, in which the growth rate are 4% per generation for A and 5% per generation for B. For clarity, the population size is plotted every second generation. The initial cell numbers are 1.6×10^5 for A and 0.4×10^5 for B. (B) Ratio of cell numbers of $A : B$. Because the B population grows faster than the A population, the proportion of A in the total population decreases.

conversely, if $a > b$, then the ratio of A cells to B cells increases without limit. Figure 1.6B shows the change in A/B for the example in part A. From a value of 4 at the beginning, the ratio declines to a value of 1.54 in 100 generations; these ratios correspond to frequencies of A of 0.80 and 0.61, respectively.

In the selection in Figure 6.1, it is not necessary to specify whether a and b differ because of survivorship or fecundity. All that matters is that they do differ. It is also important that the outcome depends only on the ratio $(1 + a)/(1 + b)$, which means that, in practice, we do not need to know the absolute growth rates of A and B but only their relative values (their ratio). In Equation 6.1, w represents the ratio $(1 + a)/(1 + b)$. The symbol w is conventionally used in discrete models of selection and, in this example, it is the **relative fitness** of genotype A to that of genotype B. In other words, in a haploid organism, the relative fitness equals the ratio of the growth rates.

Although it is sometimes instructive to do so, it is not necessary to keep track of population size in models of selection. The variable of interest is usually the allele frequency and not the population size. Therefore, let p_t and q_t represent the frequencies of genotypes A and B, respectively, in generation t, with $p_t + q_t = 1$. A method to relate the frequencies of A and B in any two successive generations is illustrated in Table 6.1. For ease of discussion, we divide each generation into three phases: birth, selection, and reproduction. In generation $t - 1$, the frequencies of A and B at birth are p_{t-1} and q_{t-1}, respectively. The genotypes A and B are assumed to survive in the ratio $w : 1$, which means that w is the probability of survival of an A genotype relative to that of an B genotype. As before, the absolute probabilities of survival of the geno-

---

TABLE 6.1   A MODEL OF SELECTION IN A HAPLOID ORGANISM, IN WHICH *w* IS THE PROBABILITY OF SURVIVAL OF AN *A* CELL RELATIVE TO THAT OF A *B* CELL

	Genotype	
	A	B
Generation $t - 1$		
Frequency before selection	p_{t-1}	q_{t-1}
Relative fitness	w	1
After selection	$p_{t-1}w$	q_{t-1}
Generation t	$\dfrac{p_{t-1}w}{p_{t-1}w + q_{t-1}}$	$\dfrac{q_{t-1}}{p_{t-1}w + q_{t-1}}$

Note: The fractions in the bottom line are expressions for the allele frequencies in generation t in terms of those in generation $t - 1$. Although this model assumes differential survival, $w : 1$ could also be the relative probability of reproduction of A and B. More generally, the relative fitness $w : 1$ represents the net output of $A : B$ for the combined effects of differential survival and reproduction.

types are not relevant. All that matters is the ratio. After selection, the ratio of frequencies of $A : B$ equals $p_{t-1} \times w : q_{t-1} \times 1$. If the surviving genotypes reproduce with equal efficiency, then the frequencies at birth in the following generation are given by the expressions across the bottom in Table 6.1; the denominators in these expressions are necessary to make the allele frequencies in generation t sum to 1.

For comparison with Equation 6.1, consider that p_t is the number of A cells in generation t divided by the total; likewise, q_t is the number of B cells divided by the total. Therefore, the ratio p_t/q_t equals the ratio of A cells to B cells in generation t because the denominators cancel. The expressions in Table 6.1 imply that the ratio of p/q in any generation equals w multiplied by the ratio of p/q in the previous generation, and so

$$\frac{p_t}{q_t} = w\frac{p_{t-1}}{q_{t-1}} = w^2\frac{p_{t-2}}{q_{t-2}} = \cdots = w^t\frac{p_0}{q_0} \qquad 6.2$$

The right-hand side of Equation 6.2 is identical to that in Equation 6.1 except that the relative frequencies p and q replace the absolute number of cells of type A and type B. Hence, to deduce the outcome of selection, we do not need to keep track of population size. All we need to know is the relative fitness w and the initial frequencies p_0 and q_0.

For application to experimental data, Equation 6.2 is often transformed by taking the logarithm:

$$\log\left(\frac{p_t}{q_t}\right) = \log\left(\frac{p_0}{q_0}\right) + t \cdot \log(w) \qquad 6.3$$

Equation 6.3 means, for example, that if the values of p_t/q_t are monitored in an experimental population of bacteria over the course of time, then a plot of $\log(p_t/q_t)$ against time (in generations) should yield a straight line with slope equal to $\log w$. This kind of experiment is examined in the following problem.

PROBLEM 6.1   In the intestinal bacterium *E. coli*, the gene *gnd* codes for the enzyme 6-phosphogluconate dehydrogenase (6PGD), which is used in the metabolism of gluconate but not in the metabolism of ribose. The data below were obtained in experiments in which otherwise genetically identical strains containing the alleles *gnd(RM77C)* and *gnd(RM43A)* were grown in competition in chemostats in which the sole source of carbon and energy was either gluconate or ribose (Hartl and Dykhuizen 1981). These *gnd* alleles are polymorphic in natural populations and code for allozymes of 6PGD. Gluconate is the experimental condition to ascertain the effects on fitness of the *gnd* alleles, and ribose is the control. In the table, p_t denotes the frequency of the strain containing *gnd(RM43A)* after t generations of competition. From the two points under each growth condition, estimate the fitness of the strain containing *gnd(RM43A)* relative to that containing *gnd(RM77C)* under the growth condition:

Growth medium	p_0	p_{35}
Gluconate	0.455	0.898
Ribose	0.594	0.587

ANSWER   In gluconate medium, log (0.898/0.102) = log (0.455/0.545) + 35 × log w, and so log w = 0.0292, or w = 1.0696. Hence, the allele *gnd(RM43A)* confers about a 7% selective advantage in competition for utilization of gluconate. In ribose medium, w = 0.999, a value that is not significantly different from 1.0, and so the alleles appear to be functionally equivalent in this environment. (There were more than two points in the original data, and the estimates of fitness were based on the slope of the linear regression; here we have quoted only two data points for computational convenience.)

Continuous Time

Bacterial populations such as those in Problem 6.1 do not reproduce in discrete generations but instead they reproduce continuously. In a continuous model, the exponential population growth of A and B are governed by the equations $dA(t)/dt = a'A(t)$ and $dB(t)/dt = b'B(t)$, where a' and b' are the growth rates. Therefore, $A(t) = A(0) \exp^{a't}$ and $B(t) = B(0) \exp^{b't}$ (Chapter 1), and so

$$\frac{A(t)}{B(t)} = \frac{A(0)}{B(0)} e^{(a'-b')t} = \frac{A(0)}{B(0)} e^{mt} \qquad 6.4$$

Equation 6.4 means that, in a continuous population, the outcome of selection depends on the difference between the exponential growth rates $a'-b'$, which is represented by the symbol m on the right-hand side. The value of m also measures the relative fitness of strain A relative to strain B, but in a continuously reproducing population. Comparing Equation 6.4 with Equation 6.1 yields the relation between m and w:

$$m = \ln w \qquad\qquad 6.5$$

In other words, the relative fitness with continuous growth m equals the natural logarithm of the relative fitness with discrete reproduction w. Selective neutrality means that $w = 1$ or that $m = 0$. For the values of w estimated in Problem 6.1, the corresponding values of m are 0.0673 and -0.001, respectively. If w is not too different from 1, then $m = w - 1$ is a reasonable approximation.

Change in Allele Frequency in Haploids

Although the discrete and continuous models are completely equivalent under the transformation in Equation 6.5, the equations for change in allele frequency look rather different. In the discrete model, the change in the frequency of strain A in generation t is given by the difference $p_t - p_{t-1}$, which can be calculated in terms of p_{t-1} from the formulas in Table 6.1. The difference $p_t - p_{t-1}$ is usually symbolized Δp and, for simplicity, the subscript $t - 1$ is suppressed. Using the expressions in Table 6.1 and the fact that $q = 1 - p$, we obtain

$$\Delta p = \frac{pw}{pw + q} - p = \frac{pq(w - 1)}{pw + q} \qquad\qquad 6.6$$

Not surprisingly, p increases if the relative fitness of A is greater than 1 and decreases if the relative fitness of A is smaller than 1. If the relative fitnesses of A and B are equal, then p does not change—provided that the population size is very large (theoretically, it has to be infinite).

The analog of Equation 6.6 in a continuous model contains the derivative dp/dt in place of Δp. This we can obtain from Equation 6.4 with a little trickery. Because $A(t)/B(t)$ equals $p(t)/q(t)$, the derivative of Equation 6.4 with respect to t must equal the derivative of $p(t)/q(t)$ with respect to t. For simplicity, we will write p and q instead of $p(t)$ and $q(t)$. The derivative of Equation 6.4 with respect to t equals mp/q and the derivative of p/q with respect to t equals $(1/q^2) \times dp/dt$. Setting these expressions equal to each other and solving for dp/dt, we obtain

$$\frac{dp}{dt} = pqm \qquad\qquad 6.7$$

Voilà! There is no denominator! What happened to it? In a technical sense, it disappeared into the difference between the discrete model and the continuous model. In a practical sense, the absence of a denominator in Equation 6.7 greatly simplifies some of the formulas to come, especially those concerned with random genetic drift in Chapter 7. Although they look very different, Equations 6.6 and 6.7 are merely different ways of saying the same thing. In this chapter, we will deal mainly with expressions analogous to Equation 6.6 because they are more easily derived for various types of selection. However, when it is necessary to dispose of a troublesome denominator, we will invoke the continuous model in Equation 6.7 and be rid of it.

Darwinian Fitness and Malthusian Fitness

The distinction between the fitness parameters in the discrete and continuous models has been incorporated into the terminology of population genetics in the terms **Darwinian fitness**, which refers to the discrete model, and **Malthusian fitness**, which refers to the continuous model. The latter is named after Thomas Malthus (1766–1834), whose views on the implications of continued population growth strongly influenced Darwin's thinking on the subject. A Darwinian fitness is conventionally represented by the symbol w, often embellished with a subscript, and Malthusian fitness is conventionally represented by the symbol m. In this book, the term *fitness*, when used without qualification, will mean Darwinian fitness unless it is clear from the context that some other meaning is intended.

SELECTION IN DIPLOID ORGANISMS

In diploid organisms, the consequences of selection are most conveniently explored under the model of random mating in Chapter 3, but incorporating selection by permitting the fitnesses of the genotypes to differ. Selection is assumed to take place on the diploid genotypes. We shall use the conventional symbols w_{11}, w_{12}, and w_{22} to represent the Darwinian fitnesses of the genotypes AA, Aa, and aa, respectively. The simplest way to interpret the fitnesses is in terms of survivorship, usually termed **viability**, which is the probability that a genotype survives from fertilization to reproductive age. If the fitness of each genotype is set equal to its probability of survivorship, then each fitness is an **absolute fitness** because its value is independent of the fitnesses of the other genotypes. In practice, we usually know only the value of the viability of each genotype relative to that of another genotype chosen as the standard of comparison. When a fitness value is expressed relative to that of another genotype, the fitness is a **relative fitness.** The relative fitness of the genotype chosen as the standard of comparison is arbitrarily assigned the value 1.

To consider a specific example, suppose that the genotypes AA, Aa, and aa have probabilities of survival from conception to reproductive age of 0.75, 0.75, and 0.50, respectively. These are the absolute viabilities of the genotypes. They can be judged realistic or not only if we specify the organism. They may be plausible values if the organism is a mammal or a bird because each offspring has a reasonable chance of survival, but implausible if the organism is an insect or an oyster because, in these organisms, most newborns are destined not to survive. Because selection depends on the relative magnitudes of the viabilities, it is usually most convenient to express the viabilities in relative terms. Taking genotype AA as the standard, the relative viabilities of AA, Aa, and aa are $0.75/0.75$, $0.75/0.75$, and $0.50/0.75$, or 1.0, 1.0, and 0.67, respectively. Equivalently, we could choose genotype aa as the standard, in which case the relative viabilities are $0.75/0.50$, $0.75/0.50$, and $0.50/0.50$, or 1.5, 1.5, and 1.0, respectively. Usually, the relative viabilities are calculated so that the largest relative viability equals 1.0. The relative viabilities are equal to the relative fitnesses of the genotypes provided that the genotypes are equally capable of reproduction. Viabilities expressed in relative terms are as valid for osprey as for oysters because the relative fitnesses are the same whether the absolute fitnesses are 0.75, 0.75, and 0.50 or 0.00075, 0.00075, and 0.00050.

Change in Allele Frequency in Diploids

If we write the allele frequencies of A and a as p_t and q_t, respectively, in generation t, then it is straightforward to derive expressions for the allele frequencies in generation t in terms of the allele frequencies p_{t-1} and q_{t-1} in the previous generation. The subscripts t and $t-1$ are rather cumbersome to carry along in equations, so we will use the symbols p and q for p_{t-1} and q_{t-1}, and the symbols p' and q' for p_t and q_t.

The relation between the allele frequencies in two consecutive generations is deduced in Table 6.2, where the fitnesses w_{11}, w_{12}, and w_{22} are the relative viabilities. In generation $t-1$, the genotype frequencies of AA, Aa, and aa

---

TABLE 6.2   DIPLOID SELECTION FOR SURVIVORSHIP (VIABILITY)

	Genotype			Total
Generation $t-1$	AA	Aa	aa	
Frequency before selection	p^2	$2pq$	q^2	$1 = p^2 + 2pq + q^2$
Relative fitness (viability)	w_{11}	w_{12}	w_{22}	
After selection	$p^2 w_{11}$	$2pq w_{12}$	$q^2 w_{22}$	$\bar{w} = p^2 w_{11} + 2pq w_{12} + q^2 w_{22}$
Normalized	$\dfrac{p^2 w_{11}}{\bar{w}}$	$\dfrac{2pq w_{12}}{\bar{w}}$	$\dfrac{q^2 w_{22}}{\bar{w}}$	

$$p' = \frac{p^2 w_{11} + pq w_{12}}{\bar{w}}$$

Generation t

$$q' = \frac{pq w_{12} + q^2 w_{22}}{\bar{w}}$$

Note: The allele frequencies p and q are those in gametes immediately prior to fertilization. The AA, Aa, and aa zygotes survive to reproductive maturity in the ratio $w_{11} : w_{12} : w_{22}$. All genotypes, as adults, are assumed to have the same reproductive capacity.

among newly fertilized eggs are given by p^2, $2pq$, and q^2, respectively, assuming random mating. By definition, newly fertilized eggs survive in the ratio $w_{11} : w_{12} : w_{22}$, and so the ratio of $AA : Aa : aa$ among surviving adults is

$$p^2w_{11} : 2pqw_{12} : q^2w_{22}$$

To proceed, we need to convert the terms in the above expression into relative frequencies by dividing each term by the sum. The value of the sum is indicated in Table 6.2 as

$$\overline{w} = p^2w_{11} + 2pqw_{12} + q^2w_{22} \qquad 6.8$$

The symbol $\overline{w}$ is the **average fitness** in the population in generation $t - 1$. Division of each term in the ratio of survivors by $\overline{w}$ yields the genotype frequencies among adults:

$$AA : \frac{p^2w_{11}}{\overline{w}} \qquad Aa : \frac{2pqw_{12}}{\overline{w}} \qquad aa : \frac{q^2w_{22}}{\overline{w}} \qquad 6.9$$

Among the surviving adults, the AA genotypes produce all A gametes, the Aa genotypes produce $\frac{1}{2}\, A$ and $\frac{1}{2}\, a$ gametes, and the aa genotypes produce all a gametes. Hence, the frequencies of the gametes that unite at random to form the zygotes of the next generation are:

$$A : p' = \frac{p^2w_{11} + pqw_{12}}{\overline{w}} \qquad a : q' = \frac{pqw_{12} + q^2w_{22}}{\overline{w}} \qquad 6.10$$

These are the relations we were after because they express the allele frequencies in any generation in terms of the allele frequencies in the previous generation. From these equations, the outcome of selection can be deduced.

As in the haploid model, it is often useful to know Δp, which is the difference in allele frequency $p' - p$ resulting from one generation of selection. Subtraction of p from the expression for p' in Equation 6.10 and a little manipulation leads to:

$$\Delta p = \frac{pq[p(w_{11} - w_{12}) + q(w_{12} - w_{22})]}{\overline{w}} \qquad 6.11$$

Equation 6.11 is the diploid analog of that in the haploid model in Equation 6.6.

At this point, an example of the use of these equations is in order. We will use data on the change in the frequency of the Cy (*Curly* wings) allele in a laboratory population of *Drosophila melanogaster*, which are plotted in Figure 6.2. The Cy allele is lethal when homozygous, so $w_{11} = 0$. The points in Figure 6.2 pertain to the frequency of Cy heterozygotes but, because Cy/Cy geno-

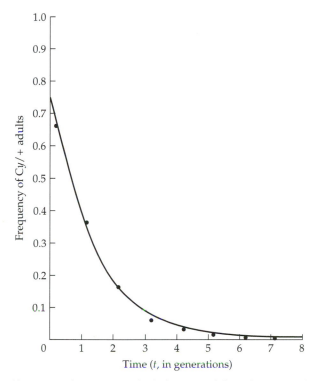

Figure 6.2   Change in frequency of adult *Drosophila melanogaster* heterozygous for the dominant mutation *Cy* (*Curly* wings) in an experimental population. The genotype *Cy/Cy* is lethal. The curve represents the theoretical change in frequency when the ratio of viabilities of *Cy/+* to *+/+* is 0.5 : 1. (Data from Teissier 1942. The fitness value of 0.5 was estimated by Wright 1977.)

types do not survive, the allele frequency p of *Cy* equals one-half the frequency of *Cy/+* adults. The points in the figure are each separated by one generation, and the initial generation has a frequency of *Cy/+* adults of 0.67, hence $p_0 = 0.335$ and thus $q_0 = 0.665$. Wright (1977) has studied these data and concluded that $w_{12} = 0.5$ for *Cy/+* genotypes, relative to a value of $w_{22} = 1.0$ for *+/+* genotypes. Substituting these values for p, q, w_{11}, w_{12}, and w_{22} into the expression for p' in Equation 6.10 yields

$$p' = \frac{0.335^2 \times 0 + 0.335 \times 0.665 \times 0.5}{0.335^2 \times 0 + 2 \times 0.335 \times 0.665 \times 0.5 + 0.665^2 \times 1} = 0.168$$

Therefore, the predicted frequency of *Cy/+* adults in the generation 1 is $2p' = 0.336$, which is reasonably close to the observed value of 0.368.

> **PROBLEM 6.2**   Assume a value of $p = 0.168$ for the frequency of the *Cy* allele in generation 1 in the population in Figure 6.2. Calculate the expected frequency of *Cy*/+ heterozygotes among adults in generation 2.
>
> ---
>
> **ANSWER**   In this case, $p' = [0.168^2 \times 0 + 0.168 \times 0.832 \times 0.5]/\overline{w}$, where $\overline{w} = 0.168^2 \times 0 + 2 \times 0.168 \times 0.832 \times 0.5 + 0.832^2 \times 1.0 = 0.832$; hence $p' = 0.0699/0.832 = 0.084$. The expected frequency of *Cy*/+ adults is $2p' = 0.168$. This result is very close to the observed value of 0.165. The theoretical curve in Figure 6.2 was calculated using the same generation-by-generation algorithm.

We make a slight digression to point out that it is sometimes convenient to think in terms of the **marginal fitnesses** of the *A* and *a* alleles. The marginal fitness equals the average fitness of all genotypes containing *A* or *a*, respectively, weighted by their relative frequency and the number of *A* or *a* alleles they contain. For example, *A* alleles are found in *AA* and *Aa* genotypes in the proportions p and q and, therefore, the marginal fitness $\overline{w}_1$ of *A*-containing genotypes equals $pw_{11} + qw_{12}$. Similarly, the marginal fitness of *a*-containing genotypes is $\overline{w}_2 = pw_{12} + qw_{22}$. The expression for p' in Equation 6.10 thus becomes $p' = p\overline{w}_1/\overline{w}$, and Equation 6.11 becomes $\Delta p = p(\overline{w}_1 - \overline{w})/\overline{w}$. This expression makes it clear that any allele increases in frequency if the marginal fitness of genotypes containing the allele ($\overline{w}_1$) is greater than the average fitness in the population ($\overline{w}$). This approach also generalizes readily to multiple alleles: for an allele with frequency p_i and marginal fitness $\overline{w}_i$, the change in frequency in one generation equals

$$\Delta p_i = \frac{p_i(\overline{w}_i - \overline{w})}{\overline{w}} \qquad 6.12$$

Time Required for a Given Change in Allele Frequency

Having derived Equation 6.11 for Δp resulting from one generation of selection, it is an appropriate next step to express p_t in terms of p_0, as we did in Chapter 5 for the analogous equations involving mutation and migration. For any specified values of the initial allele frequencies and the fitness parameters, the allele frequencies can be determined generation after generation by computer iteration, as in Problem 6.2. More generally one might want an explicit mathematical formula for p_t in terms of p_0, but Equation 6.11 does not lend itself to analytical solution.

There is an alternative approach based on a continuous model, however. If the fitnesses are expressed as Malthusian fitnesses rather than as Darwinian fitnesses, then the analog of Equation 6.11 for a continuously growing population is

$$\frac{dp}{dt} = pq[p(m_{11} - m_{12}) + q(m_{12} - m_{22})]$$ 6.13

where the values of m are the malthusian fitnesses. Note that there is no denominator in Equation 6.13 because it disappeared in the same way as the denominator in Equation 6.7. A less elegant way to derive an equation like Equation 6.13 is to suppose that the Darwinian fitnesses are all quite close to 1; then the change in allele frequency is slow enough that $\Delta p \cong dp/dt$ and, furthermore, $\bar{w} \cong 1$. Under these conditions, Equation 6.11 takes the form of Equation 6.13 with the m values replaced with w values.

To solve Equation 6.13, the terms are rearranged to isolate those in p on one side and those in t on the other, then one side is integrated over p from p_0 to p_t and the other integrated over t from 0 to t. The details are left as an exercise. The answers are most easily presented if we change the symbols. For this purpose, we rewrite the fitnesses of the genotypes as follows:

$$w_{11} = 1 \quad w_{12} = 1 - hs \quad w_{22} = 1 - s$$
$$m_{11} = 0 \quad m_{12} = -hs \quad m_{22} = -s$$

where the Malthusian fitnesses follow from the approximation $m_{ij} \cong w_{ij} - 1$ when $w_{ij} \cong 1$. Use of the h and s symbols for the fitnesses has the advantage of making the amount of selection and the degree of dominance explicit.

If s is positive and h is not negative, selection favors genotypes carrying the A allele. In this context, s is called the **selection coefficient** against the aa genotype, and h is called the **degree of dominance** of the a allele. For example, when $h = 0$, the Darwinian fitnesses of AA, Aa, and aa are 1, 1, and $1 - s$, respectively, and a is completely recessive to A. Alternatively, when $h = 1$, the Darwinian fitnesses are 1, $1 - s$, and $1 - s$, respectively, and a is completely dominant to A. In terms of the selection coefficient and the degree of dominance, dp/dt of Equation 6.13 becomes

$$\frac{dp}{dt} = pqs[ph + q(1 - h)]$$ 6.14

The following equations give p_t in terms of p_0 in three cases of importance.

- A is a **favored dominant**. In this case $h = 0$. Then $dp/dt = pq^2s$, and

$$\ln\left(\frac{p_t}{q_t}\right) + \frac{1}{q_t} = \ln\left(\frac{p_0}{q_0}\right) + \frac{1}{q_0} + st$$ 6.15

- A is favored and the alleles are **additive** in their effects on fitness. Additive effects on fitness means that the fitness of the heterozygote is exactly intermediate between the fitnesses of the homozygotes, and so $h = \frac{1}{2}$. The additive case is also referred to as **semidominance** or as **genic selection**. When $h = \frac{1}{2}$, then $dp/dt = pqs/2$, and

$$\ln\left(\frac{p_t}{q_t}\right) = \ln\left(\frac{p_0}{q_0}\right) + \left(\frac{s}{2}\right)t \qquad 6.16$$

Note that Equation 6.16 for additive alleles is similar in form to Equation 6.3 for haploid selection when $w = 1 + s/2$ and s is small. In other words, slow selection of additive alleles in a diploid species is mathematically almost equivalent to selection in a haploid species. In Problem 6.3, you will see that the precise requirement is $w_{12} = \sqrt{(w_{11}w_{22})}$.

- A is a **favored recessive**. In this case, $h = 1$, so $dp/dt = p^2qs$, and

$$\ln\left(\frac{p_t}{q_t}\right) - \frac{1}{p_t} = \ln\left(\frac{p_0}{q_0}\right) - \frac{1}{p_0} + st \qquad 6.17$$

Some of the practical implications of these equations are explored below. Problem 6.3 explores a little more deeply the relation between selection in haploid species and selection in diploid species. Figure 6.3 illustrates the changes in allele frequency for Equations 6.15 through 6.17.

PROBLEM 6.3   The discrete model of selection in a haploid species is completely equivalent to that in a diploid species if, in the diploid, the Darwinian fitness of the heterozygote equals the geometric mean of the Darwinian fitnesses of the homozygotes—that is, if $w_{12} = \sqrt{(w_{11}w_{22})}$. Show that, in this case, Equation 6.3 for Δp in a haploid species is, indeed, identical to Equation 6.11 for Δp in a diploid species. What is the equivalent value of w in the haploid in terms of the Darwinian fitnesses in the diploid?

ANSWER   Substitute $w_{12} = \sqrt{w_{11}}\sqrt{w_{22}}$ into Equation 6.11. The numerator simplifies to $pq \times (\sqrt{w_{11}} - \sqrt{w_{22}}) \times (p\sqrt{w_{11}} + q\sqrt{w_{22}})$. The denominator simplifies to $(p\sqrt{w_{11}} + q\sqrt{w_{22}})^2$. Therefore,

$$\Delta p = \frac{pq(\sqrt{w_{11}} - \sqrt{w_{22}})}{p\sqrt{w_{11}} + q\sqrt{w_{22}}} = \frac{pq\left(\sqrt{\frac{w_{11}}{w_{22}}} - 1\right)}{p\sqrt{\frac{w_{11}}{w_{22}}} + q}$$

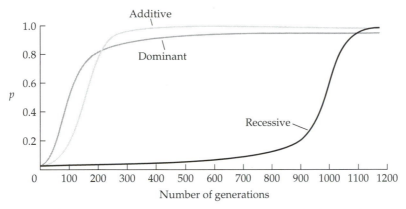

Figure 6.3   The change in frequency p of a favorable allele that is either dominant, additive, or recessive in its effect on fitness. The frequency of a favored dominant allele changes most slowly when the allele is common, and the frequency of a favored recessive allele changes most slowly when the allele is rare. In all three examples, the difference in relative fitness between the homozygous *AA* and *aa* genotypes is assumed to be five percent.

This is in exactly the same form as Equation 6.6 with $w = \sqrt{w_{11}/w_{22}}$. Taking $w_{22} = 1$ as the standard, w in the haploid model equals the fitness of the heterozygote in the diploid model. More specifically, let $w_{11} = (1 + s/2)^2$, $w_{12} = 1 + s/2$, and $w_{22} = 1$. If s is small compared to 1, then $w_{11} = (1 + s/2)^2 \cong 1 + s$, which implies that the Darwinian fitnesses are approximately additive. Furthermore, $\Delta p = pqs/2$, which has the same form as dp/dt in the additive case leading to Equation 6.17.

PROBLEM 6.4   A certain highly isolated colony of the moth *Panaxia dominula* near Oxford, England, was intensively studied by Ford and collaborators over the period 1928 to 1968 (Ford and Sheppard 1969). This colony contained a mutant allele affecting color pattern. The frequency of the mutant allele declined steadily over the period 1939 to 1968. Indeed, the accompanying steady increase in the frequency of the normal allele followed Equation 6.16 for additive genes with $s = 0.20$ (Wright, 1978, shows a graph). The species has one generation per year, and the estimated frequency of the mutant allele in 1965 was 0.008. (This value is actually the average for the seven-year period 1962 to 1968.) Estimate the frequency of the mutant allele in 1950 and in 1940.

ANSWER   Here we are given q_t and want to use Equation 6.16 to estimate q_0. Between 1950 and 1965, there were $t = 1965 - 1950 = 15$ generations. We are given $q_t = 0.008$, hence $p_t = 0.992$ and $\ln (0.992/0.008) = 4.820$. Thus, $4.820 = \ln (p_0/q_0) + (0.20/2) \times 15$, or $\ln (p_0/q_0) = 3.32$. Then $p_0/q_0 = 27.660$, or $p_0 = 0.965$ and $q_0 = 0.035$. For the year 1940, $t = 1965 - 1940 = 25$ generations, from which $p_0 = 0.911$ and $q_0 = 0.089$. (You may be interested to know that observations made at the time yielded estimates of $q_0 = 0.037$ in 1950 and $q_0 = 0.111$ in 1940.)

Application to the Evolution of Insecticide Resistance

Some of the most dramatic examples of evolution in action result from the natural selection for chemical pesticide resistance in natural populations of insects and other agricultural pests. In the 1940s, when chemical pesticides were first used on a large scale, an estimated 7% of the agricultural crops in the United States were lost to insects. Initial successes in chemical pest management were followed by gradual loss of effectiveness. Today, more than 400 pest species have evolved significant resistance to one or more pesticides, and 13% of the agricultural crops in the United States are lost to insects (May 1985). In many cases, significant pesticide resistance has evolved in 5 to 50 generations irrespective of the insect species, geographical region, pesticide, frequency and method of use, and other seemingly important variables (May 1985). Equations 6.15 through 6.17 help to understand this apparent paradox because many of the resistance phenotypes result from single mutant alleles. The resistance alleles are often partially or completely dominant, so Equations 6.15 and 6.16 are applicable. Prior to use of the pesticide, the allele frequency p_0 of the resistant mutant is generally close to 0. Use of the pesticide increases the allele frequency, sometimes by many orders of magnitude, but significant resistance is noticed in the pest population even before the allele frequency p_t increases above a few percent. Thus, as rough approximations, we may assume that q_0 and q_t are both close enough to 1 that $\ln (p_0/q_0) \cong \ln p_0$ and $\ln (p_t/q_t) \cong p_t$. Using these approximations, Equation 6.16 (additive case) implies that $t \cong (2/s) \times \ln (p_t/p_0)$ and Equation 6.15 (dominant case) implies that $t \cong (1/s) \times \ln (p_t/p_0)$. In many instances, the ratio p_t/p_0 may range from 1×10^2 to perhaps 1×10^7, and s may typically be 0.5 or greater. Over this wide range of parameter values, the time t is effectively limited to a range of 5 to 50 generations for the appearance of a significant degree of pesticide resistance. Details in actual examples depend on such factors as effective population number and extent of genetic isolation between local populations. An example of the global spread of an insecticide-resistance allele is given in Chapter 8. The evolution of resistance caused by multiple interacting alleles may be expected to take somewhat longer than single-gene resistance.

PROBLEM 6.5 In the discussion of the evolution of insecticide resistance, we used the approximation $t \cong (1/s) \times \ln(p_t/p_0)$ for the dominant case and $t \cong (2/s) \times \ln(p_t/p_0)$ for the semidominant case. Evaluate the adequacy of the approximations for the values in the accompanying table by comparing them with the more exact values calculated from Equations 6.15 and 6.16.

Example no.	p_0	p_t	s
1	1×10^{-4}	0.01	0.50
2	1×10^{-4}	0.10	0.50
3	1×10^{-4}	0.50	0.50
4	1×10^{-7}	0.10	0.50
5	1×10^{-4}	0.10	0.20

ANSWER The approximations are quite acceptable for the examples. The more exact and approximate values are as follows:

Example no.	Eqn. 6.15	Approximation	Eqn. 6.16	Approximation
1	9.3	9.2	18.5	18.4
2	14.2	13.8	28.1	27.6
3	20.4	17.0	36.8	34.1
4	28.1	27.6	55.7	55.3
5	35.6	34.5	70.1	69.1

EQUILIBRIA WITH SELECTION

An **equilibrium** value of p in a discrete model is any value for which $\Delta p = 0$. When the allele frequency is at an equilibrium in an infinite population, the allele frequency remains the same generation after generation. Because real populations are finite in size, an allele frequency is subject to chance fluctuations and so cannot usually remain exactly at an equilibrium value. For any equilibrium, therefore, it is important to consider how the allele frequency behaves when it is close, but not exactly equal, to the equilibrium value. Any equilibrium can be classified as one of several different types according to the behavior of the allele frequency when it is near the equilibrium:

- An equilibrium is said to be **locally stable** if the allele frequency, when it is already close to the equilibrium, moves progressively closer in subsequent generations. A locally stable equilibrium may also be **globally stable**. This term means that the allele frequency always moves toward the equilibrium regardless of where it starts, even if initially far away from the equilibrium. A polymorphism with a stable equilibrium is sometimes called a **balanced polymorphism**.
- An equilibrium is **unstable** if the allele frequency, initially close to the equilibrium, moves progressively farther away in subsequent generations.
- An equilibrium is called **neutrally stable** or **semistable** if the allele frequency has no tendency to change regardless of its initial value. In such a case, every allele frequency represents an equilibrium because $\Delta p = 0$ whatever the value of p. This type of equilibrium is exemplified by the Hardy-Weinberg principle in an infinite population (Chapter 3).

The concepts of stability can be applied to the case of selection governed by Equation 6.11 in which A is the favored allele. For A to be favored, we need $w_{11} \geq w_{12} \geq w_{22}$, and at least one of the strict inequalities must be true. In such a case, there are only two equilibria, namely $p = 0$ and $p = 1$. Except for $p = 0$ and $p = 1$, when $\Delta p = 0$, it is always true that $\Delta p > 0$. Hence, if p is close to 0, its value increases (moving it farther away from 0), and so the equilibrium at $p = 0$ is unstable. On the other hand, if p is near 1, it moves still closer to 1 (because $\Delta p > 0$), and so the equilibrium at $p = 1$ is locally stable. In this example, p eventually goes to 1 whatever its initial value, and so the equilibrium at $p = 1$ is globally stable also.

Overdominance

With two alleles of a gene in a diploid organism, there is the possibility that the heterozygous genotype has the highest fitness or that the heterozygous genotype has the lowest fitness. These cases illustrate equilibria in which the equilibrium value of p is between 0 and 1.

Overdominance, also called **heterozygote superiority**, is the term applied when the heterozygote has a higher fitness than both homozygotes. Symbolically, heterozygote superiority means that $w_{12} > w_{11}$ and simultaneously $w_{12} > w_{22}$. With overdominance, $p = 0$ and $p = 1$ are both equilibria because, according to Equation 6.11, $\Delta p = 0$ at these values. There is also a third equilibrium made possible by the fact that $p(w_{11} - w_{12}) + q(w_{12} - w_{22})$ can equal 0. The equilibrium frequency of A is conventionally denoted $\hat{p}$; hence the equilibrium allele frequency of a is $\hat{q} = 1 - \hat{p}$. The equilibrium can be found by solving $\hat{p}(w_{11} - w_{12}) + \hat{q}(w_{12} - w_{22}) = 0$, from which a little algebra gives

$$\hat{p} = \frac{w_{12} - w_{22}}{2w_{12} - w_{11} - w_{22}} \qquad 6.18$$

Equation 6.18 is often encountered in another form in which the fitnesses are all expressed relative to that of the heterozygote by setting $w_{11} = 1 - s$, $w_{12} = 1$, and $w_{22} = 1 - t$. (This formulation is proposed at the risk of some confusion because t is now the selection coefficient against aa rather than the time in generations.) With these substitutions, Equation 6.18 becomes

$$\hat{p} = \frac{t}{s + t}$$

This relationship makes a lot of intuitive sense because it implies that greater selection against aa increases the equilibrium frequency $\hat{p}$ of A.

The overdominance equilibrium in Equation 6.18 is globally stable whereas those at $p = 0$ and $p = 1$ are unstable. The time course is indicated in Figure 6.4A, where the arrowheads show the direction of change in allele frequency. Figure 6.4B shows the change in $\bar{w}$ with overdominance. The average fitness

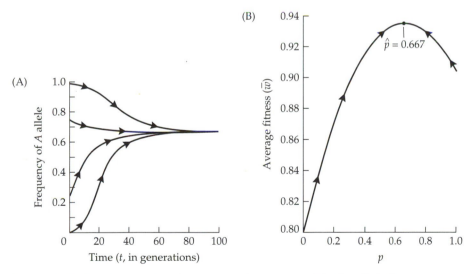

Figure 6.4 Selection when there is overdominance. (A) The allele frequencies converge to an equilibrium value irrespective of the initial frequency. In this example, $w_{11} = 0.9$, $w_{12} = 1$, and $w_{22} = 0.8$, and the equilibrium frequency of the A allele, $\hat{p}$, is 0.667. (B) Average fitness $\bar{w}$ against p for the same example. Note that $\bar{w}$ is a maximum at equilibrium.

in the population is maximized at the stable equilibrium. Maximization of average fitness is a frequent outcome of selection in random-mating populations with constant fitnesses. There are, however, many exceptions when mating is nonrandom, when the fitnesses are not constant, or when there are interactions between alleles of different genes (Ewens 1979; Curtsinger 1984). Note particularly that $\bar{w}$ is the average fitness *in* the population, not the average fitness *of* the population. The relative survivorships w_{11}, w_{12}, and w_{22} are relevant only to the differential mortality of the genotypes within a population at any given time. The average of the relative survivorships is the average "fitness" $\bar{w}$ *in* the population. However, $\bar{w}$ has no necessary relation to vernacular meanings of "fitness" such as competitive ability, population size, production of biomass, or evolutionary persistence (Haymer and Hartl 1982).

Although overdominance is one mechanism for the maintenance of polymorphisms in natural populations, it has been documented in only a few cases. The classic case is sickle-cell anemia in human beings, which is prevalent in many populations at risk for the type of malaria caused by the mosquito-borne protozoan parasite *Plasmodium falciparum* (Figure 6.5). The anemia is caused by an allele S that codes for a variant form of the β chain of hemoglobin. In persons of genotype SS, many red blood cells assume a curved, elongated shape ("sickling") and are removed from circulation. The result is a severe anemia as well as pain and disability owing to the accumulation of defective cells in the capillaries, joints, spleen, and other organs. In the absence of intensive medical care, persons of genotype SS usually do not survive. The S allele is maintained at a relatively high frequency because persons of genotype AS, in which A is the nonmutant allele, have only a mild form of the anemia but are quite resistant to malaria, perhaps because red blood cells infested with the parasite undergo sickling and are removed from circulation. Homozygous AA people are not anemic but, on the other hand, are the most sensitive to severe malaria. The result of the offsetting sickle-cell anemia and malaria resistance is that the heterozygotes have the highest fitness. In regions of Africa in which malaria is common, the viabilities of AA, AS, and SS genotypes have been estimated as $w_{11} = 0.9$, $w_{12} = 1$, and $w_{22} = 0.2$, respectively (Cavalli-Sforza and Bodmer 1971; Templeton 1982). Substitution into Equation 6.18 leads to a predicted equilibrium allele frequency for A of $\hat{p} = 0.89$. Consequently, that of S is 0.11. This value is reasonably close to the average allele frequency of 0.09 across West Africa, but there is considerable variation in allele frequency among local populations.

PROBLEM 6.6   Experimental populations of *Drosophila pseudoobscura* were periodically treated with weak doses of the insecticide DDT. One population was initially polymorphic for five different inversions

Figure 6.5   The medium gray areas show the incidence of falciparum malaria in Africa, the Middle East, and southern Europe in the 1920s before mosquito control programs were implemented. The light gray areas are regions with a high incidence of sickle-cell anemia. The extensive overlap in the distributions (darkest shade) was an early indication that there might be some causal connection. (After Cavalli-Sforza 1974.)

of the third chromosome. After 13 generations, three of the inversions had essentially disappeared from the population. The two that remained were Standard (*ST*) and Arrowhead (*AR*). Changes in frequency of each inversion were monitored and, from the values for the first nine generations, the relative fitnesses of *ST/ST*, *ST/AR*, and *AR/AR* genotypes were estimated as 0.47, 1.0, and 0.62, respectively (DuMouchel and Anderson 1968). Because the inversions undergo almost no recombination, each type can be considered as an "allele." What equilibrium frequency of *ST* is predicted? What equilibrium value of $\bar{w}$ is predicted?

ANSWER   From Equation 6.18, $\hat{p} = (1.0 - 0.62)/(2.0 - 0.47 - 0.62) = 0.42$. (The observed value after 13 generations was 0.43.) The predicted equilibrium value of $\bar{w}$, from Equation 6.8, equals $0.422^2 \times 0.47 + 2 \times 0.42 \times 0.58 \times 1.0 + 0.58^2 \times 0.62 = 0.78$.

PROBLEM 6.7   Warfarin is a blood anticoagulant used for rat control in World War II and afterward. Initially highly successful, the effectiveness of the rodenticide gradually diminished owing to the evolution of resistance among some target populations. Among Norway rats in Great Britain, resistance results from an otherwise harmful mutation R in a gene in which the normal nonresistant allele may be denoted S. In the absence of warfarin, the relative fitnesses of SS, SR, and RR genotypes have been estimated as 1.00, 0.77, and 0.46 respectively. In the presence of warfarin, the relative fitnesses have been estimated as 0.68, 1.00, and 0.37, respectively (May 1985). The reduced fitness of the RR genotype appears to result from an excessive requirement for vitamin K. Calculate the equilibrium frequency $\hat{q}$ of R in the presence of warfarin. Noting that, in the absence of warfarin, R and S are very nearly additive in their effects on fitness, estimate the approximate number of generations required for the allele frequency of R to decrease from $\hat{q}$ to 0.01 in the absence of the poison.

ANSWER   From Equation 6.18, the equilibrium frequency $\hat{p}$ of S equals $(1.00 - 0.37)/(2 - 0.68 - 0.37) = 0.66$, and so $\hat{q}$ of $R = 0.34$. Setting $q_0 = 0.34$ and $q_t = 0.01$ in Equation 6.16, with $s = 1.00 - 0.46 = 0.54$, yields $t = 14.6$ generations. (The approximation is very good even though s is large; the exact value is 14 generations.)

Local Stability

Although the curves in Figure 6.4A indicate that the interior equilibrium is locally stable when there is overdominance, an alternative approach is also applicable to the analysis of local stability in models of much greater complexity. It is based on the expression for Δp in Equation 6.11. To emphasize that Δp is a function of p, we will write it as an explicit function, $\Delta(p)$. The local stability of an equilibrium depends on the behavior of $\Delta(p)$ for a value of p close to, but not equal to, the equilibrium, as illustrated in Figure 6.6. It is convenient to write $\Delta(p + \varepsilon)$ as the change in allele frequency when the starting point is a small deviation, ε, from any allele frequency p. The function $\Delta(p + \varepsilon)$ can be expanded term by term into an infinite sum:

$$\Delta(p + \varepsilon) = \Delta(p) + \frac{d\Delta(p)}{dp}\varepsilon + \frac{d^2\Delta(p)}{dp^2}\frac{\varepsilon^2}{2!} + \frac{d^3\Delta(p)}{dp^3}\frac{\varepsilon^3}{3!} + \cdots$$

The mathematical basis of this type of expansion is beyond the scope of the book. If you are unfamiliar with it and want to look it up, you will find it under the heading the Taylor series in most textbooks of calculus. It is named after the mathematician Brook Taylor (1685–1731).

The value of the Taylor series expansion is that, when ε is sufficiently small, then all terms in ε^2 and higher can be ignored. Therefore, for any value

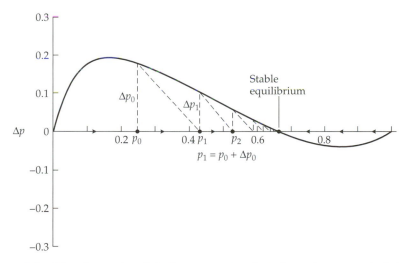

Figure 6.6 The change in allele frequency Δp plotted as a function of allele frequency p for a case of overdominance in which $w_{11} = 0.6$, $w_{12} = 1$, and $w_{22} = 0.2$. Starting with an allele frequency p_0, smaller than the equilibrium value, the positive value of Δp_0 indicates that the allele frequency in the next generation, p_1, will be greater than p_0 because $p_1 = p_0 + \Delta p_0$. At an allele frequency of p_1, the value of Δp_1 is also positive, and so p_2 is greater than p_1 because $p_2 = p_1 + \Delta p_1$. The steady increase continues until the population arrives at the equilibrium point $\hat{p}$. The same logic shows that, starting with an initial allele frequency greater than $\hat{p}$, the allele frequency decreases in each succeeding generation and ultimately converges to the equilibrium from the other side.

of p, we can approximate $\Delta(p + \varepsilon)$ in terms of $\Delta(p)$ itself and its first derivative. Furthermore, if p is one of the equilibrium points, then $\Delta(p) = 0$ by definition, and so the sign of $\Delta(p + \varepsilon)$ depends on the sign of first derivative of $\Delta(p)$ evaluated at the equilibrium in question. By definition, an equilibrium is locally stable if the allele frequency, starting at a point near the equilibrium, moves ever closer to the equilibrium. In symbols, this means that $\Delta(p + \varepsilon) < 0$ if $\varepsilon > 0$ and $\Delta(p + \varepsilon) > 0$ if $\varepsilon < 0$. Therefore, any equilibrium point, denoted generically as $\hat{p}$, is locally stable if, and only if,

$$\left. \frac{d\Delta(p)}{dp} \right|_{\hat{p}} < 0 \qquad\qquad 6.19$$

where the vertical line and $\hat{p}$ mean that the derivative should be evaluated at the equilibrium in question.

In practice, calculating the derivative of $\Delta(p)$ can be quite tedious without the use of computer software like Mathematica to do the algebraic manipulations. The result of differentiating Equation 6.11 is that

$$\frac{d\Delta(p)}{dp} = \frac{pqw}{\overline{w}} + \frac{(q - p)(p - \hat{p})w}{\overline{w}} - \frac{2pq(p - \hat{p})^2 w^2}{\overline{w}^2}$$

where $w = w_{11} - 2w_{12} + w_{22}$. With overdominance, $w < 0$. Note that, when $d\Delta(p)/dp$ is evaluated at $p = 0$ or $p = 1$, both the first and last terms equal 0; when it is evaluated at $p = \hat{p}$, the second and last terms equal 0. The stability analysis proceeds as follows:

- At $p = 0$, sign $[d\Delta(p)/dp] = -$sign $(w) > 0$;
- At $p = \hat{p}$, sign $[d\Delta(p)/dp] = $ sign $(w) < 0$;
- At $p = 1$, sign $[d\Delta(p)/dp] = -$sign $(w) > 0$.

Therefore, as is already clear from Figure 6.4A, the equilibrium points at 0, $\hat{p}$, and 1 are unstable, locally stable, and unstable, respectively. This stability analysis is predicated on the assumption of heterozygote superiority, which implies that $w < 0$. Exactly the same equilibrium points are present when there is heterozygote inferiority, but then $w > 0$, which means that the stability property of each equilibrium point is reversed. This situation is discussed next.

Heterozygote Inferiority

Heterozygote inferiority means that the fitness of the heterozygous genotype is smaller than that of both homozygotes: $w_{12} < w_{11}$ and $w_{12} < w_{22}$. An interior equilibrium, given by Equation 6.18, exists in this case also. The analysis in the previous section indicates that this equilibrium is unstable, whereas the equilibria at $p = 0$ and $p = 1$ are both locally (but not globally) sta-

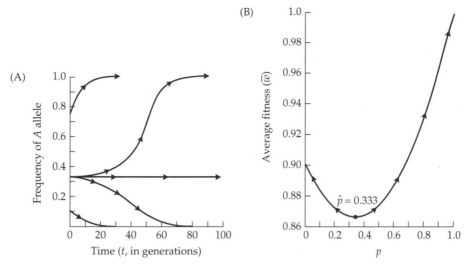

Figure 6.7 Selection when there is heterozygote inferiority. (A) The allele frequency goes to 0 or 1 depending on the initial frequency. In this example, $w_{11} = 1$, $w_{12} = 0.8$, and $w_{22} = 0.9$, and there is an unstable equilibrium when the frequency of the A allele is $\hat{p} = 0.333$. An infinite population with $p = \frac{1}{3}$ maintains this frequency, but any slight upward change in the frequency of A results in eventual fixation, and any slight downward change in the frequency of A results in ultimate loss. (B) Average fitness $\bar{w}$ against p for the same example. The unstable equilibrium represents the minimum of $\bar{w}$.

ble. An example of heterozygote inferiority is depicted in Figure 6.7A, where the arrows again denote the direction of change in allele frequency. If the initial allele frequency is exactly equal to the equilibrium value (in this example, $\hat{p} = \frac{1}{3}$), then the allele frequency remains at that value. In all other cases, p goes to 1 or 0 depending on whether the initial allele frequency was above or below the equilibrium value.

Figure 6.7B shows the change in average fitness. The unstable equilibrium at $\hat{p} = \frac{1}{3}$ is the minimum average fitness. The shape of the $\bar{w}$ curve has an important implication that carries over to more complex examples. Imagine a population with an allele frequency near 0, at which $\bar{w} = 0.9$. In terms of average fitness in the population, the population would be better off if the allele frequency were near 1, because then $\bar{w} \cong 1.0$. However, as shown by the direction of the arrows, the population cannot evolve toward $p = 1$. It cannot get through the "valley" because $p = 0$ is a locally stable equilibrium. The population has no way to escape from the equilibrium even though, in doing so, it would eventually end up with a greater average fitness. This consideration would seem to limit the ability of natural selection to increase average fitness in such cases, but one way out of the impass is suggested in the next section.

The Adaptive Topography and the Role of Random Genetic Drift

Any graph of $\bar{w}$ against allele frequency is called an **adaptive topography**. The simplest example is Figure 6.7B. In order to generalize the example, try to imagine an adaptive topography in many dimensions with $\bar{w}$ a function of the allele frequencies at many loci. In many dimensions, the adaptive topography is a complex surface upon which there may be "peaks" and "pits" and even "saddle-shaped" regions. The peaks represent locally stable equilibria. Even if natural selection changes the allele frequencies so as to move $\bar{w}$ to the top of some peak, the peak it perches on may not be the highest peak that exists on the whole surface. However, as illustrated in Figure 6.7B, the population may become stuck there because the peak is a locally stable equilibrium.

By what process can a population stranded on a submaximal fitness peak get off the peak? To do so, it has to travel through a nearby valley to a place where natural selection can carry it to the top of an even higher fitness peak. This is something that natural selection acting alone cannot accomplish because it entails a temporary reduction in fitness. There is, however, a process that can accomplish the task—random genetic drift. In a sufficiently small population, the allele frequencies can change by chance, even producing a reduction in average fitness. Theoretically, random genetic drift can shift a population from a locally stable equilibrium, through a nearby valley, and into a region where it is attracted by another locally stable equilibrium toward a higher fitness peak. Random genetic drift can therefore play a crucial role in evolution by allowing a population to explore the full range of its adaptive topography. This role of random genetic drift has been particularly emphasized by Wright (1977 and earlier) in his proposed **shifting balance theory** of evolution. Additional discussion of the theory is found in this chapter's section on interdemic selection; see also Hartl (1979), Provine (1986), and Coyne et al. (1997).

MUTATION-SELECTION BALANCE

You may recall from Chapter 4 that outcrossing species typically contain a large amount of hidden genetic variability in the form of recessive, or nearly recessive, harmful alleles, each present at a low frequency. Now we can explain why harmful alleles are not completely eliminated. Selection cannot eliminate them because they are continually created anew through recurrent mutation. To be specific, suppose that a is a harmful allele of the wildtype A and that mutation of A to a takes place at the rate μ per generation. Because the allele frequency of a, which we call q, remains small, reverse mutation of a to A can safely be ignored. The calculation of p' carried out to obtain

Equation 6.10 is still valid, except that a proportion μ of A alleles mutate to a in each generation. Therefore,

$$p' = \frac{\left(p^2 w_{11} + pq w_{12}\right)\left(1 - \mu\right)}{\overline{w}} \qquad 6.20$$

To proceed further, it is convenient to write the relative fitnesses as

$$w_{11} = 1 \qquad w_{12} = 1 - hs \qquad w_{22} = 1 - s$$

The value of s is the selection coefficient against the homozygous aa genotypes and h is the degree of dominance of the a allele. If $h = 0$, then a is a complete recessive because AA and Aa have an identical fitness. If $h = 1$, then a is dominant because Aa and aa have an identical fitness. Semidominance means that $h = \frac{1}{2}$. In mutation-selection balance, we are concerned with harmful alleles that are near the recessive end of the spectrum, and so h will usually be substantially smaller than 0.5.

Equilibrium Allele Frequencies

When selection is balanced by recurrent mutation, there is a globally stable equilibrium at an allele frequency of $\hat{p}$, which is the value of p in Equation 6.20 for which $p' = p$. The equilibrium frequency of the harmful a allele is therefore $\hat{q} = 1 - \hat{p}$. There are two important cases:

- When the harmful allele is a **complete recessive** ($h = 0$), then

$$\hat{q} = \sqrt{\frac{\mu}{s}} \qquad 6.21$$

- When the harmful allele shows **partial dominance** ($h > 0$), then, to an excellent approximation for realistic values of μ, h, and s,

$$\hat{q} = \frac{\mu}{hs} \qquad 6.22$$

Use of these equations is exemplified by Huntington disease in human beings. This severe inherited disorder is characterized by a degeneration of the neuromuscular system that typically appears after age 35. Although the disease itself results from a dominant mutation, the effects on fitness show only partial dominance owing to the late age of onset of the disease. Relative to a value of $w_{11} = 1$ for the homozygous nonmutant genotype, the fitness of the heterozygous genotype has been estimated as $w_{12} = 0.81$ (Reed and Neel 1959). Homozygous mutant genotypes also have the disease, but they are so rare that the equilibrium frequency of the mutant allele is determined by the fitness of the heterozygote. Equation 6.22 with $hs = 0.19$ is appropriate in this example. If we knew either μ or $\hat{q}$, we could estimate the other. In a Michigan population, $q = 5 \times 10^{-5}$ for the Huntington allele (Reed and Neel 1959). Assuming that the population is in equilibrium, we can estimate μ from Equation 6.22 as $\mu = 5 \times 10^{-5} \times 0.19 = 9.5 \times 10^{-6}$. This use of Equation 6.22 illustrates one of the common indirect methods for the estimation of mutation rates in human beings.

The degree of dominance of a harmful allele is a primary factor in determining its equilibrium frequency. Harmful alleles held in mutation-selection balance are rare. Thus the great majority of harmful alleles are present in heterozygous genotypes. Because there are so many heterozygous genotypes, relative to homozygous mutant genotypes, even a small reduction in fitness in the heterozygote has a large effect in decreasing the equilibrium allele frequency. This effect is shown quantitatively in Figure 6.8, which depicts $\hat{q}$ as a function of μ/s and h. Note how the surface bends sharply upward at the far-right corner where $h = 0$. The increase indicates that, for a given value of μ/s, a completely recessive allele is maintained at a higher equilibrium frequency than a partially dominant allele. Furthermore, the surface drops sharply as h increases from 0, which means that even a small

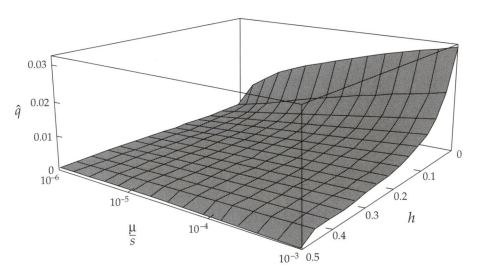

Figure 6.8   Allele frequencies maintained at equilibrium by mutation-selection balance. At each point on the surface, the height is the equilibrium frequency $\hat{q}$ of a harmful allele, given as a function of the mutation rate μ (expressed in multiples of the selection coefficient s) and the degree of dominance h. Note that the surface bends sharply upward toward $h = 0$, a characteristic that means that even a small degree of dominance results in a substantial decrease in the equilibrium frequency of the harmful allele. The μ/s axis is easiest to interpret when the harmful allele is a lethal ($s = 1$).

degree of dominance can cause a large reduction in equilibrium frequency. In general, for realistic values of μ, s, and h, the value of $\hat{q}$ is typically less than 0.01. Therefore, although mutation-selection balance can account for low-frequency deleterious alleles, it cannot readily account for a harmful allele with a frequency greater than 0.01.

PROBLEM 6.8   To confirm for yourself that a small amount of dominance can have a major effect in reducing the equilibrium frequency of a harmful allele, imagine an allele that is lethal when homozygous ($s = 1$) in a population of *Drosophila*. Suppose that the allele is maintained by mutation-selection balance with $\mu = 5 \times 10^{-6}$. Calculate the equilibrium frequency of the allele for a complete recessive and for partial dominant when $h = 0.025$.

ANSWER   For a complete recessive, $\hat{q} = \sqrt{\mu/s} = \sqrt{(5 \times 10^{-6})} = 2.24 \times 10^{-3}$. For partial dominance, $\hat{q} = \mu/hs = (5 \times 10^{-6})/0.025 = 2.00 \times 10^{-4}$. With partial dominance, the equilibrium allele frequency is reduced more than tenfold, and the frequency of homozygous recessive genotypes at equilibrium is reduced more than a hundredfold. It is of interest that $h = 0.025$ is near the average degree of dominance estimated for "recessive" lethals in *Drosophila* (Simmons and Crow 1977).

The Haldane-Muller Principle

The Haldane-Muller principle, named after the geneticists J. B. S. Haldane (1892–1964) and H. J. Muller (1890–1967), deals with the effect of mutation-selection balance on the average fitness of a population. Ignoring recurrent mutation, selection would be able to rid a population completely of a harmful allele. Then, $\hat{q} = 0$, and $\bar{w} = 1$. Because of recurrent mutation, the equilibrium frequency is greater than 0. When $h = 0$, the average fitness in the population at equilibrium equals $1 - \hat{q}^2 s = 1 - (\mu/s)s = 1 - \mu$. The reduction in average fitness due to mutation therefore equals $1 - (1 - \mu) = \mu$, which is called the **mutation load**. When a is partially dominant, the mutation load is approximately 2μ because the average fitness at equilibrium is $1 - 2\hat{p}\hat{q}hs - \hat{q}^2 s \cong 1 - 2\mu$. This result is obtained by ignoring terms in $\hat{q}^2$ because they are so small. With or without partial dominance, therefore, the effect of recurrent mutation in reducing the average fitness in the population is independent of how harmful the mutation is. That the effect of recurrent mutation on average population fitness depends only on the mutation rate is the **Haldane-Muller principle**. The implication is that the harmful effect of an increase in the mutation rate is the same irrespective of whether the mutations produced are mildly detrimental or severely harmful. The effects of severe and mild mutations balance out because a more harmful mutation comes to a lower equilibrium frequency.

MORE COMPLEX TYPES OF SELECTION

Although the two-allele model of viability selection illustrates the possible outcomes of selection, it ignores many potential complications. For example, when the genotypes differ in fertility rather than survivorship, then the model of viability selection is inadequate except in special cases. Most mutations have **pleiotropic effects**; that is, they affect more than one phenotypic attribute of the organism. For example, a gene affecting embryonic growth rate may also affect age at first reproduction. When the pleiotropic effects act in opposing directions (for example, increasing viability but reducing fertility), the net effect on fitness may be quite small. As a result, mutations with offsetting effects on different components of fitness may remain segregating in a population for many generations.

Additional complications arise because fitness is determined by many genes that interact with each other. Simple models of selection are valid only when the alleles interact in such a way that their effects on fitness are additive or multiplicative across genes. Other complications result when the fitnesses of the genotypes are not constant but variable in time or space. In this section we briefly examine a sample of more complex models. Many of the models are of interest because they can maintain genetic polymorphisms. Although the list is extensive, it is by no means complete. You should not try to memorize all the different types of selection. They are collected here only for ease of reference.

Frequency-Dependent Selection

Frequency-dependent selection takes place when fitness is a function of either allele frequencies or genotype frequencies. There is no restriction on the type of frequency dependence except that each Darwinian fitness must be nonnegative. A simple example that illustrates frequency dependence is one in which the fitness of each genotype decreases in proportion to its frequency with a constant of proportionality equal to c:

$$AA: \ w_{11} = 1 - cp^2 \qquad Aa: \ w_{12} = 1 - 2cpq \qquad aa: \ w_{22} = 1 - cq^2$$

In this example, $\Delta p = cpq(q - p)(p^2 - pq + q^2)/\overline{w}$, and so there are equilibria at $p = 0$, $\frac{1}{2}$, and 1. (The factor $p^2 - pq + q^2$ does not have a root for p in the range [0, 1].) A curious feature of this type of frequency-dependent selection is that, at equilibrium, w_{12} is smaller than either w_{11} or w_{22}, so there is heterozygote inferiority; yet $p = \frac{1}{2}$ is a globally stable equilibrium and $\bar{w}$ is a maximum at this equilibrium. The peculiarities of this example are illustrative of frequency-dependent selection in general. Because the fitnesses can be any functions of allele or genotype frequency, nearly anything can happen.

Density-Dependent Selection

Density-dependent selection means that the fitnesses are functions of the population size. Models of density-dependent selection must explicitly include population size and population growth. With logistic growth of two haploid genotypes whose numbers at time t are $A(t)$ and $B(t)$, Equation 1.11 in Chapter 1 becomes

$$\frac{dA(t)}{dt} = r_1 A(t) \left(\frac{K_1 - [A(t) + B(t)]}{K_1} \right) \qquad \frac{dB(t)}{dt} = r_2 B(t) \left(\frac{K_2 - [A(t) + B(t)]}{K_2} \right)$$

Each genotype has its own intrinsic rate of increase (r_1 or r_2) and its own carrying capacity (K_1 or K_2), but they affect each other's growth through the total population size $A(t) + B(t)$. At any time, the outcome of selection depends on the total population size. When the population size is much smaller then either K_1 or K_2, then the right-hand factor in each growth equation equals approximately 1, and so the selection is determined by the relative values of r_1 and r_2. When the population size becomes approximately equal to the smaller of K_1 or K_2, then the genotype with the smaller carrying capacity stops growing while the other continues, and so the selection is determined by the relative values of K_1 and K_2. Interesting events happen when the selection for r favors one genotype and the selection for K favors the other, especially in situations in which stochastic factors also affect population size or there is a time lag between population size and its affect on growth rate. For further information on these types of models, see Roughgarden (1979), May (1981), Bulmer (1994), and Cohen (1995).

Fecundity Selection

In **fecundity selection**, differences in fitness between the genotypes result from the differing abilities of mating pairs to produce offspring. Because both genotypes in a mating pair contribute to the total number of offspring, the number of fitness parameters potentially equals the number of distinct kinds of mating pairs. For two alleles of one gene, there are nine possible types of mating because reciprocal matings may differ in the expected number of offspring; for example, the expected number of offspring from the mating $Aa \female \times aa \male$ may differ from that from the mating $Aa \male \times aa \female$. The presence of so many fitness parameters complicates the mathematical analysis. An analysis of selection based on individual genotypes, analogous to viability differences, is not possible unless the overall fecundity of any mating pair can be written as either the product or the sum of two parameters, one for each genotype in the mating pair. When this strong simplification does not hold, models of selection with fertility differences become rather complex (Ewens, 1979; Clark and Feldman 1986). Models in which differences in fecundity are combined with differences in survivorship can retain genetic polymorphisms even if there is directional selection in one or the other component of fitness.

Age-Structured Populations

Age-structured populations with overlapping generations present problems even more formidable than those caused by fecundity and survivorship differences in populations with discrete, nonoverlapping generations. In each short interval of time, a new cohort of newborns comes into existence and, as it ages, the fate of each organism in the cohort is governed by the functions $l(x)$, which is the probability of survival from birth to age x, and $b(x)$, which is the probability that an organism of age x (actually in the infinitesimal age interval x to $x + dx$) reproduces. If the functions $l(x)$ and $b(x)$ maintain the same form over time, then it can be shown that the population eventually reaches a stable age distribution in which the number of organisms in each age group increases or decreases at a constant rate. At the stable age distribution, the overall growth rate of the population is the value of m that satisfies the equation:

$$1 = \int_0^\infty e^{-mx} l(x) b(x) dx$$

(See Crow and Kimura, 1970, for a derivation.) For this value of m, $dN/dt = mN$, where N is the total population size. In an age-structured population, m corresponds to the intrinsic rate of increase denoted r_0 in Equation 1.7 in Chapter 1.

So far so good, but genetics complicates this situation enormously. If the $l(x)$ and $b(x)$ functions differ for different genotypes, then the allele frequencies change through time. As the allele frequencies change, so does the age structure, and the genotype frequencies in each age class may be different. The result is that the age structure may not become stable until selection reaches some equilibrium (possibly fixation). The sorts of complexities that can arise have been examined by Charlesworth (1980).

Heterogeneous Environments and Clines

Heterogeneous environments refer to models in which the relative fitnesses change according to the environment. The environmental heterogeneity may be spatial or temporal or both. Selection of this type can maintain polymorphisms in the absence of overdominance. If each homozygous genotype is favored in a different subset of environments, then there can be **marginal**

overdominance, in which the heterozygous genotype has the highest fitness when averaged across all the environments, even though it is not the most fit genotype in any particular environment.

In some cases, the relative fitnesses of the genotypes vary geographically across a more or less smooth environmental gradient, for example, according to latitude, altitude, aridity, or salinity. If sufficiently stable in time, a gradient of selection across a region can result in a gradient of allele frequency across the region. A geographical trend in an allele frequency is called a **cline**. An unusually extreme example of a cline is found in the *hemoglobin–I¹* allele in the eelpout fish *Zoarces viviparus*, the allele frequency of which drops from a value of nearly 1 in the North Sea to a value of nearly 0 in the Baltic Sea (Christiansen and Frydenberg 1974). In human aboriginal populations, there is a cline of increasing frequency of the allele I^B in the ABO blood groups from Southwest to Northeast Europe.

Although clines can result from selection—for example, when one genotype is favored at one extreme of the environmental gradient but disfavored at the other extreme—clines can also result from other processes. Migration is one possibility: differences in allele frequency in local populations at the extremes of the range may result from chance processes (for example, different founding populations), and migration of organisms from the extremes into the intermediate zone produces the cline.

The strongest evidence that a cline results from selection is when a cline is reproduced in different locations along a similar environmental gradient. A example of parallel clines played out on a grand scale is found in the electrophoretic polymorphism of alcohol dehydrogenase (the *Adh* gene) in *D. melanogaster*. In Eastern North America, the frequency of the Adh^F allele increases as one goes north, whereas DNA polymorphisms flanking *Adh* show no such geographic trend (Berry and Kreitman 1993). The cline is shown in the upper part of Figure 6.9. The frequency of Adh^F is correlated with cooler temperatures and less rainfall in the more northern latitudes. In Australia, as shown in the lower part of Figure 6.9, the frequency of the Adh^F allele increases as one goes south (Oakeshott et al. 1982). This pattern is in apparent contradiction to that in Eastern North America but, because Australia is in the Southern Hemisphere, the clines are actually parallel. Both show an increase in the frequency of Adh^F as one proceeds from the equator toward the polar cap—the North Pole in the Northern Hemisphere and the South Pole in the Southern Hemisphere. On a much smaller geographical scale, in mountainous regions, the frequency of the Adh^F allele shows a clinal increase with altitude, which is again correlated with cooler temperature and less rainfall. Data from the Caucasus Mountains (Grossman et al. 1970) have been discussed in Problem 4.2; parallel clines have also been studied in the mountains of Mexico (Pipkin et al. 1976).

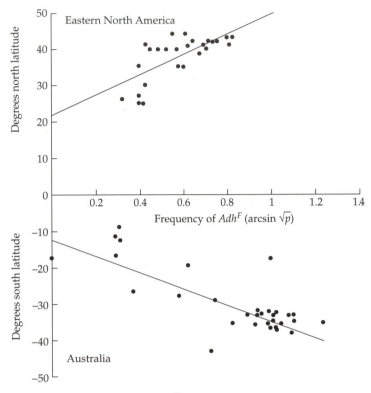

Figure 6.9   Parallel clines of the *Adh^F* (*alcohol dehydrogenase fast*) allele in Eastern North America and in Australia. The allele frequency is given as arcsin($\sqrt{p}$), where p is the allele frequency of *Adh^F*. The angular transformation stretches the scale near the extreme values of p: for values of $p = 0.1, 0.5,$ and 0.9, the values of arcsin($\sqrt{p}$) are 0.322, 0.785, and 1.249, respectively, where the angles are measured in radians. The angular transformation is often used for proportions because it separates the variance of an estimate from the estimate itself: for a binomial proportion p based on n observations, the variance of p is $p(1-p)/n$, whereas the variance of arcsin($\sqrt{p}$), with the angle expressed in radians, is approximately $1/4n$. (North American data from Berry and Kreitman 1993; Australian data from Oakeshott et al. 1982.)

Diversifying Selection

The term **diversifying selection** refers narrowly to selection that favors extreme phenotypes. In a normal distribution of phenotypes, for example, diversifying selection means that organisms in the tails of the distribution are favored relative to those in the middle. More generally, diversifying selection refers to any type of selection in which genotypes are favored merely because they are different. Genes under diversifying selection tend to maintain a relatively large number of alleles. Examples include genes of the major histocompatibility complex in mammals, in which the selective agent is thought to be through resistance to parasitic microorganisms (Satta et al. 1993) and bacterial genes that produce toxins (colicins) that kill other bacteria, in which the selective agent is the destruction of competitors (Riley 1993; Ayala et al., 1994).

Some plants have genes for **gametophytic self-incompatibility**, in which a pollen grain that carries any self-incompatibility allele is unable to pollinate a plant that carries the same allele. Self-incompatibility of this type implies that no plant can fertilize itself. Because a plant of genotype S_iS_j can produce only S_i and S_j pollen, the pollen cannot fertilize S_iS_j plants. Furthermore, homozygous genotypes are not normally found because their formation would require that S_i pollen fertilize an S_iS_j plant. It is easy to show that there is positive selection for new self-sterility alleles and that, at equilibrium, every allele has the same frequency. For n alleles, if S_i has frequency p_i, then the frequency of S_iS_j genotypes with random mating is $2p_i(1 - p_i)/(1 - \Sigma p_i^2)$. The denominator is necessary because of the absence of homozygous genotypes. The probability that an S_i pollen can be successful in fertilization is therefore the probability of genotypes other than S_iS_j, which equals $1 - 2p_i(1 - p_i)/(1 - \Sigma p_i^2)$. At equilibrium, we must have $p_i(1 - p_i) = p_j(1 - p_j)$. From these expressions follow some important conclusions summarized in Problem 6.10. For more information on gametophytic self-incompatibility systems, see Ioerger et al. (1991) and Uyenoyama (1995).

PROBLEM 6.9   Show that $p_i(1 - p_i) = p_j(1 - p_j)$ for all i and j implies that $p_i = p_j = 1/n$, where n is the number of self-incompatible alleles and $n \geq 3$. Use these equilibrium allele frequencies to show that the probability that a pollen grain lands on a compatible style equals $(n - 2)/n$. Finally, show that the probability of successful fertilization by a new mutant S allele, relative to that of any preexisting allele, equals $n/(n - 2)$.

ANSWER $p_i(1 - p_i) = p_j(1 - p_j)$ implies that $p_i - p_j = p_i^2 - p_j^2 = (p_i - p_j)$ $(p_i + p_j)$ so that either $p_i = p_j$ for all i and j or $p_i + p_j = 0$. Because $n \geq 3$, $(p_i + p_j) \neq 1$. Because there are n alleles, we must have $\Sigma p_i = 1$, and so $p_i = 1/n$. The probability of a pollen grain landing on a compatible style is $1 - 2p_i(1 - p_i)/(1 - \Sigma p_i^2) = 1 - 2/n = (n - 2)/n$. A pollen grain containing a newly arising S allele will always land on a compatible style, and so its probability of fertilization, relative to that of a preexisting allele, equals $1/[(n-2)/n] = n/(n-2)$. If effect, this is the relative fitness of a new mutation. For $n = 3, 4, 5, 10, 50$, and 100, it equals $3, 2, 1.67, 1.25, 1.04$, and 1.02, respectively.

Differential Selection in the Sexes

Some genes may have different effects in the two sexes. If the fitnesses of genotypes differ between the sexes, then genotypes that are disfavored in one sex may be favored in the other. The offsetting effects increase the opportunity for a balanced polymorphism. The survivorship model of selection can be extended to include this case by supposing that the relative viabilities of the genotypes AA, Aa, and aa are given by w_{11}, w_{12}, and w_{22} in females and by v_{11}, v_{12}, and v_{22} in males. One of the w's and one of the v's can be set arbitrarily to 1, which leaves four fitness parameters rather than two. A more serious complication is that the allele frequencies in gametes are no longer the same in males and females. Letting p_f and p_m be the allele frequency of A in female and male gametes, respectively, then the genotype frequencies of AA, Aa, and aa in the zygotes are $p_f p_m$, $p_f q_m + q_f p_m$, and $q_f q_m$, respectively, where $q_f = 1 - p_f$ and $q_m = 1 - p_m$. One of the consequences of differential selection in the sexes is that, with an appropriate choice of fitnesses, it is possible to have more than one stable polymorphic equilibrium. A stable equilibrium is also possible with heterozygote inferiority in one sex or with incomplete dominance when selection works in opposite directions in the two sexes.

X-linked Genes

Genes located in the X chromosome can have the same sort of complications as differential selection in the sexes, but the possibilities for polymorphism are not quite so numerous because there are only three fitness parameters instead of four. If A and a are alleles of an X-linked gene, then there are three genotypes in females (AA, Aa, and aa) and two genotypes in males (either A or a along with the Y chromosome). One fitness parameter in each sex can be set arbitrarily to 1. As with differential selection in the sexes, the allele frequencies differ in eggs and sperm. However, in any generation, the frequency of A in male zygotes equals the frequency of A in female gametes of the preceding generation. If you do not understand why, think about the parental origin of the X chromosome in a male.

Gametic Selection

Many plants go through a life cycle in which both haploid products of meiosis and the diploid products of fertilization are exposed to selection. In mosses and vascular plants, for example, a diploid organism (the *sporophyte*) produces spores each of which germinates to form a haploid organism (the *gametophyte*) that reproduces asexually by mitosis. The gametophytes give rise to haploid male and female gametes, which undergo fertilization creating a new diploid generation. In mosses, the prominent stage of the life cycle is the gametophyte whereas, in higher plants, the prominent stage is the sporophyte.

When the haploid phase of the life cycle is exposed to selection, the selection is called **gametic selection**. As a concrete model, suppose that the relative survivorships of A and a gametophytes (the haploid phase) are given by v_1 and v_2, respectively. In the sporophytes (the diploid phase), the survivorships can be written as before as w_{11}, w_{12}, and w_{22}. If p and q are the allele frequencies of A and a at the beginning of the haploid phase, then after the differential haploid mortality has taken place, the frequencies will be $p^* = pv_1/\bar{v}$ and $q^* = qv_2/\bar{v}$, where $\bar{v} = pv_1 + qv_2$. With random fertilization among the gametes, the diploid genotypes AA, Aa, and aa are formed in the proportions p^{*2}, $2p^{*2}q^{*2}$, and q^{*2}, and these survive in the relative proportions w_{11}, w_{12}, and w_{22}. You may verify for yourself that, at the beginning of the haploid phase of the next generation, the allele frequency of A is

$$p' = \frac{p^2 w_{11} v_1^2 + pq w_{12} v_1 v_2}{p^2 w_{11} v_1^2 + 2pq w_{12} v_1 v_2 + q^2 w_{22} v_2^2}$$

This equation has the same form as the equation for p' in Equation 6.10 except that w_{11} is replaced with $w_{11} v_1^2$, w_{12} with $w_{12} v_1 v_2$, and w_{22} with $w_{22} v_2^2$. The conditions for fixation or for a stable or unstable equilibrium are therefore determined by the relative magnitude of the composite "fitness" of the heterozygous genotype relative to those of the homozygous genotypes.

Meiotic Drive

A situation analogous to, but distinct from, gametic selection takes place when there is non-Mendelian segregation in the heterozygous genotype. In females, unequal recovery of reciprocal products of meiosis can be caused by nonrandom segregation of homologous chromosomes to the functional egg nucleus, which is why non-Mendelian segregation is known generically as **meiotic drive**. In other cases, the unequal recovery is caused by a gene or genes that act to render gametes carrying the homologous chromosome nonfunctional. Examples include "sperm killers" such as *segregation distortion* in *Drosophila melanogaster* (Charlesworth and Hartl 1978) and the *t* alleles in the house mouse (Hammer and Silver 1993) as well as "spore killers" described in filamentous fungi (Raju 1994).

Because meiotic drive acts only in the heterozygous genotype, its effect is to alter the term $pq w_{12}$ in Equation 6.10 for p'. This term comes from the expression $\frac{1}{2} \times 2pq w_{12}$ for the proportion of A-bearing gametes from surviving Aa genotypes, and the $\frac{1}{2}$ is the Mendelian segregation ratio. If the ratio of $A : a$ gametes from Aa heterozygotes is $k : 1 - k$ instead of $\frac{1}{2} : \frac{1}{2}$, then the expression for p' becomes

$$p' = \frac{p^2 w_{11} + 2kpq w_{12}}{\overline{w}} \qquad\qquad 6.23$$

where $\overline{w}$ is the average survivorship in the population defined in Equation 6.8. Since A is the driven allele, $k > \frac{1}{2}$. Equation 6.23 is illustrative of meiotic drive even though it requires that the non-Mendelian segregation affect both sexes equally, a case that is not generally found in practice. One implication of the equation is that, unless selection counteracts the meiotic drive, the driven allele goes to fixation. In particular, if the relative viabilities are equal, then $p' = p^2 + 2kpq$ and $\Delta p = pq(2k - 1)$, so that $p \rightarrow 1$ because $k > \frac{1}{2}$.

In some examples of meiotic drive, including *segregation distortion* and the t alleles, the driven allele is lethal when homozygous (Hartl 1970). Assuming that the lethality is completely recessive, the survivorships are $w_{11} = 0$, $w_{12} = 1$, and $w_{22} = 1$. Equation 6.23 implies that $p' = 2kp/(1 + p)$ and so $\Delta p = p[(2k - 1) - p]/(1 + p)$. There is an interior equilibrium at $\hat{p} = 2k - 1$, which intuition suggests (correctly) is locally stable. It is also globally stable (Figure 6.10). Note that $\hat{p}$ is between 0 and 1 for any value of k between $\frac{1}{2}$ and 1. The calculations for a recessive-lethal driven allele are a special case of the slightly more general model discussed in Problem 6.11.

PROBLEM 6.10   Suppose that the AA genotype is not completely lethal but that its survivorship is given by $1 - s$ relative to a value of 1 for both Aa and aa genotypes. Show that $\Delta p = pq[(2k - 1) - ps]/(1 - p^2 s)$. Find $\hat{p}$ and define the conditions, in terms of k and s, for which $\hat{p}$ is between 0 and 1. Show also that the equilibrium is locally stable.

ANSWER   Equation 6.23 implies that $p' = [p^2(1 - s) + 2kpq]/(1 - p^2 s)$. $\Delta p = p' - p$ simplifies to the formula given. Setting $\Delta p = 0$ yields equilibria at 0, 1, and $\hat{p} = (2k - 1)/s$. For $\hat{p} > 0$, we need $(2k - 1)/s > 0$, or $k > \frac{1}{2}$. For $\hat{p} < 1$, we need $(2k - 1)/s < 1$, or $k < (s + 1)/2$. Note that, as the selection against the A allele becomes smaller (s closer to 0), more values of k result in fixation of the unfavorable A allele and fewer result in an interior equilibrium. The stability of $\hat{p}$ can be deduced by evaluating the derivative in Equation 6.19. For this purpose, it is

continued on page 250

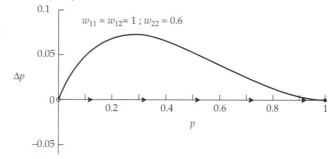

(A) Viability only

$w_{11} = w_{12} = 1 \; ; w_{22} = 0.6$

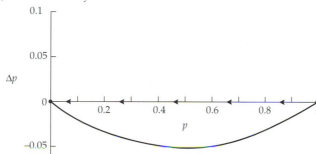

(B) Meiotic drive only

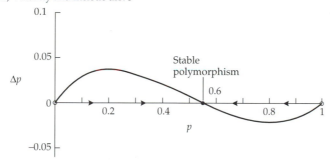

(C) Viability and meiotic drive

Stable polymorphism

0.6

Figure 6.10 The balance between meiotic drive and viability selection. (A) Δp versus p for viability alone, when the fitnesses are $w_{11} = w_{12} = 1$ and $w_{22} = 0.6$. With these fitnesses, viability selection would eliminate the a allele. (B) Meiotic drive alone, where the heterozygous genotype Aa produces 40% A-bearing gametes and 60% a-bearing gametes. With meiotic drive alone, the A allele would be lost. (C) Δp versus p when both viability selection and meiotic drive are operating at the same time, using the same fitness and meiotic drive parameters as above. In this example, when both processes operate simultaneously, their offsetting effects create a stable polymorphism.

convenient to write Δp as $pqs(\hat{p} - p)/(1 - p^2 s)$. In taking the derivative, remember that any term containing $\hat{p} - p$ becomes 0 when $p = \hat{p}$, so these terms can be neglected. The derivative, evaluated at $\hat{p}$, equals $-\hat{p}\hat{q}\, s/(1 - \hat{p}^2 s)$, where $\hat{q} = 1 - \hat{p}$. The sign of this number must be negative, and so the equilibrium at $\hat{p}$, when it exists, is locally stable.

Multiple Alleles

The presence of multiple alleles complicates the analysis of selection because the number of fitness parameters increases. With n alleles, there are $n(n + 1)/2$ possible genotypes, each with its own fitness. Furthermore, simple generalizations from two-allele theory do not necessarily carry over to multiple alleles. Consider the example of heterozygote superiority. Intuitively, one might expect that fitnesses yielding stable, multiple-allele polymorphisms would be easy to generate by requiring that each heterozygous genotype have a greater fitness than the homozygous genotypes formed from the constituent alleles. This is not the case, however. If, for n alleles, the fitnesses of the genotypes are assigned at random between 0 and 1, subject to the condition that, for each i and j, $w_{ij} > \max(w_{ii}, w_{jj})$, then only a relatively small proportion of systems with four or more alleles yields a stable polymorphism with all alleles present. For four, five, and six alleles, the percentage of fitness sets yielding a stable equilibrium is 12.6, 1.2, and 0.03, respectively (Lewontin et al. 1978). The reason for the low percentages is that, even if a heterozygote is more fit than its constituent homozygotes, there might be a different homozygote more fit than all three. All right, how about requiring that each heterozygote be better than *every* homozygote? Surprisingly, this requirement does not help matters much. In this case, for four, five, and six alleles, the percentage of fitness sets yielding a stable equilibrium is 34.3, 10.4, and 1.3, respectively (Lewontin et al. 1978). The point is that polymorphisms with greater than three or four alleles are extremely unlikely to be maintained by selection for simple heterozygous advantage with constant survivorship. If selection is implicated in such a case, models of selection such as diversifying selection or heterogeneous environments are much more plausible. On the other hand, the fitnesses of genotypes in nature are not chosen simultaneously by a random number generator. Each new allele that arises is tested against the resident alleles, and the new allele is able to invade the population if its marginal fitness exceeds the mean fitness of the population. By this process, multiple allele polymorphisms can be accumulated, and the order in which the mutations appear makes a difference (Spencer and Marks 1988).

The possibility of multiple alleles also creates surprising situations in which the outcome of natural selection depends on the order in which the alleles are introduced into the population. Earlier in this chapter we mentioned the sickle-cell hemoglobin polymorphism in Africa and its relation to malaria resistance. People who are homozygous AA for the normal allele are susceptible to falciparum malaria, those who are heterozygous AS for the sickle-cell allele are resistant to malaria and have a mild anemia, and those who are homozygous SS for the sickle-cell allele have a life-threatening anemia. This is a classic case of heterozygote superiority. There is another allele, C, found at low frequency in populations in which the S allele is prevalent. The C allele is also protective against malaria, but the allele is recessive, and so only the CC genotypes are resistant. Unlike the S allele, the C allele does not cause anemia.

The relative survivorship of each of the various hemoglobin genotypes has been estimated based on studies of more than 32,000 people in 72 populations in West Africa (Cavalli-Sforza and Bodmer 1971). The survivorships are given in the following table, which indicates the genotypes that are resistant and those that have severe hemolytic anemia. The survivorships were estimated in a geographical region where malaria was common. Note that the S allele causes a severe anemia in the heterozygous SC genotype, but not so serious as that in the homozygous SS genotype.

Genotype	AA	AS	SS	AC	SC	CC
Survivorship	0.9	1.0	0.2	0.9	0.7	1.3
Health status		Resistant	Anemic		Anemic	Resistant

Inspection of these survivorships reveals a paradox. The CC genotype has the highest fitness, yet the C allele is not fixed. The reason is found in the historical order in which the S and C mutations took place. The A allele is the ancestral type and undoubtedly predated the human settlement of regions subject to malaria. In such a region, the appearance of an S allele creates a heterozygous advantage, and natural selection quickly attains a stable equilibrium at which the ratio of $A : S$ alleles is approximately 8 : 1. At this equilibrium, the average fitness in the population is $\bar{w} = 0.911$. Now suppose that mutation or migration were to introduce a small number of C alleles. Because C alleles are rare, each is present in either the AC genotype, with probability $8/9$, or in the SC genotype, with probability $1/9$. The average fitness of genotypes heterozygous for C is therefore 0.878, which is smaller than the average fitness in the population. Hence, the frequency of C decreases, and C goes extinct. The C allele has no chance of invading an A/S polymorphism unless the initial frequency of C is sufficiently large. Figure 6.11 illustrates this phenomenon. With the survivorships given in this example, the critical initial frequency of C that allows invasion is 0.073. Once C can get established in the population, it eventually becomes fixed.

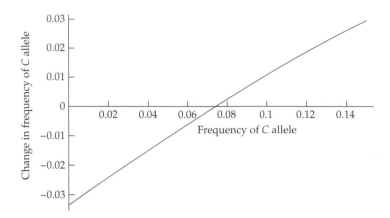

Figure 6.11   Change in frequency of the hemoglobin *C* allele in a population in which the *A* and *S* alleles are present in their equilibrium proportions of 8 : 1. When the initial frequency of *C* is small, the change in frequency is negative, and so *C* is eliminated even though *CC* genotypes have the highest fitness. The *C* allele is unable to invade unless its initial frequency is greater than 0.073, and in that case *C* goes to fixation. The plot is based on the survivorship values given in the text.

Multiple Loci and Gene Interaction: Epistasis

With multiple loci, as many types of gametes are possible as there are combinations of alleles. The simplest example is the two-locus, two-allele case, in which the possible gametes are *AB*, *Ab*, *aB*, and *ab*. In the absence of recombination ($r = 0$), each type of gamete can be regarded as an "allele" of one locus with four alleles. The principles of multiple-allele selection then apply, and some of the "alleles" may be eliminated by selection. The presence of recombination complicates matters because each gametic type is continually recreated by recombination even if it is disfavored by selection. The influence of recombination on the outcome of selection is determined by the recombination fraction and by the degree of interaction between the loci. When selection acts on the phenotype produced by the joint effects of multiple loci, there are two general situations:

- Changes in allele frequency are driven primarily by the selection coefficients and recombination plays a minor role.
- Selection and recombination are about equally important in determining the outcome.

The former is usually the case with weak epistasis and moderate or loose linkage; the latter is more prevalent with strong epistasis and tight linkage. The term **epistasis** is often used in population genetics as a synonym for gene interaction; it applies to any situation in which the genetic effects of different loci that contribute to a phenotypic trait are not additive. In the two-

---

TABLE 6.3   TWO-LOCUS FITNESSES (SURVIVORSHIPS)

		Genotype at B locus			
		BB	*Bb*	*bb*	
Genotype at A locus	*AA*	w_{11}	w_{12}	w_{13}	w_{AA}
	Aa	w_{21}	$w_{22} = 1$	w_{23}	w_{Aa}
	aa	w_{31}	w_{32}	w_{33}	w_{aa}
		w_{BB}	w_{Bb}	w_{bb}	

Note: The table assumes that the two types of double heterozygotes, AB/ab and Ab/aB, have the same fitness, w_{22}.

---

locus, two-allele example, the fitnesses (survivorships) of the genotypes can be written as shown in Table 6.3, where it is assumed that the two types of double heterozygote (AB/ab and Ab/aB) have the same fitness; for convenience, this value is often set at $w_{22} = 1$. For each single-locus genotype, the average survivorship is equal to the weighted average across each genotype at the other locus. In Table 6.3, these averages are denoted w_{AA}, w_{Aa}, and so on. Additivity across loci means that $w_{11} = w_{AA} + w_{BB}$, $w_{12} = w_{AA} + w_{Bb}$, and so forth for all genotypes, including $w_{22} = w_{Aa} + w_{Bb} = 1$. If additivity does not apply across all nine genotypes, then epistasis is said to be present. A discussion of epistasis from a statistical point of view appears in Chapter 9.

When there is strong epistasis and tight linkage, complications abound. With two loci and two alleles at each, there are as many as 15 equilibria. Most of them are unstable, but examples are known in which four interior equilibria are simultaneously stable. Figure 6.12 is one example that shows the average fitness in the population $\bar{w}$ as a function of the allele frequencies of A and B. At any point in time, the gametic frequencies in the population are determined not only by the allele frequencies of A and B but also by the linkage disequilibrium parameter, which was denoted by the symbol D in the section on linkage disequilibrium in Chapter 3. In this example, all 15 equilibria are realized. There are four *corner equilibria* in which one gametic type is fixed, namely, AB, Ab, aB, or ab; there are also four *edge* equilibria in which one allele of either locus is fixed, namely, A, a, B, or b. With the survivorships as in Figure 6.12, all of the corner and edge equilibria are unstable. There are also three unstable interior equilibria, each of which has $p_A = p_B = \frac{1}{2}$ and so is located at the position of the open circle on the saddle in Figure 6.12; these equilibria have the same allele frequencies but differ in the degree of linkage disequilibrium. The positions of the stable equilibria are indicated by the solid circles, each of which represents two equilibrium points with the same

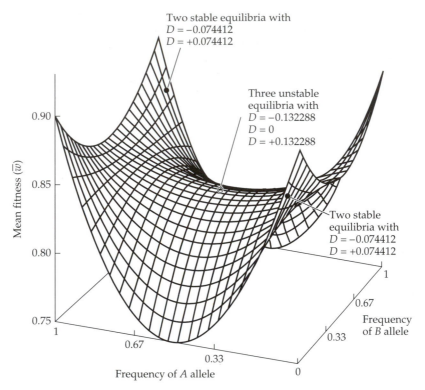

Two stable equilibria with
$D = -0.074412$
$D = +0.074412$

Three unstable
equilibria with
$D = -0.132288$
$D = 0$
$D = +0.132288$

Two stable
equilibria with
$D = -0.074412$
$D = +0.074412$

Mean fitness ($\bar{w}$)

Frequency
of B allele

Frequency of A allele

Figure 6.12   An example of two-locus, two-allele survivorship selection in which there are four stable interior equilibria, the positions of which are indicated by the dots near two of the corners. Each dot represents two stable points differing in the sign of the linkage disequilibrium. This example also includes three unstable interior equilibria (represented by the open circled point in the center), four unstable edge equilibria, and four unstable corner equilibria. The survivorship parameters, in the notation of Table 6.3, are $w_{11} = w_{13} = w_{31} = w_{33} = 0.9$, $w_{21} = w_{23} = 0.8$, and $w_{12} = w_{32} = 0.6$; the recombination fraction is $r = 0.09$. (Example from Hastings 1985.)

allele frequencies but differing in their value of D. In this case, the equilibria are symmetrical.

Figure 6.12 is a plot of $\bar{w}$ to emphasize that the average fitness in the population is not necessarily a maximum at equilibrium. In this example, none of the four stable equilibria is a point of maximum average fitness. The maximum fitness is found at either of the four corners, and these equilibria are unstable. Furthermore, in the vicinity of each stable equilibrium, as the population moves toward the equilibrium from certain directions, the average fitness must decrease as the equilibrium is approached. Hence, not only is average fitness not necessarily a maximum at equilibrium, natural selection can cause a decrease in average fitness.

In models in which fitness depends on multiple interacting loci, do we really have to give up the attractive generalization from one-locus theory that selection acts in such a way as to increase average fitness? Not altogether. Although even the two-locus, two-allele model of survivorship selection is beyond present techniques of mathematical analysis, an important generalization has come from approximate solutions as well as from computer simulations (Ewens 1979): if epistasis is not too strong, and linkage is not too tight, then the average fitness in the population usually increases.

This statement is multiply qualified ("not too strong . . . not too tight . . . usually increases") because exceptions can rather easily be constructed. However, the generalization is observed in most generations in most numerical examples when the survivorships are chosen at random (Karlin and Carmelli 1975). To the extent that it is true, the generalization supports the powerful metaphor that natural selection tends to increase average fitness. If one can imagine a complex surface of hills and valleys corresponding to regions of high and low average fitness, then one can speak metaphorically of a population as a sort of "hill climber" moving across this surface and scaling a fitness "peak." This picturesque analogy is a central concept in Wright's shifting balance theory of evolution, which is discussed in the last section of this chapter. However, there are enough exceptions to the hill-climbing generalization that maximization of average fitness cannot be used as a guide to predicting the outcome of any particular set of fitness values. Each model must be considered in detail on its own.

Sexual Selection

It seems that, wherever you look in nature, animals have physical adornments or behavioral displays to help them in obtaining mates. In some cases, there is direct competition between animals, usually males, as exemplified by the contests of antler bashing in moose or head butting in bighorn sheep. In other cases, there is indirect competition, as seen in the behavioral displays of male peacocks in full plumage strutting their stuff. These are dangerous activities. A bighorn sheep can get his skull fractured or fall off a cliff. The male peacock is conspicuous, burdened, and preoccupied—vulnerable to any predator.

Darwin (1871) was the first to draw attention to competition for mates as a source of selection not necessarily related to adaptation of the organism to its environment. This type of selection he called **sexual selection**. In the case of direct competition for mates, it is easy to understand that a successful male leaves more progeny than an unsuccessful male, and so alleles promoting the physical adornments, strength, and aggressiveness needed for successful competition for mates are perpetuated even though they may occasionally be detrimental. The example of indirect competition is considerably more subtle because the male is merely advertising. The female does the choosing. One theory for the evolution of male sexual displays is that, in the early stages of their evolution, the displays take advantage of a female preference. The origin of the initial preference is unclear. Darwin suggested that female choosiness and offspring number are both associated with superior nutrition, hence choosy females may, at the beginning, have had more offspring. Whatever the cause, given an initial choosiness among females, males with more effective displays are chosen preferentially as mates, and their offspring receive alleles that create both the displays in the males and the preferences in the females. If these traits are genetically correlated—as, for example, through common hormonal or neurological pathways or through linkage disequilibrium—then selection becomes a self-accelerating process promoting increasingly elaborate displays and increasingly greater choosiness. According to Fisher (1930):

> The two characteristics affected by such a process, namely plumage development in the male, and sexual preference for such developments in the female, must thus advance together, and so long as the process is unchecked by severe counterselection, will advance with ever-increasing speed. In the total absence of such checks, it is easy to see that the speed of development will be proportional to the development already attained. There is thus, in any situation in which sexual selection is capable of conferring a great reproductive advantage, the potentiality of a runaway process which will, however small the beginnings from which it arose, must, unless checked, produce great effects, and in the later stages with great rapidity.

The ever-accelerating process is called **runaway sexual selection**, and the conditions under which it takes place have been studied theoretically (Lande and Arnold 1985; Kirkpatrick and Barton 1995; Iwasa and Pomiankowski 1995).

KIN SELECTION

One alternative type of selection, called kin selection, makes use of an extended concept of "fitness." In **kin selection**, a positive selection for certain alleles takes place indirectly through enhanced reproduction of the genetic relatives of carriers of the alleles rather than directly through an increased fitness of the carriers themselves. Kin selection has been postulated in attempts to account for the evolution of altruism. A behavior is regarded as **altruism** if it increases the fitness of other organisms at the expense of one's own fitness. Altruistic behavior is exhibited most dramatically by social insects such as termites, ants, and bees, in which certain worker castes exert their labors for the care, protection, and reproduction of the queen and her offspring but do not reproduce themselves. Other, less dramatic examples of altruistic behavior include phenomena such as the care of offspring by their parents.

A central consideration in kin selection is that relatives have genes in common. Therefore, a gene that causes altruistic behavior can increase in frequency if the increase in the recipient's fitness as a result of altruism is sufficiently large to offset the decrease in the altruist's own fitness. The essentials of the situation can be made clear by considering the case of identical twins. Because identical twins are genetically identical, the reproduction of one's twin is genetically equivalent to reproduction by oneself. Thus, it makes no difference if an altruistic organism decreases its own fitness for the sake of an equal increase in fitness of an identical twin; from an evolutionary point of view, it is an even trade because the combined number of offspring from both twins remains unchanged. By the same token, if an altruistic act decreases the fitness of an organism by an amount less than the increase gained by an identical twin, then the altruism results in a net increase in the combined number of offspring. One would, therefore, expect altruism between identical twins to be favored by natural selection as long as the risk to the altruist is no greater than the benefit to the recipient.

These considerations of identical twins can be extended to other degrees of relationship as well, but the risk to the altruist must be correspondingly smaller than the benefit to the recipient because other types of relatives share fewer genes than identical twins. The break-even points for altruism toward various degrees of relationship have been trenchantly summarized by J. B. S. Haldane, who is said to have quipped that he would lay down his life for two brothers, four nephews, or eight cousins. In any case, fitness considerations that take into account not only an organism's own fitness but also the fitness of relatives (other than direct descendants) constitute what is called the **inclusive fitness** of the organism.

To be concrete, suppose that altruism results in a decrease in fitness c of the altruist that is offset by an increase in fitness b in the recipient. The gene for altruism increases in frequency if the ratio of cost to benefit is great enough, relative to the genetic relationship between the altruist and the recipient; that is, the gene for altruism increases in frequency if

$$\frac{c}{b} < r \qquad\qquad 6.24$$

as shown first by Hamilton (1964) and discussed in detail by Cavalli-Sforza and Feldman (1978) and Uyenoyama and Feldman (1980). In this context, r is a measure of genetic relationship between the altruist X and the recipient of the altruism Y, defined as

$$r = \frac{2F_{XY}}{(1 + F_X)} \qquad\qquad 6.25$$

where F_X is the inbreeding coefficient of the altruist X, and F_{XY} is the inbreeding coefficient of a hypothetical offspring of X and Y. As illustrated in Figure 6.13, r equals the probability that two gametes from X and Y contain alleles that are identical by descent, F_{XY}, relative to the probability that two gametes from X contain alleles that are identical by descent, $(1 + F_X)/2$. The cost-benefit tradeoff in Equation 6.24 is generally valid for weak selection when $F_X = 0$ and valid for additive alleles even when $F_X \neq 0$ (Aoki 1981).

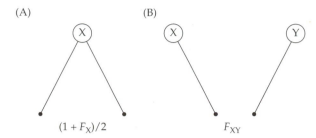

Figure 6.13 Definition of the genetic relationship between an altruist X and the recipient of the altruism Y. (A) Two alleles chosen at random from an organism X are identical by descent with probability $(1 + F_X)/2$ (see Figure 4.13). (B) Two alleles chosen at random, one from X and the other from Y, are identical by descent with probability F_{XY}, which is the inbreeding coefficient of a hypothetical offspring of X and Y. The ratio of F_{XY} to $(1 + F_X)/2$ is the appropriate measure of genetic relationship in the consideration of kin selection.

PROBLEM 6.11   For the illustrated pedigrees (A) and (B) of full siblings shown in the accompanying figure, calculate the break-even value of the benefit b to the recipient of altruism Y, relative to a cost value $c = 1$ to the altruist X, in order to ensure an increase in frequency of an additive gene for altruism. Why are the answers different in the two cases?

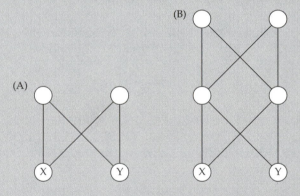

ANSWER   In case (A), a hypothetical offspring of X and Y has an inbreeding coefficient of $F_{XY} = (\frac{1}{2})^3 + (\frac{1}{2})^3 = \frac{1}{4}$, and $F_X = 0$. Therefore, $r = 2 \times \frac{1}{4} = \frac{1}{2}$, and the break-even value of $c/b = \frac{1}{2}$. Hence, for $c = 1$, the break-even value of $b = 2$. (This calculation is the theoretical basis of Haldane's quip about laying down his life for two brothers.) In pedigree (B), $F_{XY} = 4 \times (1/2)^5 + 2 \times (1/2)^3 = 3/8$, and $F_X = 2 \times (1/2)^3 = 1/4$. Therefore, $r = 2(3/8)/(1 + 1/4) = 3/5$. For a cost of $c = 1$, the break-even value of b equals $5/3$. The values differ in the two cases because of the differing inbreeding. In case (B), even though X is inbred, the break-even value of b is smaller because of the closer genetic relationship between X and Y.

INTERDEME SELECTION AND THE SHIFTING BALANCE THEORY

Another alternative type of selection arises in the context of **interdeme selection**, which takes place between semi-isolated subpopulations (*demes*) of the same species. If subpopulations composed of certain genotypes are more likely to become extinct and have their vacated habitats recolonized by migrants from other subpopulations composed of other genotypes, then the more successful subpopulations can, in some sense, be considered as having a greater "fitness" than the less successful ones. Since this concept of **population fitness** is a characteristic of the entire population and not merely the average fitness of the genotypes within it ($\overline{w}$), interdeme selection is outside the realm of most conventional models of selection. Interdeme selection is one type of group selection (Wilson 1983).

Interdeme selection plays an essential role in the shifting balance theory of evolution of Wright (1977 and earlier). In the **shifting balance theory**, a large population that is subdivided into a set of small, semi-isolated subpopulations (demes) has the best chance for the subpopulations to explore the full range of the adaptive topography and to find the highest fitness peak on a convoluted adaptive surface. If the subpopulations are sufficiently small, and the migration rate between them is sufficiently small, then the subpopulations are susceptible to random genetic drift of allele frequencies, which allows them to explore their adaptive topography more or less independently. In any subpopulation, random genetic drift can result in a temporary reduction in fitness that would be prevented by selection in a larger population, and so a subpopulation can pass through a "valley" of reduced fitness and possibly end up "climbing" a peak of fitness higher than the original. Any lucky subpopulation that reaches a higher adaptive peak on the fitness surface increases in size and sends out more migrants to nearby subpopulations, and the favorable gene combinations are gradually spread throughout the entire set of subpopulations by means of interdeme selection.

The shifting balance process includes three distinct phases:

1. An exploratory phase, in which random genetic drift plays an important role in allowing small subpopulations to explore their adaptive topography.

2. A phase of mass selection, in which favorable gene combinations created by chance in the random drift phase become rapidly incorporated into the genome of local subpopulations by the action of natural selection.

3. A phase of interdeme selection, in which the more successful demes increase in size and rate of migration; the excess migration shifts the allele frequencies of nearby subpopulations until they also come under the control of the higher fitness peak. The favorable genotypes thereby become spread throughout the entire population in an ever-widening distribution. Where the region of spread from two such centers overlaps, a new and still more favorable genotype may be formed and itself become a center for interdeme selection. In this manner, the whole of the adaptive topography can be explored, and there is a continual shifting of control from one adaptive peak to control by a superior one.

The shifting balance theory has played an important role in evolutionary thinking, in part because of its use of mountain-climbing terms as tropes for stages in the evolutionary progress: "exploration" of the adaptive topography, chance "discovery" of a route to a higher adaptive peak, and ultimately the "conquest" of the highest adaptive peak by the whole species. However, as a comprehensive theory of evolution, many aspects of the theory remain untested. For the theory to work as envisaged, the interactions between alleles must often result in complex adaptive topographies with many peaks and valleys. The population must be split up into smaller subpopulations, which must be small enough for random genetic drift to be important, but large enough for mass selection to fix favorable combinations of alleles. Although migration between demes is essential, neighboring demes must be sufficiently isolated for genetic differentiation to take place, but sufficiently connected for favorable gene combinations to spread. Because of uncertainly about the applicability of these assumptions, the shifting balance process remains a picturesque metaphor that is still largely untested. However, computer simulations have been carried out to investigate the range of magnitudes of the key parameters that are necessary for the shifting balance process to be effective; these parameters include the size of the subpopulations, the rate of migration and range of dispersal of the migrants, the degree of epistasis between genes, and the rate of recombination (Bergman et al. 1995). Some empirical studies have also explored the partitioning of genetic variance within and between groups for traits associated with fitness (Wade and Goodnight 1991).

One important implication of interdeme selection is that alleles that are harmful in themselves may nevertheless be favored because they are beneficial to the group. This principle is illustrated in the model in Table 6.4, where the allele A' is harmful to organisms within demes but favorable to the deme as a whole. Equation 6.11 implies that, within the ith deme, $\Delta q_i = -cq_i(1 - q_i)$ (assuming that $\bar{w} \cong 1$). Averaging across all of the subpopulations, the change in allele frequency resulting from selection within subpopulations, Δq_w, equals $-c\bar{q}(1 - \bar{q})(1 - F)$, where F is the fixation index F_{ST} discussed in Chap-

TABLE 6.4   MODEL OF INTERDEME SELECTION

Genotype	AA	AA′	A′A′
Frequency in deme i	p_i^2	$2p_iq_i$	q_i^2
Within-population fitness	1	$1-c$	$1-2c$
Between-population fitness of deme i		$1+2(b-c)q_i$	

ter 4. At the same time, within-subpopulation selection takes place, interdeme selection favors demes containing A', and the change in allele frequency resulting from between-subpopulation selection, Δq_b, equals $2(b-c)\bar{q}(1-\bar{q})F$, as shown by Crow and Aoki (1982). Putting the within-subpopulation and between-subpopulation selection together, the total change in the frequency of A' is

$$\Delta q = \Delta q_w + \Delta q_b = -c\bar{q}(1-\bar{q})(1-F) + 2(b-c)\bar{q}(1-\bar{q})F \qquad 6.26$$

The terms on the right-hand side can be interpreted by considering the extremes of $F = 0$ and $F = 1$. When $F = 0$, there is no population substructure, which means that all subpopulations have the same allele frequency $\bar{q}$; in this case, the change in allele frequency is just $-c\bar{q}(1-\bar{q})$. At the other extreme, when $F = 1$, each subpopulation is fixed for either A or A', and the proportion fixed for A' equals $\bar{q}$. The between-subpopulation selection is therefore analogous to selection between alleles in a haploid organism in which the fitnesses of A and A' demes are in the ratio $1 : 2(b-c)$. In this case, therefore, the change in allele frequency is $2(b-c)\bar{q}(1-\bar{q})$ (from Equation 6.6, assuming that $\bar{w} \cong 1$).

Equation 6.26 implies that $\Delta q > 0$ if

$$\frac{b-c}{c} > \frac{1-F}{2F} \qquad 6.27$$

This is the condition necessary for selection between demes to override selection within demes, and the formulation is quite general (Crow and Aoki 1982). A biological interpretation of the inequality in Equation 6.27 can be inferred by comparison with the break-even point for kin selection given in Equations 6.24 and 6.25. Expressing 6.27 in terms of $r = 2F/(1+F)$, which means that $F = r/(2-r)$, yields $c/b < r$; this condition is identical to Equation 6.24. In these models, the equivalence between kin selection and interdeme selection results from the shared remote ancestry of the members of each subpopulation caused by random genetic drift among the subpopulations. The members of each subpopulation are related by kinship, and so interdeme selection is the same phenomenon as kin selection; the break-even point is that at which the benefit b to one's kin through interdeme selection equals the cost to one's self c through direct selection against the A' allele.

If there are a large number of subpopulations, each of size N, that exchange migrants in such a way that m is the proportion of genes in each deme that are exchanged each generation for genes chosen at random from the other demes, then the approximate value of F at equilibrium is given by Equation 5.17 as $F = 1/(1 + 4Nm)$. Consequently, the right-hand side of Equation 6.27 becomes $2Nm$. In other words, $(1 - F)/2F$ equals the number of migrant diploid organisms per generation. We therefore conclude from Equation 6.27 that selection between demes overrides selection within demes only when the benefit to the group $(b - c)$, relative to the cost to the individual organism (c), is greater than the average number of migrant organisms per generation. This principle defines a rather stringent limit above which migration among demes cancels any possible effects of interdeme selection.

SUMMARY

Natural selection can take place in many different ways. The simplest case is that in a haploid organism in which the relative fitnesses of the alternative genotypes are constant. Models of discrete generations and of continuous exponential growth are presented. In the discrete model, the relative fitnesses are called Darwinian fitnesses; in the continuous model, they are called Malthusian fitnesses. The relationship is that $\ln w = m$, where w and m are the Darwinian and Malthusian fitnesses, respectively.

In a diploid organism, continuous population growth is difficult to model when the genotypes differ in their rates of reproduction. The "standard" diploid model is that of discrete generations in which the genotypes may differ in the probability of survival from fertilization to adulthood (survivorship or viability selection) but are equal in fertility. In such a model with two alleles, A and a, and constant fitnesses of the diploid genotypes, four outcomes of selection are possible: A becomes fixed; a becomes fixed; there is a globally stable equilibrium; or there is an unstable equilibrium. Fixation of A or a results from directional selection in which either AA or aa is favored and the fitness of the heterozygous genotype is intermediate between the homozygous genotypes (or possibly equal to one of them). The stable equilibrium results from heterozygote superiority (overdominance), in which the fitness of the heterozygous genotype exceeds that of both homozygous genotypes. At the stable equilibrium of allele frequency, the average fitness in the population $\bar{w}$ is maximized. An unstable equilibrium arises when the fitness of the heterozygous genotype is smaller than that of both homozygous genotypes. The outcome of selection then depends on the initial conditions; fixation of either A or a takes place according to whether the initial frequency of A is greater than or less than the unstable equilibrium frequency.

Mutation-selection balance refers to the maintenance of a harmful allele in a population at a low equilibrium frequency because, in every generation, the elimination of preexisting harmful alleles by selection is offset by the introduction of new harmful alleles by mutation. For a completely recessive allele in which the relative fitness of the homozygous recessive genotype is $1 - s$, the equilibrium frequency of the harmful allele is given by $\hat{q} = \sqrt{\mu/s}$, where μ is the rate of mutation per generation of the wildtype allele to the harmful allele. For a partially dominant allele, the relative fitnesses of the heterozygous and homozygous genotypes carrying the harmful allele are $1 - hs$ and $1 - s$, where h is the degree of dominance. In this case, the equilibrium allele frequency is given approximately by $\hat{q} = \mu/hs$. An important implication of these formulas is that a small degree of dominance of a "recessive" allele has a disproportionate effect in decreasing the allele frequency at equilibrium. Another important implication is the Haldane-Muller principle, which states that, at mutation-selection balance, the total genetic load (measured as the product of the genotype frequency times the decrease in fitness of the genotype) is independent of the fitnesses and depends only on the rate of recurrent mutation.

In nature, selection must often be expected to have a more complex mechanism than that of differential survivorship envisaged in the standard model. Among the more complex types of selection are frequency-dependent selection, density-dependent selection, fecundity selection, selection in age-structured populations, selection when there are heterogeneous environments, diversifying selection favoring rare alleles or genotypes, differential selection in the sexes, selection for X-linked genes, gametic selection, meiotic drive (non-Mendelian segregation), multiple-alleles selection, multiple-loci selection, and sexual selection. Multiple loci are a particularly important source of complexity even when the fitness differences are entirely due to survivorship. In particular, with strong epistasis and tight linkage, there may be multiple stable interior equilibria and the equilibria may not coincide with points of maximum average fitness. With weak epistasis and loose linkage, however, the average fitness in the population usually does tend to increase.

Extended concepts of fitness can include the effects of selection acting on groups of relatives or on subpopulations. Kin selection invokes the concept of inclusive fitness, which embraces not only an organism's own fitness but also the fitness of its relatives (exclusive of direct descendants). Kin selection has been invoked to explain the evolution of many behavioral traits that appear to be detrimental to the individual organism but beneficial to its relatives. The most dramatic examples are found in social insects, in which certain organisms are reproductively sterile and devote their lives to the care and feeding of the queen and the protection of the colony. Generally speaking, alleles for altruistic behavior can increase in frequency if the loss in fitness of the altruist is offset by the increase in inclusive fitness to the beneficiaries of the altruism. More precisely, for additive alleles, the condition for increase in frequency of an allele predisposing to altruism is $c/b < r$, where c and b are the fitness cost to the altruist X and benefit to the relative Y, respectively, and $r = 2F_{XY}/(1 + F_X)$.

Interdeme selection plays an important role in the shifting balance theory of evolution. According to this theory, adaptive topographies are highly complex surfaces with many peaks and valleys. In small, partially isolated subpopulations, random genetic drift promotes the random exploration of the topography. When, by chance, a subpopulation comes under the control of a higher fitness peak, mass selection takes precedence and rapidly multiplies the favored gene combinations. Excess migration from the successful subpopulation shifts the allele frequencies in surrounding subpopulations and, through repetition of the selection process, the favored gene combinations progressively spread in waves throughout the entire population. Influential as metaphor, the shifting balance theory has not yet been adequately evaluated as an accurate description of the principal mechanism of evolutionary change.

PROBLEMS

1. Suppose that in the ith generation of a haploid population the fitnesses of A and a are $1 : s_i$. Show that $p_n/q_n = (p_0/q_0)(s_0 s_1 s_2 \cdots s_{n-1})$. If this is written as $p_n/q_n = (p_0/q_0)s^n$, then how can s be interpreted?

2. If the fitnesses of AA, Aa, aa are 1.0, 0.9, 0.6, and $p_0 = 0.7$, calculate p_1, p_2, and p_3, the allele frequencies after 1, 2, and 3 generations of selection.

3. Calculate the equilibrium allele frequency with overdominance when the fitnesses of AA, Aa, and aa are, respectively:
   a. 0.300, 1, 0.700.
   b. 0.930, 1, 0.970.
   c. 0.993, 1, 0.997.

4. Calculate $\bar{w}$ for $w_{11} = 0.9$, $w_{12} = 1$, $w_{22} = 0.6$, and $p = 0.8$, assuming random mating. Does any other p give a larger $\bar{w}$? Why or why not?

5. If a rare allele that is lethal when homozygous decreases in frequency by 1% each generation (i.e., $q' = 0.99q$), then what is the selection coefficient against heterozygotes? (*Hint:* Assume that qh is small compared to 1.)

6. If selection is not too intense, an additive gene giving fitnesses $1 + s$, $1 + s/2$, and 1 in AA, Aa, and aa will increase in frequency approximately according to $\ln(p_t/q_t) = \ln(p_0/q_0) + (s/2)t$. Calculate the approximate number of generations required to evolve significant insecticide resistance in an insect population when $s = \frac{1}{2}$ and $p_0 = 10^{-5}$. Significant resistance in the population may be taken as $p_t = 10^{-1}$. Show that, when $p_t/p_0 << 1$, $t = (2/s)\ln(p_t/p_0)$.

7. Show that a random mating diploid population with fitnesses 1, $1 - s$, and $(1 - s)^2$ for AA, Aa, and aa gives the same change in the allele frequency p of A as a haploid population with fitnesses 1 and $1 - s$ of A and a.

8. If selection is not too strong, the time required for the allele frequency of a favored dominant allele to change from p_0 to p_t is given by

$$\ln(p_t/q_t) + (1/q_t) = [\ln(p_0/q_0) + (1/q_0)] + st$$

Use this equation to derive the analogous equation for a favored recessive.

9. The following equation has equilibria at $\hat{p} = 0$, $\frac{1}{2}$, and 1. Classify the equilibria as to stability. If there is a stable equilibrium, is it locally or globally stable?

$$\Delta p = p(\frac{1}{2} - p)(1 - p)$$

10. Show that the allele frequency of a recessive lethal in generation n is given by $q_n = q_0/(1 + nq_0)$. (*Hint:* It is easiest to derive an expression first for $1/q_n$.) How many generations are required to reduce the allele frequency by half?

11. The mutation rate to a dominant gene for neurofibromatosis is approximately 9×10^{-5} and the reproductive fitness of affected individuals is estimated as $\frac{1}{2}$. What is the expected equilibrium frequency of affected individuals at birth?

12. What is the equilibrium frequency of a recessive gene arising with a mutation rate of 4×10^{-6} and a reproductive fitness in homozygotes of 0.8? What would it be if the gene were partially dominant with $h = 0.05$?

13. What is the equilibrium frequency of a recessive gene arising with a mutation rate of 10^{-6} with a fitness of 0.4 in homozygotes? How much would this be reduced if the homozygotes did not reproduce at all?

14. For a rare allele maintained at an equilibrium frequency of $q = \mu/h$, where h is the selection coefficient against heterozygotes, show that the proportion of heterozygous zygotes resulting from new mutations is approximately equal to h.

15. A polymorphism is said to be *protected* if all of the fixation states are unstable equilibria. Suppose the viabilities of males and females are as follows:

	AA	Aa	aa
Females	0.9	1	0.8
Male	1.0	v_{12}	0.5

What is the smallest value of v_{12} that ensures a protected polymorphism? (*Hint:* Some algebra shows that a condition for polymorphism is $w_{12}/w_{11} + v_{12}/v_{11} > 2$ and $w_{12}/w_{22} + v_{12}/v_{22} > 2$.)

16. If allele a is a recessive lethal in zygotes and the relative fitness of $A : a$ gametes is $1 - s : 1$, then what is the equilibrium allele frequency of a?

(*Hint:* The recursion simplifies greatly for the case of a recessive lethal, and equilibrium is given by $\hat{p} = v_1 w_{12} / [(v_1 + v_2) w_{12} - v_1 w_{11}]$.)

17. In a *Drosophila* population cage containing a meiotically driven chromosome known as segregation distorter, the equilibrium frequency of the driven chromosome was approximately 0.125 and the segregation ratio in heterozygotes was about $k = 0.75$ (Hiraizumi, Sandler, and Crow 1960). The meiotic drive chromosome is homozygous lethal in both sexes. The equilibrium between viability and meiotic drive in this case is $\hat{p} = 2(k-1)w_{12} / (1 - 2w_{12})$. Use this equation to estimate the approximate value of w_{12} consistent with these data.

18. In a multiple allele system in which each heterozygote is superior to the homozygotes for the alleles it contains, why are all alleles not maintained by selection?

19. The viabilities of genotypes A'A', A'A, and AA are 0.5, 1, and 0.7, respectively. If the initial frequency of allele A' is .05, what will the frequency be when the population comes to equilibrium? If a mutation occurs, introducing a novel allele A'', such that the fitnesses of A''A'', A''A', and A''A are all 0.8, determine whether this allele will increase in frequency.

20. Suppose alleles A_1, A_2, A_3, and A_4 are additive in their effects, and the homozygote fitnesses are $A_1 A_1 : 0.8$, $A_2 A_2 : 0.6$, $A_3 A_3 : 0.4$, and $A_4 A_4 : 0.2$. What are the heterozygote fitnesses? If all alleles are equally frequent, what is the mean fitness for this locus?

CHAPTER 7

Random Genetic Drift

RANDOM GENETIC DRIFT • BINOMIAL SAMPLING • WRIGHT-FISHER MODEL
COALESCENCE

OR EACH GENERATION there is an element of chance in the drawing of gametes that will unite to form the next generation. Chance alone can result in changes in allele frequency, and because the allele frequencies do not change in any predetermined way by this sampling process, the process is known as **random genetic drift**. In Chapter 5 we looked at some of the basic principles of how random genetic drift affects levels of variation in populations, but the subtlety and importance of drift are such that we will now devote this chapter to the subject.

RANDOM GENETIC DRIFT AND BINOMIAL SAMPLING

Consider a large population at Hardy-Weinberg equilibrium with alleles A and a at equal frequencies $p = q = \frac{1}{2}$. In this population, the genotype frequencies are $\frac{1}{4}$ AA, $\frac{1}{2}$ Aa and $\frac{1}{4}$ aa. Suppose four individuals are drawn at random from this population to start a colony. It is possible, by chance alone, that the sample will consist of 4 AA individuals. (This chance is $(\frac{1}{4})^4 = 1/256$.) Similarly, it is possible that all four will be aa. Any other possible sample could have been drawn, and it is not difficult to work out the probability for each type of sample. If the colony remains at just four individuals, this same kind of random sampling occurs each generation. At each generation, there is an opportunity for a large change in gene frequency caused purely by this process of sampling. One consequence of drift soon becomes clear—eventually the population will have either all A alleles or all

267

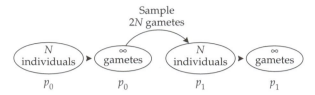

Figure 7.1   The gene frequencies and sampling that occur in the Wright-Fisher model. Initially there are N diploid adults with a gene whose frequency is p_0. The adults make an infinite number of gametes having the same allele frequency. From this pool, $2N$ gametes are drawn at random to constitute the N diploid individuals for the next generation.

a alleles. Once the population reaches such a "fixation" state, it is stuck. Only new mutations or migrants into the population can reintroduce variation.

In the example above we sampled four diploid individuals each generation. For our purposes, this is equivalent to drawing eight gametes at random from a pool of gametes. For example, if eight gametes are drawn from a population with $p = \frac{1}{2}$, there are nine possible outcomes, having 0, 1, 2, 3, . . . , 8 copies of the A allele and the remaining copies being the a allele. The probability of each of the nine possibilities is given by the **binomial distribution**, first introduced in Chapter 1. For the case of fixation, we need to find the probability of drawing eight copies of the A allele. Each draw is considered independent of the other draws, and each has a chance of $\frac{1}{2}$ of yielding an A. This means that the probability of drawing eight consecutive A alleles is $(\frac{1}{2})^8 = 1/256$. It is no coincidence that this is the same as the probability of drawing four AA genotypes as described above.

In sampling gametes from a finite population, the sampling process is depicted in Figure 7.1. In each generation there are N diploid individuals in the population. Regardless of the way fertilization occurs, we can imagine the sampling process to be one of sampling with replacement, such that the diploid individuals contribute to an essentially infinite gamete pool whose allele frequency is the same as the allele frequency in the adults. From this infinite gamete pool, $2N$ gametes are drawn and unite at random to form the next generation. Under this kind of sampling process, the distribution of frequencies of gametes is expected to be binomial.

PROBLEM 7.1   Suppose there are a thousand round pea seeds and a thousand wrinkled pea seeds in a soup pot. Enumerate all possible samples of four seeds drawn from the pot, and calculate the probability of each.

ANSWER The chance of drawing a round seed is $\frac{1}{2}$ (as is the chance of drawing a wrinkled seed). The chance of drawing four round seeds is roughly $(\frac{1}{2})^4 = \frac{1}{16}$, since the fraction of round seeds remains fairly close to $\frac{1}{2}$ even after a few are drawn. The chance of drawing all four wrinkled seeds is also $\frac{1}{16}$. There are four ways to get three round and one wrinkled seed: RRRW, RRWR, RWRR, and WRRR, and each of these has chance $\frac{1}{16}$. Similarly, there are four ways to get three wrinkled and one round seed: WWWR, WWRW, WRWW, and RWWW, and again, each of these four possibilities had probability $\frac{1}{16}$. Finally, we could draw two round and two wrinkled seeds, and such a sample could be drawn in any of six possible orders: RRWW, RWRW, RWWR, WRRW, WRWR, and WWRR. Each order has chance $\frac{1}{16}$, so the total chance of getting two round and two wrinkled seeds is $\frac{6}{16} = \frac{3}{8}$. We have exhaustively enumerated all 16 possible samples of four seeds, and since each of the 16 possibilities has chance $\frac{1}{16}$, the sum of the probabilities of the events is 1. This is a check that we have considered all possibilities. Note that the binomial distribution (Equation 7.1) makes these calculations much easier.

To take a specific example, a population of nine diploid organisms arises from a sample of just 18 gametes, but the gametes can be thought of as being sampled from an essentially infinite pool of gametes. Because small samples are frequently not representative, an allele frequency in the sample may differ from that in the entire pool of gametes. In fact, if the number of gametes in a sample is represented as $2N$ (in this example, $2N = 18$), the probability that the sample contains exactly i alleles of type A is the binomial probability

$$\Pr(i) = \binom{2N}{i} p^i q^{2N-i} \qquad 7.1$$

where $\binom{2N}{i}$ means $(2N)!/i!(2N-i)!$; p and q are, respectively, the allele frequencies of A and a in the entire pool of gametes ($p + q = 1$); and i takes on any integer value between 0 and $2N$. The new allele frequency in the population (call it p') is therefore $i/2N$ because, by definition, the allele frequency of A equals the number of A alleles (in this case i) divided by the total (in this case $2N$). In the next generation, the sampling process occurs anew, and the new probability of a prescribed number of A alleles occurring in the $2N$

gametes is given by the binomial probability above, with p now replaced by p' and q by $1 - p'$. Thus, the allele frequency may change at random from generation to generation. Computer-generated examples based on random numbers are shown in Figure 7.2. Each line in Figure 7.2A gives the number of A alleles in 20 successive generations of random genetic drift in a population of size $N = 9$ (so $2N = 18$). As you can see, individual populations behave very erratically. In seven populations, the A allele became **fixed** (that is, $p = 1$); in five populations, A became **lost** (that is, $p = 0$). The other eight populations remained **unfixed** (A was neither fixed nor lost), but the final allele frequency among the unfixed populations was as likely to be one value as another. Figure 7.2B shows the same kind of simulation, except now $2N = 100$. With a larger population size, the rate at which populations go to fixation is evidently slower. The principal conclusion from Figure 7.2 is that allele frequencies behave so erratically in any one population that prediction is virtually impossible.

Although changes in allele frequency due to random genetic drift in any individual population may defy prediction, the average behavior of allele frequencies in a large number of populations can be predicted. Consider a large number of populations all starting at the same time with the same allele frequency and same population size N. Each of these populations is assumed to undergo drift independently of the other populations. Except for their finite size, the subpopulations are assumed to satisfy all the assumptions of the Hardy-Weinberg model, with the additional stipulations that (1) the number of males and females is equal, and (2) each individual has an equal chance of contributing successful gametes to the next generation. The key point illustrated in Figure 7.3 is that we can describe how these populations change in allele frequency by considering time slices through the graph, and tallying a histogram of the counts of populations having each specified allele frequency. Initially, the populations will all be close to the starting allele frequency. As time passes, the populations "drift" apart, and eventually they are spread over all possible allele frequencies. Finally, as we will see, each population must go to fixation for one allele or the other.

The trick in understanding drift is to learn how to deduce the distributions of allele frequencies plotted in Figure 7.3. We just described what would happen after one generation—the set of populations would have a range of allele frequencies as described by the binomial distribution. The binomial distribution gives us the probability that a population has allele frequency p' after one generation of drift. If we consider 1000 populations all starting at p, the binomial distribution gives us the fraction of those populations with allele frequency p'. What about the following generation? For each population, one can imagine the whole sampling process as starting over again. The population does not remember where it was the previous generation, and so the binomial sampling occurs again. But this time, the allele frequency is p',

(A)

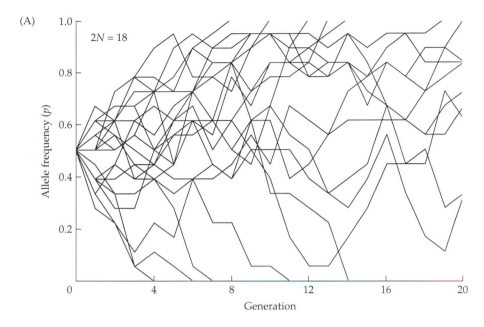

(B)

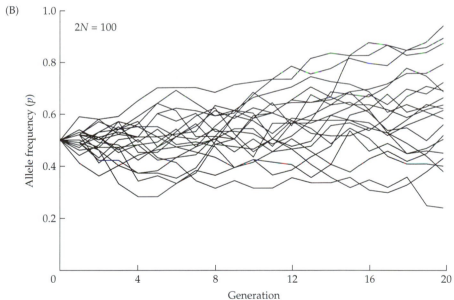

Figure 7.2   Computer simulations of the Wright-Fisher model of random genetic drift. Each line represents a population of size (A) $2N = 18$ or (B) $2N = 100$, simulated for 20 generations. Each generation alleles are sampled with replacement as described in the text. An allele frequency of $p = 0.5$ in A implies that there are nine copies of the A allele, and nine copies of the a allele. In B, an allele frequency of 0.5 implies 50 copies of each allele. Note that the larger population size in B results in smaller oscillations of allele frequency, and a slower rate of fixation.

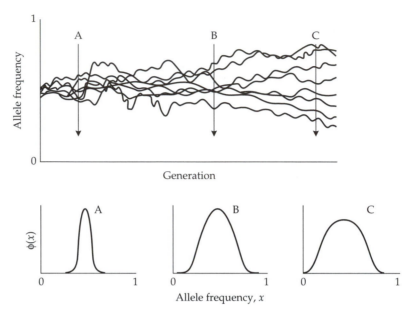

Figure 7.3   The model of random genetic drift can be seen by imagining a large collection of populations undergoing the process of repeated sampling. As the top part of the figure indicates, the populations' allele frequencies change erratically, and tend to drift apart. At time intervals, a snapshot of the populations would produce distributions of allele frequencies whose variance increases over time.

and this value must be the frequency used in Equation 7.1. Each of the 1000 populations may have a different p' after one generation of drift, so to get the second generation, we need to calculate the binomial distribution 1000 times and add the values up. Fortunately, R.A. Fisher and Sewall Wright figured out an easier way to do this, which is described in the next section.

An experiment designed along the same lines as the one given in Figure 7.3 is shown in Figure 7.4. In this study, the history of 19 generations of random genetic drift in 107 subpopulations of *Drosophila melanogaster* was followed. Each population was initiated with 16 bw^{75}/bw (bw = brown eyes) heterozygotes and maintained at a constant size of 16 individuals by randomly choosing eight males and eight females to produce the next generation. Each histogram in Figure 7.4 gives the number of populations containing 0, 1, 2, . . . , 32 bw^{75} alleles. The pattern of change in allele frequency in Figure 7.4 may at first appear to be complicated, but in reality a simple thing is happening. The initially humped distribution of allele frequency gradually becomes flat as populations fixed for bw^{75} or bw begin to pile up at the boundaries. The piling up occurs because, once an allele has been fixed or

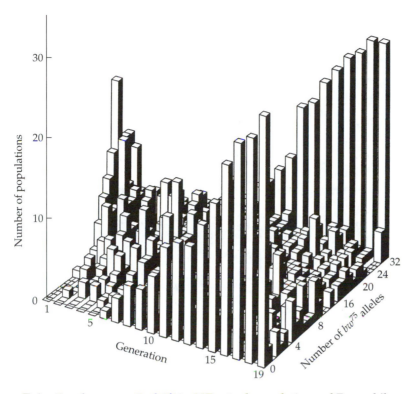

Figure 7.4   Random genetic drift in 107 actual populations of *Drosophila melanogaster*. Each of the initial 107 populations consisted of 16 bw^{75}/bw heterozygotes ($N = 16$; bw = brown eyes). From among the progeny in each generation, eight males and eight females were chosen at random to be the parents of the next generation. The horizontal axis of each curve gives the number of bw^{75} alleles in the population, and the vertical axis gives the corresponding number of populations. (Data from Buri 1956.)

lost, it remains fixed or lost since mutation is negligible over such a small number of generations in small populations. After 19 generations, most of the populations are fixed for one allele or the other, and among the unfixed populations, the distribution of allele frequencies is essentially flat.

PROBLEM 7.2   Consider a self-pollinating plant population consisting of a single heterozygous (*Aa*) individual on a small barren island. Suppose the plant reproduces and dies, so that the generations are discrete, and the population can only consist of a single plant. What is the probability that the population is homozygous at this genetic locus by the second generation?

ANSWER   The chance that the first generation offspring is AA is $\frac{1}{4}$ and the chance that it is aa is also $\frac{1}{4}$, so the chance of fixation in one generation is $\frac{1}{2}$. If the first generation offspring is Aa, then the probability of fixation in the second generation (given that the population is not fixed in the first generation) is again $\frac{1}{2}$. The probability of not fixing in generation 1 and then fixing in generation 2 is $\frac{1}{2} \times \frac{1}{2} = \frac{1}{4}$. Add to this the chance of fixing in one generation and we get $\frac{3}{4}$ as the probability of fixation by two generations. Note that the probability of *not* going to fixation each generation is $\frac{1}{2}$, and so the chance of not fixing for two generations is $\frac{1}{2} \times \frac{1}{2} = \frac{1}{4}$.

Consider an infinitely long bowling alley with minor imperfections that displace the ball one way and the other. The gutters represent the fixation states of $p = 0$ and $p = 1$. Once the ball goes in the gutter, it cannot get out again. The imperfections keep the ball from rolling in a straight line, and eventually it rolls into the gutter. In this analogy, the size of the population corresponds to the width of the bowling alley; a larger population implies a wider alley. The imperfections still deflect the ball but, in proportion to the width of the alley, the ball's zigs and zags are of a smaller magnitude. Consequently, the ball remains out of the gutter for a longer time, analogous to the longer time to fixation for a larger population. But just as certainly, the ball will eventually land in the gutter.

THE WRIGHT-FISHER MODEL OF RANDOM GENETIC DRIFT

Fisher (1930) and Wright (1931) both considered the consequences of the sort of binomial sampling that occurs in small populations when the sampling occurs repeatedly over many generations. This model, known as the Wright-Fisher model, derives the distribution of allele frequencies among populations undergoing random genetic drift. Although neither Fisher nor Wright formulated the problem in terms of matrices, as used here, this approach makes the problem much simpler and gives the same results. If a population has $2N$ genes, and there are two alleles (A and a) that may be segregating, then the **state** of the population can be described by the number of A alleles in the population. The possible states are then $0, 1, 2, \ldots, 2N$. The states 0 and

$2N$ are special in that these are fixations, and once the population gets into these states, it cannot leave. The states 0 and $2N$ are called **absorbing states**. From any other allele frequency, it is possible for the population to drift to a different allele frequency. However, to use an example from Figure 7.4, if $2N = 32$, then the chance of drifting in one generation from 30 copies of gene A to 29 copies of gene A is greater than the chance of drifting to two copies. The probability of the population drifting from the state having i copies to j copies of allele A is known as the **transition probability**. The transition probability for the Wright-Fisher model is obtained directly from the binomial distribution. If a population has i copies of allele A, then the allele frequency is $p = i/2N$, and the frequency of allele a is $q = 1 - i/2N$. The probability of going from i copies of A to j copies of A in one generation is:

$$T_{ij} = \binom{2N}{j} p^j q^{2N-j} \qquad\qquad 7.2$$

The transition probabilities can be put in a square matrix $\mathbf{T}$, with elements T_{ij} giving the transition probability from state i to state j for $i, j = 0,1,2, \ldots , 2N$. The matrix $\mathbf{T}$ contains everything that is needed to predict the expected distribution of populations like those in Figure 7.4 over a series of generations. This type of model, expressed in terms of discrete states with fixed probabilities of going from one state to another, is known as a **Markov chain**, and it has some very elegant mathematical properties. Iterations of the Wright-Fisher model give the expected outcome of a pure drift process (Figure 7.5). We will use the Wright-Fisher model only to show one aspect about fixation probabilities.

PROBLEM 7.3   Consider a population of four diploid individuals. Calculate the probability that a population with four copies of allele A (allele frequency $p = \frac{1}{2}$) drifts in one generation to having three copies. What is the probability that the population has four copies of A? Five copies? Now consider a population of the same size, but initially with two copies of A. What is its probability of drifting to one, two, or three copies?

ANSWER   Applying Equation 7.2, we get $T_{4,3} = [8!/(5!3!)](\frac{1}{2})^8 = 7/32 = 0.219$. $T_{4,4} = [8!/(4!4!)] (\frac{1}{2})^8 = 70/256 = 0.273$. $T_{4,5} = T_{4,3} = 0.219$. (Note that the binomial distribution is symmetric when $p = \frac{1}{2}$, so there is equal probability for samples that are symmetrically divergent from

$p = \frac{1}{2}$.) In the case when the initial frequency is $\frac{2}{8}$, we get $T_{2,1} = [8!/1!7!)] \, (\frac{1}{4})(\frac{3}{4})^7 = 0.267$, $T_{2,2} = [8!/(2!6!)] \, (\frac{1}{4})^2(\frac{3}{4})^6 = 0.311$, and $T_{2,3} = [8!/(3!5!)] \, (\frac{1}{4})^3(\frac{3}{4})^5 = 0.208$.

The above problem illustrates an important feature of the Wright-Fisher model. The magnitude of change in allele frequency is greater when the allele frequency is $\frac{1}{2}$ than it is when the allele frequency is more skewed. The changes are greater because the variance in the binomial sampling distribution is greatest when $p = \frac{1}{2}$. (The formula for the binomial variance is $pq/2N$.) The variance drops to zero at $p = 0$ and $p = 1$. The variance formula makes it clear that a large population will change allele frequency more slowly than a smaller population because the sampling variance varies as the reciprocal of population size. Furthermore, the probability of an increase in allele frequency is the same as the probability of a decrease in allele frequency, regardless of the allele frequency. The process of drift does not recognize when a population is close to a fixation. The chance of drifting up in frequency is always equal to the chance of drifting down in frequency, regardless of the current population allele frequency.

Fisher and Wright also addressed the expected time to fixation. Since another approach yields the solution to this problem much more easily, we will consider times to fixation in the next section.

PROBLEM 7.4 Simulating random drift can be a very time-consuming proposition. If one wants to simulate a population of 1000 individuals for 1000 generations, one has to draw 10^6 random numbers and for each decide whether to accept or reject each genotype. Kimura (1980b) came up with a shortcut that relates very closely to how the diffusion approximation works (see the next section). The trick is to use the recursion: $p' = p + (2U - 1)\sqrt{(3pq/2N)}$, where U is a random number uniformly distributed between 0 and 1. Each generation, one picks one random number U, and a sample realization of the next generation's allele frequency is gotten from the above recursion. Why does this approach work? (*Hint:* The variance in a uniform distribution is the square of the range divided by 12.)

ANSWER   The expression $2U - 1$, where U is a number between 0 and 1, gives a value from -1 to $+1$, or a range of 2. The range of $(2U - 1)\sqrt{(3pq/2N)}$ is therefore $2\sqrt{(3pq/2N)}$. Squaring this expression and dividing by 12, the variance of this uniform random variable is thus $pq/2N$, just what we get from a binomial sampling distribution. Each generation the allele frequency has an equal chance of increasing or decreasing, and the variance in the allele frequency change is $pq/2N$. Even though the distribution of change in allele frequency is uniform in the pseudosampling simulation instead of binomial (as it is in the Wright-Fisher model), this process can reproduce most of the results of the complete brute-force simulation at a tiny fraction of the computer time.

THE DIFFUSION APPROXIMATION

The pattern of change in allele frequency shown in Figure 7.4 is very nearly that expected theoretically for an ideal population, and although the full-blown theory of random genetic drift requires mathematics beyond the scope of this book, some background might be of interest (see Kimura 1955, 1964, 1976; Wright 1969; Crow and Kimura 1970; Kimura and Ohta 1971). The representation of random drift by a differential equation was first applied by Fisher (1922), who noted that the equation describing the diffusion of heat through a solid bar applies to random genetic drift. The distribution of populations with allele frequencies ranging from 0 to 1 is called $\phi(x,t)$, where x represents the allele frequency and t indicates time. Figure 7.5 shows a particular realization of $\phi(x,t)$ changing through time. The theoretical problem is to formulate an equation that describes how $\phi(x,t)$ changes under random genetic drift, and to solve the equation.

The parallel between the physical process of diffusion and what is actually going on in a finite population is a bit abstract, but it is not particularly difficult. Consider an axis of allele frequency extending from 0 to 1. The number of populations whose frequency is between x and $x + \partial x$ at time t is our probability density $\phi(x,t)$. Populations may enter this range of allele frequencies by drifting in from a lower frequency, which occurs with a probability flux $J(x,t)$. Populations may leave this range of allele frequencies by drifting out, which occurs with probability flux $J(x + \partial x,t)$. The rate of change in $\phi(x,t)$ is the difference in these fluxes, which we can write

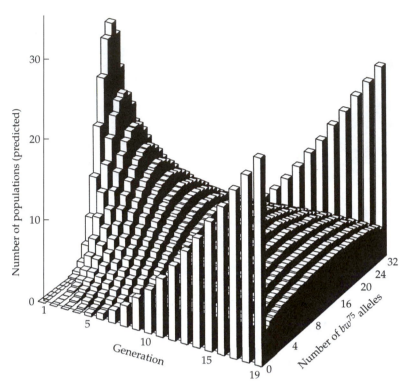

Figure 7.5   Prediction of the Wright-Fisher model for the distribution $\phi(x,t)$ of populations of size $N = 16$ with allele frequency x at generation t, for 20 generations after an initial frequency of 0.5. The values of $\phi(x,t)$ were generated using the Markov transition probability matrix, whose terms are given by the binomial distribution. The model with $2N = 32$ predicts that fewer populations have fixed by generation 19 than actually did go to fixation in the experiment in Figure 7.4. This is because the effective population size is smaller than the observed count (see Figure 7.12).

$$\frac{\partial}{\partial t}\phi(x,t) = -\frac{\partial}{\partial x}J(x,t)$$ 7.3

because $J(x,t) - J(x + \partial x, t) \approx -\dfrac{\partial}{\partial x}J(x,t)$ when ∂x is very small.

The probability flux is

$$J(x,t) = M(x)\phi(x,t) - \frac{1}{2}\frac{\partial}{\partial x}V(x)\phi(x,t)$$ 7.4

where $M(x)$ is the average change in allele frequency in a population whose current allele frequency is x, and $V(x)$ is the variance in change in allele frequency.

$M(x)$ is zero unless there is some force, like mutation or selection, driving the allele frequency to change in a particular direction. (Remember that with pure drift, allele frequency increases and decreases with equal chance.) $V(x)$ tells how fast allele frequencies change for a population with frequency x. Under the Wright-Fisher model, $V(x) = x(1-x)/2N$, which means that the binomial sampling variance describes the magnitude of allele frequency change. But the probability flux depends on the *difference* in rates of change from x to $x + \partial x$, and so, just as in classical physical models of diffusion, the flux depends on the gradient in whatever is diffusing. In the case of chemical diffusion, this gradient would be the gradient in concentrations, and a greater difference in concentrations would yield higher flux. In the case of our population genetic model, the gradient is the change in sampling variance as x is varied, or $\partial V(x)/\partial x$.

Substituting Equation 7.4 into Equation 7.3, we get

$$\begin{aligned}\frac{\partial}{\partial t}\phi(x,t) &= -\frac{\partial}{\partial x}\left[M(x)\phi(x,t) - \frac{1}{2}\frac{\partial}{\partial x}V(x)\phi(x,t)\right]\\ &= -\frac{\partial}{\partial x}\left[M(x)\phi(x,t)\right] + \frac{1}{2}\frac{\partial^2}{\partial x^2}\left[V(x)\phi(x,t)\right]\end{aligned}\tag{7.5}$$

Equation 7.5 is known as the diffusion equation, the forward Kolmogorov equation, or, in the context of the physics of heat diffusion, the Fokker-Planck equation. For the Wright-Fisher model, $M(x) = 0$ and $V(x) = x(1-x)/2N$ so we get

$$\frac{\partial}{\partial t}\phi(x,t) = \frac{1}{4N}\frac{\partial^2}{\partial x^2}\left[x(1-x)\phi(x,t)\right]\tag{7.6}$$

Many aspects of this problem were explored by Wright (1931), and the formal solution to this equation, found by Kimura (1955), required some heavy mathematics. For our purposes, some graphs will illustrate the important properties of the diffusion equation. The two families of curves in Figure 7.6 are the theoretical distributions $\phi(x,t)$ of allele frequency among unfixed populations after various times (t) measured in units of N generations. In Figure 7.6A, all populations have an initial allele frequency of $\frac{1}{2}$, as in the actual populations in Figure 7.4; after about $t = 2N$ generations, the distribution of allele frequency is essentially flat, and by this time about half the populations are still unfixed. The distributions in Figure 7.6 refer only to those populations that are unfixed; as time goes on, more and more of the populations become fixed, and the distributions progressively pile up at 0 and 1, as in the histograms in Figure 7.4. Indeed, in Figure 7.6, the area under each curve is equal to the proportion of unfixed populations, which becomes progressively smaller. In particular, the rate at which the height of the distribution decreases once it becomes flat is about $1/2N$ per generation.

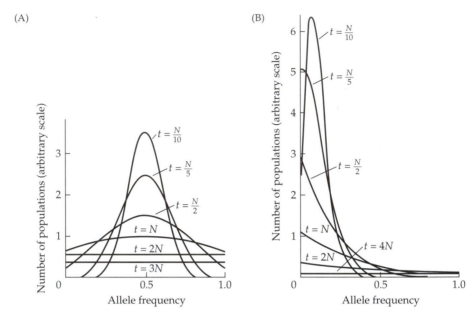

Figure 7.6   Theoretical results of random genetic drift. (A) Initial allele frequency = $\frac{1}{2}$. (B) Initial allele frequency = 0.1. The curves have been scaled so that the area under each curve is equal to the proportion of populations in which fixation or loss has not yet occurred. The curves are therefore the distributions of allele frequencies among segregating populations. (From Kimura 1955.)

Figure 7.6B shows what happens when the initial allele frequency is 0.1; here the distributions are highly asymmetrical, and the distribution of allele frequency does not become flat until about $t = 4N$ generations, by which time only about 10% of the populations remain unfixed. Once a flat distribution of allele frequency is reached, the distribution remains flat, but random drift continues until fixation or loss has occurred in all populations.

PROBLEM 7.5   Demonstrate from Equation 7.6 that the fixation states, $x = 0$ and $x = 1$, are equilibria of the diffusion process.

ANSWER   A condition for equilibrium is that $\phi(x,t)$ remains stationary, so that $\frac{\partial}{\partial t}\phi(x,t) = 0$. Substituting into Equation 7.6, we get

$0 = \dfrac{1}{4N} \dfrac{\partial^2}{\partial x^2} \big[x(1-x)\phi(x,t) \big]$. If $x = 0$ or $x = 1$, this equality is clearly

satisfied, because $\dfrac{\partial^2}{\partial x^2}[0] = 0$. It requires a bit more work to show that

these are the only equilibria.

To illustrate that the diffusion approximation and the Wright-Fisher model give very similar results, Figure 7.7 shows the diffusion approximation for the data in Figure 7.4, with $2N = 32$, $x_0 = \frac{1}{2}$, and t running from generation 1 through generation 19.

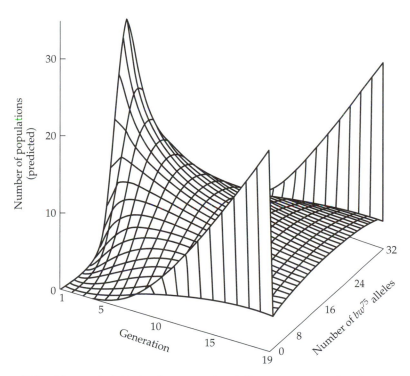

Figure 7.7 Kimura's (1955) solution to the diffusion equation for the particular case of $N = 16$. This is the three-dimensional view of Figure 7.6, and represents the diffusion approximation to the exact solution obtained by the Wright-Fisher model in Figure 7.5.

Absorption Time and Time to Fixation

One useful application of the diffusion approximation has been to determine expressions for the expected time for a neutral allele to go to fixation. Assuming that the allele starts at frequency p, Kimura and Ohta (1969) showed that the mean time (in generations) until the allele is fixed (ignoring cases where the allele is lost) is

$$\bar{t}_1(p) = -\frac{4N}{p}\left[(1-p)\log(1-p)\right] \qquad 7.7$$

Similarly, they showed that the mean time to loss of the allele is

$$\bar{t}_0(p) = -\frac{4N}{1-p}\left[p\log(p)\right] \qquad 7.8$$

Combining Equations 7.7 and 7.8, the mean persistence time of an allele is

$$\bar{t}(p) = -4N\left[p\log(p) + (1-p)\log(1-p)\right] \qquad 7.9$$

where $\bar{t}(p)$ is the average time that an allele remains segregating in a population (that is, until its frequency is either 0 or 1).

The average times for fixation, loss, and persistence of a neutral allele are shown graphically in Figure 7.8. An allele is expected to remain in a population for the longest time when its initial frequency is $\frac{1}{2}$. When

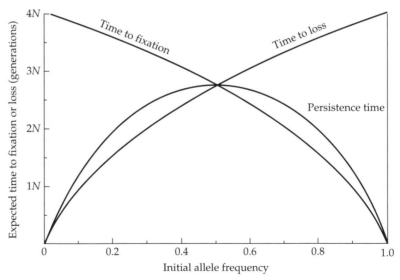

Figure 7.8   Average persistence of a neutral allele in an ideal diploid population of size N, plotted against initial allele frequency.

$p_0 = \frac{1}{2}$, the average time that a population remains unfixed is about $2.77N$ generations.

PARALLELISM BETWEEN RANDOM DRIFT AND INBREEDING

Consider a set of four subpopulations each started with allele frequency $p = \frac{1}{2}$, and each undergoing random drift independently following binomial sampling (Figure 7.9). Within any particular subpopulation (call it subpopulation

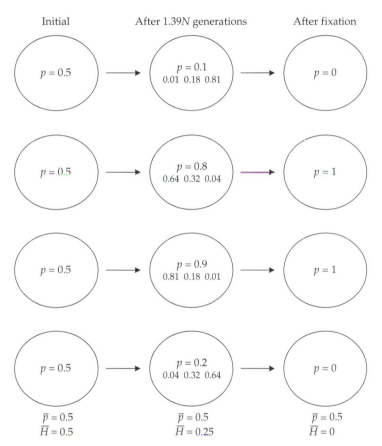

Initial   After 1.39N generations   After fixation

$p = 0.5$ → $p = 0.1$ / 0.01 0.18 0.81 → $p = 0$

$p = 0.5$ → $p = 0.8$ / 0.64 0.32 0.04 → $p = 1$

$p = 0.5$ → $p = 0.9$ / 0.81 0.18 0.01 → $p = 1$

$p = 0.5$ → $p = 0.2$ / 0.04 0.32 0.64 → $p = 0$

$\bar{p} = 0.5$ $\bar{p} = 0.5$ $\bar{p} = 0.5$
$\bar{H} = 0.5$ $\bar{H} = 0.25$ $\bar{H} = 0$

Figure 7.9 A schematic diagram showing a set of four populations undergoing the process of drift. Initially the allele frequency is $\frac{1}{2}$ in all four populations, and the average heterozygosity is $\frac{1}{2}$. As the populations drift in allele frequency, the average is expected to remain the same (indicated by $\bar{p}$ remaining $\frac{1}{2}$) but the average heterozygosity decreases. (Genotype frequencies are given for the intermediate generation.) Finally, all populations go to fixation, half fix one allele and half fix the other, so the average allele frequency is still $\frac{1}{2}$, but the heterozygosity is zero.

number i), mating is random because all the assumptions in Table 7.1 hold true. If the allele frequencies of A and a in the ith subpopulation are denoted p_i and q_i, then the genotype frequencies of AA, Aa, and aa are given by the familiar Hardy-Weinberg principle as p_i^2, $2p_iq_i$ and q_i^2. Furthermore, picture the situation in Figure 7.9 at a time so advanced that all subpopulations are fixed for one allele or the other. Within the ith subpopulation, therefore, either p_i equals 0 or p_i equals 1. The genotype frequencies of AA, Aa, and aa in that subpopulation are either 0, 0, and 1 (if $p_i = 0$), or 1, 0, and 0 (if $p_i = 1$). These genotype frequencies, though extreme, still satisfy the Hardy-Weinberg principle. Thus, within any one subpopulation in Figure 7.9, the frequency of heterozygotes is that expected with random mating.

The situation regarding the total population in Figure 7.9 is very different, however, as there is an overall deficiency of heterozygotes. Suppose that we sample the four subpopulations, but that we are unaware of the existence of the four subpopulations, and instead we think that the sample contains a single randomly mating population. Considering the four populations at the right side of Figure 7.9 (after all variation is lost), if we were to calculate the allele frequency we would obtain $p = \frac{1}{2}$. We would then naively expect a fraction $2pq = \frac{1}{2}$ of the genotypes to be heterozygous. In fact, we would have no heterozygotes at all in our sample! This rather paradoxical result—that there is a deficiency of heterozygotes in the total population even though random mating occurs within each subpopulation—is a consequence of the random genetic drift of allele frequencies among subpopulations due to their finite size. This extreme case when each subpopulation is fixed is easy to understand: a population with allele frequency $\frac{1}{2}$ could only be made up of two subpopulations each fixed for A, and two subpopulations each fixed for a. The entire population has no heterozygotes whatsoever, but the average allele frequency is $\frac{1}{2}$. The total population has a deficiency of heterozygotes, much as if there were inbreeding. This inbreeding-like effect of population subdivision is known as the Wahlund principle (Chapter 4), and we are now

---

TABLE 7.1 ASSUMPTIONS OF MODEL OF RANDOM GENETIC DRIFT

---

(1) Diploid organism
(2) Sexual reproduction
(3) Nonoverlapping generations
(4) Many independent subpopulations, each of constant size N
(5) Random mating within each subpopulation
(6) No migration between subpopulations
(7) No mutation
(8) No selection

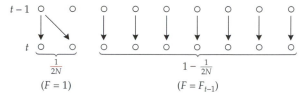

Figure 7.10   Diagram illustrating the reasoning behind the recursion for F in a finite population. When the gametes are drawn to make up the population at generation t, there is a chance $1/2N$ that any pair of alleles will be drawn in generation $t-1$. If this happens, the probability of identity is 1. For the allele pairs drawn in generation t from two distinct alleles at generation $t-1$ (the probability of this is $1-1/2N$), the probability of identity is F_{t-1}. Adding the probabilities of these two events, we get $F_t = 1/2N + (1-1/2N)F_{t-1}$.

in a position to quantify the manner in which subpopulations diverge in allele frequency under random genetic drift.

In Chapter 4 we measured the extent of inbreeding with the inbreeding coefficient, F. F is the probability of **autozygosity**, or the probability that an individual carries a pair of alleles that are identical by descent (derived from a common ancestor). Even though random mating occurs within each subpopulation in Figure 7.9, because gametes do combine at random, any two alleles in a subpopulation may be identical by descent due to the limited population size. Thus F_t does not equal zero. The value of F_t can be calculated as in Figure 7.10. This figure shows the $2N$ alleles in a breeding population of generation $t-1$. In sampling alleles for generation t, the first chosen allele may be any of those present in generation $t-1$ with equal chance. The probability that the second chosen allele is of the same type as the first is $1/2N$, because this is the frequency of each allelic type in the gametic pool; the probability that the second chosen allele is of a different type from the first is accordingly $1-1/2N$. In the first case, the probability of identity-by-descent is 1; in the second case it is F_{t-1}. Altogether the recursion is

$$F_t = \frac{1}{2N} + \left(1 - \frac{1}{2N}\right)F_{t-1} \qquad\qquad 7.10$$

Multiplying both sides by –1 and adding 1 leads to

$$1 - F_t = 1 - \frac{1}{2N} - \left(1 - \frac{1}{2N}\right)F_{t-1} = \left(1 - \frac{1}{2N}\right)(1 - F_{t-1})$$

and so

$$1 - F_t = \left(1 - \frac{1}{2N}\right)^t (1 - F_0) \qquad\qquad 7.11$$

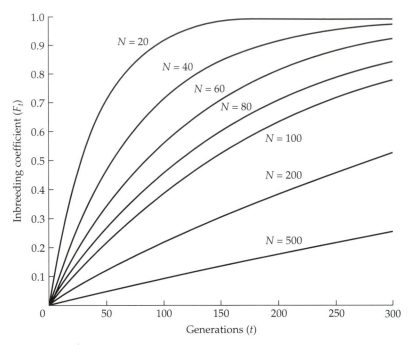

Figure 7.11 Increase of F_t in ideal populations as a function of time and effective population size N.

or, when $F_0 = 0$,

$$F_t = 1 - \left(1 - \frac{1}{2N}\right)^t \qquad \text{7.12}$$

Figure 7.11 shows the rapid increase of F_t in small populations. Another aspect of the same phenomenon can be appreciated by the probability of drawing a pair of alleles that are not identical by descent. This probability is the same as the heterozygosity, and it can be written

$$H_t = 1 - F_t \qquad \text{7.13}$$

By substitution for F_t we obtain the rate of change in heterozygosity from random genetic drift

$$H_t = \left(1 - \frac{1}{2N}\right)H_{t-1} \qquad \text{7.14}$$

and so

$$H_t = \left(1 - \frac{1}{2N}\right)^t H_0 \approx H_0 e^{-t/2N} \qquad 7.15$$

Recall again that a single population undergoing random drift remains in approximate Hardy-Weinberg proportions, and that the symbol H_t represents a sort of "virtual heterozygosity" averaged across many subpopulations. The above equations show that pure random drift should result in the heterozygosity decreasing at a geometric rate, since H_t is multiplied by the constant $(1 - 1/2N)$ each generation. Experimental tests of this prediction are shown in Figure 7.12. Figure 7.12A shows how the heterozygosity averaged across the populations in Figure 7.4 declines over generations, but the theoretical curve when $N = 16$ does not fit the data very well. In fact, the rate of decline of heterozygosity is greater than the theoretical expectation, as though the population size were smaller than $N = 16$. On the other hand, the allele frequency, averaged across populations, is not expected to change, and the data agree with this aspect of the theory quite well (Figure 7.12B).

PROBLEM 7.6   Use Equation 7.15 to determine how long it takes for a finite population to halve in heterozygosity.

ANSWER   Set $\frac{1}{2}H_0 = H_0 \, e^{-(t/2N)}$. Dividing out the H_0, and taking logarithms, we obtain

$$\ln\left(\frac{1}{2}\right) = \frac{-t}{2N}$$

so $t = 1.39N$. In other words, it takes $1.39N$ generations to halve the heterozygosity, regardless of its initial value. Fisher expressed this result by saying that it takes $1.39N$ generations to halve the *genic variance* in the population. Since the variance of a binomial sample is $pq/2N$, and the variance in allele frequency among subpopulations is proportional to the heterozygosity, it follows that both the variance and the heterozygosity decrease at the same rate.

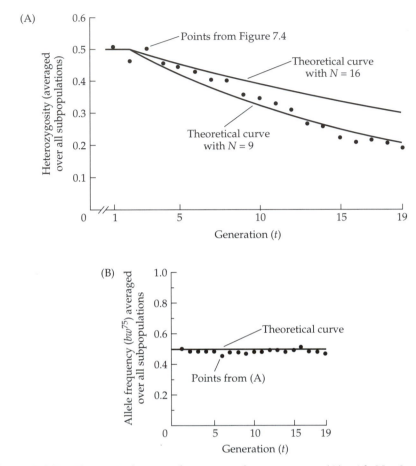

Figure 7.12 Theoretical curves for average heterozygotes (A) with $N = 9$ or $N = 16$, along with actual values (plotted as points) from the experiment in Figure 7.4. In (B) the observed and expected allele frequencies (averaged across the 107 subpopulations) are plotted. (Data from Buri 1956.)

Several important consequences of the population structure in Figure 7.9 can now be summarized. First, although each subpopulation is finite in size, we can imagine so many of them that the size of the total population is effectively infinite. For an infinite population that obeys the assumptions in Table 7.1, the allele frequencies must remain constant. That is, even though the allele frequency in any individual subpopulation may change willy-nilly due to random genetic drift, the overall average allele frequency of A among subpopulations remains p_0, where p_0 represents the allele frequency of A in the base population. Figure 7.12B gives an experimental demonstration of the constancy of average allele frequency. Since F_t is the probability of autozygosity of a gene in an individual in generation t, the probability of allozygosity (obtaining a pair of alleles that are not identical by descent) is $1 - F_t$. Because p_0 is the overall allele frequency of A, the probability that a randomly chosen individual will be genotypically AA is $p_0^2(1 - F_t)$ [for the case of allozygosity] $+ p_0F_t$ [for the case of autozygosity]. Similarly, the probability that the individual will be Aa equals $2p_0q_0(1 - F_t)$; and the probability that the individual will be aa equals $q_0^2(1 - F_t) + q_0F_t$. Note that the genotypic frequencies in the total population are different from the standard Hardy-Weinberg proportions, because there is an apparent excess of homozygotes. However, within any one subpopulation, the genotypic frequencies still obey the Hardy-Weinberg principle because of random mating. Substituting for F_t in Equation 7.12 implies that the average heterozygosity among subpopulations at time t equals $2p_0q_0(1 - F_t) = 2p_0q_0(1 - 1/2N)^t$; this is the theoretical curve plotted in Figure 7.12A (with $p_0 = q_0 = 1/2$).

Since F_t eventually goes to 1, all subpopulations eventually become fixed for one allele or the other. Because the average allele frequency of A remains p_0 even when all subpopulations have become fixed, the proportion of subpopulations that eventually become fixed for A must be p_0 (and the proportion that eventually become fixed for a must be q_0). Stated another way, the probability of ultimate fixation of an allele in any ideal subpopulation is equal to the frequency of that allele in the initial population. This point is illustrated by the actual example in Figure 7.4, where $p_0 = 1/2$; by generation 19, a total of 58 populations have become fixed, 30 for the bw allele and 28 for bw^{75}.

EFFECTIVE POPULATION SIZE

As we saw in the *Drosophila* experiments in Figure 7.12, populations generally fluctuate in allele frequency by an amount greater than $pq/2N$. The reason is that no real population obeys all the assumptions in Table 7.1 exactly. In any actual case, there must be corrections for such complications as fluctuations in population size, unequal numbers of males and females, age structure, and skewed distributions in family size (see Crow and Kimura 1970). The degree to which genetic drift can change allele frequencies, and the rates of allele fixation by drift, can be approximated under these complicating circumstances by calculating the **effective size** of the population and using this value in the theory for an ideal population. That is, the effective population size of an actual population is the number of individuals in a theoretically ideal population having the same magnitude of random genetic drift as the actual population. There are three kinds of effective population size based on how we choose to measure "magnitude," namely: (1) the change in average inbreeding coefficient, (2) the change in variance in allele frequency, or (3) the rate of loss of heterozygosity. These are called the *inbreeding effective size*, the *variance effective size*, and the *eigenvalue effective size*, respectively.

Wright (1931) first worked out the effective population size by considering the effective degree of inbreeding in various situations. As noted, the effective population size can also be calculated by determining the rate of change in variance in a population, and Kimura and Crow (1963) first applied this approach to the problem of overlapping generations. Usually, the inbreeding effective size and the variance effective size are the same, but exceptions do occur. Similarly, the variance effective size and the eigenvalue effective size can be distinct (Ewens 1982). Some of the various factors that require calculation of an effective population size will now be illustrated. We will focus on the inbreeding effective size because this concept is the most widely used.

Fluctuation in Population Size

Correction for fluctuating population size is important because natural populations actually do change in size, sometimes by a factor of 10 or more in a single generation. For the sake of simplicity, assume that the population is ideal in all respects except that its size is not constant. We will consider the situation over just two generations. Suppose that the population sizes in two successive generations are N_0 and N_1. The arguments laid out in Figure 7.10 imply that

$$1 - F_2 = \left(1 - \frac{1}{2N_1}\right)(1 - F_1) \tag{7.16}$$

and

$$1 - F_1 = \left(1 - \frac{1}{2N_0}\right)(1 - F_0) \tag{7.17}$$

Substituting from the second equation into the first leads to

$$1 - F_2 = \left(1 - \frac{1}{2N_1}\right)\left(1 - \frac{1}{2N_0}\right)(1 - F_0) \tag{7.18}$$

By analogy with the constant N case, it is appropriate to try to express this equation in the general form

$$1 - F_t = \left(1 - \frac{1}{2N}\right)^t (1 - F_0) \tag{7.19}$$

where N is now the *effective* population size. In our example $t = 2$, so

$$1 - F_2 = \left(1 - \frac{1}{2N}\right)^2 (1 - F_0) \tag{7.20}$$

Setting the two expressions for $1 - F_2$ equal to each other we obtain

$$\left(1 - \frac{1}{2N}\right)^2 = \left(1 - \frac{1}{2N_0}\right)\left(1 - \frac{1}{2N_1}\right) \qquad \text{7.21}$$

from which $1/N = \frac{1}{2}(1/N_0 + 1/N_1)$ turns out to be an excellent approximation. In general,

$$\frac{1}{N_e} = \frac{1}{t}\left(\frac{1}{N_0} + \frac{1}{N_1} + \cdots + \frac{1}{N_{t-1}}\right) \qquad \text{7.22}$$

and so the effective size N_e is the **harmonic mean** of the actual numbers—the reciprocal of the average of reciprocals. As illustrated in the problem below, the harmonic mean tends to be dominated by the smallest terms. In biological reality, this means that a single period of small population size, called a **bottleneck**, can result in a serious loss in heterozygosity. Population bottlenecks are thought to account for the very low levels of polymorphism found in extant populations of the elephant seal (Bonnell and Selander 1974) and the cheetah (O'Brien et al. 1985, 1987). A severe population bottleneck often occurs in nature when a small group of emigrants from an established subpopulation founds a new subpopulation; the accompanying random genetic drift is then known as a **founder effect** (see Holgate 1966; Nei et al. 1975; Chakraborty and Nei 1977; Neel and Thompson 1978). Founder effects in human populations have implications in medical genetics, because human populations derived from small numbers of founders may have an elevated incidence of an otherwise rare genetic disorder. Examples include Tay-Sachs diseases in Ashkenazi Jews, diastrophic dystrophy in Finns, familial hyperchylomicronemia in Quebecois, and congenital total color blindness in Pingelap Islanders. In addition to reducing the effective size, and thereby increasing F, population bottlenecks and founder effects may affect many other aspects of the genetic variation, including causing a reduced number of alleles, a distorted distribution of numbers of molecular site differences among alleles, and an increased level of linkage disequilibrium.

PROBLEM 7.7   Suppose a population went through a bottleneck as follows: $N_0 = 1000$, $N_1 = 10$, and $N_2 = 1000$. Calculate the effective size of this population across all three generations.

ANSWER   Using Equation 7.22, we get $1/N = (1/3)(1/1000 + 1/10 + 1/1000) = 0.034$, or $N = 1/0.034 = 29.4$. The average effective number over the three-generation period is only 29.4, whereas the arithmetic average number of individuals is $(1/3)(1000 + 10 + 1000) = 670$.

Unequal Sex Ratio, Sex Chromosomes, Organelle Genes

A second important case in which the effective size of a nonideal population can readily be calculated concerns sexual populations in which the number of males and females is unequal. This inequality creates a peculiar sort of "bottleneck"; because half of the alleles in any generation must come from each sex, any departure of the sex ratio from equality will enhance the opportunity for random genetic drift. This situation is important in wildlife management, where, for many game animals (pheasants and deer come immediately to mind), the legal bag limit for males is much larger than for females. Although some management goals are served by such hunting regulations (for example, the species involved are usually polygamous, so one male can fertilize many females and overall actual population size can be maintained), it must be remembered that the resultant inequality in sex ratio reduces the effective population size. Specifically, if a sexual population consists of N_m males and N_f females, the actual size is

$$N_a = N_m + N_f \qquad\qquad 7.23$$

However, the effective population size is

$$N_e = \frac{4N_m N_f}{N_m + N_f} \qquad\qquad 7.24$$

Figure 7.13 shows the relationship between sex ratio and the reduction in effective population size. To take a realistic example, if hunting is permitted to a level at which the number of surviving males is one-tenth the number of females, then the effective population size is a mere one-third of the actual number of individuals in the population.

A related problem is the effective population size for an X-linked gene. In this case, the variance effective population size is

$$N_e = \frac{9N_m N_f}{4N_m + 2N_f} \qquad\qquad 7.25$$

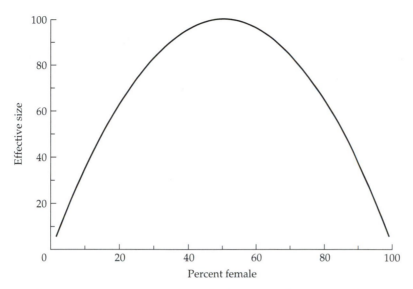

Figure 7.13   Effective size falls off rapidly in populations with a skewed sex ratio.

Equation 7.25 can be justified by noting that the sampling variance for the X chromosomes from males is $p_m q_m / N_m$, whereas the sampling variance for X chromosomes from females is $p_f q_f / 2N_f$, in which p_m and p_f are the frequencies of allele A in males and females, respectively. The frequency of an A-bearing X chromosome in the population is

$$p = \frac{1}{3} p_m + \frac{2}{3} p_f \qquad\qquad 7.26$$

and the sampling variance of p is

$$\mathrm{Var}(p) = \frac{1}{9}\left(\frac{p_m q_m}{N_m}\right) + \frac{4}{9}\left(\frac{p_f q_f}{2N_f}\right) \qquad\qquad 7.27$$

At steady state, $p_m = p_f = p$, so pq can be factored out, giving

$$\mathrm{Var}(p) = pq\left(\frac{1}{9}\frac{1}{N_m} + \frac{4}{9}\frac{1}{2N_f}\right) = \frac{pq}{2\left[\dfrac{9N_m N_f}{4N_m + 2N_f}\right]} \qquad\qquad 7.28$$

The term in the square brackets corresponds to the N_e in Equation 7.25. It shows why this is a variance effective size: the binomial sampling variance in an ideal population is $pq/2[N_e]$.

PROBLEM 7.8   What is the effective population size for mitochondrial DNA? (Assume transmission is exclusively from mothers to all offspring.) What is the effective population size for a gene on the Y chromosome, given that the population consists of N diploid individuals and all other assumptions of Table 7.1 apply? (Assume XX individuals are female and XY individuals are male.)

ANSWER   Mitochondrial DNA is transmitted essentially exclusively by females. The chance of drawing two mtDNAs that are identical by descent is $1/N_f$, where N_f is the number of females in the population. Hence the effective size is simply N_f. Similarly, the effective population size for the Y chromosome is N_m, the number of males in the population. Note that even though mtDNA is present in all individuals, while the Y is present only in males, the effective size of mtDNA is not larger. Effective size depends on the sampling properties of a gene, which depends on the gene's transmission, not just on how many individuals carry the gene.

BALANCE BETWEEN MUTATION AND DRIFT

There are many forces in population genetics that act in opposition to one another, and it is this tension that makes for interesting behavior at the population level. Mutation always increases the amount of genetic variation in a population. Random genetic drift results in the loss of genetic variation. Merely because these two forces are in opposition, it does not guarantee that there will be a stable balance between them. In order to formally ask whether the two forces do balance, we need to be careful to specify assumptions about the processes of mutation and drift. We already examined one such model in Chapter 5—the infinite-alleles model—and we saw that in this case the forces do in fact balance to provide an equilibrium level of neutral variation. Let's consider this model once again, in somewhat more detail.

INFINITE ALLELES MODEL

As we saw in Chapter 5, the infinite alleles model starts with the assumption that each mutation produces a novel allele, never before present in the population. Mutations occur such that each gene in the population has an equal, but low, chance of mutating. Random genetic drift occurs in the manner of the Wright-Fisher model—each generation the population is reconstituted by drawing a sample with replacement from the current sample of alleles. Under these assumptions we saw that the equilibrium probability of identity, F, could be approximated as

$$\hat{F} = \frac{1}{4N\mu + 1} \qquad \qquad 7.29$$

The number of selectively neutral alleles increases under mutation pressure until F satisfies this equation.

PROBLEM 7.9 Derive an expression for F in a finite population with mutation and migration.

ANSWER First assume that there is no mutation, and that new migrant alleles arrive from another population at a rate m per generation. As in the balance between mutation and drift in the infinite-alleles model, we note that alleles can be identical by descent by being drawn twice (with probability $1/2N$) or by having two different alleles drawn but having them be IBD from the previous generation. The equilibrium autozygosity can be written

$$F = \left[\frac{1}{2N} + \left(1 - \frac{1}{2N} \right) F \right] (1 - m)^2$$

because $(1 - m)^2$ is the probability that neither of the two randomly chosen alleles comes from a migrant. By analogy with the infinite-alleles model, we get (in the case of migration with no mutation)

$$\hat{F} = \frac{1}{4Nm + 1}$$

When both migration and mutation are occurring, alleles are identical by descent only if they neither mutated nor migrated, and this occurs with probability $1 - m - \mu$. Thus, the equilibrium autozygosity is

$$\hat{F} = \frac{1}{4N(m + \mu) + 1}$$

F is the probability of autozygosity, and H, which can be thought of as heterozygosity, is also the probability of drawing a pair of alleles that are not autozygous. Since $H = 1 - F$, under the infinite-alleles model, the equilibrium of H is

$$\hat{H} = \frac{4N\mu}{1+4N\mu} = \frac{\theta}{\theta+1} \qquad 7.30$$

where $\theta = 4N\mu$.

The relationship between the quantity $4N\mu$ and H was encountered in Chapter 5 and is plotted in Figure 5.7. For a per-locus mutation rate of 10^{-6} and a population size of 250,000, we get $\theta = 1$, and so $\hat{H} = 1/2$. Note that increases in population size have precisely the same effect as increases in mutation rate. Heterozygosity approaches one only if population sizes are very large (such as in microbial organisms) or if mutation rates are very high (such as at some microsatellite loci). Next we will consider how one might go about testing whether a sample from a population exhibits a pattern of genetic variation that is compatible with the infinite-alleles model.

The Ewens Sampling Formula

The infinite-alleles model has an "equilibrium" when $H = 4N\mu/(1 + 4N\mu)$. This is not an equilibrium in the usual sense. In reality, allele frequencies are always changing, new mutations continue to come into the population, and eventually they are eliminated, even perhaps, after becoming fixed for some time. The term *steady state* is probably more appropriate for this kind of behavior, since the alleles are not maintained at a constant frequency, but rather new ones are entering and old ones are leaving the population. The population remains at a steady state in the sense that the number of alleles, and the level of autozygosity, remain stationary. If the number of alleles and the level of autozygosity remain steady, then it is reasonable to assume that there is also a steady-state distribution of allele frequencies. By steady-state frequencies we mean that the most common allele always has a frequency of p_1, and the next most common has a frequency of p_2, and so on. The steady-state distribution has the curious property that, even though the most common allele is expected to have a frequency of p_1, the *identity* of the most common allele is expected to change with time. In the steady-state population, not all alleles are equally frequent, and F is greater than it would be were all alleles equally frequent.

Consider the steady-state distribution of allele frequencies from the point of view of an experimenter taking a sample from a population. Let the sample size be n genes, and suppose there are k different alleles in this sample. The sample might consist of, for example, 10 unique alleles, 3 alleles that are represented twice in the sample, 7 alleles that are present 3 times, and so on. Such a description of the sample is called the allelic **configuration** or **parti-**

tion. A remarkable finding of Ewens was that the expected configuration of a sample drawn from a population obeying the infinite-alleles model is entirely determined by the sample size, n, and the number of observed alleles, k. Ewens showed that the expected number of alleles in the sample, given θ and the sample size, is

$$E(k) = 1 + \frac{\theta}{\theta + 1} + \frac{\theta}{\theta + 2} + \cdots + \frac{\theta}{\theta + n - 1} \qquad 7.31$$

If θ is very small, $E(k) \approx 1$, whereas for very large θ, $E(k)$ approaches n, implying that for a large enough population with a high enough mutation rate, every allele that is sampled will be different. The form of Equation 7.31 suggests that, as the sample size increases, more alleles will be found, but that there is a diminishing return in finding new alleles when the sample size increases. When $E(k)$ is plotted against θ (Figure 7.14), the increase in the expected number of alleles is greatest for larger sample sizes when the population is highly diverse (large θ).

The infinite-alleles model gives a steady-state prediction of F given θ (because $F = 1/(1+\theta)$ from Equation 7.30), and a prediction of k from Equation 7.31. Combining these predictions, the expected relation between F and k is plotted in Figure 7.15. The hyperbolic relation is not surprising, because a population with many alleles will generally have a lower probability of identity of a randomly chosen pair of alleles. For $\theta = 1$, the expected F is $\frac{1}{2}$ for all

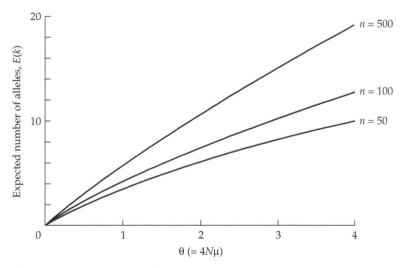

Figure 7.14 Relations between θ, the expected number of alleles, and the sample size according to the Ewens-Watterson sampling theory of a population in steady-state under the infinite-alleles model of neutral mutation.

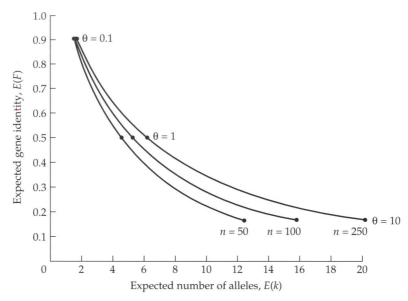

Figure 7.15   The infinite-alleles model prediction of the relation between the expected number of alleles and the expected gene identity F. The three curves represent a range of values of $\theta = 4N\mu$, starting at $\theta = 0.1$ in the upper left, and ending with $\theta = 10$ in the lower right. For the value of $\theta = 1$, the expected F, given by the relation $F = 1/(1 + \theta)$, is $\frac{1}{2}$, regardless of the sample size. Larger sample sizes always lead to larger expected numbers of alleles, but the difference is greater in more diverse populations (those with smaller F).

sample sizes, but a larger sample size should yield a greater number of distinct alleles.

The Ewens-Watterson Test

The Ewens sampling theory expressed in Equation 7.31 shows that the sample size and the number of distinct alleles observed in the sample are sufficient to give an expected configuration of allele counts. From the observed and expected configurations, a number of test statistics can be devised to determine whether the observed sample fits the expected values of the model. Figure 7.16 shows histograms of the observed and expected allele frequency configurations for alleles in a human population defined by a VNTR polymorphism. In this particular example, there appears to be a slight excess of the common allele, which is consistent with any number of causes of departure from the infinite-alleles model.

Keith et al. (1985) isolated 89 homozygous lines from a sample of *Drosophila pseudoobscura* collected at the Gundlach-Bundschu Winery in Sonoma Valley, California. Homogenized tissue from these 89 lines was then

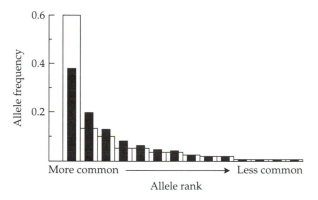

Figure 7.16   Observed (open columns) and expected (black bars) allele frequency distribution of the *HRAS-1* locus in humans, identified by Southern blotting with the *pLM0.8* probe and *Taq*I digests. Observed data are from Baird et al. (1986), and the expected distribution was generated using Ewens' sampling theory. In this sample of 490 genes there were 14 distinct alleles, four of which were present in just one individual. (From Clark 1988.)

subjected to sequential electrophoresis (a sensitive means of detecting charge and conformation differences among the protein products), and stained to reveal differences in xanthine dehydrogenase (*Xdh*) mobility. They obtained a common allele that was present in 52 of the lines, one allele that was present in nine lines, one allele that was present in eight lines, two alleles present in four lines each, two alleles that were present in two lines each, and eight singleton or unique alleles.

To test whether the observed configuration fits the expectation, a computer simulation was run to generate realizations of samples from populations that obey the infinite-alleles model, having the same number of alleles and sample size as the observed data. The algorithm to do this simulation is described by F. Stewart in the Appendix to Fuerst et al. (1977), and a listing of a program can be found in Manly (1985). From each computer-generated sample, *F* is calculated as the sum of the squared allele frequencies. Figure 7.17 shows a histogram of the computer-generated distribution of *F*, along with an arrow showing where the *Drosophila* sample fell. The sample had an observed *F* that fell in the upper tail of the distribution, and since so few values of *F* from the null hypothesis were larger than the observed *F*, Keith et al. rejected the null hypothesis and argued that the data did not fit the infinite-alleles model satisfactorily. The departure was in the direction of excess homozygosity (deficit of *H*), but since the populations were probably in Hardy-Weinberg proportions, a clearer way to state the result would be to say that there was a deficit of genetic diversity for the given number of observed alleles. The deficit means that the common allele is more common than

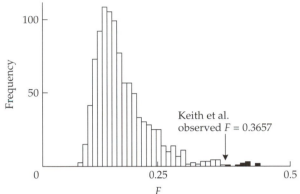

Figure 7.17   Computer-generated distribution of F obtained from 1000 samples from a population obeying the assumptions of the infinite-alleles model with $k = 15$ alleles and a sample of size $n = 89$ (as in the *Xdh* data from a sample of *Drosophila pseudoobscura* from the Gundlach-Bundschu Winery studied by Keith et al. 1985). The mean of F from the simulation was 0.168, which is well below the observed F of 0.366. A significant departure of the observed F from the predictions of the model is noted by the small area under the tail of the distribution to the right of the arrow.

expected, and there are also more singletons than expected. This pattern of frequencies is consistent with purifying selection acting to eliminate the rare, slightly deleterious alleles that continually enter the population by mutation. It is also consistent with an historical effect in which many alleles may have been previously lost and the population has not yet had time to return to equilibrium.

The results of the Ewens-Watterson test can also be reported graphically as in Figure 7.18. Each gene yields a point specified by the number of distinct alleles and the observed F. The two curves represent the 95% confidence interval generated by the Ewens sampling theory. A quick check of the concordance of the data with the model can be made by seeing whether points remain in this confidence region. Although *Xdh* in *Drosophila pseudoobscura* provides a dramatic departure from the infinite-alleles model, results like those plotted in Figure 7.18, which show an acceptable fit to neutrality, are more commonly obtained.

INFINITE-SITES MODEL

Rather than considering each mutation as generating a unique allele, with infinitely many possible alleles, we can instead consider an allele as a sequence of nucleotides with mutation altering a site in the sequence. If the mutation rate is sufficiently low, then most sites will be monomorphic, and all polymorphic sites will be segregating for just two nucleotides. Much of

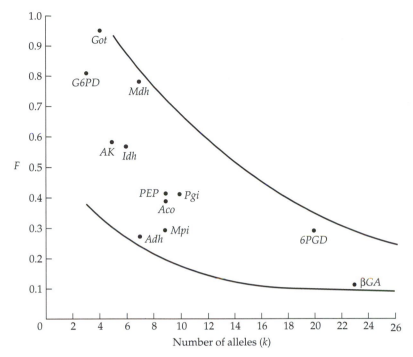

Figure 7.18   Gene identity (*F*) plotted against the observed number of alleles in a sample of 279 *E. coli*. The solid lines represent the upper 97.5% and lower 2.5% confidence limits, and the observation that all of the tested loci fall within these limits suggests good concordance with the infinite-alleles model of neutral mutation. (From Whittam et al. 1983.)

the available data on allelic variation in DNA sequence seems consistent with this view: few nucleotide sites are segregating for more than two nucleotides. If the DNA sequence is sufficiently long and the frequency of polymorphic sites low, then most of the time new mutations will occur at sites that were previously monomorphic. The **infinite-sites** model, based on these assumptions, was developed by Kimura (1969, 1971), who considered nucleotides as unlinked, and by Watterson (1975), who took account of the nearly complete linkage among sites.

The infinite-sites model is appealing because it directly addresses the type of data that molecular population geneticists can collect. Given an array of DNA sequences of alleles randomly sampled from a population, there is considerable information about the history of the alleles hidden in the patterns of similarity across alleles. The infinite-alleles model ignores this pattern and simply considers the alleles as distinct. A much more powerful treatment is to tabulate the number of sites at which all pairwise combinations of sequences differ, resulting in a so-called **mismatch distribution**. The infinite-sites model addresses the theoretically expected behavior of the mismatch distribution. Watterson (1975) considered the distribution of S_i, defined as the number of segregating sites in a sample of i genes. For the case of a random sample of two genes, Watterson showed that the steady-state probability that the sequences have i mismatches is

$$\Pr(S_2 = i) = \left(\frac{1}{\theta+1}\right)\left(\frac{\theta}{\theta+1}\right)^i \qquad 7.32$$

where $\theta = 4N\mu$, and μ is the mutation rate *per gene* (not per site). A particular case of this equation gives the probability that two sequences have no sites different, and hence are identical. Substituting $i = 0$ into Equation 7.32, we get

$$\Pr(S_2 = 0) = \frac{1}{\theta+1} \qquad 7.33$$

in agreement with the infinite-alleles model, because $\Pr(S_2 = 0) = F$, the probability that two alleles drawn at random are identical. The mean and variance in the distribution of number of segregating sites are θ and $\theta + \theta^2$, respectively.

In reality we do not sample an entire population, so it is important to determine the statistical properties of a smaller sample drawn from a population. Often the sampling properties of population genetic models are very complex and we have to resort to simulations for meaningful estimates. A few results have been obtained for samples drawn from a population obeying the infinite-sites model, and these results are very useful for testing goodness of fit to the model. The expected number of segregating sites in a sample of n alleles is

$$E(S) = \theta \sum_{i=1}^{n-1} \frac{1}{i} \qquad 7.34$$

and the variance in the number of segregating sites is

$$V(S) = \theta \sum_{i=1}^{n-1} \frac{1}{i} + \theta^2 \sum_{i=1}^{n-1} \frac{1}{i^2} \qquad 7.35$$

This expression for the variance is for the case of no intragenic recombination. It turns out that intragenic recombination does not affect $E(S)$, but it reduces $V(S)$. This is not hard to see intuitively—recombination shuffles the variation among alleles, reducing the average number of sites by which random pairs of alleles differ. The expression for the variance in the number of mismatching sites in the case of free recombination across sites is

$$V(S) = \frac{n+1}{3(n-1)}\theta + \frac{2(n^2+n+3)}{9n(n-1)}\theta^2 \qquad 7.36$$

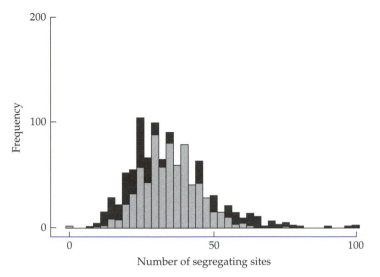

Figure 7.19   Equilibrium distribution of the number of mismatches between a pair of alleles. Note that if there is free recombination, the variance is smaller compared to the case of no recombination.

Figure 7.19 shows the mismatch distributions for a simulated set of data with free recombination (smaller variance) and with no recombination (larger variance). The relationship between the mean and the variance in the mismatch distribution can be used to make inferences about intragenic recombination (Hudson 1987).

The assumptions of the infinite-alleles model and the infinite-sites model do not seem to be entirely at odds with one another, and we saw that they predict the same steady state value for F. But the two models do make use of different aspects of the data, and so it would seem that a test of the consistency between the two models might serve as a useful test for the neutral theory. The next problem makes use of just this test, which was devised by Tajima (1989).

PROBLEM 7.10   The average heterozygosity for pairs of randomly chosen alleles under the infinite-alleles model is $E(k) = \theta$, and the expected number of sites segregating in a sample (under the infinite-sites model) is

$$E(S) = \theta \sum_{i=1}^{n-1} \frac{1}{i}$$

Two estimates of θ are therefore k, the average heterozygosity, and

$$\frac{S}{\sum_{i=1}^{n-1} \frac{1}{i}}$$

Tajima (1989) devised a test statistic to test the null hypothesis that these two estimates were identical. The test statistic is the difference between these two estimates of θ, or

$$D = E(k) - \frac{S}{\sum_{i=1}^{n-1} \frac{1}{i}}$$

If a population were growing rapidly, one might expect this to affect both the number of segregating sites and the heterozygosity. Predict the direction of change of F (probability of identity), S (the number of segregating sites), and D (Tajima's test statistic).

ANSWER   First consider a larger population at equilibrium. Since $F = 1/(4N\mu + 1)$, a larger population would have a lower F. A larger population would also have a larger number of segregating sites (S), and a higher per-site heterozygosity (k). At equilibrium, if the gene is neutral, then the Tajima statistic should be zero. In a growing population, F will decrease as added variation accumulates, S will increase and k will increase. The key point is that the increase in variation will occur in initially rare alleles, which contribute to S but only a little to k. Thus, S grows faster than k, and D will be negative. If the population stops growing, then Tajima's D statistic will return to zero at equilibrium.

GENE TREES AND THE COALESCENT

A sample of genes from a population represents more than a snapshot of counts of alleles in a population. Each gene that is sampled has an ancestral history dating back hundreds or thousands of generations. It is possible that a pair of genes sampled today may have come from identical copies of the same allele produced by the same individual just a few generations ago. Or the alleles may have had common ancestry hundreds of generations ago. The term **coalescence** refers to this process, looking backward in time, and seeing how two genes merge at times of common ancestry. Along this process, one goes from a sample of k genes, to $k - 1$ ancestors after the first coalescence, to $k - 2$ ancestors after the second coalescence, and so forth until there is a single common ancestor for the whole sample. The idea of the coalescent is to consider the ancestral history of genes in a sample by developing a model for the time to common ancestry (Kingman 1980).

To understand how the coalescent process works, consider in Figure 7.20 what happens as time moves forward. In each generation there are a number of alleles in the population, and those alleles may be reproduced and be present in the following generation (moving down the figure), or, in some cases, an allele is not reproduced and is lost from the population. By chance, some alleles may be sampled twice in constituting the next generation, and the probabilities of these events are the same as those under the Wright-Fisher model of random genetic drift. By a repetition of this process over time, eventually one of the original alleles will become "fixed" in the population. In the absence of mutation, the population would therefore be fixed for the same

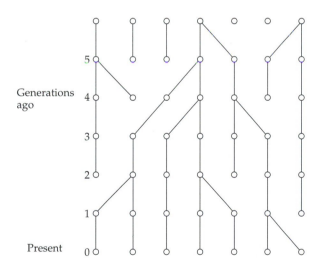

Figure 7.20   Diagram showing paths of ancestry of a set of alleles sampled at the present. The population is represented as having a constant size. Starting at the top and working down, notice that many alleles go extinct, and one allele goes to fixation. Considering this process in reverse, the current sample observed at present undergoes a series of coalescence events in which the k alleles present in the current generation had only $k - 1$ ancestors. This process continues backward in time until there is only one ancestral allele.

allele; however, because mutation may occur during the process, the alleles observed at the present will not all be identical in nucleotide sequence, even though they all descended from a single common ancestral allele.

In reality we do not have the genealogical information enabling us to follow all the alleles through time in a population. Typically what we have is a single "snapshot" represented by a small sample of alleles taken at the present time. Now consider Figure 7.20 again, but this time look at what happens when we go backwards in time. We start with the k alleles in the sample at generation 0. In going from generation 0 to generation 1 (one generation ago), we see that the two rightmost alleles "coalesced" into a single ancestral allele. As we go further back in time, the number of ancestral alleles has to either remain the same or decrease, and each reduction in the number of ancestral alleles is called a *coalescence event*. In order to show how this idea can be extended to derive expressions for the entire distribution of branch lengths of a gene tree, we next specify a model.

Consider two alleles. The probability that the two alleles came from the same allele in the previous generation is $1/2N$ (in a diploid population of size N), so the chance that they came from two distinct alleles the previous generation is $1 - 1/2N$. The probability that three alleles had three distinct ancestral alleles the previous generation is Pr(alleles 1 and 2 have distinct ancestors)Pr(allele 3 is different from both 1 and 2) = $(1 - 1/2N)(1 - 2/2N)$. In general, the probability that k alleles had k distinct parental alleles the previous generation is

$$\Pr(k) = \prod_{i=1}^{k-1}\left(1 - \frac{i}{2N}\right) \approx 1 - \frac{\binom{k}{2}}{2N} \qquad 7.37$$

Each generation the sampling process occurs independently of what happened before, and so the probability that k alleles had k distinct parental alleles two generations ago is the square of the right-hand side of Equation 7.37. Consider two alleles again. Suppose we wish to know the chance that the common ancestor of these two alleles occurred exactly t generations ago. In this case there must have been no coalescence (i.e., two distinct ancestral lineages were found) for $t - 1$ generations, and then, in the next preceding generation, a coalescence occurred. The probability of not coalescing for t generations is $(1 - 1/2N)^t$ and the chance of the two alleles coalescing in any one generation is $1/2N$. The desired probability is the product of these or

$$\Pr\left(2 \text{ alleles had common ancestor } t \text{ generations ago}\right) = \frac{1}{2N}[1 - (1/2N)]^t$$

$$\approx \frac{1}{2N}e^{-t/(2N)} \qquad 7.38$$

The exponential is an approximation that is quite good when $1/2N$ is small. This distribution has a mean of $2N$ generations and a variance of $(4N)^2$. Note that the confidence interval around the mean time is not very tight, since the standard deviation of the distribution is equal to the mean.

Returning to our sample of k alleles, the probability that the k alleles do not coalesce for t generations, then one pair coalesces to give $k-1$ alleles at $t+1$ generations ago is as follows:

Pr(k ancestors for t generations, $k-1$ ancestors at $t+1$ generations ago)

$$= \Pr(k)^t \left[1 - \Pr(k) \right]$$

$$\approx \frac{\binom{k}{2}}{2N} \exp\left[-\frac{\binom{k}{2}}{2N} t \right] \qquad 7.39$$

This approximation is valid if $k \ll N$. The distribution in Equation 7.39 has a mean of $4N/[k(k-1)]$ generations and a variance of $16N^2/[k(k-1)]^2$. Figure 7.21 shows what the gene genealogy is expected to be. Starting with five alleles, the first coalescence is expected to occur $2N/10$ generations ago, the next at $2N/6$ generations prior to that, and so on. Note that the time intervals get

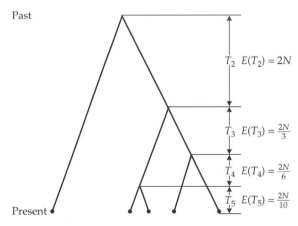

Figure 7.21   The process of coalescence can be represented by a gene tree. At each generation, if there are k alleles present, the expected time back to the next coalescence is $2N / \binom{k}{2}$. Starting with five alleles, the expected time back to the first coalescence is $2N/10$. Note that the successive times get longer. When there are only two alleles, the time back to the final coalescence is $2N$ generations.

longer and longer as the number of lineages decreases. The distribution of each of these time intervals is exponential, with ever-increasing means as one goes back in time. The time to the coalescence of all of the k alleles (i.e., the most recent time that one sample of n alleles shared a common ancestor) is

$$t = 4N(1 - 1/k)$$
7.40

with variance

$$V = 4N^2 \prod_{i=2}^{k} \frac{1}{\left(\begin{array}{c} i \\ 2 \end{array} \right)^2}$$
7.41

(Kingman 1982; Tajima 1983). As the sample size k increases toward the total population size, t approaches $4N$, which equals the expected fixation time for a newly arisen mutation. These principles allow us to generate simulated gene genealogies whose branch lengths correspond to the assumptions of the Wright-Fisher model. One thing the model still lacks is mutation, which is introduced in the next discussion.

Coalescent Models with Mutation

In order to generate simulated gene sequence data representing samples drawn from a population obeying the infinite sites model, Hudson (1990, 1993) showed that one can proceed as follows:

- Determine the sample size k and the θ for the gene region of interest;
- Draw random numbers with appropriate exponential distributions to construct a gene genealogy such that times of coalescence follow Equation 7.39;
- On each branch of this tree, distribute mutations with a Poisson distribution on each branch, such that the mean number of mutations on each branch is given by $2N\mu t$, where t is the branch length.

This procedure has been widely used in generating data sets under the neutral hypothesis for comparison to observed data sets.

From Figure 7.21, it follows that the sum of the branch lengths for the entire gene tree is

$$T = \sum_{i=2}^{k} i T_i$$
7.42

The expected number of segregating sites in the whole sample is $2N\mu T$, where T is the sum of the branch lengths, so substituting we get

$$E(S) = 2N\mu T = \frac{\theta}{2} \sum_{i=2}^{k} i E(T_i) = \theta \sum_{i=1}^{k-1} \frac{1}{i}$$
7.43

The rightmost expression agrees with Equation 7.34, which we derived for the infinite-sites model.

The coalescent approach can be used to derive many fundamental principles in population genetics. As one example, consider a population presently in mutation-drift equilibrium. In the previous generation, a pair of alleles can either coalesce, with probability $1/2N$, or failing to coalesce, one or the other allele may mutate with probability 2μ. (The factor 2 comes in because either copy can mutate.) These are the only two events that affect identity, and the sum of their probabilities is $1/2N + 2\mu$. The probability of identity is therefore the fraction of the time that the alleles coalesce:

$$F = \frac{\dfrac{1}{2N}}{\dfrac{1}{2N} + 2\mu} = \frac{1}{1+\theta} \qquad 7.44$$

We have already derived this equilibrium identity under the infinite-sites (and infinite-alleles) models. Coalescence methods are not limited to the consideration of the Wright-Fisher model. If one can develop a recursion equation for probabilities of recombination, migration, or other such phenomena in a gene tree context, then often powerful insights can be derived from coalescence approaches. For our purposes, suffice it to say that the method can generate classical results, often with much less difficulty, and the coalescence approach is especially well suited to testing hypotheses about samples drawn from populations.

PROBLEM 7.11   The probability distribution for the number of generations back to the first coalescence (in a pure drift model) in a sample of k genes taken from a haploid population of size N is approximately:

$$\Pr\left(\text{first coalescence } t \text{ generations ago}\right) = xe^{-xt}, \text{ where } x = \frac{\dbinom{k}{2}}{N}$$

From this one can show that the mean number of generations back to the first coalescence is $1/x$. The more genes in the sample, the more likely it will be that a coalescence occurred recently. Calculate the expected time to first coalescence in a population of $N = 450$ for a sample of 10 genes. How many genes would you have to sample to halve this coalescence time?

ANSWER   The expected time to first coalescence in a population of $N = 450$ for a sample of 10 genes is

$$N \bigg/ \binom{k}{2} = 450 \bigg/ \binom{10}{2} = 450/(10 \times 9/2) = 10 \text{ generations.}$$

To determine how many genes one would have to sample to halve this coalescence time, solve for

$$5 = 450 \bigg/ \binom{k}{2}$$

This is equivalent to $90 = k!/[2!(k-2!)]$. By trial and error, you will find that a sample of 14 genes will do it. Note that by increasing the sample only from 10 to 14, we expect to find a pair of alleles half as divergent from each other.

SUMMARY

Gene frequencies fluctuate at random in finite populations. The rate at which allele frequencies change varies inversely with population size. The reason for the inverse relationship is that the sampling variance, when two alleles are segregating in a population, is determined by the binomial sampling process, and the binomial variance is $pq/2N$. The **Wright-Fisher model** extended the idea of binomial sampling over multiple generations, and much of our understanding of drift has been derived from this model. In a population in which the only force acting on gene frequencies is random drift, all variation must ultimately be lost. The Wright-Fisher model shows why the probability that an allele will drift to fixation is equal to its initial frequency in the population. The **diffusion approximation** of the Wright-Fisher model is a second-order partial differential equation that yields the distribution $\phi(x,t)$, giving the number of populations with allele frequency x and time t. The diffusion approach has yielded important insights into the consequences of drift, including the expected time to fixation and loss of alleles. The expected time to fixation of a newly introduced allele is $4N$ generations, showing once again that drift happens faster in smaller populations.

A useful way to think about random drift is to consider a set of subpopulations of the same size undergoing repeated generations of sampling and drift. Within each of these subpopulations, genotypes are composed by drawing alleles at random, so that each subpopulation is always in Hardy-Weinberg equilibrium. The hypothetical population composed by pooling the subpopulations will have a deficit of heterozygotes because, as allele frequencies drift closer to fixation, the frequency of heterozygotes declines. The rate at which heterozygosity is lost in a finite population is $(1 - 1/2N)$, so that a population of size 10, say, loses 5% of its heterozygosity each generation. The allele frequencies of the subpopulations are equally likely to drift up as down, so the average allele frequency over subpopulations shows no change.

Real biological populations do not precisely fit the Wright-Fisher model. They generally exhibit changes in allele frequency that exceed the amount expected based on the actual population size. The usual reason for the discrepancy is that the drift process occurs as though there are fewer than the observed census number of individuals. The models give better correspondence to reality by calculating the effective population size. Several different factors that require consideration in calculating effective size were examined in this chapter, including unequal sex ratio, fluctuation in population size over generations, and the uniparental transmission of mtDNA and Y chromosomes.

Mutation introduces variation into populations, and random genetic drift erodes that variation. These two forces come to a steady state predicted by population genetic models. The infinite-alleles model assumes that each new mutation generates a novel allele. The steady-state balance between mutation and drift in the infinite-alleles model is given by the autozygosity, F, which is also the probability that two allelels are identical by descent. Under the infinite-alleles model for a diploid population, $\hat{F} = 1/(1 + \theta)$, where $\theta = 4N\mu$. Note that the mutation rate and population size are confounded in this model, increasing either one will decrease the autozygosity by the same amount. We can write the same equation in terms of heterozygosity $H = 1 - F$, giving $H = \theta/(1 + \theta)$. The infinite-sites model is related to the infinite-alleles model, but more specifically states that novel mutations occur at a site along the gene that has not mutated before. (If this is true, each new mutation must also generate a novel allele.) The infinite-sites model generates predictions about the number of segregating sites expected in a population at steady state. Here the result is that the expected number of segregating sites is

$E(S) = \theta \sum_{i=1}^{n-1} \frac{1}{i}$, where the mutation rate in this value of θ is the mutation rate over the entire gene in question.

Classical models of random genetic drift look forward in time, following alleles as they are lost from the population and generated anew by mutation. More recently, the **coalescent** approach has been to look backward in time, starting with the observed sample of alleles, and calculating times to common ancestry of alleles. Coalescent approaches are particularly appropriate when one wants to consider the probability that a particular observed set of molecular sequence data might have the characteristics expected of random genetic drift. Computer generation of gene trees using principles of coalescence theory makes it easy to produce a null distribution giving the full range of outcomes expected under a drift model.

PROBLEMS

1. Suppose that in one generation, in a population of size 50, the average heterozygosity (averaged across loci) is reduced from 0.50 to 0.42. Is the population mating at random?

2. In how many generations will the expected heterozygosity be 5% of the initial value in a diploid randomly mating population of size of 10? Size 100?

3. A gene in one individual in a population of 24 barn cats undergoes mutation to a new neutral allele. What is the probability that the allele eventually becomes fixed? What is the probability that it eventually becomes lost? What are the answers if the mutant gene is X-linked and the population consists of equal numbers of males and females?

4. If an isolated population of annual alpine plants decreases in heterozygosity by half every 50 years because of random genetic drift, what is its effective population size?

5. Remote Pitcairn Island in the South Pacific was settled in 1789 by Fletcher Christian and eight fellow mutineers from HMS Bounty, along with a small number of Polynesian women. Although many descendants have left the island in the intervening years, there has been essentially no immigration. Assuming an effective size of 20 in each of the eight generations since the island's settlement, what value of F_{ST} would be expected in today's population from random genetic drift?

6. In a population of effective size $N = 50$, how long is required for random genetic drift to double the value of the fixation index F from 0.01 to 0.02? From 0.05 to 0.10? Assuming that F is small, how many generations are required to double the value of F in a population of effective size N? For the latter, use the approximations that $[1 - (1/2N)]^t \approx \exp(-t/2N)$ and that, when F is small ($F < 0.10$), $\ln(1 - F) \cong -F$.

7. What is the effective population number in a population of large predatory cats in which each breeding male controls a harem of five females and the total population consists of 200 males and 200 females?

8. What is the effective population size of a herd of ten dairy cows and one bull? What is it for 40 cows and one bull? For 10 cows and two bulls?

9. What is the variance effective population size for an X-linked gene in a population consisting of 100 females and 10 males? In a population of 10 females and 100 males?

10. Among 100 restriction site differences in two inbred strains of the flour beetle *Tribolium* that are crossed and allowed thereafter to mate at random, what number of restriction sites would be expected to remain segregating after 10 generations assuming an effective population size of 80 individuals? How many would be expected to remain unfixed after 50 generations?

11. In a haploid population of constant effective size 50, what is the probability that two randomly drawn alleles shared a common ancestor exactly 100 generations ago?

12. Employing the infinite-sites model, if $\theta = 10$, how many segregating sites will one expect to find in a sample of size 10? 20? 50?

13. Consider an isolated island population with no migration, effective population size of 250,000, and a mutation rate of 10^{-6}. Calculate the expected heterozygosity under the infinite-alleles model. How much migration is necessary to increase H to $\frac{2}{3}$?

14. In a haploid population of effective size 50, how large a sample must one take to yield an expected mean coalescence time of 10 generations?

15. Show that random genetic drift requires an average of $t = 2N \ln x$ generations to reduce the heterozygosity from H_0 to H_0/x.

16. Use Equation 7.15 to show that approximately $2N$ generations of random genetic drift are required to reduce the number of segregating genes by a factor of e ($e = 2.71828 \ldots$), given initial allele frequencies close to 0.5.

17. A set of six to eight oocytes from each of three women undergoing in vitro fertilization (IVF) were recently tested for heteroplasmy (presence of more than one mitochondrial DNA type within each cell). The mtDNA from eggs of two women were all identical and matched that of somatic cells with no heteroplasmy, but the other woman produced eggs with two different mtDNA types. Densitometric scans allowed investigators to determine that the individual cells had relative frequencies of the two mtDNA types ranging from 20% to 50%. Assuming 30 cell generations from zygote to zygote in the maternal germline, and $N = 1000$ mitochondria per cell, what do you conclude from these observations? Are they consistent with neutral sampling of mtDNA types?

CHAPTER 8

Molecular Population Genetics

MOLECULAR CLOCK ⋅ SYNONYMOUS AND NONSYNONYMOUS SUBSTITUTION
CODON BIAS ⋅ GENE GENEALOGIES ⋅ ORGANELLE DNA
MOLECULAR PHYLOGENETICS

ALL THE FORCES in population genetics have an impact on the pattern of variation seen in molecular sequences of genes, including mutation, migration, selection, and random drift. A primary focus of molecular population genetics is to make inferences about the contribution of each of these evolutionary forces to produce the patterns of molecular sequence variation we see today. Usually this process involves a close interplay between mathematical model building, statistical parameter estimation, and experimental observation. Several times in the past, unexpected patterns of sequence variation have arisen which, in turn, gave rise to whole new avenues of theoretical inquiry. In many cases, inferences about evolutionary forces transcend species boundaries by making use of data on both within-species polymorphism and between-species divergence. The genetic basis for species isolation is itself amenable to analysis. But first let us begin with the basic theoretical principles that underlie molecular population genetics.

THE NEUTRAL THEORY AND MOLECULAR EVOLUTION

The first systematic application of protein electrophoretic methods to population genetics revealed extensive genetic variation within most natural populations. Typically, 15 to 50% of the genes coding for enzymes were observed to include two or more widespread, polymorphic alleles. The polymorphic alleles occurred with frequencies considered to be too high to result from equilibrium between adverse selection and mutation. Motoo Kimura suggested that most polymorphisms observed at the molecular level are selectively neutral, so that their frequency dynamics in a population are determined by random genetic drift (Kimura 1968). By extension, the hypothesis of selective neutrality would also apply to most nucleotide or amino acid substitutions that occur within a molecule during the course of evolution.

The neutral theory has been of great importance in population genetics in stimulating the collection and analysis of data in attempts to evaluate its adequacy. Mathematical investigations of its implications have resulted in one of the most complete and elegant theories in all of biology. Tests of the correspondence of sample data to the neutral theory are almost universally low in power, which means that large sets of data are needed before one has a reasonable chance of rejecting neutrality. The recent trend has been that more and more cases of departures from neutrality are being found, in part because of the expansion in available data and in part because of the increasing subtlety of tests that are applied. Regardless of the action of other forces shaping molecular sequence variation in populations, the force of random drift is always there, and for this reason the neutral theory remains useful in generating rigorous null hypotheses. The next section summarizes some of the theoretical implications of the neutral theory and some of the data bearing on it.

Theoretical Principles of the Neutral Theory

The neutral theory models the fate of mutations that are so nearly selectively neutral in their effects that their fate is determined largely through random genetic drift. A variety of mutation models have been considered, including infinite-alleles, infinite-sites, and finite-sites models. In all models, though, random drift occurs when N adult individuals produce an infinite pool of gametes from which $2N$ are chosen at random to create the N zygotes of the next generation. Much of the complexity of the mathematics of the neutral theory arises from the fact that the mutational histories of alleles are not independent, because they share an overlapping genealogical history. Before we get into the details of the predictions of the neutral theory, let us first review some of the theory's principal implications (Kimura 1983).

1. If a population contains a neutral allele with allele frequency p_0, then the probability that the allele eventually becomes fixed equals p_0. In particular, a newly arising neutral mutation occurs in just one copy, so the initial allele frequency is $p_0 = 1/2N$, and the probability of eventual fixation of the mutation is therefore $1/2N$. Figure 8.1 shows that a mutant allele arising in a smaller population has higher chance of fixation.
2. The steady-state rate at which neutral mutations are fixed in a population equals μ, where μ is the **neutral mutation rate**. It is noteworthy that the equilibrium rate of fixation does not involve the population size N. The

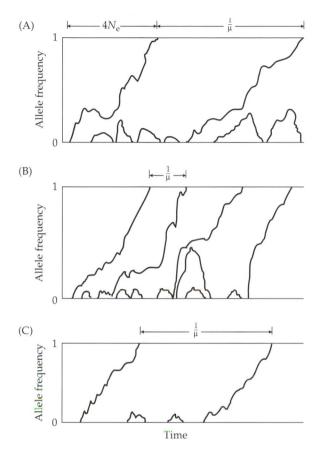

Figure 8.1 Diagram showing the trajectory of neutral alleles in a population. New alleles enter the population by mutation and have an initial allele frequency of 1/2N. Most alleles are lost, but those that go to fixation take an average of 4N generations. The time between successive fixations of neutral alleles is 1/μ generations. (A) A moderate size population. (B) The same population size; a higher mutation rate gives the same time to fixation, but less time between fixations. (C) A smaller population has alleles that go to fixation more rapidly, but the time between fixations is still 1/μ. (After Kimura 1980.)

reason is that the N cancels out: The overall rate is determined by the product of the probability of fixation of new neutral mutations ($1/2N$) and the average number of new neutral mutations in each generation ($2N\mu$), hence $(1/2N) \times (2N\mu) = \mu$.

3. The average time that occurs between consecutive neutral substitutions equals $1/\mu$. This principle follows directly from the one above. If the steady-state rate of fixation is μ per unit time, the average length of time

between substitutions will be the reciprocal, or $1/\mu$. By way of analogy, if a Swiss clock cuckoos at the rate of 24 times per day, then the average length of time between cuckoos is 1/24th of a day, or one hour. As Figure 8.1 shows, the time interval between fixations is independent of population size, and elevating the mutation rate decreases the time interval between fixations.

PROBLEM 8.1   The neutral theory makes a strong prediction about the relationship between population size and heterozygosity. Under the infinite-alleles model, we can express the prediction by the formula, $H = 4N\mu/(4N\mu+1)$, and hence small populations should have low heterozygosity and large populations high heterozygosity. Do the data support this prediction? A survey of 77 species reviewed by Nei and Graur (1984) found that species with very small populations (less than, say, 10^4) had a mean protein heterozygosity of 0.05, whereas those species with a very large population (greater than 10^9, say, which include *Drosophila* species), have heterozygosities of around 0.2. This positive correlation seems to favor the neutral theory, except that the range of H is much smaller than theory predicts in view of the enormous range in N. When these extremes of population size are excluded ($N < 10^4$ and $N > 10^9$), there is no significant correlation between population size and heterozygosity. What is going on?

ANSWER   The paradoxical result demonstrates that levels of variability in a population are determined by several forces, and that different organisms may be affected by the forces to different magnitudes. The result does not support the neutral theory insofar as it shows that population size does not, by itself, explain levels of variation. On the other hand, the discrepancy is not grounds to completely toss out the neutral theory. For one thing, the population sizes were generally roughly estimated, and effective sizes (Chapter 7), which were not estimated, are more relevant to neutral predictions of heterozygosity. There is also an implicit assumption that mutation rates are identical in all organisms, and violations of this assumption can be found.

4. Analysis of the diffusion equation has shown that, among newly arising neutral alleles that are destined to be fixed, the average time to fixation is $4N_e$ generations (where N_e is the effective population size). This too is evi-

dent in Figure 8.1: alleles that go to fixation do so in less time in the smaller population. Among newly arising neutral alleles destined to be lost, the average time to loss is $(2N_e/N)\ln(2N)$ generations. The average times required for fixation or loss apply to newly arising alleles, which are necessarily present in just one copy, so $p_0 = 1/2N$. The implication of these formulas is that, on average, neutral mutations that are going to be fixed require a very long time for this to occur, but mutations destined to be lost are lost quite rapidly.

5. If each neutral mutation creates an allele that is different from all others existing in the population in which it occurs, then, at equilibrium, when the average number of new alleles gained through mutation is exactly off-set by the average number lost through random genetic drift, the expected homozygosity equals $1/(4N_e\mu + 1)$, where μ is the neutral mutation rate. The model of mutation in which each new allele is novel is the **infinite-alleles model** of mutation. The quantity $4N_e\mu$, which shows up frequently in the neutral theory, is often denoted as θ. The equilibrium average homozygosity is therefore $1/(1 + \theta)$. Since the heterozygosity equals one minus the homozygosity, the average heterozygosity at equilibrium in the infinite-alleles model equals $\theta/(1 + \theta)$. Larger populations are expected to have a higher heterozygosity, as reflected in the greater number of alleles segregating at any one time in the larger populations in Figure 8.2.

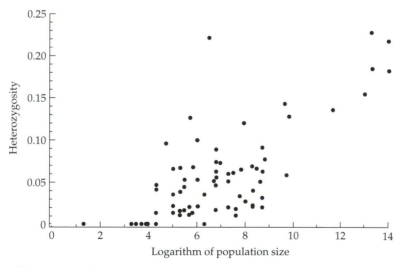

Figure 8.2 Given the enormous variation in effective population sizes, one would expect to see a wider range in variation in heterozygosity than is actually observed. The relation between population size and heterozygosity does not fit the neutral theory expectation over a wide range of intermediate population sizes. (After Nei and Graur 1984.)

ESTIMATING RATES OF MOLECULAR SEQUENCE DIVERGENCE

Rates of Amino Acid Replacement

The initial impetus for the neutral theory came from observations on the rate of amino acid replacements in proteins. When extrapolated to the entire genome, the inferred rate of evolution was several nucleotide substitutions per year. This rate was regarded as much too high to result from natural selection, because the intensity of selection must be limited by the total amount of differential survival and reproduction that occurs in the organism. Direct DNA sequencing later revealed that rates of nucleotide substitution vary according to the function (or presumed absence of function) of the nucleotides. The type of data that must be analyzed are best illustrated by example. The first 18 amino acids present at the amino terminal end of the human and mouse γ-interferon proteins constitute a signal peptide that is used in secretion of the molecules (Gray and Goeddel 1983). The sequences are:

```
Human: Met  Lys  Try  Thr  Ser  Tyr  Ile  Leu  Ala  Phe  Gln  Leu  Cys  Ile  Val  Leu  Gly  Ser
Mouse: Met  Asn  Ala  Thr  His  Cys  Ile  Leu  Ala  Leu  Gln  Leu  Phe  Leu  Met  Ala  Val  Ser
```

In order to calculate the proportion of amino acids that differ in the two signal sequences, we can simply count the number of sites that are the same and the number of sites that differ. Among the 18 amino acids there are 10 differences, so the proportion different is $10/18 = 0.56$.

To interpret these data, let us suppose that amino acid replacements occur at the rate λ per unit time. Consider two independently evolving sequences, initially identical, which at time t are found to differ in the proportion D_t of their amino acids. After the next time interval, the proportion of differences D_{t+1} is given by

$$D_{t+1} = (1 - D_t)(2\lambda) + D_t \qquad 8.1$$

In this equation, $(1 - D_t)(2\lambda)$ is the proportion of sites, previously identical, in which one or the other underwent an amino acid replacement during the time interval in question, which must be added to the already existing differences D_t in order to give the total. (The equation ignores the unlikely possibility of an amino acid replacement making two previously different amino acid sites identical.) The factor of 2 is present because the total time for evolution is $2t$ units (t units in each lineage after the split), which is illustrated in Figure 8.3. Equation 8.1 suggests the differential equation

$$dD/dt = D_{t+1} - D_t = 2\lambda - 2\lambda D_t \qquad 8.2$$

which has the solution

$$D_t = 1 - e^{-2\lambda t} \qquad 8.3$$

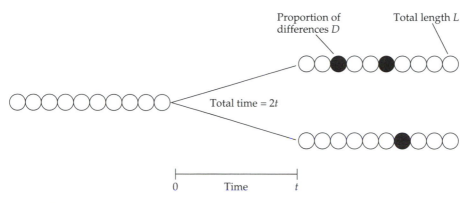

Figure 8.3   Two amino acid or nucleotide sequences that have each undergone independent evolution from a common ancestor for t time units are separated by a total time of $2t$ units because there are t units in each lineage after the split. The proportion of sites that differ in the sequence is denoted $\hat{D}$ and the total number of sites L. In this particular example, $L = 10$ and $\hat{D} = 3/10$.

An alternative argument can be used to derive Equation 8.3 without resorting to differential equations. If λ is the rate of amino acid replacement per unit time, then the probability that a particular site remains unsubstituted for t consecutive intervals along each of two independent lineages is $(1 - \lambda)^{2t}$, which is approximately equal to $e^{-2\lambda t}$, provided that λt is not too large. Thus, the probability D_t of one or more replacements occurring in t units of time after divergence is approximately $1 - e^{-2\lambda t}$, which is Equation 8.3.

Since λ is the rate of amino acid replacement per unit time, the expected proportion of differences between two sequences at any time t is

$$K = 2\lambda t \qquad\qquad 8.4$$

where the factor of 2 is again present because the total time for evolution is $2t$ units (Figure 8.3).

Substituting K from (8.4) into (8.3) and rearranging yields the following estimate $\hat{K}$ of K,

$$\hat{K} = -\ln(1 - \hat{D}) \qquad\qquad 8.5$$

where $\hat{D}$ is the observed proportion of sites in which two sequences differ. If the sequences under comparison are L amino acids in length, then the estimated variance $\text{Var}(\hat{K})$ of $\hat{K}$ is estimated from the distribution of K implied by the substitution process and is approximately

$$\text{Var}(\hat{K}) = \hat{D}/[(1 - \hat{D})L] \qquad\qquad 8.6$$

The rate of evolution at the molecular level is given by the amount of sequence divergence that occurs per unit of time. Thus, as suggested by Equations 8.4 and 8.5, if two sequences are compared, and these are known to have diverged from a common ancestral sequence an estimated $\hat{t}$ time units ago, then the rate of evolution λ may be estimated as

$$\hat{\lambda} = \hat{K}/2\hat{t} \qquad\qquad 8.7$$

The units of $\hat{\lambda}$ are usually expressed as replacements per amino acid site (or substitutions per nucleotide site) per year.

The quantity $\hat{K}$ is used in preference to $\hat{D}$ in estimating the rate of molecular evolution because $\hat{K}$ takes multiple substitutions into account. Over long periods of evolutionary time, the amino acid present at a particular site may be replaced several times, first by one alternative, then by another, then still another, and perhaps, at some stage, even return to the amino acid originally present at the site. When comparing two sequences, only the sites that are different can be identified. Sites that are identical at the present time may include some that were different in the past, and sites that are different at the present time might have undergone more than one substitution. The quantity $\hat{D}$ is determined only by the proportion of differences between the sequences observed at the present time. The estimate $\hat{K}$ makes a correction for multiple substitutions, but at the cost of introducing assumptions that the substitutions occur independently and at the same rate through time.

For relatively short intervals of evolutionary time, during which multiple substitutions remain uncommon, the correction is minor, and the value of $\hat{K}$ is close to that of $\hat{D}$. This can be seen by the fact that the initial slope of the curve plotted in Figure 8.4 is 1. As the observed sequence divergence increases, it becomes more likely that multiple hits have happened, so the slope decreases. Over longer intervals, when many multiple substitutions have occurred, the correction is important, and the assumptions on which it is based must be evaluated critically. Correction for multiple substitution events is even more important for nucleotides than it is for amino acids. With amino acids, the probability of a random replacement returning an amino acid site to its original identity is $1/20$ (assuming equal frequencies), whereas for nucleotides it is $1/4$.

PROBLEM 8.2   Use the data in the preceding example to estimate the average rate of amino acid replacement in the signal peptide of γ-interferon during the divergence of mice and humans. Based on fossil evidence, the separation of these species occurred approximately 80 million years ago.

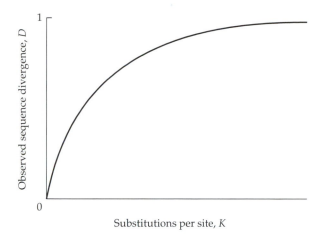

Figure 8.4   As sequences become more divergent over time, the number of substitutions per site (K) can continue to increase, but the proportion of sites that mismatch in the observed sequences (D) saturates.

ANSWER   For the signal peptide, $\hat{D} = 0.56$ and $\hat{K} = -\ln(1 - 0.56) = 0.82$. The estimated rate of evolution is therefore $0.82/[2 \times (80 \times 10^6)] = 5.1 \times 10^{-9}$ amino acid replacements per amino acid site per year. The standard deviation of $\hat{K}$ is estimated as equal to $[0.56/(0.44 \times 18)]^{1/2} = 0.27$. With such a small sample size, the estimates could ordinarily not be taken too literally. However, in this case, the average rate for the signal sequence is very close to the average rate for the molecule as a whole. For γ-interferon, among 155 amino acid sites there are 91 differences, giving $\hat{K} = 0.88 \pm 0.22$ and an average rate of 5.5×10^{-9} amino acid replacements per amino acid site per year.

Rates of amino acid replacement vary over a 500-fold range in different proteins. The rate of amino acid replacement in γ-interferon is one of the fastest rates known (Li et al. 1985). Among the slowest rates is that of histone H4, for which $\hat{\lambda} = 0.01 \times 10^{-9}$ per year. The average rate among a large number of proteins is very close to the rate found in hemoglobin, which is approximately 1×10^{-9} amino acid replacements per amino acid site per year.

To be concrete about the interpretation of the rate of amino acid replacement, consider a protein exactly 100 amino acids in length, in which the rate of amino acid replacement per amino acid site equals 1.0×10^{-9} per year. For the entire protein, the rate of replacement equals $100 \times 1.0 \times 10^{-9} = 1 \times 10^{-7}$ per year. In two different species, therefore, the protein would accumulate amino acid differences at the rate of one replacement every 5 million years since their divergence from a common ancestor [because $(5 \times 10^{6}) \times 2 \times (1 \times 10^{-7}) = 1.0$].

The simple model that we just examined makes an assumption that is violated by an abundance of data. We assumed that all amino acid replacements occur with equal likelihood. Besides the fact that real proteins violate this assumption, we might not have expected it to be true, since some amino acid changes require a single underlying nucleotide change, while others require two or even three changes. More sophisticated models for amino acid sequence evolution account for these differences by weighting amino acid changes with their observed rates of change (Dayhoff 1972; Jones et al. 1992).

Rates of Nucleotide Substitution

Nucleotide sequences are analyzed in the same manner as amino acid sequences, but the analogous equation to (8.1) is slightly more complicated because it has to correct for cases in which a substitution makes two previously different nucleotide sites identical. The correction is significant for nucleotide sequences because an expected one third of random substitutions will make two previously different nucleotides identical. The correction is usually unnecessary for proteins because only $1/19$ of random replacements make two previously different amino acids identical.

Several models of nucleotide substitution have been studied, which differ primarily in the assumptions about rates of mutation between pairs of nucleotides. The simplest model is one in which mutation occurs at a constant rate, and each nucleotide is equally likely to mutate to any other (Jukes and Cantor 1969). If α is the rate of mutating from one nucleotide to a different nucleotide, then in any time interval, A mutates to C with probability α, A mutates to T with probability α, and A mutates to G with probability α. The probability that A does not mutate in this interval is therefore $1 - 3\alpha$. The probability that a particular site is A at time $t + 1$ is

$$P_{A(t+1)} = (1 - 3\alpha)P_{A(t)} + \alpha(1 - P_{A(t)}) \qquad 8.8$$

because the first part of the equation gives the probability of having been A at time t and not mutating, and the second part is the probability of being any other nucleotide and mutating to A. From 8.8 it follows that

$$P_{A(t+1)} - P_{A(t)} = dP_{A(t)}/dt = -4\alpha\, P_{A(t)} + \alpha \qquad 8.9$$

Solving this differential equation,

$$P_{A(t)} = 1/4 + 3/4\, e^{-4\alpha t} \qquad 8.10$$

assuming that the initial state was A. This is the transition probability from A to A, which we can write as P_{AA}. If we observe two sequences that have been separated for time t, then the probability that they continue to carry the same nucleotide at a particular site is

$$P_{AA} = \frac{1}{4} + \frac{3}{4} e^{-8\alpha t} \qquad 8.11$$

because $2t$ is the total duration of time along both lineages during which changes could occur. Let d be the proportion of nucleotide sites that differ between two sequences:

$$d = 1 - P_{AA} \qquad 8.12$$

so

$$d = \frac{3}{4}(1 - e^{-8\alpha t}) \qquad 8.13$$

In the previous symbols, λ is the rate of mutation to a nucleotide different from the current nucleotide, so relating this to α, we have $\lambda = 3\alpha$. This implies that $k = 2\lambda t = 2(3\alpha t) = 6\alpha t$. Taking logarithms of both sides of Equation 8.13, we deduce

$$8\alpha t = -\ln(1 - 4\hat{d}/3) \qquad 8.14$$

and, since $k = \frac{3}{4}(8\alpha t)$,

$$\hat{k} = -\frac{3}{4} \ln(1 - 4\hat{d}/3) \qquad 8.15$$

where k is the expected proportion of nucleotide sites that differ between two sequences at a time t units after their evolutionary separation. By analogy with protein evolution, $\hat{D}$ is the observed proportion of L nucleotide sites in which the sequences differ. The variance Var $(\hat{k})$ of the estimate can be estimated as

$$\text{Var}(\hat{k}) = \hat{d}(1 - \hat{d})/[L(1 - 4\hat{d}/3)^2] \qquad 8.16$$

Figure 8.5 shows the relationship between time and d, and shows that nucleotide sequences that follow the Jukes-Cantor pattern of mutation (all nucleotides equally interchangeable) approach an asymptote, showing a divergence of $\frac{3}{4}$. This makes intuitive sense because, after sufficient time, the common ancestry of the sequences has been erased, and $\frac{1}{4}$ of the sites will match by chance.

PROBLEM 8.3   The coding region of the *trpA* genes in strains of the related enteric bacteria *Escherichia coli* strain K12 and *Salmonella typhimurium* strain LT-2 were sequenced and compared (Nichols and

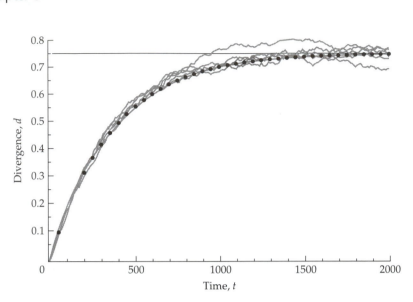

Figure 8.5   Simulations of the substitution process for nucleotide sequences show that the sequence divergence saturates at $d = 0.75$. The jagged lines are numerical simulations of a sequence of length 1000, and the dots give the prediction under the Jukes-Cantor model.

Yanofsky 1979). The *trpA* gene codes for one of the subunits of the enzyme tryptophan synthetase used in the synthesis of tryptophan. Estimate the amount of nucleotide divergence $\hat{k}$ and amino acid divergence $\hat{K}$ and their standard deviations.

K12:	GTC	GCA	CCT	ATC	TTC	ATC	TGC	CCG	CCA	AAT	GCC	GAT	GAC	GAC	CTG	CTG	CGC	CAG	ATA	GCC
	Val	Ala	Pro	Ile	Phe	Ile	Cys	Pro	Pro	Asn	Ala	Asp	Asp	Asp	Leu	Leu	Arg	Gln	Ile	Ala
LT2:	ATC	GCG	CCG	ATC	TTC	ATC	TGC	CCG	CCA	AAT	GCG	GAT	GAC	GAT	CTT	CTG	CGC	CAG	GTC	GCA
	Ile	Ala	Pro	Ile	Phe	Ile	Cys	Pro	Pro	Asn	Ala	Asp	Asp	Asp	Leu	Leu	Arg	Gln	Val	Ala

ANSWER   For the amino acid sequences, $L = 20$ and $\hat{D} = 2/20 = 0.10$; thus $\hat{K} = -\ln(0.90) = 0.105$ with standard deviation 0.074. For the nucleotide sequences, $L = 60$ and $\hat{d} = 9/60 = 0.15$; thus $\hat{k} = -\frac{3}{4}\ln(0.8) = 0.167$ with standard deviation 0.058. Assuming that *Escherichia* and *Salmonella* diverged at around the time of the mammalian radiation 80 million years ago, the rates of evolution are $0.167/(2 \times 80 \times 10^6) = 1.04 \times 10^{-9}$ nucleotide substitutions per year and $0.105/(2 \times 80 \times 10^6) = 0.66 \times 10^{-9}$ amino acid replacements per year. In the gene as a whole, the values are $\hat{k} = 0.300$ for nucleotide substitutions and $\hat{K} = 0.162$ for amino acid replacements.

The Jukes-Cantor model assumes that all possible nucleotide changes occur at an equal rate. In fact, it is generally observed from sequence comparisons that transitions, or changes either from purine to purine (G⇔A) or from pyrimidine to pyrimidine (C⇔T) are more frequent that transversions (the other possible changes). Kimura (1980a) sought to accommodate this observation by making a model with two mutation-rate parameters. Transitions occur with rate α and transversions occur with rate β. The rate matrix below shows the parameters of the Kimura two-parameter model; as you might guess, other models can also be specified by adding parameters to this table. These models can be fitted to the data in a variety of ways, including solutions of the sort we derived for the Jukes-Cantor model, as well as with more complex numerical methods.

Rate matrix for the Kimura two-parameter model:

		Ending base			
		A	C	G	T
Initial base	A	–	β	α	β
	C	β	–	β	α
	G	α	β	–	β
	T	β	α	β	–

Usually we have data on more than two sequences, and estimates of the parameters of the substitution models are sought. If the phylogeny of the organisms in the data set is known, it is possible to calculate the likelihood of the observed sequences given the phylogeny and the parameters in the model (Felsenstein 1981). Many advances have been made in recent years in applying the method of maximum likelihood for estimating parameters of the substitution process in this context (Goldman 1993; Yang 1996a).

Other Measures of Molecular Divergence

Rates of evolution and divergence times can be estimated from other kinds of molecular data if care is taken to consider carefully how the process of mutations results in differences in the data that are actually scored. For example, Randomly Amplified Polymorphic DNA (RAPD), which is analyzed by the polymerase chain reaction, can be used to estimate nucleotide divergence only if one can verify some questionable assumptions about how PCR reactions work (Clark and Lanigan 1993). VNTR loci (loci that are polymorphic due to variable numbers of tandem repeats generated by unequal exchanges) have a forward and back mutation pattern that results in very different population dynamics. In a similar manner, microsatellites, also known as STRPs (short tandem repeat polymorphisms), undergo increases and decreases in copy number such that small changes in copy number are more common than large changes. The result is a sort of stepwise mutation process, and models of this have yielded predictions about patterns of microsatellite variability that are roughly concordant with observations (Zhivotovsky and Feldman 1995).

THE MOLECULAR CLOCK

Although the rate of nucleotide substitution and amino acid replacement varies among different genes, the average rate of molecular evolution can be rather uniform throughout long periods of evolutionary time. Such uniformity in the rate of amino acid replacement or nucleotide substitution, first noted by Zuckerkandl and Pauling (1962), is known as a **molecular clock**.

An example of the approximate uniformity in amino acid substitutions is illustrated in the evolution of the α-globin gene in the organisms depicted in the phylogenetic tree in Figure 8.6. The data are summarized in Table 8.1. The numbers above the diagonal are the percent amino acid differences ($\hat{D} \times 100$) between the α-globin sequences. For example, the α-globin genes of dog and human differ in 16.3% of their amino acid sites; since mammalian α-globin contains 141 amino acids, this percentage corresponds to 23 sites in which the amino acids differ. The percentages exclude differences that result from the insertion or deletion of amino acids, which are called **gaps** in sequence comparisons. For example, the comparison between human and shark α-globin is based on 139 amino acid sites that are homologous, and excludes gaps amounting to 11 additional amino acid sites. Missing from Figure 8.6 are plants, which (remarkably) also have sequences, known as leghemoglobin, that show significant homology to vertebrate globins (Landsmann et al. 1986).

Beneath the diagonal in Table 8.1 are the estimated proportions of differences per amino acid site, calculated from Equation 8.5 as $\hat{K} = -\ln(1 - \hat{D})$. The table also gives the average value of $\hat{K}$ in all comparisons with the shark, carp, newt, chicken, echidna, kangaroo, and dog, respectively, and the divergence times from the bifurcations in Figure 8.6.

The average proportion of differences per site is plotted against divergence time in Figure 8.7. The very close fit to a straight line is evident. Since the divergence time is exactly half of the total time available for evolution (Figure 8.6), the rate of evolution λ can be estimated as one-half times the slope of the line in Figure 8.7. For these data, the slope is 1.8×10^{-9}, and therefore $\hat{K} = 0.9 \times 10^{-9}$ amino acid replacements per amino acid site per year. The good fit of the points to the straight line indicates that the actual rate of α-globin evolution has deviated little from the average for the past 450 million years.

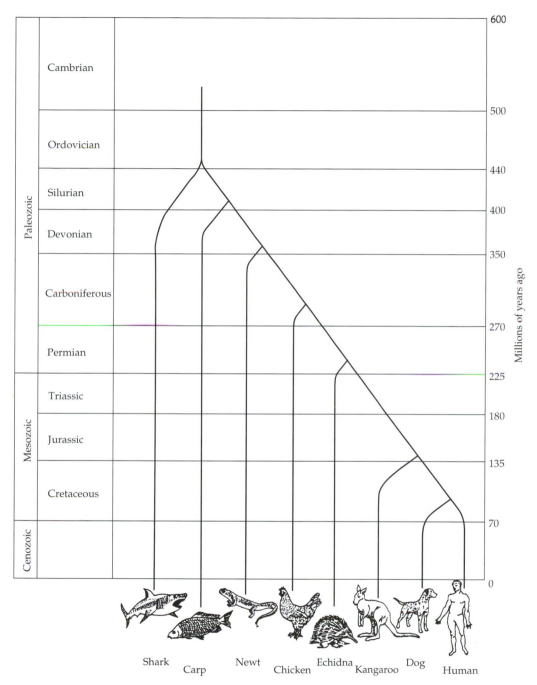

Figure 8.6 Phylogenetic relationships among eight vertebrate species and their approximate times of evolutionary divergence. (From Kimura 1983.)

TABLE 8.1 RATE OF EVOLUTION IN THE α-GLOBIN GENE

	Shark	Carp	Newt	Chicken	Echidna	Kang	Dog	Human
Shark		59.4	61.4	59.7	60.4	55.4	56.8	53.2
Carp	0.90		53.2	51.4	53.6	50.7	47.9	48.6
Newt	0.95	0.76		44.7	50.4	47.5	46.1	44.0
Chicken	0.91	0.72	0.59		34.0	29.1	31.2	24.8
Echidna	0.93	0.77	0.70	0.42		34.8	29.8	26.2
Kang	0.81	0.71	0.64	0.34	0.43		23.4	19.1
Dog	0.84	0.65	0.62	0.37	0.35	0.27		16.3
Human	0.76	0.67	0.58	0.28	0.30	0.21	.018	
Avg $\hat{K}$	0.87	0.71	0.63	0.35	0.36	0.24	0.18	
Time	450	410	360	290	225	135	80	

(Percentage data from Kimura 1983.)

Note: Values above the diagonal are the observed percent amino acid differences (*D*) between the α-globin sequences in the species, values in **boldface** are the expected amino acid differences per site [$\hat{K} = -\ln(1 - D)$]. Average values of $\hat{K}$ and the estimated times of divergence (in millions of years) are given at the bottom of the table. Abbreviation: Kang, kangaroo.

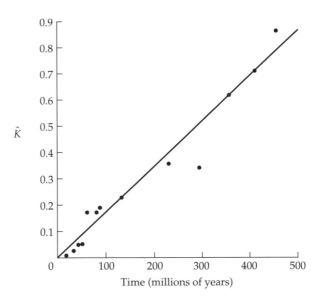

Figure 8.7 Relation between estimated number of amino acid substitutions in α-globin ($\hat{K}$) between pairs of the vertebrate species in Figure 8.6, against time since each pair diverged from a common ancestor. The straight line is expected based on a uniform rate of amino acid substitution during the entire period. (From Kimura 1983.)

PROBLEM 8.4   The β-globin molecule in primates contains 146 amino acids, and estimates of the number of amino acid differences among various primates are tabulated below (data from Kimura 1983). Calculate the average rate of evolution of β-globin molecule in primates. (*Hint:* First calculate $\hat{D}$ and $\hat{K}$ for each species pair, then plot the points with time on the x axis and $\hat{D}$ on the y axis. Finally, do a linear regression to estimate the average rate of substitution.)

Time of divergence (millions of years)	Average number of amino acid differences
85	25.5
60	24.0
42	6.25
40	6.0
30	2.5
15	1.0

ANSWER $\hat{D}$ values are obtained by dividing each number of amino acid differences by 146, and average values of $\hat{K}$ are estimated as $-\ln(1 - \hat{D})$. The average $\hat{K}$ values, from top to bottom, are 0.192, 0.180, 0.044, 0.042, 0.018, 0.007, respectively. These are the y values in the linear regression, and the x values are the divergence times. Altogether there are $n = 6$ points. In this case, $\Sigma(xy) = 3.1263 \times 10^7$, $\Sigma(x) = 2.72 \times 10^8$, $\Sigma(y) = 0.482$, and $\Sigma(x^2) = 1.5314 \times 10^{16}$. The slope of the regression is 3.15×10^{-9}, and the rate of evolution is half of this, or 1.58×10^{-9} amino acid replacements per amino acid site per year. This estimate is reasonably close to the value of 0.9×10^{-9} per year calculated for α-globin. (*Note:* Rather than calculate $\hat{K}$ from the average number of amino acid differences, it would be more accurate to calculate $\hat{K}$ for each species comparison and then take the average; however, in this example, it makes very little difference.)

Variation across Genes in the Rate of the Molecular Clock

If an organism has a particular rate of mutation in its genome, one might think at first that the rate at which the molecular clock runs would be the same for all genes. But the neutral theory predicts that the rate of molecular evolution should depend on the *neutral* mutation rate, which may be quite a bit lower than the overall mutation rate, and may vary widely across genes. Figure 8.8 shows that three different proteins in the same organisms have

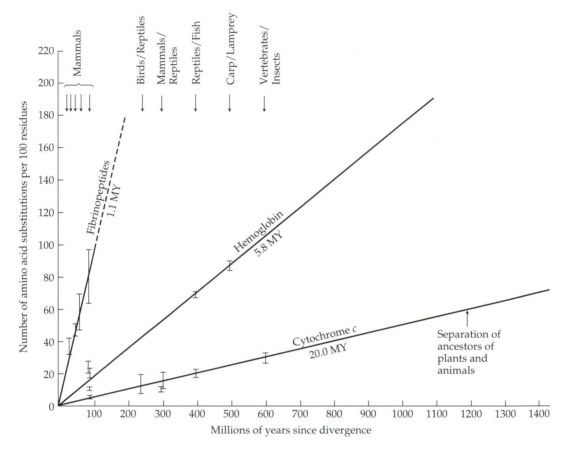

Figure 8.8   The molecular clock runs at different rates in different proteins. One reason is that the neutral substitution rate differs among proteins. Fibrinogen appears to be relatively unconstrained and has a high neutral substitution rate, while cytochrome *c* has a lower neutral substitution rate, and may be more constrained. Data are from a wide variety of organisms. (From Dickerson 1971.)

widely differing molecular clock rates. Nevertheless, within each gene, we observe reasonably uniform rates of change. The variation across genes appears to be due to the fact that some proteins are highly tolerant of substitutions, whereas others suffer deleterious effects from even one or a few minor changes. Genes whose function is well buffered from the environment generally have a slower rate of substitution than genes whose products have a premium on variability. The extremes are represented by histone H4, at the low end, and γ-interferon, at the high end, with globin proteins near the middle of the spectrum. In short, the molecular clocks for different genes "tick" at different rates.

In addition to functional constraints affecting substitution rate, the pattern of hereditary transmission also affects substitution rate. Organelle genomes are replicated and transmitted in a manner distinct from nuclear genes, so it may not be surprising that they undergo substitutions with different dynamics. Mitochondrial DNA exhibits wide variation in substitution rates across its relatively tiny genome, but in animals the substitution rate is generally much higher than the substitution rate of chromosomal genes. In plants, on the other hand, comparisons among nucleotide substitution rates of nuclear DNA, chloroplasts, and mitochondria reveal clear differences, with mtDNA showing less than one-third the substitution rate of chloroplast DNA, which in turn has about half the substitution rate of nuclear genes (Wolfe et al. 1987). In general, genes on the X chromosome have a lower rate of substitution than do genes on autosomes (Miyata et al. 1987). A higher rate of mutation in males (Shimmin et al. 1993) would lower the X-chromosome rate because the X chromosome spends more time in females. But the substitution rate for Y-linked genes is indistinguishable from that of autosomes (McVean and Hurst 1997), which suggests that the mutation rate is equal in both sexes but lower in X-linked genes than in autosomal genes.

Not only do substitution rates vary from one gene to another, but they also vary widely across sites within each gene! If all sites did undergo substitution at the same rate, then the number of substitutions per site should have a Poisson distribution. Fitch and Margoliash (1967) noticed that the cytochrome c data did not fit this model unless invariant and hypervariable sites were excluded. The models that we have developed so far assume that all sites evolve in the same way, so to accommodate this variability (and to test for how different the rates are) models that specifically incorporate rate variation must be developed. One convenient model is to assume that the rates vary according to a gamma distribution (Golding 1983; Wakeley 1993). Yang (1996b) reviews estimates of the rate-variation parameter of the gamma distribution, and finds that all 17 cases examined show significant among-site variation in substitution rate.

Variation across Lineages in Clock Rate

The neutral theory predicts that the rate of the molecular clock should run at different rates for different organisms having different neutral mutation rates. The range of mutation rates is impressive. Figure 8.9 shows the number of nucleotide differences observed in the influenza NS genes, plotted against the year of isolation of the virus containing them. The rate of gene substitution averages $\hat{\lambda} = 1.94 \pm 0.09 \times 10^{-3}$ nucleotide substitutions per nucleotide site per year. Although the rate of gene substitution is about 10^{6}-fold faster than observed in germline genes in eukaryotes, it is neverthe-

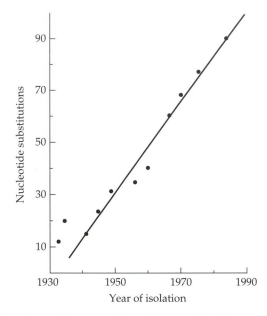

Figure 8.9   Molecular evolution in the *NS* genes of influenza virus determined from strains isolated and stored during the past 60 years. The total rate of evolution in the 890-nucleotide sequence averages 1.73 ± 0.08 nucleotide substitutions per year, and the rate is remarkably uniform. (From Buonagurio et al. 1986.)

less approximately constant during the period available for study. The extraordinary rate of evolution in influenza virus is thought to be related to a high rate of spontaneous mutation resulting from errors in replication (Holland et al. 1982). As in many other RNA-based viruses, the RNA replicase enzyme that replicates the influenza genome lacks a proofreading function. Rapid rates of gene substitution can be of immense medical significance. Yokoyama et al. (1988) estimated the rate of substitution in the *pol* gene of the human immunodeficiency virus as 0.5×10^{-3} per nucleotide site per year. The time of divergence between HIV1 and HIV2 was estimated at just 200 years ago, and the bulk of the genetic variability among recently isolated strains of HIV1 has been generated in the last 20 years.

The rate of the molecular clock also varies among taxonomic groups (Britten 1986). For example, the insulin gene evolved much more rapidly in the evolutionary line leading to the guinea pig than in other evolutionary lines (King and Jukes 1969), and the C-type viral sequences integrated into the primate genome evolved at twice the rate in Asian primates as in African primates (Benveniste 1985). Figure 8.10 illustrates another example of a retardation in the clock in one lineage. Such departures from constancy of the clock rate pose a problem in using molecular divergence to date the times of

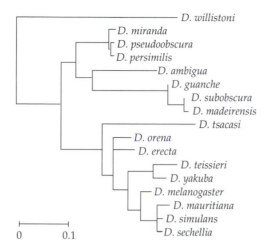

Figure 8.10   Gene genealogy of *Drosophila Adh* sequences showing a significant slow-down of substitutions in the *pseudoobscura* clade. (After Takezaki et al. 1995.)

existence of most recent common ancestors. Before this inference can be justified, one needs to know that the set of species one is examining have a uniform clock.

PROBLEM 8.5   The simplest way to test whether substitutions have occurred at the same rate in different organisms is to consider a tree like that in Figure 8.11. We expect that the divergence between A and C should be the same as the divergence between B and C if the clock is uniform on all branches. Tests of this hypothesis are known as **relative rate tests**. Any site that underwent a substitution along the branch from X to C (but not on the other branches) will have the property that A = B ≠ C. Sites that underwent a substitution on the branch from X to B (but not the other branches) will show A = C ≠ B. Tajima (1993) showed that a simple and robust relative rate test could be performed by simply doing a chi-square test of the null hypothesis that the numbers of these two kinds of sites are equal. Suppose we observe sequences as follows:

```
A    ATG  CTA  GCA  TGC  ATG  CTA  GC
B    ATC  CTA  GCA  TCC  ATG  GTA  GT
C    ATG  CTA  TCA  TGC  TTG  GTA  GC
```

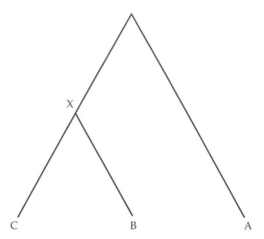

Figure 8.11   A simple tree for illustrating the relative rate test of Tajima (1993).

Calculate the observed and expected numbers of sites in the two categories ($A = B \neq C$ and $A = C \neq B$), and calculate the chi-square statistic to determine whether they are equal.

ANSWER   The observed number of sites for which $A = B \neq C$ is 2, and for $A = C \neq B$ there are 3 sites. Sites where $A = B = C$ or $A \neq B \neq C$ are ignored in this test. The expected number of sites of the two types is each $(2 + 3)/2$, so the chi-square tests gives $(2 - 2.5)^2/2.5 + (3 - 2.5)^2/2.5 = 0.2$, which is clearly not significant. This example had insufficient data for an adequate test, but it provides an example in which there is no evidence for significant difference in rates. A more flexible but more involved test, based on maximum likelihood, can be found in Muse and Weir (1992).

The Generation-Time Effect

One observed feature of molecular evolutionary clocks is that their rate is approximately constant in a time scale measured in years. This is quite unexpected because mutation rates are thought to be more nearly constant when measured in generations. However, the appropriate time scale of molecular evolution is not completely settled (Easteal 1985), as there is some evidence that the rate of synonymous substitution in genes in the rodent lineage (short generation time) might be about two times as rapid as occurs in the same genes in the human lineage (Wu and Li 1985; Li and Wu 1987). Evidence from immunoglobulin genes further suggests that among mammals, the primate lineage has the slowest rate of nucleotide substitution (Sakoyama et al. 1987).

Even if true, a nearly constant rate of gene substitution per year is not necessarily in conflict with a constant rate of neutral mutation per generation. The reason is that organisms with short generation times tend to be small and to maintain large population sizes. In such organisms, the proportion of nearly neutral mutations will be reduced because effective neutrality requires that $Ns \ll 1$, where s is the selection coefficient against the mutation. However, the smaller proportion of nearly neutral mutations in these organisms is offset against the occurrence of more mutations per unit time than in larger organisms, because the generation time is shorter. Thus, the effects of short generation time and larger population size act in opposite directions and tend to cancel out (Crow 1985).

Does the Constancy of Substitution Rates Prove the Neutral Theory?

The possibility that gene substitutions might occur at an approximately constant rate gave some credence to the simplest version of the neutral theory. Theoretical principle 2, discussed earlier in this chapter, states that the expected rate of substitution of neutral alleles equals the rate of mutation μ to neutral alleles. Therefore, on the face of it, the occurrence of molecular clocks would seem to support the neutral theory. But when we dig a bit deeper into the predictions of the molecular clock, we find that things are not necessarily so simple.

In a theoretically perfect molecular clock driven by a random process identical to that of radioactive decay (a Poisson process), the variance in the rate of ticking would be equal to the average rate of ticking. Tests based on the number of substitutions between pairs of species in three proteins showed that the variance was significantly larger than the mean (Ohta and Kimura 1971). Langley and Fitch (1974) backed this up by an analysis in which they estimated the number of substitutions on each branch of the phylogenetic tree, and compared the mean and variance of these counts for each branch. Again, there was a highly significant excess variance. Gillespie (1989) examined the ratio R of the variance to the mean number of substitutions in a set of four nuclear and five mitochondrial genes in mammals, and found that R ranged from 0.16 to 35.55. (The value of 35.55 is for cytochrome oxidase II, which shows 65 amino acid differences between human and mouse, 61 differences between human and cow, and only 21 differences between mouse and cow.) Gillespie argued that the large range of R implies a sixfold difference among mammalian lineages in rates of nucleotide substitution. This excess variance in substitution rate has been called an "episodic clock," characterized by periods of stasis alternating with periods of rapid substitution.

Why does the clock appear to be episodic? One possible reason is that the substitution process is not really a simple Poisson process. If instead the rate itself changes in a random or stochastic manner, the data could be fitted much better. Such a process, where the substitution rate for a Poisson process is itself stochastic, is called a doubly stochastic process, and it does indeed seem to fit the data better (Gillespie 1991). Such a compound Poisson process ought to show clusters of rapid change separated by periods of relative quiescence, a pattern that is generally supported by the data (Gingerich 1986; Gillespie 1989, 1991). One means of causing variation in the substitution rate is natural selection in a stochastically varying environment, and such models can also fit the data satisfactorily (Gillespie 1986). Takahata (1987) has argued that the variance can be inflated by a "fluctuating neutral space" model, in which changes in selective constraints among lineages result in variation in substitution rate among lineages. The dynamics of substitutions are sufficiently complicated that a wide range of models can fit the data, but for now, one thing we are sure of is that the simplest Poisson process is not adequate.

PATTERNS OF NUCLEOTIDE AND AMINO ACID SUBSTITUTION

We have now seen several examples illustrating the general principle that nucleotide substitutions occur at a greater rate than amino acid replacements. The difference in rates, sometimes much greater than in these data, results from redundancy in the genetic code. As illustrated in Table 8.2, the codons for eight amino acids contain N (standing for any nucleotide) in their third position, seven terminate in Y (any pyrimidine, which means T or C), and five terminate in R (any purine, which means A or G). Coding sites containing an N are called **fourfold degenerate sites** because any of the four nucleotides will do, and those containing a Y or R are **twofold degenerate sites** (Li et al. 1985). Because of degeneracies, nucleotides in a gene can change without affecting the amino acid sequence. These changes are called **synonymous** or silent nucleotide substitutions. Nucleotide substitutions that do change amino acids are **nonsynonymous** substitutions.

Calculating Synonymous and Nonsynonymous Substitution Rates

In calculations involving synonymous and nonsynonymous nucleotide sites, the total number of synonymous sites is calculated as the number of fourfold degenerate sites plus one-third of the number of twofold degenerate sites. The total number of nonsynonymous sites in a coding region is defined as the number of nondegenerate sites (nucleotides in which any change results in an amino acid substitution), plus two-thirds of the number of twofold degenerate sites (the latter because, with random mutation at twofold

TABLE 8.2 **DEGENERACY IN THE GENETIC CODE**

		Second nucleotide in codon		
	T	**C**	**A**	**G**
T	TTY Phe TTR Leu	TCN Ser	TAY Tyr TAR Stop	TGY Cys TGA Stop TGG Trp
C	CTN Leu	CCNPro	CAY His CAR Gln	CGN Arg
A	ATH Ile ATG Met	ACN Thr	AAY Asn AAR Lys	AGY Ser AGR Arg
G	GTN Val	GCN Ala	GAY Asp GAR Glu	GGN Gly

First nucleotide in codon (left axis label)

Note: In this representation of the standard genetic code, the symbol N stands for any nucleotide (T, C, A, or G), the symbol Y for any pyrimidine (T or C), and the symbol R for any purine (A or G). The H in the set of codons for isoleucine (Ile) stands for "not-G" (T, C or A). Degeneracies are as follows: N represents a fourfold degenerate site, Y and R represent twofold degenerate sites. The H in the set of codons for isoleucine is considered as twofold degenerate, as are the first nucleotides in four leucine codons (TTA, TTG, CTA, and CTG) and four arginine codons (CGA, CGG, AGA, and AGG). All other nucleotides are nondegenerate.

degenerate sites, two-thirds of the mutations are expected to result in amino acid changes). These conventions are illustrated above.

PROBLEM 8.6   For the sequences of the region of the *trpA* gene given earlier, calculate the synonymous and nonsynonymous substitution rates. Start by using Table 8.2 to assign degeneracy classes to each site. For each difference between *E. coli* and *Salmonella*, the difference is synonymous either if the site is fourfold degenerate or if it is twofold degenerate and the change is a transition (that is, A to G or the reverse, or T to C or the reverse). The difference is nonsynonymous either if the site is nondegenerate or if it is twofold degenerate and the change is a transversion (that is, A or G to T or C). Equation 8.15 is used to estimate the proportion of nonsynonymous nucleotide substitutions per nonsynonymous site and the proportion of synonymous substitutions per synonymous site. The degeneracy assignments are therefore as follows:

	*		*								*				*	*			*	*	*
	004	004	004	002	002	002	002	004	004	002	004	002	002	002	204	204	004	002	002	004
K12:	GTC	GCA	CCT	ATC	TTC	ATC	TGC	CCG	CCA	AAT	GCC	GAT	GAC	GAC	CTG	CTG	CGC	CAG	ATA	GCC
	Val	Ala	Pro	Ile	Phe	Ile	Cys	Pro	Pro	Asn	Ala	Asp	Asp	Asp	Leu	Leu	Arg	Gln	Ile	Ala
LT2:	ATC	GCG	CCG	ATC	TTC	ATC	TGC	CCG	CCA	AAT	GCG	GAT	GAC	GAT	CTT	CTG	CGC	CAG	GTC	GCA
	Ile	Ala	Pro	Ile	Phe	Ile	Cys	Pro	Pro	Asn	Ala	Asp	Asp	Asp	Leu	Leu	Arg	Gln	Val	Ala
	N	S	S								S			S	S				N	N	S

ANSWER   The stars above indicate differences with *Salmonella* and the letters below indicate which changes are nonsynonymous (N) and which are synonymous (S). Altogether there are 38 nondegenerate sites, 12 twofold degenerate sites, and 10 fourfold degenerate sites. The total number of nonsynonymous sites is $38 + (2/3)12 = 46$, and the total number of synonymous sites is $10 + (1/3)12 = 14$. There are three nonsynonymous changes ($\hat{D} = 3/46 = 0.065$) and six synonymous changes ($\hat{D} = 6/14 = 0.429$). Now we use Equation 8.15 to estimate the proportion of nonsynonymous nucleotide substitutions per nonsynonymous site and the proportion of synonymous substitutions per synonymous site. The number of nonsynonymous nucleotide substitutions per nonsynonymous site is $\hat{k} = 0.068$, and the number of synonymous nucleotide substitutions per synonymous site is $\hat{k} = 0.635$.

Estimates of synonymous and nonsynonymous substitution rates for a mammalian protein-encoding gene are plotted in Figure 8.12. A striking observation is that the synonymous rates are generally much greater than the rates of substitution at nonsynonymous sites. These rates are scaled, so that if all mutations were equally likely to go to fixation, the rates would be equal. The depression in nonsynonymous substitution rate is interpreted as being caused by natural selection eliminating those changes that are deleterious. There also appears to be greater variability of nonsynonymous rates than there is in the synonymous rates, although even the latter vary by more than twofold. Figure 8.13 shows that the two rates are correlated, suggesting that either the mutation rates vary from gene to gene or that the constraints on nonsynonymous sites are somehow correlated with those on synonymous sites. We shall see how this correlation might arise at the end of this section.

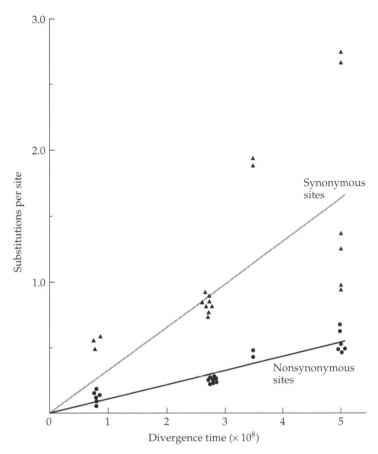

Figure 8.12 Synonymous sites and nonsynonymous sites in β-globin undergo substitutions at different rates, but to a first approximation, both may appear to exhibit a clocklike substitution process. (From Li et al. 1985a.)

One problem that may be apparent with the above method for counting synonymous and nonsynonymous sites is that the status of a particular site may change during evolution. The reason is that changes elsewhere in the codon may make a site that was formerly four-fold degenerate now become two-fold degenerate. In fact, the way the sites are tallied depends on the order in which they are considered. Another way to calculate nonsynonymous and synonymous substitution rates is to consider each codon and count the number of changes that occurred. For codons that changed at a single site, the change is scored as synonymous if there was no alteration in the

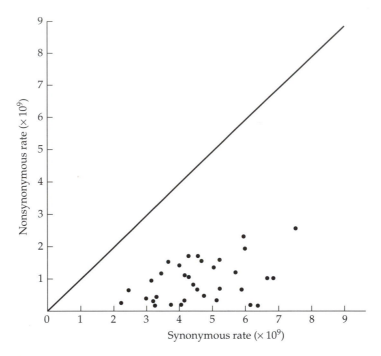

Figure 8.13   Plotting the data of Figure 8.12 in another way, the relative rates of synonymous and nonsynonymous substitutions vary somewhat, but in all cases synonymous rates are lower. (Data from Li, et al. 1985a.)

resulting amino acid sequence, and nonsynonymous if there was an alteration. When there are two differences in a codon, then it is necessary to consider both orders of occurrence, and if we have no reason to assume one order is more likely, then both are considered equally likely. The two orders may have differing numbers of synonymous and nonsynonymous changes. For example, if a codon changes from CCG (proline) to AGG (arginine), it could have done so either through CCG→ACG (threonine)→AGG or through CCG→CGG(arginine)→AGG. The first possibility entails two nonsynonymous changes, whereas the second entails only one. If there are three changes in a codon, there are six possible orders in which they might have occurred. This all-possibilities method, by Nei and Gojobori (1986), seems like an improvement, but actually the estimates come out to be very similar to the method in Problem 8.6. Furthermore, even this method does not avoid the problems of sites changing status due to flanking changes. Far more complicated models are needed to fully avoid this problem, but in the end, the estimates that they give are also very similar to the simplest method outlined in Problem 8.6 (Muse and Gaut 1994; Goldman and Yang 1994).

Paralleling the evolutionary rates for amino-acid-changing substitutions, the rates of nonsynonymous nucleotide substitution vary tremendously among different proteins. Among the slowest rates is that of histone H4, for which $\hat{k} = 0.004 \times 10^{-9}$ substitutions per nonsynonymous nucleotide site per year, and among the fastest is that of γ-interferon, for which $\hat{k} = 2.80 \times 10^{-9}$ substitutions per nonsynonymous nucleotide site per year. The average rate among a large number of proteins is very close to the rate found in hemoglobin, which is 0.87×10^{-9} substitutions per nonsynonymous nucleotide site per year (Figure 8.14). As in the examples given here, rates of nonsynonymous nucleotide substitution are usually quite similar to the rates of amino acid replacement in the same genes.

In contrast with the highly variable rates of nonsynonymous nucleotide substitutions among proteins, the rates of synonymous substitution are much more uniform. For example, in mammalian genes, the fastest rate of synonymous substitution is only 3 to 4 times greater than the slowest rate (see Figure 8.14). However, the average rate, $\hat{k} = 4.7 \times 10^{-9}$ substitutions per synonymous site per year, is not only greater than the average rate of nonsynonymous substitutions, but it is greater than the fastest known rate of nonsynonymous substitutions (for γ-interferon).

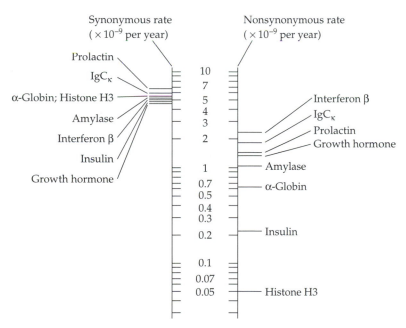

Figure 8.14 Comparison of rates of synonymous and nonsynonymous nucleotide substitutions. Synonymous rates are generally much faster and much more uniform than nonsynonymous rates. (From Kimura 1986.)

The great variability among proteins in the rate of nonsynonymous nucleotide substitution, when contrasted with the much smaller variability found in the rate of synonymous substitutions, is illustrated graphically in Figure 8.14. This disparity has been used as evidence in favor of the neutral theory. Interpreted according to the neutral theory, the variation in rates occurs because there are selective constraints on amino acid substitutions that do not operate as strongly on synonymous nucleotide substitutions. Not just any amino acid will serve at a particular position in a protein molecule, because each amino acid must participate in the chemical interactions that fold the molecule into its three-dimensional shape and give the molecule its specificity and ability to function. The need for proper chemical interactions and folding constrains the acceptable amino acids that can occupy each site. Although some amino acid replacements may be functionally equivalent or nearly equivalent, many more are expected to impair protein function to such an extent that they reduce the fitness of the organisms that contain them. Thus, the constraints on acceptable amino acids are **selective constraints** because unacceptable amino acid replacements are eliminated by selection.

If an amino acid replacement does occur, its effect on the function of the protein product will depend on many factors, but one of the most important determinants of protein conformation is the charge of the amino acid. Different amino acid replacements give different numbers of charge changes, and in most cases the smallest change in charge might be expected to result in the smallest conformational change. Peetz et al. (1986) examined the charge changes in the evolution of seven proteins, and found that hemoglobin α, hemoglobin β, myoglobin, and insulin all accumulated charge changes at a rate slower than expected by random substitution. This finding is consistent with constraints on the conformation of these proteins that limit permissible charge changes. On the other hand, cytochrome c and fibrinogens A and B accumulate charge changes at the expected neutral rate.

For comparison of rates, it would be useful to study rates of nucleotide substitution in stretches of DNA wholly devoid of function and therefore subject exclusively to the whims of mutation and random drift. A likely candidate is found in a class of genes called **pseudogenes**, which are DNA sequences that are homologous to known genes but that have undergone one or more mutations eliminating their ability to be expressed. Pseudogenes are thought to be completely nonfunctional relics of mutational inactivation, and, in fact, their extremely rapid rate of nucleotide substitution is offered in support of this view. The average rate of nucleotide substitution in pseudogenes is faster than the average rate found in intervening sequences, flanking regions, and fourfold degenerate (synonymous) sites. Pseudogenes evolve at the fastest rates known, which may correspond to rates of substitution when DNA is completely unconstrained by natural selection. The fact that fourfold degenerate sites evolve more slowly than pseudogenes may be a suggestion that these sites are not totally lacking in constraint, an idea we shall return to shortly.

Rates of nucleotide substitution also vary within protein molecules. Human insulin is a good illustration. The A and B polypeptide chains found in the mature insulin molecule are created by post-translational cleavage of a longer polypeptide known as preproinsulin. Preproinsulin contains a signal peptide for secretion and an internal C-peptide, neither of which are present in the active molecule. The rates of nucleotide substitution in these three regions are 0.16 for the A and B chains, 0.99 for the C peptide, and 1.16 for the signal peptide. [As in Li et al. (1985), rates are expressed in terms of nonsynonymous nucleotide substitutions per nonsynonymous site per billion years.] In insulin, while there is a sevenfold difference between the maximum and minimum rates of nonsynonymous substitution in different regions of the molecule, the rates of synonymous substitution differ only twofold. Moreover, there is a negative correlation between functional importance and rate of nonsynonymous substitution within the insulin molecule. Many diverse amino acid sequences can serve as signal peptides provided they are hydrophobic, which suggests that selective constraints on signal peptides may be reduced in comparison with sequences in mature polypeptides. In insulin, as expected, the rate of nonsynonymous substitution is fastest in the signal peptide and slowest in the functional subunits of the mature molecule. This kind of negative correlation between selective constraint and substitution rate has also been observed in several other proteins (Li et al. 1985).

Within-Species Polymorphism

So far we have talked only about differences between nucleotide sequences of genes from distinct species. DNA sequence differences between alternative alleles of the same gene in a single species may also be synonymous or nonsynonymous, and it is instructive to compare levels of within-species polymorphism at synonymous vs. nonsynonymous sites. In this case we do not generally talk about substitution rate, but rather quantify the variability with the **nucleotide diversity**. Nucleotide diversity, often symbolized with the Greek letter π, is the probability that a sample of a particular nucleotide site drawn from two individuals will differ. It is essentially the heterozygosity at the nucleotide level.

Figure 8.15 illustrates the first systematic study of DNA sequence variation in a set of 11 alcohol dehydrogenase alleles of *Drosophila melanogaster* (Kreitman 1983). Of the 2659 nucleotides sequenced, 52 were variable across the 11 alleles. The nucleotide diversity over the entire gene was 0.0065 ± 0.0017, meaning that 99.4% of the time, pairs of alleles will match at a site. The level of nucleotide diversity differs in different regions of genes. Figure 8.16 illustrates the estimates of nucleotide diversity found in different parts of the *Drosophila Adh* gene. The different parts are the 5′ (upstream) flanking region, the 5′ transcribed but untranslated region, the coding region (nonsynonymous substitu-

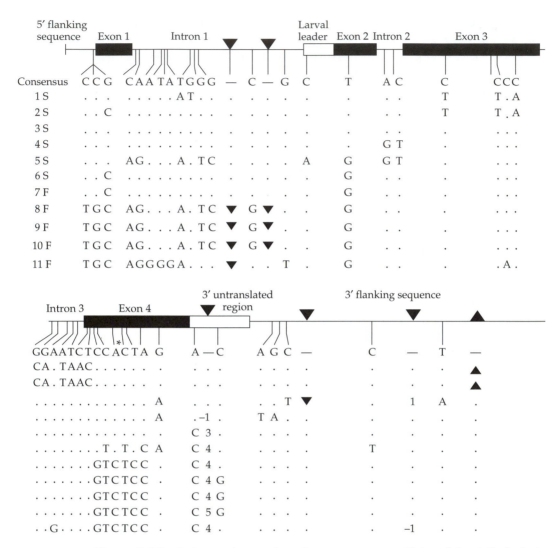

Figure 8.15 Polymorphic nucleotide sites among 11 alleles of the *Adh* alcohol dehydrogenase gene of *D. melanogaster.* The first line gives a consensus sequence for *Adh* at sites that vary; subsequent lines give the nucleotides from each copy for the polymorphic sites. A dot indicates that the site is identical to the consensus sequence. The triangles indicate sites of insertion or deletion relative to the consensus sequence. The star in exon 4 indicates the site of the amino acid replacement (threonine-to-lysine) responsible for the Fast-Slow mobility difference in the Adh protein. (After Kreitman 1983.)

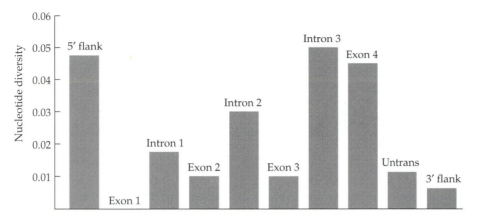

Figure 8.16   Nucleotide diversity in *Adh* of *Drosophila melanogaster*.

tions only, with both the slowest and fastest rates shown), intervening sequences, the 3' (downstream) transcribed but untranslated region, and the 3' untranscribed region. On the average, the fastest rates of substitution occur in intervening sequences and the 3' flanking regions, but the average rates in the 5' flanking regions and the 3' untranslated region are all substantially faster than 0.88×10^{-9}, which is the average rate of nonsynonymous substitution in coding regions (see Figure 8.14). Neutralists would argue that the high rates of substitution in noncoding regions and variation among different parts of the coding region result from varying degrees of selective constraints on different parts of the gene. It is to be emphasized that Figure 8.16 depicts the results for just one gene, and in individual instances, especially in comparisons of closely related species, there may be fewer substitutions observed in flanking sequences than in coding sequences, or fewer changes in synonymous sites than in nonsynonymous sites.

Comparison of nucleotide diversity in different functional regions of a single gene can reveal features of the gene's evolutionary history. For example, in 11 sequenced *Adh* genes of *Drosophila melanogaster* (Kreitman 1983), among 14 substitutions that were observed in the coding region, 13 were silent substitutions. Considering the genetic code and the codon usage in the *Adh* gene, it is possible to calculate what portion of the substitutions would be silent if all substitutions occurred with equal frequency. This figure is about 30% in the case of the *Adh* gene in *Drosophila*, which implies that about 70% of the substitutions would be expected to cause amino acid replacements. Since only one out of 14 observed substitutions was an amino acid replacement, such substitutions are greatly underrepresented. This finding is consistent with the view that most amino acid replacements are eliminated from the population by purifying selection. The same logic can be extended to argue that sequences that are conserved are likely to be functionally important; this type of reasoning led to the identification of a new open reading frame in the HIV (AIDS) virus genome (Miller 1988).

The action of natural selection can sometimes be inferred from levels of synonymous and nonsynonymous polymorphism. For genes that determine surface antigens of pathogens or those that determine the major histocompatibility antigens of mammalian cells, the rates of nucleotide substitution can be quite high. One way to address whether the high rate of substitution is driven by selection is to examine the levels of synonymous and nonsynonymous diversity in these genes. For example, Hughes and Nei (1988) found that in the regions coding for the antigen recognition sites in the class I *MHC* (major histocompatibility complex) genes of humans and mice, the rate of nonsynonymous substitution exceeded the rate of synonymous substitution by a ratio of 3 : 1. This ratio is the reverse of that found in the usual situation and in other regions in the same genes, where silent substitutions are present in excess. The excess of amino acid replacements is consistent with a model in which mutations that generate diversity are often advantageous, and hence natural selection accelerates the substitution process. Endo et al. (1996) developed software to scan the gene sequence databases for cases in which the nonsynonymous rate significantly exceeded the synonymous rate, and they recovered 17 cases. Nine of these 17 cases were cell surface antigens or immune system genes—proteins for which one can easily imagine scenarios in which high levels of diversity are advantageous. High rates of nonsynonymous substitution are also found in protein toxins called colicins that certain bacteria produce to kill potential competitors in their immediate vicinity (Riley 1993; Ayala et al. 1994).

Implications of Codon Bias

Synonymous substitutions occur at a greater rate than nonsynonymous substitutions, implying that they face weaker selective constraints. But are synonymous changes completely neutral, or do they too face some form of constraint? One potential type of constraint occurs through codon preferences, which are correlated with the relative abundance of tRNA molecules that interact with and translate the codons. In bacteria and yeast, for example, highly abundant proteins tend to use codons for abundant tRNA molecules, whereas proteins produced in small amounts tend toward codons for less abundant tRNA molecules (Ikemura 1985). A plot of the frequency of use of the synonymous codons that code for leucine shows that CUG is much more frequent than the others, corresponding to an increased abundance of this tRNA. A second potential constraint on synonymous substitutions occurs through possible secondary structures that the RNA might form, in which certain nucleotides must undergo base pairing (see the next section for an elaboration). Pre-messenger RNA secondary structure may influence the speed or accuracy of intron splicing, rate of transport, or stability. A third potential constraint on synonymous substitutions is related to the fact that, during translation, the probability of misincorporation of the wrong amino acid increases if there is a pause while the translation machinery waits to find a rare tRNA. Such translation errors are known to occur (in fact, mistranslation of an mRNA that bears a frameshift mutation can yield an active protein). Pausing during translation may also be of importance to the folding of the protein into its proper three-dimensional structure.

If synonymous codons are neutral, then one would expect their frequencies of use to correspond to the product of the nucleotide frequencies. If all four bases were equally frequent, all synonymous codons should be used equally frequently. A more subtle way to test for departure from equal codon use is to count the incidence of polymorphisms and substitutions toward or away from the most abundant codon. If the most abundant codon became the most abundant by chance, then the substitutions toward and away from this codon should show no bias. But if the most abundant codon is "preferred," then there will be a deficit of substitutions away from this codon. Application of this kind of approach for codon bias in *E. coli* suggested an average selection coefficient against disfavored codons of about $s = 7.3 \times 10^{-9}$ (Hartl et al. 1994). Even *Drosophila*, whose effective size might be around 10^6, appears to exhibit significant codon preference (Akashi 1995), suggesting that selective constraints on synonymous codons must be greater than 10^{-6} in this organism (Figure 8.17).

From the selectionist viewpoint, while granting that substitutions in pseudogenes may be neutral, and synonymous substitutions may be constrained by natural selection only weakly, it is nevertheless maintained that nucleotide substitutions that change amino acid sequences are inevitably subject to the action of natural selection of an intensity that is sufficient to counteract the effects of random genetic drift. Thus, selectionists would argue that amino acid substitutions that have occurred in a protein during the course of evolution became fixed by natural selection because they increased the fitness of the carriers through improvement in function of the molecule. However, neutralists argue back, the selectionist viewpoint cannot easily explain the negative correlation between functional importance and rate of substitution within proteins. Furthermore, a neutralist might add, even a slightly detrimental mutation has some chance of being fixed unless a population is very large (Chapter 7).

POLYMORPHISM AND DIVERGENCE IN NUCLEOTIDE SEQUENCE DATA

The effects of varying the neutral mutation rate on levels of polymorphism within a species and the interspecific divergence in nucleotide sequences are plotted in Figure 8.18. The theory is consistent with the idea that genes with a high rate of nucleotide substitution, as indicated by a large number of

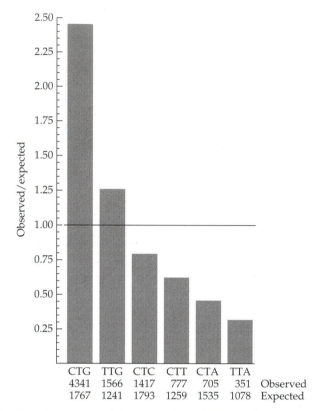

Figure 8.17   The frequency of the six codons that encode leucine in *Drosophila melanogaster* is not uniform. This kind of codon bias, in which one codon is present in excess, is commonly observed. (Data from FlyBase, http://cbbridges.harvard.edu:7081.)

interspecific sequence differences, should also have a high level of intraspecific polymorphism. Polymorphism depends only on the product of the neutral mutation rate and the effective size, through the formula $H = \theta/(1 + \theta)$ that we encountered in Chapter 7. For strictly neutral genes, interspecific divergence does not depend on the population size, but instead follows the formula $k = 2\mu t$. If we compare two genes, the level of intraspecific polymorphism would let us estimate a θ value for each gene. Given the θ value for gene A and the observed interspecific divergence, an estimate of the divergence time could be estimated. For gene B, we would also have a θ estimated from the level of polymorphism, and we could use the divergence time estimated from gene A to determine a predicted value of divergence in gene B.

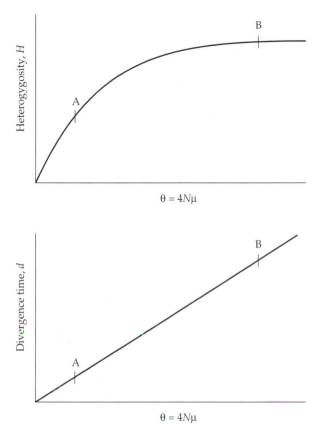

Figure 8.18 Reasoning behind the HKA test. Consider two genes, A and B, that differ in neutral substitution rate. θ can be estimated for each gene based on observed levels of nucleotide heterozygosity (top panel). Given the observed divergence between two species in gene A (determined by the neutral mutation rate and time), the divergence in gene B can be predicted based on its neutral substitution rate, and the divergence time obtained from gene A. The HKA test is a goodness-of-fit test to the observed levels of intraspecific diversity and interspecific divergence under a model whose parameters are population sizes, neutral mutation rates, and times of divergence.

The above reasoning has been formalized in a popular test of neutrality based on nucleotide sequence data within and among species (Hudson et al. 1987). Sequences of at least two genes from a number of individuals of each of two species are needed to apply the test. Define S_i^A and S_i^B as the number of polymorphic nucleotide sites in gene i in species A and B, respectively, and d_i as the number of differences in gene i between a pair of alleles sampled randomly, one from species A and one from species B. The expected values of these parameters are obtained from the infinite-sites neutral model, assuming that the two species diverged t generations ago, that the population sizes are $2N$ and $2Nf$, and that each gene has an associated $\theta_i = 4N\mu_i$. Estimates of θ_i, f, and t are obtained by a least-squares method that gives the best fit of the expressions for the expected values and variances of S_i^A, S_i^B, and d_i to the data, and goodness-of-fit is tested with an appropriate chi-square test. Using data from the *Adh* coding and 5′ flanking regions in *D. melanogaster* and *D. sechellia*, Hudson et al. (1987) found that the observed values deviated significantly from the neutral model in a direction consistent with the operation of balancing selection acting on the coding region of *Adh*. This finding is consistent with Kreitman's (1983) observation of an excess of silent substitutions in *Adh*, except that the test of Hudson et al. makes use of the genetic variation observed within and among species. The "HKA" test has seen many applications in molecular population genetics (Kreitman and Hudson 1991; Aguadé et al. 1992; Begun and Aquadro 1993; Gaut and Clegg 1993).

PROBLEM 8.7   In a set of 12 *Adh* sequences in *Drosophila melanogaster*, McDonald and Kreitman (1991) observed 42 silent (synonymous) polymorphisms and two replacement (nonsynonymous) polymorphisms. As had been concluded by Kreitman (1983), this suggests that most replacement mutations are deleterious and are eliminated from the population. When they examined fixed differences between *melanogaster* and either *D. simulans* or *D. yakuba*, they found that seven of the fixed differences were replacements and seventeen were silent. What is the significance of this observation?

ANSWER   A null hypothesis might be that the effects on fitness of a mutation would be the same whether within a species or at any time along the ancestral history of two species back to the common ancestor. If this is true, then we would expect the ratio of silent to replacement polymorphisms to be the same as the ratio of silent to replacement fixed differences. A simple test of this is to do a 2×2 contingency chi-square:

	Fixed	Polymorphic
Replacement	7	2
Silent	17	42

For this table we get $\chi^2 = 8.20$, and with one degree of freedom, $P < 0.01$. (A correction is often applied to the chi-square for tables with counts less than 5, but it does not make much difference in this case.) The low probability means we reject the null hypothesis and conclude that, within species, there is a tendency to avoid replacement polymorphisms; however, between-species replacement differences are much more likely to occur. McDonald and Kreitman (1991) argue that this pattern is consistent with adaptive fixation of amino acid replacements, since they are relatively more frequent in interspecific comparisons, and such adaptive polymorphisms would be less common than neutral polymorphisms because adaptive differences would not remain polymorphic for as long a duration. This simple test is useful in assessing the relative importance of neutral drift versus selection in interspecific differences.

Impact of Local Recombination Rates

Recall from Chapter 5 that the level of polymorphism in *Drosophila* shows a striking correlation to the local rate of recombination. Regions of low recombination rate are nearly devoid of variation, whereas regions with high rates of recombination are highly polymorphic. The idea of comparing polymorphism and divergence makes this pattern even more striking and allows us to eliminate a possible cause. One possible reason for the correlation is that recombination itself is mutagenic, or that somehow the two processes are related mechanistically. (That is, perhaps when mutations occur, the DNA configuration is altered to increase the chance of recombination.) If this were the case, then the regions of low recombination rate should also have a low mutation rate, and hence lower interspecific divergence. Figure 8.19 shows that a lower divergence is not observed. Levels of interspecific divergence are independent of local recombination rates. The conclusion is that the correlation between recombination rates and levels of polymorphism observed by Aquadro et al. (1994) must be due to more rapid elimination of the variation in regions of low recombination.

Two known mechanisms that remove variation faster in regions of low recombination are selective sweeps and background selection. Background selection is thought to be the primary mechanism for the reduced variation (discussed in Chapter 5 in the section on linkage and recombination), but this does not mean that sweeps do not occur. Selective sweeps occur when a favorable mutation takes place, and selection rapidly increases its frequency. Such sweeps can have a dramatic effect on levels of variation in the selected

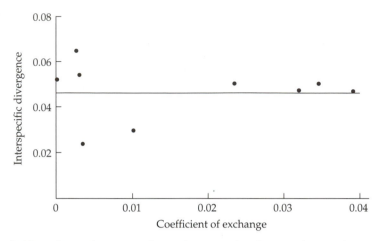

Figure 8.19   The striking correlation between local rates of recombination and levels of intraspecific nucleotide diversity cannot be explained by a lower mutation rate in regions of low recombination. If regions of low recombination had low rates of mutation, the interspecific divergence would be lower in these regions. That it is not is shown by these data. (From Aquadro et al. 1994.)

gene and the region around it. The size of the "swept" region depends on the rate of recombination and is larger for regions of low recombination. This means that the chance that a particular site has been swept free of variation is greater in regions of low recombination, assuming the density of selective sweeps is uniform across the genome. An example of a selective sweep is an *esterase B* allele in the mosquito that is associated with pesticide resistance (Figure 8.20). The resistant allele has apparently undergone a nearly global sweep, judging from the near monomorphism of the *esterase B* gene (Raymond et al. 1991; Ffrench Constant et al. 1991). We do not know how frequent such sweeps are, but one possible means of identifying them is to score many highly polymorphic markers in many populations and look for regions of reduced variation. Schlötterer et al. (1997) performed such a survey and found several cases of individual genes in single populations that were depauperate in variation, perhaps due to a local sweep event.

GENE GENEALOGIES

There is an important distinction between the construction of trees from sequences of genes from different species and from sequences of alleles from a single species. The former yields a customary phylogenetic **gene tree**, while the latter produces what is called a **gene genealogy**. The relationships among species result from macroevolutionary processes, whereas allelic differences result from a number of microevolutionary processes, including aspects of genetic transmission. Once the nucleotide sequences of alleles are

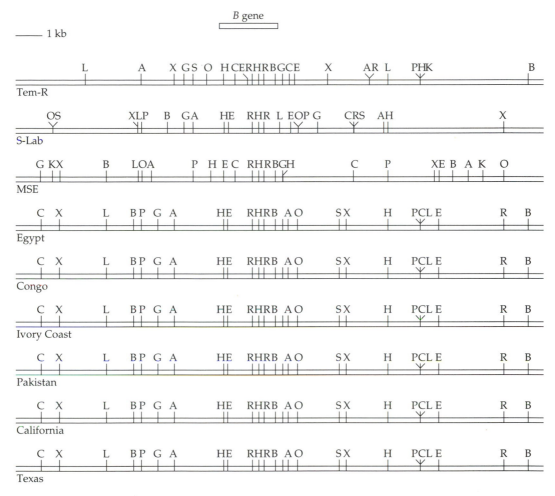

Figure 8.20   Restriction maps of the *esterase B* gene from global samples of the mosquito *Culex pipiens*. Note the identity of a haplotype from Egypt through Texas. This haplotype is associated with insecticide resistance, and probably underwent a global sweep in the face of strong selection. (From Raymond et al. 1991.)

known, the different alleles can be treated like genes in different species in applying standard methods for inferring a phylogenetic tree. However, great care is needed in constructing gene genealogies, because recombination among the sequences results in a gross violation of the assumptions of most tree-building methods. Provided the rate of recombination is not too high, localized blocks of sequence can be identified in which there appears to have been no recombination in the ancestral history of the sampled alleles. With this caveat, gene genealogies can be of great use in inferring the evolutionary

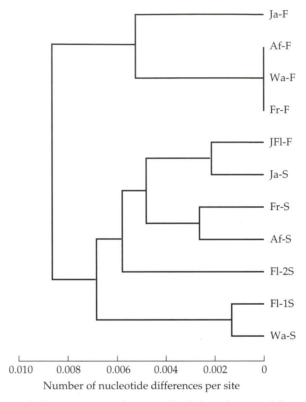

Figure 8.21 A phylogenetic tree for 11 *Adh* alleles of *Drosophila melanogaster* based on 43 nucleotide differences. The scale is the number of nucleotide differences per site. Ja: Japan; Af: Africa; Wa: Seattle, Washington; Fl: Southern Florida; Fr: France. S and F refer to the slow and fast electrophoretic forms. (Data from Kreitman 1983.)

history of a polymorphism. For example, they can reveal which of a group of alleles is older, or which alleles are more closely related to each other. Figure 8.21 shows the gene genealogy from Kreitman's (1983) *Adh* sequence data, and the higher diversity of the *S* allele clearly makes it appear to be older.

Hypothesis Testing Using Trees

Beyond the descriptive approach to showing relationships among alleles, gene genealogies can be used to test fundamental forces of population genetics, including natural selection. For example, consider a phylogenetic tree based purely on neutral variation. As illustrated in Figure 8.22A, when the substitution rate is μ, the expected time to coalescence to a common ancestor for a randomly chosen pair of alleles is 4N generations (Chapter 2). Under a model like Ohta's (1973), where many mutations are slightly deleterious, the tree is not changed very much because the alleles included in a sample are

(A) No selection

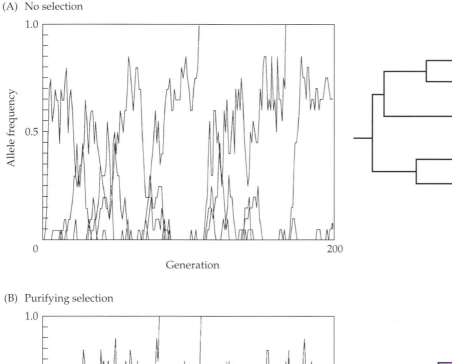

(B) Purifying selection

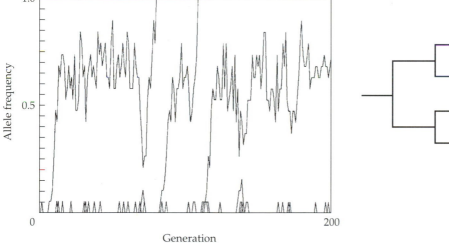

Figure 8.22 Computer simulations of the infinite-allele model of molecular evolution. (A) With strict neutrality, the expected time from mutation to fixation of alleles that will go to fixation is $4N_e$ generations. (B) Purifying selection (in this case with half of the mutations having a fitness of 0.5) results in less polymorphism at any given time. (C, next page) Stabilizing selection (overdominance or frequency dependence) can retain alleles in a polymorphic state for much longer times. Representative trees are plotted to the right of each panel.

(C) Stabilizing selection

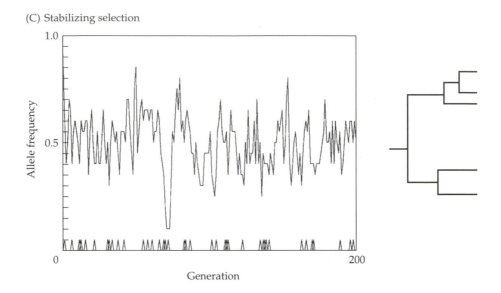

the subset of mutations that occurred that were nearly neutral. On the other hand, in the case of adaptive mutations, the rate of fixation would be much faster than with neutrality, so that sites of adaptive mutation would have shorter coalescence times than flanking neutral sites (Figure 8.22B). Finally, with balancing selection (heterozygote advantage), polymorphisms would be maintained for a longer time than under the pure drift model (Figure 8.22C). The number of statistical methods for inference of population genetic forces from gene genealogies is increasing rapidly, and there is ample opportunity for exciting progress in this area.

PROBLEM 8.8   A study of variation in the gene encoding superoxide dismutase in *Drosophila melanogaster* (Hudson et al. 1994b) revealed 63 polymorphic sites in three *slow* alleles and 22 *fast* alleles (where fast and slow refer to the mobility of the protein product in an electrophoretic gel). An additional 16 *slow* alleles were separately scored, giving a total of 19 *slow* alleles that were found to be identical in nucleotide sequence. The *fast* allele broke into 10 distinct haplotypes, and the most common was *FastA* with nine copies. The partial table of pairwise counts of numbers of sites that differ between alleles is:

	FastA	FastH	FastB	FastJ	FastK
Slow	1	3	4	9	16
FastA		2	3	8	15
FastH			3	10	17
FastB				11	18
FastJ					7

How would you address the question of whether this sample is typical of a sample from a neutral gene?

ANSWER   The aspect of the pattern of variation that is unusual is that the *fast* alleles appear to be quite variable, whereas all 19 *slow* alleles are identical. A gene genealogy of the *fast* alleles would look like a typical neutral tree with roughly exponentially distributed branch lengths, but the complete tree would then have 19 identical *slow* alleles placed one substitution away from *FastA*. The suspicion is that the *slow* allele must have arisen recently and is being pulled to high frequency by selection. An observed increasing trend in the *slow* allele frequency supported this conjecture. To make a formal test out of this observation, Hudson et al. (1994b) used the coalescent procedure described in Chapter 7 to generate simulated data sets with a sample size of 25 and having 63 polymorphic sites. For each of the 10,000 simulated samples, they asked, how often is there a set of 12 alleles that differ by 0 or 1 substitutions? (The 9 *FastA* alleles and 3 slow alleles in the original observed sample differ at just 1 site.) The answer was 81 of the 10,000 cases, giving a probability of 0.0081. The observed sample is not a likely occurrence under neutrality.

It is instructive to note that these data were consistent with neutrality by the Fu and Li (1993) test, the Tajima (1989) test, and the HKA test (Hudson et al. 1987), demonstrating that even strong departures from neutrality may be missed by these standard tests. This problem illustrates a common principle in molecular population genetic analysis, which is that ad hoc approaches tailored to particular observations often are necessary.

The topology of gene trees affords an opportunity for yet another test of goodness of fit of data to the neutral theory. We saw in Chapter 7 that the coalescent approach provides a description for the expected topology of a gene tree under the infinite-sites model. In particular, the expected time back to the next preceding coalescent event is exponentially distributed with parameter $1/\binom{k}{2}$ where k is the current number of distinct alleles. A test of Fu and Li (1993) makes use of the fact that the model predicts a relationship between θ and the number of "external mutations." An external mutation is a mutation that occurs on a branch of the gene genealogy that terminates in an observed allele (an external or terminal branch). The remarkable observation that Fu and Li made was that the expected number of external mutations is θ, *independent of sample size*. The test is based on the idea that selection will affect the number of external branches more than it will affect internal branches, and Fu and Li devised test statistics for goodness of fit between observed and expected numbers of external mutations. The test has some advantages over the Tajima test, but Simonsen et al. (1995), after extensive simulations to test the power of various neutrality tests, conclude that the Tajima test (see Problem 7.10) is generally the most powerful against alternative hypotheses of selective sweeps, population bottlenecks, or population subdivision (but see Problem 8.8).

Inferences about Migration Based on Gene Trees

Data from a panmictic population obeying the infinite-site model will have a characteristic gene tree topology. If the population is divided into two semi-isolated groups, the alleles within each group will, on average, be more similar to one another than comparisons between groups. This would mean that a gene tree in such a subdivided population would have two major clades corresponding to the two populations. For higher levels of migration, the gene tree will be somewhere between these two extremes. Slatkin and Maddison (1989, 1991) devised means for estimating the number of migrants per generation, Nm, from the inferred gene genealogy. In essence, the approach uses parsimony to obtain a direct count of migration events compatible with the tree, and Nm is estimated from this count.

With sufficient DNA sequence data, one can be confident that identical alleles are truly identical by descent, an important aspect of inference of population history and migration. In their analysis of *D. pseudoobscura Adh* sequences, Schaeffer and Miller (1991) found that geographically distant populations had identical alleles, and more generally, that the gene tree did not partition geographically, as though the population were panmictic. This was an exciting result, given the extraordinary level of population subdivision in *D. pseudoobscura* third chromosome inversions. It implies that the latter subdivision is not just a historical accident, but is being maintained in the face of sufficient migration to homogenize other sorts of genetic variation.

The data of Bowcock et al. (1994) show yet another aspect of very high resolution molecular data. After constructing a tree based on 30 microsatellite loci, they observed that human samples showed a significant tendency to cluster by continent. Although lower-resolution methods had shown some degree of dissimilarity among groups of humans, this was the first study to show that reduced intercontinental migration was sufficient to partition human genetic variation.

MITOCHONDRIAL AND CHLOROPLAST DNA EVOLUTION

We already saw in Chapter 5 that mitochondrial DNA can be highly informative about the geographic structure of populations. Some of the advantages of using DNA sequence variation from this organelle genome include:

- The DNA molecule (in most animals) is relatively small and easy to isolate.
- It is present in multiple copies per cell; therefore, older and less well preserved samples are still likely to yield useful information.
- The mitochondrial genome does not undergo recombination, so it is more likely to show a clean branching structure to its gene trees.
- It evolves rapidly.

The primary problems with mtDNA are:

- The absence of recombination means that the gene tree constructed from any mitochondrial DNA gene will reflect just a single realization of the genealogical process. As such, the data will not be as informative about species or population trees as, say, a dozen nuclear genes.
- Much of the work with mtDNA has been on the control region sequence. While this region is highly variable, the variability occurs at a subset of sites that are so mutable that multiple substitutions often occur.

In animals, mitochondria are usually inherited through the egg cytoplasm (maternal inheritance) and are genetically uniform within an individual. The mitochondrial genome consists of a single circular DNA molecule, denoted mtDNA, the size of which varies over a remarkably narrow range in different species of vertebrates (15.7–19.5 kb), averaging about 16 kb. Human mtDNA is fairly typical, containing a control region for the initiation of DNA replication, genes for two ribosomal RNA molecules, 22 transfer RNA molecules, and 13 proteins. Twelve of the proteins are subunits of enzyme complexes that carry out electron transport and ATP synthesis. The genetic code of mammalian mitochondria differs from the standard code in that ATA codes for Met, TGA codes for Trp, and AGR codes for End (termination of protein synthesis); thus, every codon in the mitochondrial code can be written as either NNY or NNR. Animal mitochondria also contain several hundred enzymes used in metabolic functions, but these are coded for by nuclear genes, and the enzymes are transported into the mitochondria.

At the nucleotide level, the rates of substitution in mammalian mtDNA are typically 5 to 10 times greater than occur in single-copy nuclear genes, averaging approximately 10×10^{-9} substitutions per nucleotide site per year. The reason for the high rate of substitution is thought to be either a high rate of nucleotide misincorporation or a low efficiency of repair of the DNA polymerase. Support for the latter view comes from the observation that, unlike

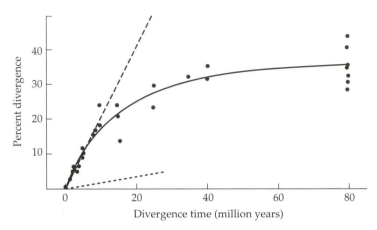

Figure 8.23   Relationship between percent sequence divergence (100*d*) and divergence time. The points represent estimates from pairwise comparisons of restriction endonuclease cleavage maps. The initial rate of mtDNA sequence is shown by the longer dashed line and the rate of divergence of single-copy nuclear DNA by the shorter dashed line. (From Brown et al. 1979.)

the nuclear DNA polymerase, the mitochondrial DNA polymerase lacks the proofreading function. In protein-coding mitochondrial genes, the rate of synonymous substitution is about five times greater than the rate of nonsynonymous substitution, which is comparable with the ratio found in nuclear genes. Mitochondrial tRNA genes in mammals evolve approximately 100 times as rapidly as their nuclear counterparts (Brown 1985; Avise 1986). One result of this faster rate of nucleotide substitution is that the divergence between two sequences saturates relatively soon, so that the linearity of divergence over time (the molecular clock) is an accurate approximation only for species that have diverged less than about 10 million years (Figure 8.23). Exceptions to the elevated rate of mtDNA divergence have been found, notably in *Drosophila* (Powell et al. 1986).

PROBLEM 8.9   The mitochondrial DNA of 21 humans of diverse geographic and racial origin were digested with 18 restriction enzymes, 11 of which exhibited one or more fragments in which size polymorphism occurred (Brown 1980). All restriction site polymorphisms could be explained by single-nucleotide differences, thus there was no evidence for insertions, deletions, or other mtDNA rearrangements. Altogether, 868 nucleotide sites were assayed for differences among individuals, and the average number of differences per nucleotide site per individual was estimated at 0.0018. Assuming that mammalian DNA undergoes sequence divergence at the rate of 5 to

10×10^{-9} nucleotide substitutions per site per year, and that the rate is uniform in time, calculate the length of time since all of the 21 contemporary mtDNA molecules last shared a common ancestor. Calculate the effective size of the population from the level of mtDNA variability.

ANSWER   Given an average number of differences per nucleotide site per individual of 0.0018 and an average rate of divergence of 5 to 10×10^{-9} per site per year, the time of the most recent common ancestor would be between $0.0018/(10 \times 10^{-9})$ and $0.0018/(5 \times 10^{-9})$ or 180,000 to 360,000 years. Assuming a generation time of 20 years, this means that all mtDNA in the diverse sample could have been from a single female in the population between 9,000 to 18,000 generations ago. To estimate the long-term effective size of the population, recall that the expected time to fixation of a newly arisen neutral mutation is $4N_e$ generations. This result applies to an autosomal gene in a diploid species. For mitochondrial genes, only females transmit them, and they are effectively haploid, so the corresponding fixation time for mtDNA is just N_e generations. If we argue that the one mtDNA type went to fixation in 9,000 to 18,000 generations, this is equivalent to saying that the long-term population size has been $N_e = 9,000$ to 18,000. This sounds like a low number, but modern anthropologists find it reasonable, given the population structure of ancient humans and the rapid, nearly starburst-like growth since the adoption of agricultural methods.

One of the most dramatic claims in the history of population genetics was that human genetic variation in mtDNA indicates a recent African origin of modern humans (Cann et al. 1987). This claim was based on restriction site variation among mtDNA of 147 humans in five populations. The 12 restriction enzymes sampled an average of 370 restriction sites per individual, equivalent to assaying 9% of the mtDNA genome per individual. A total of 195 polymorphic sites were found in the genome, and the precise location on the mtDNA sequence of all polymorphic sites was identified. When the 133 distinct mtDNA haplotypes were assembled into a phylogenetic tree, a clade was found in which the most ancient branch pointed to a group of people of African ancestry (Figure 8.24). Given the observed number of differences between the two most divergent mtDNA types, and assuming there is 2 to 4% divergence in mtDNA sequences per million years (estimated from the human–chimp split at 5 MYA), the common ancestor to all of the observed haplotypes was estimated to have existed 140,000 to 280,000 years ago.

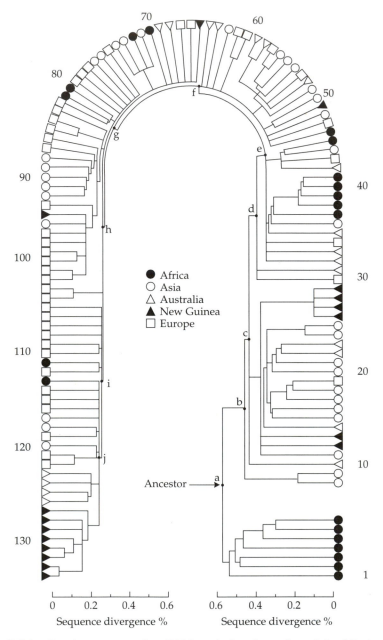

Figure 8.24   Parsimony tree of mtDNA variation from the original "mitochondrial Eve" paper. Much was made of the observation that there is an isolated clade consisting only of Africans. (From Cann et al. 1987.)

Sequences of the control region, which diverge at a rate of 12 to 15% per million years, produce a date for the common ancestor of 166,000 to 249,000 years ago (Vigilant et al. 1991). Several other data sets have been collected to address the issue of date, and all have produced estimates of the date of the common ancestor of human mtDNA of between 100,000 and 400,000 years ago (Hasegawa and Horai 1991; Pesole et al. 1992; Ruvolo et al. 1993). These figures, and their interpretation, have launched a controversy centering on: (1) the best way to infer the time of the common ancestor, (2) the meaning of higher African diversity, (3) the confidence in an African root, (4) the neutrality of human mtDNA variation, and (5) the implications for human evolution. Whether modern humans migrated out of Africa in the past 200,000 years may not be supported with statistical rigor by mtDNA alone (Templeton 1993), but when haplotypes of nuclear genes (Tishkoff et al. 1996), or when many nuclear genes are considered in addition, the case for African origin is strong (Nei and Roychoudary 1993).

We must be careful to realize, however, that the fact that Africa has the greatest genetic diversity today does not by itself guarantee that modern humans originated in Africa. If the African population has had a long-term effective size much larger than other populations, or if the other populations suffered a bottleneck that Africa did not, then Africa would be more diverse no matter where humans originated (Relethford 1995). In addition, just because a gene genealogy appears to have a root that coincides with an African allele does not mean that modern humans came from an expansion of the African population to cover the earth. It only means that the one gene has the observed ancestral history. Other genes may trace back to other origins.

Inferences about human origins from extant patterns of genetic variation require an understanding of nonequilibrium models, where populations grow in size, new colonies are founded, and populations remained connected by some level of migration. Recently there has been much attention paid to the influence of past changes in population size on patterns of variation. It was observed that a growing population produces a gene genealogy that has a more starlike shape than does a stationary population, and this in turn produces a peak in the distribution of pairwise counts of mismatches (Slatkin and Hudson 1991; Rogers and Harpending 1992). The use of patterns of human genetic variation to make inferences about our ancestral history is an active and lively area of inquiry.

Chloroplast DNA and Organelle Transmission in Plants

Chloroplasts are cellular organelles that also have their own genome and also are transmitted in a non-Mendelian fashion. Chloroplast DNA (cpDNA) ranges in size from 135 to 160 kb, and it occurs in multiple copies in each chloroplast. Its structural organization is conserved in higher plants, and the rate of synonymous nucleotide substitution is approximately 1×10^{-9} substitutions per site per year. Thus, the evolution of cpDNA is conservative in

TABLE 8.3 RATES OF SEQUENCE AND STRUCTURAL EVOLUTION IN ORGANELLE DNA

Genome	Rate of nucleotide substitution	Rate of structural evolution
Angiosperm cp DNA	Slow	Slow
Angiosperm mtDNA	Slow	Rapid
Mammalian mtDNA	Rapid	Slow
Fungal mtDNA	Rapid	Rapid

regard to both sequence and structure (Table 8.3). The opposite extreme, with a very fast rate of evolution, is found in the mtDNA of fungi, which changes rapidly in both sequence and structure.

The mtDNA of angiosperm plants has the opposite pattern of evolution as found in animal mtDNA. In sequence evolution the rate in angiosperms is slow, but in structural evolution it is fast. In plants, the mtDNA genome is large and highly complex. In some instances, a single molecule can resolve itself into smaller circles and even linear molecules. For example, in the turnip (*Brassica campestris*), a 218 kb molecule undergoes an internal recombination event that produces smaller circles of 135 kb and 85 kb. Maize mtDNA contains six pairs of repeated sequences that can undergo recombination and create a variety of structural derivatives. The *Arabidopsis* mtDNA genome was recently sequenced, and although it is 366 kb, nearly all the increase in size compared to mammalian mtDNA is noncoding (Unseld et al. 1996). Many plant mitochondria also contain autonomously replicating plasmid DNA molecules, and mtDNA is also capable of incorporating segments of cpDNA. Why plant mtDNA genomes are so large, complex, and variable in size is not understood.

Maintenance of Variation in Organelle Genomes

Organelle genomes have unusual population genetics because of their (typically) uniparental transmission and because many copies are passed from the mother to the progeny through the egg. Uniparental transmission has important implications in the operation of natural selection, since it is equivalent to a haploid clonal population structure, and pure selection models can maintain polymorphism in such populations only if the fitnesses are frequency dependent. From the outset, then, uniparental transmission makes it less likely for polymorphisms to be maintained by natural selection, even if epistatic effects with the nuclear genome are allowed (Clark 1984). The widespread polymorphisms observed in mtDNA must then be attributed largely to high mutation rates, just as the rapid substitution rate was attributed to a high mutation rate. Polymorphisms can also be maintained by interspecific hybridization, and it is possible to obtain estimates of rates and directions of interspecific matings from nuclear and mtDNA data (Asmussen et al. 1987). Unusual forms of transmission, such as the doubly uniparental transmission of the mussel *Mytilus edulis*, results in separate male and female lineages, which are highly divergent (Skibinski et al. 1994; Stewart et al. 1995).

The theory of random genetic drift for organelles is more complex than that for nuclear genes because individual cells have many organelles that are apportioned among daughter cells; thus there is an additional level of sampling when heteroplasmic cells divide. Models of the dual sampling process have been examined in some detail (Birky et al. 1983; Takahata 1983, 1984). These models predict some level of heteroplasmy, and although early empirical studies did not detect heteroplasmy, it has now been described in crickets (Harrison et al. 1987), *Drosophila* (Hale and Singh 1986; Solignac et al. 1983, 1984, 1987), lizards (Densmore et al. 1985), mice (Boursot et al. 1987), cattle (Hauswirth and Laipis 1982), frogs (Monnerot et al. 1984), treefrogs, and bowfin fish (Bermingham et al. 1986). Heteroplasmy can be maintained by a steady-state balance between the forces of random genetic drift and mutation, but heteroplasmy is most frequently observed in restriction length polymorphisms, in which variants differ in the number of copies of a small repeat. Simple deterministic models show that heteroplasmy can be stably maintained by infrequent paternal transmission (leakage), by natural selection, or by bi-directional mutation, such as the gain/loss events one would expect for changes in copy number of a small repeat (Clark 1988). Distributions of heteroplasmy in the field cricket are consistent with a model of mutation-selection balance, with smaller genomes favored by selection (Rand and Harrison 1986).

Evidence for Selection in mtDNA

There are several clear examples of nonneutrality of mtDNA mutations. For example, many forms of cytoplasmic male sterility are caused by defects in mtDNA (Grun 1976; Levings 1983). Similarly, cytoplasmically transmitted drug resistance genes have been shown to be associated with the mitochondrial genome of yeast. The potential importance of mtDNA variation in human health was revealed in the implication of mitochondrial DNA defects in the muscle diseases known as mitochondrial myopathies. The celebrated bicycle racer Greg Lemond, a three-time winner of the Tour de France, was forced into early retirement by a defect in mitochondrial oxidative metabolism. Effects of natural selection also have left their mark on extant patterns of mtDNA sequence variation, as revealed by the discordance between levels of polymorphism and divergence in synonymous vs. nonsynonymous sites (Ballard and Kreitman 1994; Rand and Kann 1996). The strongly skewed distribution of frequencies of segregating sites also suggests that human mtDNA has faced selection pressure (Hey 1997).

If a cytoplasmically related factor of any sort is associated with a particular mtDNA type, then the mtDNA will "hitchhike" along with the other cytoplasmic factor. A striking example of this mode of evolution in action was caught by Turelli et al. (1992), when they noticed that a cytoplasmically transmitted *Wolbachia* infection in *Drosophila simulans* was rapidly spreading north in California, and as it did so, it propelled a single mtDNA type to high frequency. While the mtDNA genome may seem small, its uniparental transmission makes it susceptible to any cytoplasmic factor that may carry a particular cytoplasmic type to fixation. However, most populations have fairly high levels of mtDNA variation, suggesting that such sweep events are not very common.

MOLECULAR PHYLOGENETICS

The use of techniques of molecular biology, particularly those for determining amino acid or nucleotide sequences, has added a new dimension to phylogenetic inference. For example, the analysis of 5S RNA sequences in a broad variety of microorganisms has led to a reclassification at the deepest of phylogenetic levels, resulting in a new kingdom, the Archaea (Woese 1981). In addition to the satisfaction of understanding the history of relationships among living things, the application of comparative molecular analysis to infer robust and accurate phylogenetic relationships has spawned interest in the application of those phylogenetic trees for testing hypotheses about evolutionary mechanisms. The problem of inferring the correct branching topology for a tree that relates a set of organisms is a challenge in part because of the enormous number of possible bifurcating trees. If there are n species to be placed, there are $(2n - 3)!/2^{n-2}(n - 2)!$ rooted trees that describe possible ancestral histories. For five species this number is 105, and for 10 species it is 34,459,425. For many data sets of 30 or more species, the number of possible trees is so enormous that it is not possible to examine all topologies and assess the fit of the data to each tree, even with the very fastest computers. Fortunately, the trees are not all independent of one another, and the key to many of the algorithms that try to find the best fitting tree is to eliminate whole classes of trees based on the observed data. Let us consider a few of these tree-building methods.

Algorithms for Phylogenetic Tree Reconstruction

If a gene in a pair of species or populations evolves in clocklike fashion, and if the degree of divergence between two genes implies that they have been diverging for t generations, then we can infer that the genes separated from a common ancestor $t/2$ generations ago. This reasoning provides a group of methods of tree construction based on measures of genetic distance. One such method is the unweighted pair-group method with arithmetic mean

(UPGMA) or average distance method. This method requires that all sequences evolve at the same rate, an assumption that other methods can relax to some degree, but the ease of understanding UPGMA still gives it heuristic appeal. With a matrix of all pairwise distances, a tree is built up by first grouping the two species with the smallest distance. A new distance matrix is then constructed, with the grouped species now considered as one unit. If the grouped species were indexed i and j, then for all $k \neq i,j$ the distance from k to the group $\{i,j\}$ is $d_{k(ij)} = \frac{1}{2}(d_{ik} + d_{ij})$. In words, the distance from each other species k to the group $\{i,j\}$ is the average of the distances from species k to each of species i and j in the group. The new distance matrix is again searched for the smallest element, and the appropriate grouping again occurs. This process is repeated until all species are clustered into a tree.

Tree-building methods can not only produce a tree topology, but they generally also give estimates of branch lengths of the tree. An example of one method for branch-length estimation is the method of Fitch and Margoliash (1967). Suppose the number of substitutions distinguishing sequence i and j is d_{ij}. If the tree relating sequences 1, 2, and 3 has branch lengths A, B, and C (Figure 8.25), then the branch lengths can be estimated from

$$A = \frac{1}{2}(d_{12} + d_{13} - d_{23})$$

$$B = \frac{1}{2}(d_{12} + d_{23} - d_{13}) \qquad\qquad 8.17$$

$$C = \frac{1}{2}(d_{13} + d_{23} - d_{12})$$

These relations were found by solving the equations $d_{12} = A + B$, $d_{13} = A + C$, and $d_{23} = B + C$. With more than three sequences, the tree is built up by considering three units at a time, beginning with the two most closely related sequences and grouping the remaining sequences. If sequences 1 and 2 are the most similar, then the distances from sequence 1 to the remaining group is the average of the distances from sequence 1 to each member of the group.

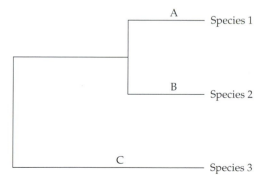

Figure 8.25   A simple phylogenetic tree. A, B, and C represent branch lengths from the most recent common ancestor.

In this way, only three distances are considered at a time, and Equations 8.17 allow branch lengths to be estimated. This method is known as **least squares**, and it turns out that Equations 8.17 minimize the sum of squared deviations from the model, much like linear regression.

Another algorithm for tree construction is particularly well suited to the situation in which one does not know whether rates of substitution are constant across clades of the tree. This method is known as **neighbor-joining**, because it groups species having the property "neighbors" (Saitou and Nei 1987). Begin by assuming that the sequences are all related to one another by a star phylogeny (Figure 8.26). For a star phylogeny with N sequences, the sum of the branch lengths is

$$S_0 = \sum_{i \neq j} d_{ij} / (N-1)$$

(It may help to draw a star phylogeny to see that each branch gets counted $N - 1$ times.) Next we begin a procedure that groups certain sequences together. For each possible pair of sequences, a tree like that in step 1 in Figure 8.26 is constructed. Branch lengths for this tree are estimated by least squares, and the sum of the branch lengths for the entire tree (S_{ij}) is calculated. We consider as neighbors that pair of sequences i and j that give the minimum of the S_{ij}'s. After the first pair of neighbors is found, that pair is considered as a single entity (joined neighbors), and the process of considering all possible pairings is repeated. The distance from any one sequence k to this pair of neighbors (i and j) is the average of the two distances, or $\frac{1}{2}(d_{ik} + d_{jk})$. The process ends when there are just three neighbors left, and at this point we have a finished neighbor-joining tree complete with branch lengths. The criterion for neighbor-joining is to minimize the sum of branch lengths, and sometimes it is possible to find tree topologies that are even shorter, using a method called **minimum evolution** trees (Rzhetsky and Nei 1992).

PROBLEM 8.10   Consider a sample of one allele drawn from each of three species. Suppose that the tree that one gets from these alleles may be represented ((A,B),C), implying that A and B are most closely related, and C is the outgroup. What are the possible relationships among the species bearing these alleles?

ANSWER   This problem bears on an important issue in phylogeny reconstruction, namely that any one gene tree does not necessarily reflect the true pattern of splitting of species. The easiest way to see

	B	C	D	E
A	0.53	0.99	1.02	0.82
B		0.80	0.93	0.73
C			0.65	0.81
D				0.94

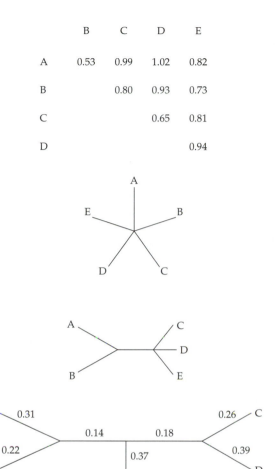

Figure 8.26 Illustration of the neighbor-joining method for phylogeny reconstruction. Given a distance matrix, one starts with a star phylogeny and tests all trees having different pairs separated from the rest. The tree with A–B joined is the shortest such tree. The process of testing all pairs of "neighbors", where a neighbor may be either a single allele or a cluster of alleles, is repeated until no more joining can be done. (See Saitou and Nei 1987.)

this is to consider ancestral populations as being polymorphic, in which case the speciation process may sort out the alleles in various ways. It turns out that the possible species trees include ((A,B),C), ((A,C),B), and ((B,C),A). In other words, the gene tree does not eliminate the possibility of *any* of the species trees.

Distance Methods versus Parsimony

There is no universal theory that provides a single optimal way to construct phylogenetic trees, and as basic as the distance matrix seems, it is not required by all methods. Another method, known as **maximum parsimony**, uses the smallest number of mutational events necessary to account for the evolution of a set of sequences from a common ancestor to construct the trees. There are a number of such parsimony methods based on trees with the smallest number of substitutions, but none guarantee that the most parsimonious tree is the correct tree. For example, when rates of substitution differ in different branches of the tree, the parsimony method often fails to give the correct topology (Felsenstein 1978). Methods for constructing phylogenetic trees have been reviewed by Felsenstein (1981, 1982) and more recently by Nei (1996). Massive simulation studies have been done to test the statistical reliability of tree-constructing methods (Rohlf and Wooten 1988; Sourdis and Nei 1988; Hillis 1996). Results of these simulations are easy to summarize: if the data allow one method to assign a topology with good statistical confidence, generally all the popular methods work pretty well. But if the data have many apparent reverse mutations, variable rates among branches, or wide variation in rates across sites, then none of the methods works very well.

Bootstrapping and Statistical Confidence in a Tree

Because there are so many possible tree topologies, it is important to assess how much statistical confidence one can place in a particular tree. One cannot assign a numerical standard error to a tree; by its geometrical nature a tree is actually a complicated statement of phylogenetic relationships, such that we might have high confidence in some branches, and low confidence in others. A widely used method of assessing confidence in the nodes of a tree is the **bootstrap test** (Felsenstein 1985). The basic idea is quite simple: a subset of the original data is drawn with replacement and, from this new data set, a tree is drawn. For each node in the original tree, we ask whether the new tree has the same cluster of sequences. The whole operation of resampling the data, drawing a tree, and tallying up nodes that are in the original tree is repeated perhaps 1000 times, and the final result is displayed graphically as a number next to each node indicating the percentage of time that cluster is present among the resampled trees. If that fraction is high, then one gains confidence that the given cluster actually belongs together.

Another means of testing the statistical confidence in a tree is to test the null hypothesis that each interior branch has length zero. From distance methods, we often obtain estimates of all branch lengths in the tree, along with their standard errors. If we fail to reject the null hypothesis of zero length for an interior branch, then we lose confidence in the nodes surrounding that branch.

Shared Polymorphism

One might intuitively expect that all the alleles of a species should cluster together on a gene tree, implying that the common ancestor of all the alleles is an ancestral allele within the same species. A few gene trees have been found to have the unexpected property that alleles in two or more species appear to be interdigitated on the tree. This pattern, known as **shared polymorphism** or **trans-species polymorphism**, has been observed in major histocompatibility alleles in primates (Lawlor et al. 1988), in self-incompatibility alleles of plants (Ioerger et al. 1991), and in several genes in the *melanogaster* species subgroup of *Drosophila* (Hey and Kliman 1993). Figure 8.27 shows the probable means by which shared polymorphism arises, namely, that the ancestral *species* was polymorphic, and two or more alleles remain in the descendant species ever since the time of the common ancestor. Recall that the expected fixation time for a new mutation, given that it goes to fixation, is $4N$ generations. This means that neutral alleles are quite unlikely to remain polymorphic for much longer periods. Consequently, observation of shared polymorphism implies that either strong selection is retaining the alleles in the population, or that the species have diverged relatively recently. In the first two examples above, there is good evidence that selection has maintained the polymorphisms, while in

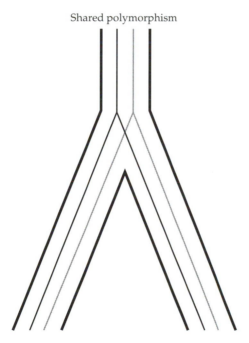

Shared polymorphism

Figure 8.27   Trans-species or shared polymorphism may occur if the ancestor was polymorphic for two or more alleles and if alleles persist to the present in both species.

the third example, the *Drosophila* species are recently enough diverged that some shared neutral polymorphisms are expected.

Interspecific Genetics

Phylogenetic inference from molecular sequences is a descriptive goal in the sense that the primary objective is to obtain an accurate representation of the ancestral history of the species. Population genetics can also address the genetic basis for species differences, particularly in the case of species in which some hybrids are at least partially fertile. Although these studies do not directly address the genetic causes for species origination, they are relevant to the genetic causes of barriers to interspecific gene flow. Investigation of the genetic basis for hybrid infertility and inviability among species in the *Drosophila melanogaster* species subgroup (comprising the species *melanogaster, simulans, sechellia,* and *mauritiana*) is a very active area. One focus in this work has been an investigation of the genetic basis for Haldane's rule, which states that, in interspecific hybrids in which only one sex is sterile or inviable, the sex likely to be affected is the heterogametic sex (Coyne 1985; Coyne et al. 1991). Rather than one or two genes of large effect, interspecific hybrid sterility appears to be caused by many genes that also have a complex pattern of interaction, so that some particular combinations are sterile and other combinations are fertile (Palopoli and Wu 1994). A powerful tool for studying the genetic basis of hybrid sterility has been to introgress small pieces of the genome from one species into the other. By doing this for many regions distributed all over the genome, one can learn about the relative roles of the X chromosome and autosomes, the relative incidence of male vs. female infertility, and so forth (True et al. 1996). Other features of interspecific differences are amenable to genetic analysis by either introgression methods (applied to differences in cuticular hydrocarbons by Coyne 1996) or by scoring an array of anonymous markers in many backcross individuals (applied to genital arch morphology by Liu et al. 1996).

MULTIGENE FAMILIES

Genes increase in number through duplication. Several successive rounds of duplication result in a family of homologous genes with related functions, a **multigene family**, the members of which are often arrayed in tandem along the chromosome. Among genes that normally exist in tandemly arrayed multigene families are the rRNA genes and the histone genes. Analysis of the sequences of members of multigene families has led to some interesting surprises. Figure 8.28 shows a scenario whereby a gene underwent a duplication that ultimately became fixed in the population either through drift or selection. Subsequently, sufficient sequence divergence occurred that the two genes could be distinguished. Later a speciation event produced two differ-

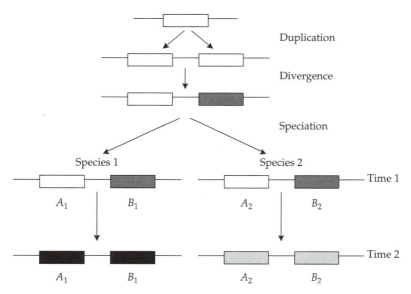

Figure 8.28   Multigene families originate by a process of gene duplication. After the duplication the genes may retain very similar functions (like rRNA genes), or they may diverge (like globin genes). If the species splits into two species, then time 1 and time 2 depict the relationship between the genes shortly after speciation and long after speciation (see Figure 8.29).

ent species sharing this pair of genes. Figure 8.29 shows the gene genealogies at two time points in the evolution of this gene family. At time 1, the *A* genes in species 1 and 2 have a more recent common ancestor than do genes *A* and *B* within species 1. At time 2 the pairs of genes present in the same species are more similar. This is the pattern that is observed in some multigene families. The close resemblance of A_1 with B_1, and of A_2 with B_2, seems paradoxical, since both species have the duplication, and Figure 8.28 makes it appear that genes A_1 and A_2 in the two species have a more recent common ancestor than do genes A_1 and B_1. Genes A_1 and B_1, as well as A_2 and B_2, may have more similar sequences because the genes evolve together, in concert, under the influence of mechanisms that operate to homogenize their sequences. This tendency toward homogenization is known as **concerted evolution**.

Causes of Concerted Evolution

Two important mechanisms of concerted evolution are gene conversion and unequal crossing-over. **Gene conversion** is a process in which nucleotide pairing between two sufficiently similar genes is accompanied by the excision of

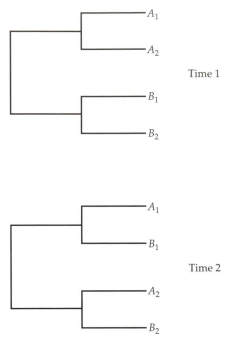

Figure 8.29 Referring to Figure 8.28, at Time 1, genes A_1 and A_2 in the two species are more similar to each other than either is to gene B, and likewise B_1 and B_2 are closest neighbors. This tree reflects the fact that the common ancestor of A_1 and A_2 is more recent than that of A_1 and B_1. If at Time 2 a tree like the bottom panel is observed, then sequences of A_1 and B_1 have become more similar, possibly by the process of gene conversion. The bottom tree illustrates the phenomenon known as concerted evolution.

all or part of the nucleotide sequence of one gene and its replacement by a replica of the nucleotide sequence from the other gene. Formally, the result is that the sequence in one gene "converts" the sequence in the other gene to be exactly like itself. In **unequal crossing-over**, meiotic pairing between the tandem repeats in homologous chromosomes is out of register, and crossing-over results in an increase in the number of copies in one chromosome and a corresponding decrease in the number of copies in the other chromosome. Repeated rounds of unequal crossing-over can result in the disproportionate representation of certain sequences among members of the multigene family, a result that is formally equivalent to gene conversion.

A theoretical model of concerted evolution has been studied by Ohta (1982). In this model, a tandemly arranged multigene family consists of a fixed number of n members, and λ is the probability that a particular member of the gene family becomes converted by another member in any one generation. (Equivalently, λ is the probability of completion of a cycle of unequal crossing-over resulting in the replacement of one sequence in the family by another.) The mutation rate per copy is μ, and the population number is N.

In a tandemly arrayed multigene family, there are three distinct types of identity by descent (IBD) among the gene copies (Figure 8.30):

1. Genes at different positions in the same chromosome may be IBD (probability c_1).
2. Genes at different positions in different chromosomes may be IBD (probability c_2).
3. Genes at the same position in different chromosomes may be IBD (probability f).

Complex formulas for the equilibrium values of c_1, c_2, and f have been derived by Ohta (1982), but they are greatly simplified when recombination within the gene cluster is ignored. In such a case, the equilibrium values are approximately

$$c_1 = c_2 = \frac{\lambda}{\lambda + (n-1)\mu}$$

$$f = \frac{4N\lambda c_2 + 1}{4N\lambda + 4N\mu + 1}$$

8.18

In Equations 8.18, the quantity $(n-1)\mu$ is very nearly equal to $n\mu$ if n is reasonably large. Because n is the number of copies of the gene in each tandem array, $n\mu$ is the total rate of mutation in the multigene family, summed across all copies. Thus, the implication of Equation 8.18 is that there is a delicate balance between the rate of gene conversion λ and the total mutation rate $n\mu$. If the rate of gene conversion is much greater than the total mutation rate, then the probability of IBD of genes at different positions within the

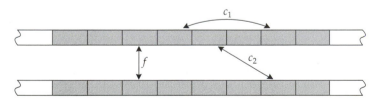

Figure 8.30   Three types of identity by descent in multigene families. They are the identity between genes at homologous sites (probability f), between genes at nonhomologous sites in the same chromosome (probability c_1), and between gene at nonhomologous sites in different chromosomes (probably c_2). (After Ohta 1982.)

family (c_1 and c_2) is close to 1.0. On the other hand, if λ is much smaller than the total mutation rate, then the probability of IBD of genes at different positions within the family is close to zero.

Concerted evolution does not homogenize all multigene families. Depending on the balance of the forces of mutation, gene conversion, and unequal crossing-over, the pair of genes may remain active and very similar, or they may diverge in function (such as different tissue-specific forms of amylase or lactate dehydrogenase), or one gene may lose function and become a pseudogene. Multigene families can avoid the accumulation of mutations when there is sufficiently strong natural selection, and positive selection is necessary for genes to evolve new functions. Walsh (1988) addressed the question of genes within a family escaping from gene conversion, and he showed that higher mutation rates and lower conversion rates lead to greater likelihood for a gene escaping conversion. Once a gene is sufficiently divergent to have escaped conversion, it can either lose function and become a pseudogene or it can acquire a new function. Simple models of such a duplicated gene show that very little selection is needed in a large population to avoid a pseudogene fate (Walsh 1995).

Multigene Family Evolution through a Birth and Death Process

Duplicate genes can evolve in separate ways under the influence of natural selection, mutation, and random genetic drift. In time, some members of a multigene family may diverge to a greater or lesser degree in their function. This process of **duplication and divergence** is thought to be the major mechanism by which genes with novel functions are created. Some multigene families retain a tandemly arrayed structure and similarity in function across members despite the fact that the differences between individual members is of functional significance. This pattern is particularly true of genes in the immune system, including immunoglobulin genes and major histocompatibility genes. Interspecific comparisons of genes in families of this sort exhibit some genes that are clearly homologous, and others that are more distantly related. In addition, the rate of duplication, loss of function through pseudogenes, and loss by deletion, may be fairly high. This kind of pattern of multigene family evolution is different from concerted evolution, because the differences between the genes can be high enough that intergenic conversion is very rare. Figure 8.31 illustrates the distinctness of this pattern of gene evolution, called a birth-and-death process by Ota and Nei (1994).

Figure 8.32 illustrates the result of duplication and divergence in two related multigene families in mammals that code for the α-like and β-like polypeptide chains of hemoglobin. The genes are specialized for different periods of life. The ϵ (epsilon) genes are expressed in embryos; the $^G\gamma$ and $^A\gamma$ genes and the α genes in the fetus; and the α, β, and δ genes in the adult. The inference from differences in nucleotide sequence is that the original

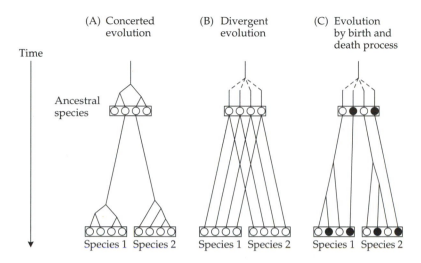

Time

(A) Concerted evolution

(B) Divergent evolution

(C) Evolution by birth and death process

Ancestral species

Species 1 Species 2

Species 1 Species 2

Species 1 Species 2

Figure 8.31 In addition to concerted evolution and simple divergent evolution, multigene families frequently exhibit the phenomenon of genes being added and lost to families by a "birth and death process." (From Ota and Nei 1994.)

α-β duplication took place approximately 500 million years ago, when vertebrates were represented by the bony fishes, and the β-γ duplication took place about 80 million years ago, during the mammalian radiation. More recent duplications have also occurred, for example those leading to the two functional α genes, the cluster of three α-like pseudogenes, and the two γ genes. There are several models for the sequence of duplication, deletion, and conversion events that could have led to the current array of globin genes (Goodman et al. 1984; Hardies et al. 1984; Hardison 1984; Margot et al. 1988), but it appears well substantiated that the ancestral cluster that predated the mammalian radiation was 5'–εγηδβ –3'. Within the mammalian radiation, the different orders of mammals evolved along different routes. In prosimian primates, such as lemurs, there was a fusion of η and δ. In higher primates, including humans, there was a δ-β conversion and a γ duplication. In rodents, β and γ both duplicated, η was deleted, and there was a δ-β fusion, mediated probably by an unequal crossover. In rabbits, η was deleted and there was a δ-β conversion. Finally, in goats, γ was deleted, there was a δ-β conversion, and the remaining four gene array was then triplicated!

The evolutionary history of the fetal globin genes in humans reveals that the ^{G}γ and ^{A}γ genes originated as part of a relatively recent 5 kb tandem duplication (Shen et al. 1981). Furthermore, evidence from nucleotide

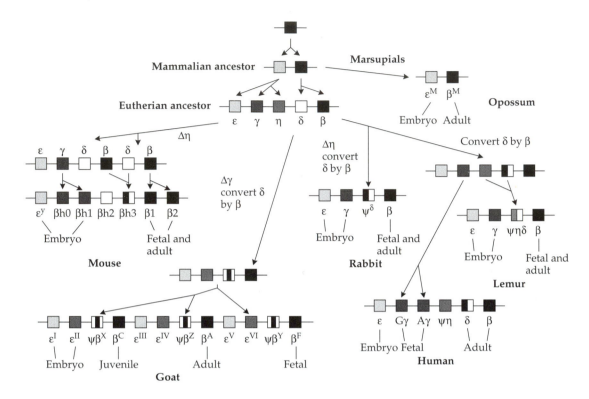

Figure 8.32 Reconstruction of the β-globin sequences in a series of mammals illustrates the complexity of duplication, loss, and gene conversion in this multigene family. (After Hardison 1984.)

sequences strongly suggests that a gene conversion event also occurred, which converted part of one particular $^A\gamma$ allele into a $^G\gamma$ allele (Slightom et al. 1980). The converted $^A\gamma$ allele is very similar to a $^G\gamma$ allele for about 1550 bp on the upstream (5′) side of a putative recognition signal for gene conversion (a stretch of repeating TG and CG dinucleotides); but on the downstream (3′) side of the putative signal, the converted $^A\gamma$ allele is typical of other $^A\gamma$ alleles in the human population. The $^A\gamma$ to $^G\gamma$ gene conversion occurred much more recently than the duplication resulting in the close sequence similarity of the $^A\gamma$ and $^G\gamma$ genes.

The estimate of the time of occurrence of the $^A\gamma$–$^G\gamma$ duplication can be improved by using the nucleotide sequence data from the entire duplicated 5 kb region. In the entire region, 14% of the nucleotide sites differ, which translates into $\hat{k} = 0.155 \pm 0.006$; this suggests a time for the duplication of $0.155 \times 100 \times 2.2 \times 10^6 = 34$ million years (Shen et al. 1981).

Unequal crossing-over in multigene families can result in a decrease in the number of genes as well as an increase. It is therefore not surprising that deletions of one or more of the hemoglobin genes are found in most parts of the world. Although usually very rare, in a few places the frequency of the deletions reaches levels too great to be accounted for by chance, especially in view of the observation that the carriers are mildly to severely anemic. Although a deletion of the β-gene results in death when homozygous, a β deletion and other mutations that decrease the abundance of the β-hemoglobin chain are relatively common in the Mediterranean Sea basin where malaria is endemic. For this reason, the decreased-β-chain diseases are called β-thalassemias (literally translated as "sea-anemias"). The well-established link between sickle-cell anemia and malaria, along with the geographical correlation between the β-thalassemias and malaria, provides a strong circumstantial case for malarial parasites being an important selective agent. Deletion of one or more of the α-globin genes results in another form of anemia called α-thalassemia, whose frequency in populations is also correlated with the incidence of malaria.

Red-green colorblindness is a common X-linked disorder with a frequency of about 8% in Caucasian males. The genes for the red and green visual pigments match at 98% of their nucleotides, indicating that they arose by a relatively recent duplication. Individuals with normal color vision have one copy of the red pigment gene and varying numbers of copies of the green pigment gene. When genomic DNA from colorblind males was analyzed by Southern blotting, those defective in green vision were lacking fragments of the green pigment gene. Further analysis showed that 24 of 25 colorblind individuals had lost one or the other pigment gene through gene rearrangements that were due either to unequal crossing-over or gene conversion. In this example, the high sequence similarity of the red and green pigments works to human disadvantage by greatly increasing the likelihood of exchange events that lead to loss of color vision (Nathans et al. 1986). The relationship between the molecular basis of light absorption and perception was made particularly clear when it was found that a normal polymorphism in red pigments, which confers a difference in the absorption peak of the protein product, also confers a measurable difference in the perception of color balance (Merbs and Nathans 1992).

Duplication of genes also occurs in plants, including a particularly important gene in plants that encodes the carbon fixing enzyme ribulose-1,5-bisphosphate carboxylase (RBC) (Clegg et al. 1997). The functional RBC holoenzyme consists of eight large and eight small subunits. Early in plant evolution, both the large and small subunits of RBC were encoded by the chloroplast genome, but the small subunit gene was transferred to the nuclear genome at an early stage and has now been lost from the chloroplast genome. Diploid angiosperms contain from two to eight copies of the gene for the small RBC subunit (*rbcS*). All copies of *rbcS* appear to be functionally equivalent, and sequence analysis shows that the genes that are closest together in the genome are also generally more similar in sequence. In sequence comparisons among *rbcS* genes of tobacco and tomato, homologous genes compared between the two species are more similar than within species comparisons of gene copies. This finding is not the pattern expected under concerted evolution. The variable number of loci across angiosperms suggests that gain and loss of gene copies occurs to give a pattern like the birth-and-death process described above.

Structural RNA Genes and Compensatory Substitutions

Transfer RNA and ribosomal RNA molecules derive their biochemical properties from the secondary structure into which they fold. We are still learning the chemical rules by which such macromolecules attain their final folded configuration, but one thing that is very clear is that complementary base pairing is important. The stems of tRNAs are critical to maintaining the tightly folded structure of these essential molecules. Substitutions that occur in stems will weaken the stability of the stem unless there is a compensatory change on the other strand that maintains base pairing. Kimura (1985) realized that one could obtain evidence for such compensatory changes. More recently, such compensatory changes have been demonstrated in an intron, demonstrating that the folding structure of introns may also be important to regulating gene expression (Kirby et al. 1995).

Further evidence of the importance of secondary structure of rRNA comes from analysis of rRNA pseudogenes in plants (Buckler et al. 1997). One attribute of secondary structure is measured as the difference in free energy attributable to complementary base pairing in the folded vs. unfolded state. Computer predictions of the best folding structure of the rRNA pseudogenes suggested that the difference in free energy decreases as the sequences accumulate substitutions. Tests of randomly permuted sequences showed that the functional rRNA sequences are significantly more stable than would be obtained by chance, whereas predicted pseudogene RNAs are not. Some introns have a significantly open secondary structure, such that random substitutions in their sequences result in more stable structures (Leicht et al. 1995). The reason some introns retain an open structure may be for access to regulatory proteins. This possibility has been indirectly demonstrated by showing that stable stems inserted into introns in yeast can disrupt normal splicing.

The ribosomal RNA gene cluster in *Drosophila melanogaster* consists of about 200 copies of a repeated unit on both the X and the Y chromosome, with each repeated unit containing an 18S and a 28S rRNA gene separated by an intergenic sequence (IGS) (Glover and Hogness 1977). The rRNA genes provide a clear example of concerted evolution because of great interspecific differences in spite of a high degree of sequence conservation within species (Coen et al. 1982). Furthermore, within individuals of *D. mercatorum*, there appears to be little sequence variation, yet there are clear differences between individuals due to length variation in the intergenic sequence (Williams et al. 1985). This finding suggests the operation of a strong homogenizing force maintaining sequence fidelity within individuals. In humans, the rDNA repeat consists of a 13 kb transcribed portion and a 31 kb spacer (Wellauer and Dawid 1979). This repeated unit is present in about 300 copies located near the tips of the short arms of five nonhomologous chromosomes. Despite the dispersed locations, concerted evolution still occurs as evidenced by much less variation among sequences within an individual than among species. Interchromosomal exchange events would lead to conservation of sequence distal to the rDNA cluster on each chromosome, and evidence for this conservation has been found (Worton et al. 1988).

Multigene Superfamilies

In some cases, several sets of multigene families and single-copy genes may share recognizable homology, implying a common ancestry, but they have undergone major divergence in function and relocation of position within the genome. These sets of historically related but functionally distinct genes constitute a **multigene superfamily**.

The remarkable similarities found among portions of genes in related gene families has suggested that many proteins have functional modules that can be combined in various ways in what is called **exon shuffling**. One example of shuffling is found in tissue plasminogen activator (TPA), which has portions of three other proteins, including plasminogen, epidermal growth factor, and fibronectin. The striking finding is that the junctions of these protein segments fall precisely at intron–exon junctions. The epidermal growth factor shares exon similarity with several other proteins, including blood clotting factors IX and X, urokinase, and complement C9 (Doolittle 1985). The gene for the low-density lipoprotein (LDL) receptor in human beings extends over 45 kilobases and contains 18 exons that show similarity to a bewildering variety of other proteins, including epidermal growth factor and blood clotting factors (Südhof et al. 1985). Just as a computer programmer recognizes the value of reusing subroutine modules in different programs, nature has capitalized on the efficiency of modular gene organization.

One extensively studied multigene superfamily that serves diverse functions in immunity is illustrated in Figure 8.33 (Hood 1985; Hunkapiller and Hood 1986). The primordial single-copy gene may have coded for a cell-surface receptor containing the basic homology unit of the superfamily, which is about 110 amino acids in length with a strategically placed disulfide bridge and folding characteristics enabling it to combine with other similar units. An early duplication and divergence of the primordial gene resulted in the

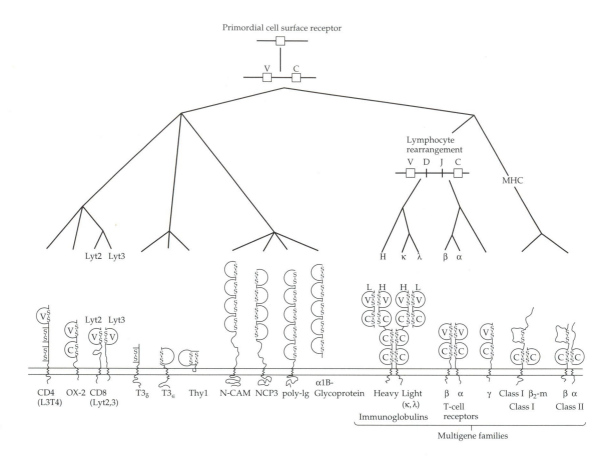

Figure 8.33   Proposed evolution of the immunoglobulin multigene superfamily from a primordial gene coding for a cell-surface receptor. Details of the evolutionary relationships are speculative. The superfamily has diversified into 12 single-gene representatives (all of those at the left, plus β_2-microglobulin—β_2-m—at the right), and eight multigene families (remaining representatives at the right). These include genes for antibodies, T-cell receptors, major histocompatibility antigens, and other functions. The single-gene members include T-cell molecules implicated in MHC recognition (CD4 and CD8) and possibly ion channel formation (T3δ, T3ϵ), an immunoglobulin-transport protein (poly-Ig), a plasma protein ($\alpha_1\beta$-glycoprotein), two molecules restricted to lymphocytes and neurons (Thy-1 and OX-2), two brain-specific proteins (N-CAM and NCP3), and β_2-microglobulin. The multigene families include the heavy (H) and light (κ, λ) components of antibody molecules, the α, β, and γ chains of T-cell receptors, and the Class I and Class II molecules from the major histocompatibility complex (HLA). (Adapted from Hood et al. 1985 and Hunkapiller and Hood 1986.)

variable (V) and constant (C) domains that have been so versatile in their diversification for specialized immune functions. In some members of the immunoglobulin superfamily, shown at the left in Figure 8.33, the functional products are usually individual polypeptide chains, sometimes containing internal duplications of the primordial folding unit. These products include the poly-Ig receptor that mediates the transport of immunoglobulin molecules across cell membranes.

In the other main branch of the superfamily, shown at the right, the functional products are usually aggregates of polypeptide chains. In this branch, there occurred multiple duplications of the V regions and specialization of D (diversity) and J (joining) regions during the evolution of the DNA splicing mechanism in lymphocytes, which today results in the tremendous diversity of antibodies and T-cell receptors. During the formation of heavy-chain antibody genes in the lymphocytes, any one of a large number of DNA sequences coding for the variable part of the molecule can become spliced with any one of a small number of DNA sequences coding for the constant part, with diversity and joining regions incorporated in between. The many possible V-D-J-C combinations enables enormous numbers of different possible antibodies to be formed, which is increased still further by slight variation in the exact positions of the splice junctions. An analogous type of splicing process occurs in the formation of antibody light-chain genes and T-cell receptor genes.

In yet another offshoot of the immunoglobulin superfamily, shown at far right in Figure 8.33, the C region underwent duplication and specialization to form molecules of the major histocompatibility complex (MHC), which, among other functions, are necessary for the T cells of the immune system to recognize foreign antigens. Complete sequencing of a 100 kb region of the T-cell receptor gene family has revealed a spectacular degree of sequence conservation between human and mouse (Koop and Hood 1994). The opportunities for exceptionally detailed analysis of multigene family evolution have enlarged with genomic sequencing methods already producing the complete sequence of entire arrays of genes (Rowen et al. 1996). Although many aspects of the immunoglobulin superfamily tree in Figure 8.33 are speculative, the molecules are undoubtedly related because comparison of the relevant units gives 15 to 40% homology at the amino acid level, and at the DNA level each homology unit is encoded in a separate exon. The immunoglobulins thus demonstrate the immense evolutionary potential of repeated rounds of duplication and divergence through specialization of function.

Dispersed Highly Repetitive DNA Sequences

A second major class of highly repetitive DNA in eukaryotes is not localized in clusters of tandemly repeating units, but is dispersed throughout the genome with single-copy sequences. The importance of dispersed repetitive elements to the human genome project is made clear by the realization that they constitute 35% of our genome (Smit 1996). In vertebrates, this dispersed highly repetitive DNA occurs primarily in two categories, denoted SINEs and LINEs (Singer 1982). SINEs (short **in**terspersed **e**lements) are sequences typically shorter than 500 base pairs which occur in 10^5 or more copies in the genome. Like tRNA genes, they contain internal transcriptional start sites and are transcribed by RNA polymerase III. LINEs (long **in**terspersed **e**lements) are sequences typically greater than 5000 base pairs that occur in 10^4 or more copies in the genome. They are processed pseudogenes (see below) and, when transcribed, are transcribed by RNA polymerase II. Marked differences in the particular array of subfamilies of SINEs and LINEs or both are frequently observed among even closely related species (Figure 8.34). The mechanisms and possible significance of such massive and rapid changes in repetitive DNA in the genome are very obscure.

One example of SINEs in human DNA is the *Alu* family, named because the sequence contains a characteristic restriction site for the restriction enzyme *Alu*I. The *Alu* sequence is about 300 nucleotides in length. *Alu* sequences are present in approximately one million copies in the human genome and constitute approximately ten percent of the total DNA (Smit 1996). Sequences closely related to *Alu* are found in other primates, and more distantly related sequences occur in rodents and probably in all placental mammals. Two randomly chosen human *Alu* sequences differ, on the average, at 15 to 20% of their nucleotide sites, which calculates to a time of divergence of between 16.7 and 23.3 million years. In the human genome there is an *Alu* element an average of every 3 to 5 kb, but the distribution is not uniform. For example, the β-tubulin and thymidine kinase gene regions have about 10 times the average density of *Alu* repeats (Slagel et al. 1987), and *Alu* repeats show a preference for integrating into oligo-dA runs (Daniels and Deininger 1985).

PROBLEM 8.11   The third chromosome of *Drosophila pseudoobscura* is polymorphic for more than a dozen inversions that result in different gene orders. Polymorphisms of this sort are different from nucleotide site substitutions because they retain some information about the order of events. Consider, for example, the sequences A-B-C-D-E and C-E-A-D-B. Can you deduce the order of the events that connect them?

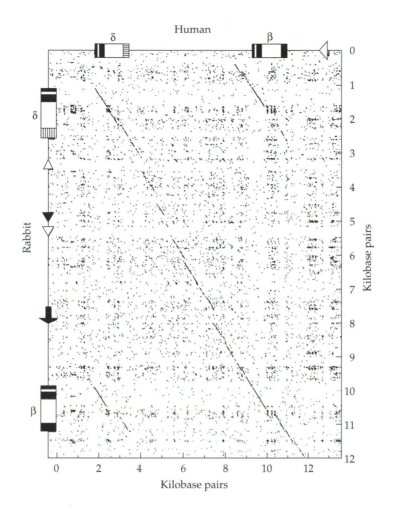

Figure 8.34   A dot plot comparison of the human and rabbit sequences spanning δ- and β-globins. Each dot represents a small bit of sequence similarity, much of the background due solely to chance, and the regions of extended similarity stand out as diagonal line segments. The scales are in kilobases, and the rectangles indicate the location and organization of the globin genes. The solid arrows show the location of a rabbit L1 repeat, and open triangles indicate human *Alu* sequences and rabbit *OcC* repeats (a rabbit SINE). The major diagonal line indicates that there is noticeable homology retained through the δ-β intergenic region, and the sequence similarity of human β-globin with rabbit δ-globin (and vice versa) is evident. (From Margot et al. 1988.)

ANSWER   From A-B-C-D-E, the first inversion must have been the segment A-B-C, giving the sequence C-B-A-D-E. Next, the segment A-D-E inverted to give C-B-D-A-E. Finally, the segment B-D-A-E inverted to give C-E-A-D-B. Much more elaborate problems of inference have arisen to determine the ancestral series of inversions and number of events needed to go from one gene order to another. Computer scientists refer to this problem as "sorting by reversals." You can see that given any random ordering of integers, a finite number of inversions or reversals will put them into the correct order. Motivated by the biological problem, an algorithm for finding the minimum number of reversals to go from one order to another was recently implemented (Bafna and Pevzner 1996). As more genomes are fully mapped and sequenced, this is likely to be an area of considerable excitement. Ehrlich et al. (1997) recently estimated that the number of rearrangements that were required to connect the human and mouse genetic maps as about 180.

An example of LINEs in the human genome is the *L1* family of sequences (also called LINE-1 or *Kpn*, because of a characteristic restriction site). The *L1* sequences average about 2,000 nucleotides, and the 50,000 copies of the sequence in the human genome account for about 4% of the total DNA. As with the *Alu* family, sequences related to *L1* are found in other mammals, including the mouse (Hardies et al. 1986) and the rabbit (Demers et al. 1986). Not all insertions of *L1* sequences are innocuous. Kazazian et al. (1988) found two cases of hemophilia A that were caused by *de novo* insertions of an *L1* sequence into exon 14 of the factor VIII gene, whose function is necessary for normal blood coagulation. This insertional mutation event was evidently mediated by an RNA intermediate and provides a mechanism for natural selection to operate on *L1* elements. Another deleterious mutation caused by a transposable element in humans was an insertion of an *L1* sequence into the *myc* oncogene in a human breast cancer (Morse et al. 1988).

In their molecular organization, LINE sequences strongly resemble a class of pseudogenes known as *processed pseudogenes*. Processed pseudogenes are thought to result from the reverse transcription of an RNA molecule into DNA, followed by insertion of the DNA into the genome. The reverse transcription and integration process can be carried out by an enzyme called **reverse transcriptase**, which is coded in the genome of a class of RNA-containing viruses called **retroviruses**. In cells infected with retrovirus, the reverse transcriptase makes a DNA copy of the viral RNA, and another enzyme inserts the DNA into the chromosome. When reverse transcription and integration happen to a processed RNA molecule, the result is a dispersed duplicate copy that is generally transcriptionally inactive due to loss of regulatory sequences. Such a sequence is known as a **processed pseudogene**. Many genes are known to have processed pseudogene counterparts, including the genes for human κ-immunoglobulin and β-tubulin, rat α-tubulin and cytochrome *c*, and mouse α-globin. Not all genes that have been processed through an RNA intermediate are pseudogenes. Human phosphoglycerate kinase (*PGK*) occurs as an active X-linked gene, a processed X-linked pseudogene, and an autosomal gene with remarkable properties. The normal *PGK-1* gene contains 11 exons and 10 introns, but the autosomal gene has no introns and has remnants of a poly-A tail, strongly implying that it was reverse transcribed from an RNA transcript. The intron-free autosomal gene (*PGK-2*) is expressed in human testes (McCarrey and Thomas 1987).

The processed pseudogene model of dispersed repeated DNA evolution is illustrated in Figure 8.35 (Hardies et al. 1986). The functional, transcribed copies of the gene family are shown at the top, and the horizontal arrows represent gene conversion, which promotes concerted evolution of the functional genes. The gene in the center is a preferred donor for gene conversion (*biased gene conversion*). Emanating from the functional genes are numerous

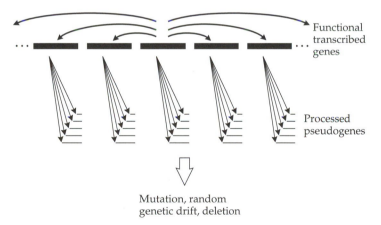

Functional transcribed genes

Processed pseudogenes

Mutation, random genetic drift, deletion

Figure 8.35   Model for the evolution of a dispersed highly repetitive family of processed pseudogenes. A small number of functional genes (top), which undergo concerted evolution by means of gene conversion, are transcribed under conditions that favor reverse transcription and integration into numerous dispersed chromosomal locations. The resulting nonfunctional genes undergo mutation and random genetic drift, and are ultimately eliminated by deletion or other mechanisms. (From Hardies et al. 1986.)

copies of processed pseudogenes distributed throughout the genome. These copies are essentially functionless and undergo sequence divergence promoted by mutation and random genetic drift, which is offset in part by gene conversion and other homogenizing processes among the pseudogenes. Eventually the pseudogene sequences are cleared from the genome by deletion or extreme sequence rearrangement or divergence.

One implication of the model in Figure 8.35 is that, eventually, a balance is reached in which the clearance of old pseudogenes from the genome is equaled by the creation and insertion of new ones. In the equilibrium state there is a steady turnover among sequences in the family, but the total number neither grows nor shrinks. Studies of a dispersed repeated sequence in the mouse related to human *L1* suggest a turnover with a half-life of approximately two million years. That is, after two million years, half the members of the gene family will have been removed and replaced with new ones. However, the *L1* family may evolve more rapidly than is typical.

The very abundance of pseudogenes implies that many unrelated genes may have pseudogenes in the same vicinity, as is the case with *Alu* sequences interspersed in the β-globin cluster. Some fraction of these linked pseudogenes may alter the level, timing, or tissue distribution of transcription of the genes to which they are linked, or they may have subtle effects on chromatin structure that affect gene expression. Through any of a diversity of mechanisms, pseudogene copies of dispersed highly repeated gene families could, in principle, have effects on phenotype and thus be subject to the influence of natural selection. While true in principle, such effects have not yet been demonstrated. To the extent that such effects can safely be ignored, the evolutionary mechanism of highly dispersed repeated DNA sequences is that of selfish DNA, subject to the conflicting forces of neutral mutation/random drift and the diverse homogenizing processes of concerted evolution.

SUMMARY

The discipline of molecular population genetics has as its theoretical foundation the neutral theory, which provides a rich set of testable hypotheses about the mechanisms that modify patterns of sequence divergence and sequence polymorphism. We saw that underlying models must be specified even to do seemingly straightforward things like estimating rates of substitution. The reason substitution rate estimates are not trivial is that, with greater divergence, subsequent mutations may not further increase the divergence if the site has already been substituted. From observed counts of amino acid or nucleotide differences, we usually want to estimate numbers of changes per site. The model for amino acid substitution is not very difficult because there are 20 amino acids, but even the simplest nucleotide substitution model of Jukes and Cantor is subtle. More complicated models account for differences in rates of transition and transversion substitutions, and it immediately becomes apparent that both the process of mutation and of substitution can be of any imagined degree of complexity.

Out of sequence analyses there emerges the pleasing generalization that many sequences appear to diverge at an approximately clock-like rate. This molecular-clock concept should be interpreted somewhat loosely, because rigorous statistical tests have identified significant irregularities in its rate. In addition, there are dramatic differences in rate of evolution across genes, because the neutral substitution rate differs from one gene to the next. Some lineages appear to have accelerated or decelerated clock rates, and one cause for the variation is a change in generation time (for example, from rodents to primates).

Synonymous and nonsynonymous substitutions have different effects on the protein product, so estimating the rates of these two kinds of substitution independently can be informative about the causes of evolutionary change. For example, most genes, like *Drosophila Adh*, have a large excess of synonymous changes, an observation that is accounted for by the presumed deleterious effect of most amino acid replacements. All synonymous codons are not used with equal frequency, and the bias in codon usage implies that even synonymous substitutions may not be selectively neutral. The most sensitive tests for selection make use of comparisons between intraspecific polymorphism and interspecific divergence. Under strict neutrality, these two quantities should be related to one another, and departures in either direction can be detected through heterogeneity among genes.

The neutral theory also makes predictions about the shape of gene trees, and there has been a great deal of excitement about the possibility of testing hypotheses about evolutionary forces based on inferred gene genealogies. (Problem 8.8 gives one example.) Gene trees have been used to test hypotheses about selection, recombination, homogeneity of mutation, and even migration. The ability to account for the patterns of correlation built up by the ancestral history of genes has been a major advance in statistical population genetics.

Organelle genome evolution occupies an important position in the development of molecular population genetics, in part because of the numerous studies of mtDNA and cpDNA variation. Of particular interest and controversy was the work on human mtDNA variation, which raised many intriguing problems about human origins. This work stimulated a huge amount of theoretical study concerning the statistical inferences that could be made from sample data, including times of common ancestry, inference of past demographic histories, and so forth. Several recent studies have shown that mtDNA exhibits patterns consistent with the past operation of natural selection, in violation of many of these models.

Molecular phylogenetics seeks to reconstruct the ancestral history of extant organisms, and shares many analytical procedures with molecular population genetics. There are several widely used algorithms for reconstructing a tree from sequence data, and we examined in some detail the UPGMA method, least-squares, neighbor-joining, and parsimony methods. One of the more intriguing patterns of variation to emerge from such interspecific comparisons is that of shared polymorphism, in which two or more species share a number of alleles in common. It is unlikely that shared polymorphism would be maintained for long by chance, so it is not surprising that cases of shared polymorphism are generally found in genes known to be under strong selection or in species that have recently diverged.

When multiple copies of similar genes exist in the genome, they can exchange sequences through unequal recombination and gene conversion. Such exchanges can result in concerted evolution, a process whereby genes in a multigene family are very similar to one another within a species, even though the duplication events that gave rise to the family occurred far in the past. Not all multigene families undergo concerted evolution. A more common finding is that many multigene families exist as groups of genes with related function that have diverged enough in sequence to escape gene conversion. In this case, new genes appear by duplication and old ones disappear by deletion, sometimes preceded by inactivating mutations that generate pseudogenes. This birth-and-death process gives rise to complex patterns of relationships among genes within gene families.

PROBLEMS

1. Suppose that you have sequences of gene A and gene B from each of two species. The fraction of sites that differ in gene A is 0.7 and the fraction of sites that differ in gene B is 0.05. Apply the Jukes-Cantor formula to obtain the estimate of the number of substitutions per site for each gene. Which gene do you think would have a smaller estimate of variance of substitution rate? Why?

2. Suppose you discover a community of deep sea creatures that have very unusual DNA that has not four bases but six. Adenine and thymine pair, and guanine and cytosine pair just like most DNA, but there are also nitidine and liondine, which also pair. You obtain sequences from two of these creatures and determine that 20% of the sites mismatch in aligned sequences. From this figure, estimate the number of substitutions per site that have occurred since the common ancestor of the two species. (*Hint:* You know that the number is higher than 0.20, because back mutations could have occurred. Derive an expression like the Jukes-Cantor formula.)

3. The following is a small portion of the gene coding for 6-phosphogluconate dehydrogenase in two natural isolates of *E. coli*.

```
1 CTG ACC AAA ATC GCC GCC GTA GCT GAA GAC GGT GAA CCA TGC GTT ACC TAT ATT GGT GCC
2 CTG AAG CAG ATC GCG GCG GTT GCT GAA GAC GGT GAG CCG TGT GTG ACT TAT ATA GGT GCC
```

Infer the correct translational reading frame of the sequences and estimate:

a. the number of amino acid differences/site.
b. the number of nucleotide differences/site.
c. the number of nonsynonymous substitutions per nonsynonymous site
   (regarding codon sites 1 and 2 as nonsynomyous).
d. the number of synonymous substitutions per synonymous site
   (regarding codon site 3 as synonymous).

4. In the human immunodeficiency virus HIV, which causes acquired
   immune deficiency syndrome (AIDS), the rate of nucleotide evolution
   has been estimated at about 0.01 substitutions per synonymous site per
   year. Two viruses isolated in 1983 in Zaire and San Francisco differ in
   approximately one third of their synonymous sites. Estimate the year in
   which the viruses last shared a common ancestor. (Data from Li et al.,
   1988.)

5. The data below give the proportion of nucleotide sites that differ in a
   gene in four RNA viruses (Yokoyama et al. 1988). HIV1 and HIV2 are two
   rather distinct types of human immunodeficiency viruses, VISNA is a
   lentivirus, and MMLV is a mouse cancer-causing virus. Estimate the
   number of nucleotide substitutions per site using these data. What do the
   numbers imply about the evolutionary relationships among the viruses?

	HIV2	VISNA	MMLV
HIV1	0.34	0.54	0.62
HIV2		0.52	0.63
VISNA			0.63

6. What inference would you make regarding the selective constraints on a
   region of DNA in which the rate of evolution was 5×10^{-9} nucleotide substitutions per site per year?

7. What might you infer about the evolutionary forces affecting a coding
   region in which the rate of amino acid replacement was greater than the
   rate of synonymous nucleotide substitution?

8. Ribsomal RNA forms a complex secondary structure in which many
   regions of the molecules are folded back and undergo base pairing with
   complementary nucleotide sequences elsewhere in the same molecule.
   What pattern of nucleotide sequence evolution might be expected in
   these paired regions?

9. What is the largest value of d that makes sense in Equation 8.15 and what
   does it mean?

10. If the rate of nucleotide evolution along a lineage is 0.5% per million years, what is the rate of substitution per nucleotide per year? What is the total rate of divergence of two lineages?

11. While analyzing the DNA sequences of two copies of a gene, you find that there are a total of 34 synonymous substitutions and 16 nonsynonymous substitutions. Using the method of Nei and Gojobori, you find that there were 310 synonymous nucleotide sites and 633 nonsynonymous sites. If possible, estimate the rates of synonymous and nonsynonymous substitution, and interpret the result.

12. If the effective size of a diploid population is N with respect to autosomal genes, what is it with respect to
    a. X-linked genes?
    b. Y-linked genes?
    c. mtDNA?

13. Analysis of mtDNA in humpback whales (Baker et al., 1990, *Nature* 344:238–240) has shown that not only do the Atlantic and Pacific populations show differences, but there are clear geographic subpopulations within oceans despite the lack of geographic barriers. Such a pattern may be observed if either: (1) there were a low rate of migration and a low rate of mtDNA sequence divergence, or (2) a higher rate of migration with a higher rate of mtDNA sequence divergence. Can you distinguish these two possibilities? Can you separately estimate the rate of neutral mutation and the rate of migration in a subdivided population?

14. Suppose the phylogeny of five species is {(A,B)R(C(D,E))}, where R designates the root. Can you ascribe the substitution events of the following data uniquely to branches on this genealogy? Number the sites 1–10 and label the substitutions by site number on the tree.

```
Species A   TAG  CTG  ATC  A
Species B   TAG  CCG  AGC  A
Species C   TAC  CCG  ATT  G
Species D   TAC  CCT  ATC  A
Species E   TGC  CCT  ATC  A
```

15. For an ideal population of effective size N, the average time to loss of a new mutation destined to be lost is $2\ln(2N)$, and the average time to fixation of a new mutation destined to be fixed is $4N$. For what values of N does
    a. Fixation time $= 10 \times$ loss time?
    b. Fixation time $= 100 \times$ loss time?

16. For the model of gene conversion with gene identities given in Equation 8.18, what value of λ makes the organization of the gene family irrelevant in the sense that $f = c_1 = c_2$? What is the common value in this case? (λ is

the probability that a particular member of the gene family becomes converted in any one generation.)

17. For the model of gene conversion with gene identities given in Equation 8.18, what are the values of f and $c_1 = c_2$ when $\lambda = \mu$? (The equations assume $4N\mu \ll 1$.)

18. In a repetitive gene family being eliminated from the genome by deletion, if the fraction of sequences present at time 0 that are still present at time t equals $\exp(-ht)$, show that the half life of the sequences equals $-\ln(\frac{1}{2})/h$.

19. For a repetitive gene family eliminated as described in Problem 18, show that the average persistence of an element is $1/h$.

CHAPTER 9

Quantitative Genetics

ARTIFICIAL SELECTION • HERITABILITY • COMPONENTS OF GENETIC VARIANCE
GENOTYPE × ENVIRONMENT INTERACTION • THRESHOLD TRAITS
GENETIC CORRELATION • EVOLUTIONARY QUANTITATIVE GENETICS • QTL MAPPING

M ANY IMPORTANT problems in evolutionary biology begin with observations of phenotypic variation. Darwin formulated his ideas about evolution by natural selection based on observations of phenotypic variation. He struggled for many years to explain the cause of the phenotypic variability, but he was unsuccessful at one level because he did not know about Mendelian genetics. Darwin did, however, appreciate the importance of the observation that offspring resemble their parents. Continuously varying traits, like body size, are influenced by both genetic and environmental factors. Crossing experiments demonstrate that the genetic components of these traits are not determined by single genes because the offspring do not fall into discrete classes with simple Mendelian ratios. Instead, what is observed is a general resemblance between parents and offspring, suggesting that there is an underlying genetic basis to the trait, but that the genetic transmission is complex.

A wealth of statistical tools have been developed for analyzing such **polygenic** traits that do not show simple Mendelian transmission. These approaches allow not only a description of the genetic basis of observed phenotypic distributions, but they also provide a means of predicting the distributions of phenotypes among offspring from observation of the parental phenotypes. Most polygenic traits are influenced by the environment to varying degrees, and they are often called **multifactorial** traits to emphasize their determination by multiple genetic and environmental factors. For example, variation in human weight is partly due to genetic differences among individuals and partly due to environmental factors such as exercise and level of nutrition. The study of polygenic inheritance goes beyond an oversimplified nature-versus-nurture dichotomy because it is concerned with specifying, in precise quantitative terms, the relative importance of nature, nurture, and their interactions, in accounting for variation in phenotype among individuals. Another compelling reason to study polygenic inheritance is that natural selection occurs at the level of the composite phenotype, and so *fitness is a multifactorial trait*.

Since natural selection operates on phenotypes, there arises an immediate problem in understanding how phenotypic evolution is reflected in changes that occur at the molecular level. One of the great challenges facing population genetics is to unify the principles of molecular evolution with those governing evolution at the phenotypic level.

TYPES OF QUANTITATIVE TRAITS

Multifactorial traits may be considered as resulting from the combined effects of many quantities, some genetic in origin and some environmental, and for this reason they are often called **quantitative traits**. The study of quantitative traits constitutes **quantitative genetics**.

Three types of quantitative traits may be distinguished:

1. Traits for which there is a continuum of possible phenotypes are **continuous** traits; examples include height, weight, milk yield, and growth rate. The distinguishing feature of continuous traits is that the phenotype can take on any one of a continuous range of values. In theory, there are infinitely many possible phenotypes, among which discrimination is limited only by the precision of the instrument used for measurement. However, in practice, similar phenotypes are often grouped together for purposes of analysis.

2. Traits for which the phenotype is expressed in discrete, integral classes are **meristic** traits; examples include number of offspring or litter size, number of ears on a stalk of corn, number of petals on a flower, and number of bristles on a fruit fly. The distinguishing feature of meristic traits is that the phenotype of an individual is given by an integer that equals the number of elements of the trait that the individual displays. For example, a popular meristic trait used in experimental studies of quantitative genetics in *Drosophila* is the number of bristles that occur on the abdominal segments or sternites. Normally there are 14 to 24 bristles per sternite. A male with 19 bristles on the fifth abdominal sternite therefore has a phenotype of 19. The distribution of numbers of abdominal bristles in a sample of *Drosophila* appears in Figure 9.1. When the number of possible phenotypes of a meristic trait is large (as it is with abdominal bristle

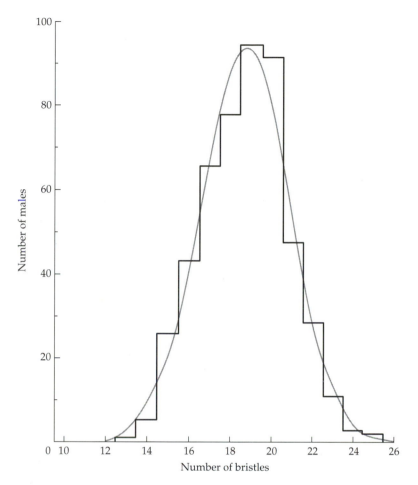

Figure 9.1   Number of bristles on the fifth abdominal sternite in males of a strain of *Drosophila melanogaster*. The smooth curve is that of a normal distribution with mean 18.7 and standard deviation 2.1. (Data from T. Mackay.)

number) then the line between continuous traits and meristic traits becomes indistinct.

3. The third category of quantitative traits consists of **discrete traits**, which are either present or absent in any one individual. In these cases, the multiple genetic and environmental factors combine to determine an underlying risk or **liability** toward the trait. Liability values are not directly observable. However, an individual that actually expresses the trait is assumed to have a liability value greater than some threshold or

triggering level. Traits of this type are called **threshold** traits, and examples in human genetics include diabetes and schizophrenia. With threshold traits, studies of affected individuals and their relatives permit inferences to be made about the underlying values of liability. These methods are discussed later in this chapter.

Quantitative traits are of utmost importance to plant and animal breeders, because agriculturally important characteristics such as yield of grain, egg production, milk production, efficiency of food utilization by domesticated animals, and meat quality are all quantitative traits. Even as modern methods of genetic engineering are applied to animal and plant improvement, quantitative genetics continues to play an important role because commercially desirable traits result from complex interactions among many genes. In addition to being essential ingredients in plant and animal improvement programs, the principles of quantitative genetics, appropriately modified and interpreted, can be applied to the analysis of quantitative traits in humans and natural populations of plants and animals.

RESEMBLANCE BETWEEN RELATIVES AND THE CONCEPT OF HERITABILITY

For Darwinian evolution to be possible, a necessary feature of the transmission of traits is that offspring must tend to resemble their parents. Even before the rediscovery of Mendel's work, Francis Galton was collecting detailed statistical data on resemblance between parents and offspring (Chapter 2). We will demonstrate the central ideas of the transmission of quantitative traits, using some of the concepts that Galton developed. Then we will show how models of Mendelian inheritance can account for these features of hereditary transmission. Calculation of the degree of resemblance among relatives in terms of underlying Mendelian genetics was first provided by Fisher (1918). Fisher's paper, notoriously difficult, was of great historical importance to population genetics, because it provided the first demonstration that multiple Mendelian genes could account for the observed patterns of transmission of multifactorial traits.

Figure 9.2 shows a plot of the mean of male offspring for a quantitative trait (y values) against the phenotypic value of the father (x values), displayed in the way Galton devised. The line is the best-fitting straight line, called the **regression line**, of offspring on parent. Regression is relevant to one of the primary aims in animal and plant breeding, namely to be able to improve attributes of the stock. An essential part of genetic improvement is to be able to predict what sort of offspring would be obtained from a given pair of parents. For quantitative traits, prediction cannot be done exactly, but a statistical description of the most likely offspring can be obtained by the procedure of plotting the parent-offspring regression. For reasons that will

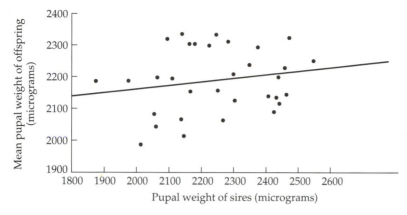

Figure 9.2   Mean weight of male pupae of the flour beetle *Tribolium castaneum*, against pupal weight of father (sire). Each point is the mean of about eight male offspring. The regression coefficient of male offspring weight on sire's weight is $\hat{b} = 0.11$, and h^2 is estimated as $2\hat{b}$. (Courtesy of F.D. Enfield.)

become clear in a moment, we are interested in the slope of the regression line. The slope is most easily expressed in terms of the **covariance of x and y,** defined as $\mathrm{Cov}(x,y) = [\Sigma(x - \bar{x})(y - \bar{y})]/n = (\overline{xy}) - (\bar{x})(\bar{y})$, where the bar over a symbol means the average. This quantity is the sample covariance of x and y. The slope of the line through a cluster of points having the smallest summed squared distance to the points is the regression coefficient: $b = \mathrm{Cov}(x,y)/\mathrm{Var}(x)$. A related quantity that also arises in quantitative genetics is the product-moment correlation coefficient, often simply referred to as the **correlation**: $r = \mathrm{Cov}(x,y)/\sqrt{\mathrm{Var}\,(x)\mathrm{Var}\,(y)}$.

An important concept in statistics is the distinction between **parameters** and **estimators**. Descriptors that are calculated from a set of data to describe a sample (such as the sample mean and sample variance) are considered as *estimates* of the parameters that determine the true distribution. The sample is thought of as having been drawn from some perfect distribution (whose parameters we can never know), and the sample statistics give us a best guess at what that true distribution is. In statistics, the distinction is generally made by unadorned Greek symbols for parameters and circumflexes for estimates. The usual symbols are: μ for the parametric mean, σ^2 for the variance, σ_{xy} for the covariance of x and y, and ρ for the correlation. Using the circumflex notation for estimates, $\hat{\mu}_x$ denotes the sample mean of x, so that, $\hat{\mu}_x = \bar{x}$. Similarly, $\hat{\sigma}_x^2 = \mathrm{Var}(x)$, and $\hat{\sigma}_{xy} = \mathrm{Cov}(x,y)$ are the sample estimates of the variance of x and the covariance of x and y. When describing models of quantitative genetics, it is the true distributions that are of interest, and so the parameters are used. When describing the results of an experiment, it is more appropriate to use the notation for estimates. The covariance and the correlation coefficient are convenient measures of the degree of association between x and y. If x and y are independent, then σ_{xy} and ρ are both zero. Since the covariance between any two variables measures their degree of association, the covariance may be positive or negative. Positive covariance means that values of x and y tend to increase or decrease together; negative covariance means that, as one variable increases, the other tends to decrease. The limiting values of the covariance are $-\sigma_x\sigma_y$ on the negative side, and $\sigma_x\sigma_y$ on the positive. The limits are achieved only when the variables demonstrate a perfect linear relationship with each other.

Returning now to Figure 9.2, if Cov(x,y) represents the covariance between phenotypic values of fathers (sires) and those of their male offspring, and Var(x) represents the variance of phenotypic values of the fathers, then the slope of the regression line is equal to the regression coefficient, Cov(x,y)/Var(x), which can be seen as follows. Suppose that the equation of the line is represented as

$$y = c + bx \qquad\qquad 9.1$$

where c and b are constants, b being the slope. Taking means of both sides yields

$$\bar{y} = c + b\bar{x} \qquad\qquad 9.2$$

subtracting the second equation from the first yields

$$y - \bar{y} = (c + bx) - (c - b\bar{x}) = b(x - \bar{x}) \qquad\qquad 9.3$$

Now multiply through by $x - \bar{x}$ to obtain

$$(x - \bar{x})(y - \bar{y}) = b(x - \bar{x})^2 \qquad\qquad 9.4$$

Taking means of both sides produces

$$\text{Cov}(x,y) = b\text{Var}(x) \qquad\qquad 9.5$$

In other words, the slope b of the regression line equals

$$b = \text{Cov}(x,y)/\text{Var}(x) \qquad\qquad 9.6$$

As noted, the slope is called the **regression coefficient** of offspring on one parent.

A graphical interpretation of regression is illustrated in Figure 9.3, which shows the distribution, in two dimensions, of the variables x and y. The variables may represent, for example, the phenotypic values of parents (x) and offspring (y). When there is no association between x and y, the distribution is a random scatter of points, and any line through the points fits equally badly. Figure 9.3 shows the appearance of the scatter of points for different

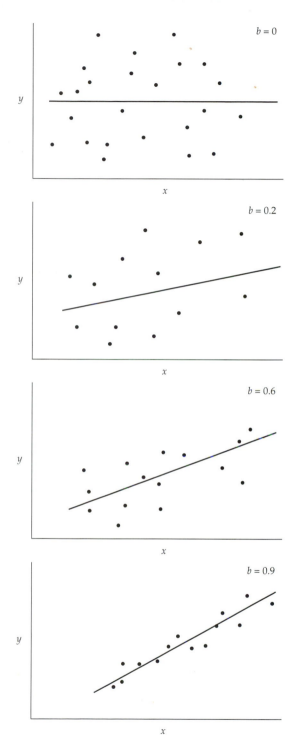

Figure 9.3 Plots of random scatters of points having the same variance on the x axis but a range of covariances. With zero covariance (top), the regression coefficient is zero. A stronger linear trend results in a higher regression coefficient.

values of association between the two variables. Note that, while each parameter measures an aspect of association between x and y, the covariance, the regression coefficient, and the correlation coefficient are different things. For example, the covariance and the regression coefficient are unbounded, whereas the correlation coefficient must be between –1 and 1.

Two extreme examples may help clarify parent-offspring regression. At one extreme, if there were no genetic contribution to the trait, then the scattergram might appear as a random scatter as in the top panel of Figure 9.3 with no tendency to follow a line. In such a case, knowing the phenotype of the parents would not help to predict that of the offspring, because there would be no parent-offspring resemblance. On the other hand, even with no genetic variation, the points might nevertheless show a substantial tendency to follow a line. To see why this is so, consider families living in different environments. In favorable environments with plenty of food and resources, parents and offspring might all be big and strong, while in unfavorable environments, parents and offspring might be small and sickly. A parent-offspring plot would show that big strong parents have big strong offspring, while small sickly parents have small sickly offspring, even though there is absolutely no genetic basis for the trait. The tendency of points to follow a line in a parent-offspring scattergram tells us nothing about the genetic basis of the trait, unless we are willing to make some claims (which hopefully can be tested experimentally) about the **environmental covariance** (the tendency of parents and offspring to resemble one another due to shared environments). Only if there is no environmental covariance will the parent-offspring regression indicate a degree of genetic influence on the resemblance. The possibility of environmental covariance is absolutely critical in human quantitative genetics, where the influence of shared environments can be very subtle and very strong.

Assuming now that the environmental covariance is zero, the regression coefficient b of offspring on one parent can be calculated for any random-mating population, and it indicates the degree to which the variance in the trait is determined by genetic variation. It is for this reason that the regression coefficient is related to an important quantity in quantitative genetics called **heritability**. There are two types of heritability that will be distinguished shortly, but for now, we note that the "narrow-sense" heritability (h^2) can be estimated from the relationship

$$b = \tfrac{1}{2}h^2 \qquad\qquad 9.7$$

The $\tfrac{1}{2}$ occurs in Equation 9.7 because the regression involves only a single parent (the father, in the case of Figure 9.2), and only half of the genes from any one parent are passed on to the offspring. In Figure 9.2, $\hat{b} = 0.11$, so $\hat{h}^2 = 0.22$. Notice the considerable scatter among the points in the figure, which represents data from 32 families. Because this sort of scatter is typical, heritability estimates tend to be quite imprecise unless based on data from several hundred families. Note however, that even with an enormous sample, there would be no less scatter to the points—we would merely have a more accurate measure of how much scatter there is. One further point about Figure 9.2: in organisms such as mammals, the regression is better performed on the father's phenotype, rather than on the mother's, in order to avoid potential bias in the estimate of heritability caused by such maternal effects as intrauterine environment. In organisms where nurturing does not impart significant maternal effects, scattergrams can be constructed with the x axis being the average of the two parents (the **midparent**) and the y axis the offspring phenotypes. From this sort of plot the regression coefficient is equal to the heritability: in symbols, when the x axis is the midparent, $b = h^2$.

PROBLEM 9.1   This example of calculating h^2 from parent-offspring regression uses data from Cook (1965), who studied shell breadth in 119 sibships of the snail *Arianta arbustorum*. For computational convenience, the data have been grouped into six categories. Estimate the heritability of shell breadth from these data.

Number of sibships	Midparent value (mm)	Offspring mean (mm)
22	16.25	17.73
31	18.75	19.15
48	21.25	20.73
11	23.75	22.84
4	26.25	23.75
3	28.75	25.42

ANSWER   Letting x refer to the midparent value and y refer to the offspring mean, then, $\bar{x} = 20.2626$, $\bar{y} = 20.1786$, $\Sigma x_i^2 = 49{,}823.4375$, $\Sigma y_i^2 = 49{,}267.1875$, $\hat{\sigma}_{xy} = 5.1826$, $\hat{\sigma}_{xy}^2 = 8.1801$, and $\hat{b} = \hat{h}^2 = 0.63$. (In actual practice we might not want to group the data into categories, because there is some loss of accuracy from grouping. The regression coefficient for the ungrouped data is $b = 0.70$. In addition, it should be noted that there is substantial assortative mating for shell breadth, and so the heritability estimate is artificially large.)

To this point we have shown that heritability can be used to measure the degree of resemblance between parents and offspring. Although the definition of heritability in terms of the regression coefficient between midparents and offspring is reasonable, heritability defined in this manner is merely a descriptive, empirical quantity because it makes no assumptions about genetics. In the next section we show how heritability in this purely statistical sense can be used to predict the result of artificial selection.

ARTIFICIAL SELECTION AND REALIZED HERITABILITY

The deliberate choice of a select group of individuals to be used for breeding constitutes **artificial selection**. The most common type of artificial selection is **directional selection**, in which phenotypically superior animals or plants are chosen for breeding. Although artificial selection has been practiced successfully for thousands of years (for example, in the body size of domesticated dogs), only during this century have the genetic principles underlying its successes become clear. Understanding the genetic principles of artificial selection permits prediction of the rapidity and amount by which a population can be altered through artificial selection in any particular generation or small number of generations. The theory of artificial selection is also strongly motivated by the idea that natural selection may operate in a similar way. For example, if only those individuals with greater than a certain amount of body fat survive, or only those individuals with less than a critical rate of evaporative water loss survive, then natural selection acts on the distribution of phenotypes in much the same way that breeders select characters of agricultural importance.

Artificial selection in outcrossing, genetically heterogeneous populations is usually successful in that the mean phenotype of the population changes over generations in the direction of selection (provided the population has not previously been subjected to long-term artificial selection for the trait in question). In experimental animals, the mean of almost any quantitative trait can be altered in whatever direction desired by artificial selection. For example, in *Drosophila*, body size, wing size, bristle number, growth rate, egg production, insecticide resistance, and many other traits can be increased or decreased by selection. In domesticated animals and plants, birth weight, growth rate, milk production, egg production, grain yield, and countless other traits respond to selection. Figure 9.4 shows the results of a long-term selection program involving oil content in corn. Amazingly, the line selected for high oil content is still responding after more than 90 generations (Dudley and Lambert 1992).

The general success of artificial selection in outcrossing species indicates that a wealth of genetic variation affecting quantitative traits exists. On the other hand, in a genetically uniform population, the mean phenotype of the population cannot usually be changed through artificial selection, because genetic variation is required for progress under artificial selection. For example, in experiments with the Princess bean, Johanssen (1909) found that artificial selection consistently resulted in failure when practiced within essentially homozygous lines. He obtained this result because, in genetically

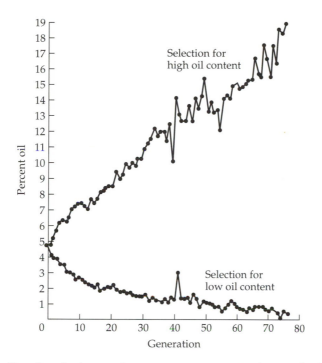

Figure 9.4   Results of a famous long-term experiment selecting for high and low oil content in corn seeds. Begun in 1896, the experiment has the longest duration of any on record and still continues at the University of Illinois. Note the steady, linear rise in oil content shown by the upper curve. The lower curve started on a roughly linear path and continued so for about ten generations, but then the response tapered off, presumably because zero percent oil is an absolute lower limit for the trait. (After Dudley and Lambert 1992.)

homozygous populations, the only source of genetic variation comes from new mutations. In contrast, since genetically variable populations usually respond to artificial selection, and genetically uniform populations do not respond, the response to artificial selection might be used as a measure of the extent of genetic variation in the trait. This notion of selection response reflecting genetic variation will be formalized in the next section.

Prediction Equation for Individual Selection

When individuals are selected for breeding based solely on their own individual phenotypic values, the type of artificial selection is called **individual selection**. Figure 9.5 illustrates a variety of individual selection called **truncation selection**. The curve in panel A represents the normal distribution of a quantitative trait in a population, and the shaded part of the distribution to

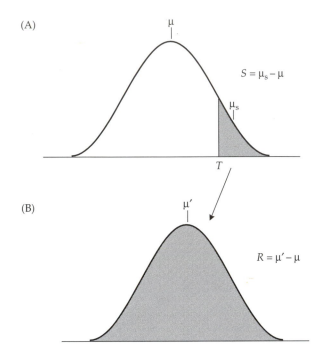

Figure 9.5   Diagram of truncation selection. (A) Distribution of phenotypes in the parental population, mean μ. Individuals with phenotypes above the truncation point (T) are saved for breeding the next generation. The selected parents are denoted by the shading and their mean phenotype by μ_s. (B) The mean of the distribution of phenotypes in the progeny is denoted μ′. Note that μ′ is greater than μ but less than μ_s. The quantity S is called the selection differential, and R is called the response to selection.

the right of the phenotypic value denoted T indicates those individuals selected for breeding. The value T is called the **truncation point**. The mean phenotype in the entire population is denoted μ, and that of the selected parents is denoted μ_s. When the selected parents are mated at random, their offspring have the phenotypic distribution shown in panel B, where the mean phenotype is denoted μ′.

An example of truncation selection for seed weight in edible beans is shown in Figure 9.6. In this example, $T = 650$ mg, $\mu = 403.5$ mg, $\mu_s = 691.7$ mg, and $\mu′ = 609.1$ mg. In this case—as is typical of truncation selection—the offspring mean μ′ is greater than the previous population mean μ but less than the parental mean μ_s. The reason μ′ is greater than μ is that some of the selected parents have favorable genotypes and therefore pass favorable genes

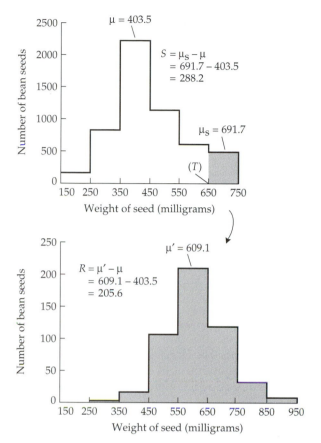

Figure 9.6 Truncation selection experiment for seed weight in edible beans of the genus *Phaseolus*, laid out as in Figure 9.5. The truncation point (*T*) is 650 mg. The selection differential *S* is the difference in means between the selected parents and the whole population. The response *R* is the difference in means between the progeny generation and the entire population in the previous generation. The quantity *R/S* is the realized heritability. (Data from Johannsen 1903.)

on to their offspring. At the same time, μ' is generally less than μ_S for two reasons:

1. Because some of the selected parents do not have favorable genotypes; rather, their exceptional phenotypes result from chance exposure to exceptionally favorable environments.
2. Because alleles, not genotypes, are transmitted to the offspring, and exceptionally favorable genotypes are disrupted by Mendelian segregation and recombination.

The difference in mean phenotype between the selected parents and the entire parental population is the **selection differential** and is designated S. In symbols,

$$S = \mu_S - \mu \qquad\qquad 9.8$$

The difference in mean phenotype between the progeny generation and the previous generation is the **response** to selection and is designated R. Symbolically,

$$R = \mu' - \mu \qquad\qquad 9.9$$

In quantitative genetics, any equation that defines the relationship between the selection differential S and the response to selection R is known as a **prediction equation**. Since selection can be applied to a population in many different ways (others will be discussed later in this chapter), the prediction equation may differ corresponding to the different modes of selection. A general prediction equation that applies to many forms of selection, including truncation selection (the type of selection illustrated in Figure 9.5), is

$$R = h^2 S \qquad\qquad 9.10$$

where h^2 is the **realized heritability**. Later in this chapter, we will show that the realized heritability is identical to the narrow-sense heritability defined by regression, provided the phenotypes and the magnitudes of genetic effects follow a bell-shaped Gaussian distribution. These assumptions are necessary in order to apply regression to the problem. This equivalence emphasizes again that heritability can be understood at several different levels. Equation 9.10 implies that the realized heritability of a trait can be interpreted as a mere description of what happens when artificial selection is practiced. In Figure 9.6, for example, $\hat{S} = 288.2$ and $\hat{R} = 205.6$, so $\hat{h}^2 = \hat{R}/\hat{S} = 205.6/288.2 = 71.3\%$. When estimated like this from empirical data, h^2 is the realized heritability, and it simply summarizes the observed result.

PROBLEM 9.2   Below are data on the number i of sternital bristles in samples from two consecutive generations G_1 and G_2 of an experiment in directional selection for increased bristle number. In the G_1 generation, individuals with 22 or more bristles (enclosed in brackets) were mated together at random to form the G_2 generation. Estimate the realized heritability of the number of sternital bristle in this experiment. (Data kindly provided by Trudy Mackay. In order to make the sexes comparable, the value of 2 has been added to the bristle number in males.)

i	G_1	G_2	i	G_1	G_2	i	G_1	G_2
15	0	2	20	20	13	25	[1]	3
16	21	4	21	12	14	26	0	2
17	5	7	22	[13]	12	27	0	0
18	18	16	23	[3]	6	28	0	2
19	17	17	24	[5]	3			

ANSWER   Estimates of the means are $\hat{\mu} = 2220/15 = 19.3$, $\hat{\mu}_S = 22.7$, $\hat{\mu}' = 2035/11 = 20.1$. The selection differential $\hat{S} = 22.7 - 19.3 = 3.4$ (Equation 9.8) and the response $\hat{R} = 20.1 - 19.3 = 0.8$ (Equation 9.9). The realized heritability estimated from Equation 9.10 is $\hat{h}^2 = 0.8/3.4 = 0.235$.

Data from experiments by Mackay (1985) demonstrate the potential significance of new mutations in quantitative genetics. The base population on which selection was performed was created by a cross that mobilizes the transposable element P that results in new P-element insertions in the germline and a syndrome of partial infertility and other reproductive abnormalities known as hybrid dysgenesis. As a control, a genetically identical base population was formed by the reciprocal cross, in which the P element is not mobilized and hybrid dysgenesis does not occur. In the dysgenic cross, the realized heritability in abdominal bristle number was increased by 40% as compared with the nondysgenic control. More strikingly, the phenotypic variance of bristle number in the selected dysgenic lines increased by a factor of three over the course of eight generations. These results demonstrate that the genetic variation affecting quantitative traits may even include insertions of transposable elements. On the other hand, other comparable experiments using hybrid dysgenesis have not given such dramatic results.

Selection Limits

Progress under artificial selection does not continue forever. Any population must eventually reach a selection limit, or **plateau**, after which it no longer responds to selection. One of the reasons that a population eventually reaches a plateau is exhaustion of genetic variance, such that all alleles affecting the selected trait have become fixed, lost, or are otherwise unavailable for selection. With no genetic variance, no progress under individual selection can be achieved. However, many experimental populations that have reached a selection limit readily respond to reverse selection (selection in the reverse direction of that originally applied), so genetic variance affecting the trait is still present. Indeed, in such populations, the phenotype may change in the direction of its original value if continuing artificial selection is simply suspended (**relaxed** selection). The consequences of relaxed selection for one example in *Drosophila* are illustrated in Figure 9.7.

One frequent reason for the occurrence of selection limits in populations with considerable genetic variation is that artificial selection is opposed by natural selection. In mice, for example, response to selection for small body size ultimately ceases because small animals are less fertile than larger ones, and the smallest animals are sterile (Falconer and Mackay 1996). Selection for small body size gradually becomes less effective due to the opposing effects of natural selection until, eventually, no further progress is possible. When selection is relaxed, the natural selection is unopposed and results in a retrogression in the artificially selected trait. Some backward slippage with relaxed selection also results from diminution in the linkage disequilibrium that usually builds up during the course of long-term artificial selection. If natural selection opposes the artificial selection, then when artificial selection is relaxed, natural selection results in at least a partial return to the initial phenotypic mean.

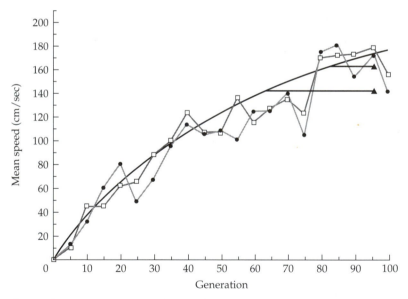

Figure 9.7   Response to selection for wind tunnel flight speed in *Drosophila melanogaster*. One line was maintained without selection for 30 generations starting at generation 65, and another was maintained without selection for 10 generations starting at generation 85 (triangles). In these examples, the flight performance did not degrade after selection was relaxed. Apparently the selection response occurred with little correlated response on fitness. (After Weber 1996.)

TABLE 9.1 **SELECTION LIMITS AND DURATION OF RESPONSE FOR VARIOUS TRAITS IN LABORATORY MICE**

Character selected	Direction of selection	Total response[a]	Half-life of response[b]
Weight (in strain N)	Up	3.4 σ_p	0.6N
	Down	5.6 σ_p	0.6N
Weight (in strain Q)	Up	3.9 σ_p	0.2N
	Down	3.6 σ_p	0.4N
Growth rate	Up	2.0 σ_p	0.3N
	Down	4.5 σ_p	0.5N
Litter size	Up	1.2 σ_p	0.5N
	Down	0.5 σ_p	0.5N

Source: From Falconer 1977.

[a] Total response is expressed as a multiple of the initial phenotypic standard deviation, σ_p.
[b] Half-life of response is the number of generations taken to progress halfway to the selection limit; here the half-life is expressed in multiples of effective population number (N).

In most genetically heterogeneous populations, artificial selection can change the phenotype well beyond the range of variation found in the original population. Pertinent data for populations of mice are presented in Table 9.1. As can be seen, a total selection response of three to five times the original phenotypic standard deviation is not unusual, and for selection to change a population of effective size N halfway to its selection limit typically requires about $1/2N$ generations.

In some cases the total response to artificial selection is very large. For example, in a long-term selection experiment for pupal weight in *Tribolium*, in which the base population consisted of the progeny of a cross between two inbred lines, 100 generations of selection resulted in a population in which the mean pupal weight in the selected population was 17 standard deviation units greater than the mean in the base population (Enfield 1980). The ability to select a population in which virtually every phenotype is greater than the maximum in the original population strikes many students as paradoxical. It does seem plausible to argue that, if all of the alleles eventually selected are already present in the original population, then all possible favorable genotypes should be present also, though perhaps at low frequency. The fallacy in the argument is that real populations subjected to artificial selection are actually small in size, consisting of at most a few hundred organisms. Therefore, if the favored alleles are rare, then the frequency of the favored genotypes may be so small that the expected number of such genotypes will be much smaller than one, and so the superior genotypes, while theoretically possible, do not actually exist in the original population.

Some traits consistently fail to respond to artificial selection, suggesting a lack of suitable genetic variation. Bilateral symmetry is an example of a trait that has not been amenable to change by artificial selection. The failure of Maynard-Smith and Sondhi (1961) to create bilateral asymmetry in *Drosophila* by selecting for an excess of dorsal bristles on the left side is typical. The apparent lack of genetic variation determining bilateral asymmetry is of interest in regard to embryonic development, for it implies that the genetic control of development of symmetrical structures specifies patterns that are common to the left and the right sides of the body. That is, rather than left-bristle genes and right-bristle genes, there appear to be generic bristle genes whose spatial expression is determined symmetrically. Of course asymmetrical structures do exist (such as the vertebrate heart) and recently inroads have been made in understanding the molecular genetic basis for this asymmetry (Isaac et al. 1997). Genes that affect left-right asymmetry do not do so in a continuous manner—rather they either successfully establish the asymmetry or they do not; absence of symmetry is fatal.

Not all traits with heritable variation obey the prediction equation and show a simple linear change in the mean. Sometimes a trait responds to directional selection for a few generations, then ceases to respond, but later responds again as selection is continued. One possible mechanism for this stop-and-start response is that the population at a plateau is in linkage disequilibrium, and it takes time for recombination to break up the allelic associations and release the latent genetic variation. This phenomenon was observed in a long-term study of the quantitative genetics of wing veins in *Drosophila* (Scharloo 1987). In this case a bimodal phenotypic distribution was also generated during selection (Figure 9.8), which was proposed to reflect a nonlinear mapping from genetic and environmental factors to the determination of phenotype.

As we have seen, heritability can be interpreted in purely statistical terms with no genetic content. However, if we postulate that there are Mendelian genes underlying the phenotypes, then the genetic underpinning allows us to do more than merely describe statistical relations among individuals. By bringing Mendelian genetics into the picture, we will see why the response to any kind of artificial selection is determined by the magnitude of the heritability. In particular, the genetic basis of response to artificial selection comes from changes in gene frequencies and sometimes also to changes in linkage disequilibrium.

GENETIC MODELS FOR QUANTITATIVE TRAITS

When h^2 is interpreted as realized heritability, then Equation 9.10 is hardly a "prediction equation" inasmuch as it merely describes what has already happened in one generation of selection. Of course, the equation could be used to predict the result of the next generation of selection, but artificial selection

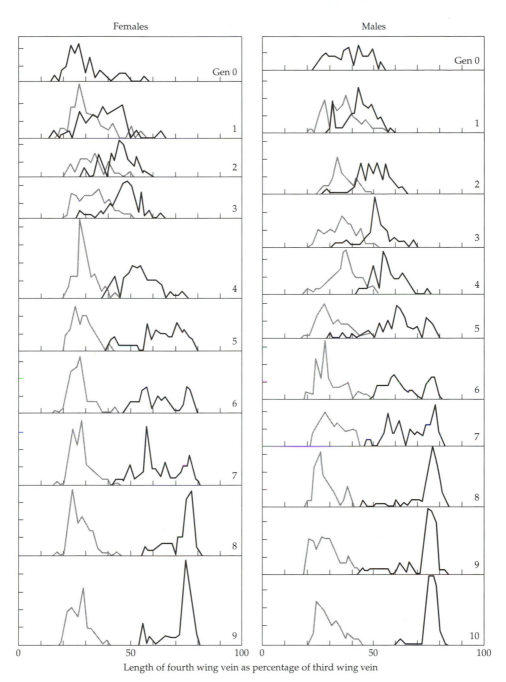

Figure 9.8 Frequency distributions in females (left) and males (right) of a line of *Drosophila melanogaster* selected for fourth wing vein length. The broken lines represent selection for a short vein, and solid lines represent selection for a long vein. In the line selected for long veins, both sexes displayed a bimodal frequency distribution when the relative vein length was approximately 60–80%. (From Scharloo 1987.)

is impossible in many natural populations and is time consuming and expensive in many domesticated plants and animals. It would therefore be useful if one could estimate heritability without actually performing any artificial selection. If the heritability h^2 could be estimated in such a manner, then Equation 9.10 would be a true prediction equation in the sense that the response R could be predicted for any selection differential S, based on the estimated value of h^2. Such an estimate of h^2 is indeed possible, but it involves an understanding of heritability at a level that includes the underlying genetic basis of quantitative traits.

An understanding of the genetics behind Equation 9.10 requires three items: (1) a concept of how alternative alleles of a gene affect a quantitative trait; (2) a determination of how selection changes the allele frequencies; and (3) a calculation of how much the mean of the trait increases as a result of the change in allele frequency. Some detail is required to establish these three items, but the detail is necessary in order to understand the genetic meaning of heritability.

Nilsson-Ehle (1909) was the first to show that a trait with a nearly continuous distribution of phenotypes could result from the joint effects of several genes. The trait of interest is the intensity of red pigment in the glume of wheat *Triticum vulgare*, which Nilsson-Ehle found to result from three unlinked genes, each with two alleles. The situation is exceptionally simple for a quantitative trait; the environment has a negligible effect on phenotype, because the alleles of each gene are additive (i.e., heterozygotes have a phenotype that is exactly intermediate between homozygous phenotypes), and because the genetic effects are also additive across genes (i.e., the total genetic effect of any three-locus genotype is just the sum of the separate effects of each gene). To simplify matters, consider just two of the genes, and let their alleles be denoted (A, a) and (B, b). With additivity within and across genes, we may assume that the genotype *aabb* has a color score of 0 (white) and that each A or B allele in the genotype contributes one unit of red pigment. Figure 9.9A shows the nine possible two-gene genotypes, their frequencies with random mating when the allele frequencies of A and B are both $1/2$, and the color score of each genotype assuming additivity. The mean color score of the population is 2. Indeed, when the allele frequencies of A and B are both p, then the mean of a population with random mating can be shown to equal $4p$. To connect this trait with the prediction, Equation 9.10, suppose that the two lowest phenotypic classes (i.e., 0 and 1) are selected as parents of the next generation. We first calculate μ_S, μ', S, R, and $h^2 = R/S$ using the allele frequency of A and B among selected parents; then we use the mean = $4p$ formula to obtain the mean of the offspring with random mating.

In this example, $\mu = 4(1/2) = 2$ is given. The selected parents consist of genotypes *Aabb*, *aaBb*, and *aabb* with respective frequencies $2/5$, $2/5$, and $1/5$, and the mean of parents = $\mu_S = (2/5)(1) + (2/5)(1) + (1/5)(0) = 4/5$. The allele frequency of A and B among parents = $(1/2)(2/5) = 1/5$, and therefore the mean among off-

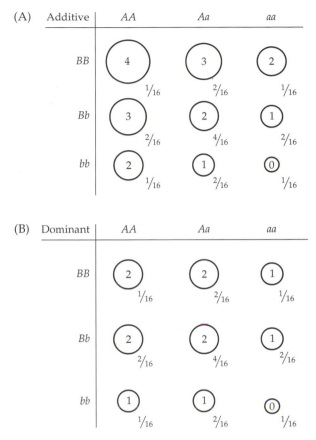

Figure 9.9 Frequencies of two-locus genotypes (outside circles) and respective phenotypes (within circles) in a population with allele frequency $\frac{1}{2}$ for each locus. Panel A illustrates the case of additivity of effects at each locus and across loci. In panel B, A and B are each dominant to a and b respectively, but the effects of the two loci are additive.

spring is $\mu' = 4(\frac{1}{5}) = \frac{4}{5}$. Then $S = (\frac{4}{5}) - 2 = -\frac{6}{5}$ and $R = (\frac{4}{5}) - 2 = -\frac{6}{5}$, so $h^2 = R/S = 1.0$. As demonstrated in the next paragraph, this high heritability is due to the additivity within and across genes and not merely to the fact that environmental effects are negligible.

Figure 9.9B refers to a hypothetical situation in which the A and B alleles are dominant but still additive across genes. Thus, genotypes AA, Aa, BB, and Bb each add one unit of red pigment to the phenotype. In this case, it can be shown that the mean of a random-mating population with allele frequencies of A and B both equal to p is given by $2p(1 + q)$, where $q = 1 - p$. If the two lowest phenotypic classes (i.e., 0 and 1) are selected as parents of the next generation, then the mean of parents = μ_S = $(\frac{1}{7})(1) + (\frac{2}{7})(1) + (\frac{1}{7})(1) + (\frac{2}{7})(1) + (\frac{1}{7})(0) = \frac{6}{7}$. The allele frequency of A and B among parents is $p = (\frac{1}{7}) + (\frac{1}{2})(\frac{2}{7}) = \frac{2}{7}$, and the mean of the offspring is therefore μ' = $2(\frac{2}{7})[1 + (\frac{5}{7})] = \frac{48}{49}$. Thus, $S = (\frac{6}{7}) - (\frac{3}{2}) = -\frac{9}{14}$ and $R = (\frac{48}{49}) - (\frac{3}{2}) = -\frac{51}{98}$. In the case where A and B are dominant, so $h^2 = \frac{51}{63} = 0.81$. Although environmental effects on seed color are still negligible in the dominance case, the heritability has become less than 1.0. This perhaps surprising result occurs because certain genetic effects (such as those resulting from dominance or, in other examples, nonadditivity across genes) are not useful in changing a population by means of the type of individual selection discussed here.

To see how an underlying genetic model can be formulated for continuous characters, refer to Figure 9.10, which shows the normal distribution of a trait in a hypothetical random mating population. In truncation selection, all individuals with phenotypes above the truncation point T are saved for breeding, and the shaded area B of the distribution represents the proportion of the population selected. (The total area under any normal density equals 1.) The height of the normal density at the point T is denoted Z, and, as before, the mean phenotype among the selected individuals is called μ_S. One of the special properties of the normal distribution to be used below is that

$$(\mu_S - \mu)/\sigma^2 = Z/B \qquad\qquad 9.11$$

To determine the amount of increase in mean phenotype in a population resulting from one generation of truncation selection, we first imagine a gene

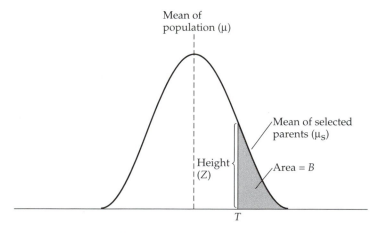

Figure 9.10   Normal distribution of a quantitative trait in a hypothetical population, showing some important symbols used in quantitative genetics. Here μ is the mean of the population, T the truncation point, Z the height (ordinate) of the normal density at the point T, B is the shaded area under the normal curve to the right of T, and μ_S is the mean among selected parents.

that affects the trait in question and that has alleles A and A' with respective allele frequencies p and q. Because of random mating, genotypes AA, AA', and $A'A'$ are present in the population with frequencies p^2, $2pq$, and q^2, respectively, but the individual genotypes cannot be identified through their phenotypic values because of the variation in phenotype caused by environmental factors and genetic differences in other genes. If the genotypes could be identified, their individual distributions of phenotypic value might appear as shown in Figure 9.11. Each distribution is normal and has the same variance, but the means are very slightly different. The mean phenotypes of AA, AA', and $A'A'$ genotypes are denoted $\mu^* + a$, $\mu^* + d$, and $\mu^* - a$, respectively. The symbols a and d serve as convenient representations of the effects of the alleles in question on the quantitative trait. The difference between means of homozygotes is $(\mu^* + a) - (\mu^* - a) = 2a$, and d/a serves as a measure of dominance. The relationship $d = a$ means that A is dominant, $d = 0$ implies additivity (heterozygotes exactly intermediate in phenotype between the homozygotes), and $d = -a$ means that A' is dominant. (Use of a and d in this manner simplifies some of the subsequent formulas.) Calculation of a and d for an actual example involving two alleles that affect coat coloration in guinea pigs is illustrated in Table 9.2. In this case, $\hat{a} = 0.127$, $\hat{d} = -0.016$ (the negative sign on d means that the c^d allele is partially dominant), and

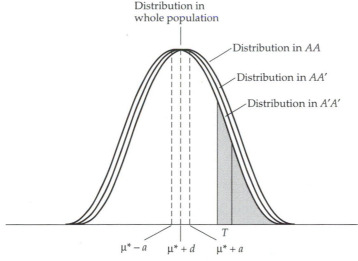

Distribution in whole population

Distribution in AA

Distribution in AA'

Distribution in $A'A'$

T

$\mu^* - a$ $\mu^* + d$ $\mu^* + a$

Figure 9.11   Same distribution as in Figure 9.10, showing the slightly different distribution of phenotypic value among the three genotypes (AA, AA', and $A'A'$) for a gene with two alleles that contributes to the quantitative trait. The means of the distributions of AA, AA', and $A'A'$ are symbolized $\mu^* + a$, $\mu^* + d$, and $\mu^* - a$, respectively.

TABLE 9.2   CALCULATION OF μ*, *a*, AND *d* FOR ALLELES AT A LOCUS AFFECTING COAT COLORATION IN GUINEA PIGS[a]

Genotype	Amount of black coloration[b]
$c^r c^r$ (AA)	$1.202 = \mu^* + a = 1.075 + 0.127$
$c^r c^d$ (AA')	$1.059 = \mu^* + d = 1.075 - 0.016$
$c^d c^d$ (A'A')	$0.948 = \mu^* - a = 1.075 - 0.127$
	$\hat{\mu}^* = (1.202 + 0.948)/2 = 1.075$
	$\hat{a} = 1.202 - 1.075 = 0.127$
	$\hat{d} = 1.059 - 1.075 = -0.016$

Source: Data from Wright 1968.

[a] The calculations to be carried out first are those beneath the data; then the right-hand column is completed.

[b] Here the amount of black coloration is measured as arcsin $(\sqrt{x})$, where x is the percentage of black coloration on the animal. For $c^r c^r$, $c^r c^d$, and $c^d c^d$ genotypes, the corresponding x values are 87%, 76%, and 66%, respectively.

$\hat{d}/\hat{a} = -0.126$. Assuming Hardy-Weinberg genotype frequencies, the mean phenotype in the entire population is

$$\mu = p^2(\mu^* + a) + 2pq(\mu^* + d) + q^2(\mu^* - a) \qquad 9.12$$

PROBLEM 9.3   Crosses between the Danmark (P_1) and Red Currant (P_2) tomato gave the following mean fruit weights and their log transforms. P_1 and P_2 are the parental means, F_1 and F_2 are the first and second hybrid generation, and B_1 and B_2 are the progeny of the backcross of $F_1 \times P_1$ and $F_1 \times P_2$, respectively.

	Expected mean	Mean weight	Log (weight)
P_1	$\mu + a$	10.36 ± 0.581	0.98 ± 0.03
P_2	$\mu - a$	0.45 ± 0.017	-0.36 ± 0.02
F_1	$\mu + d$	2.33 ± 0.130	0.33 ± 0.03
F_2	$\mu + \frac{1}{2}d$	2.12 ± 0.105	0.27 ± 0.01
B_1	$\mu + \frac{1}{2}(a + d)$	4.82 ± 0.253	0.64 ± 0.02
B_2	$\mu + \frac{1}{2}(d - a)$	0.97 ± 0.045	-0.05 ± 0.01

Use this information to calculate μ, a, and d for both the weights and the log transformed weights. Do the simple weights or the log transformed weights fit the model better? (Data from Powers 1951.)

ANSWER   The difference between the two parental means is $2a$, so $\hat{a} = (10.36 - 0.45)/2 = 4.96$. This gives $\hat{\mu} = 5.4$. The F_1 has a mean $\mu + d = 2.33$, so $\hat{d} = 2.33 - 5.4 = -3.07$. The F_2 should have mean $(1/4)(\mu + a) + 1/2 (\mu + d) + 1/4 (\mu - a) = \mu + 1/2 d = 5.4 + 1/2(-3.07) = 3.86$. The B_1 refers to backcrosses of the F_1 to P_1, which should yield one-half of genotypes like P_1 and one-half of genotypes like the F_1, so the mean should be $1/2(\mu + a) + 1/2(\mu + d) = \mu + 1/2(a + d) = 6.34$. Similar reasoning gives an expected mean for B_2 of 1.38. The estimates for the means of the F_2, B_1, and B_2 do not fit very well at all. Trying again with the log transformed data, we get $\hat{a} = 0.67$, $\hat{\mu} = 0.31$, and $\hat{d} = 0.02$. The expected means for the F_2, B_1, and B_2 are then $0.31 + 1/2(0.02) = 0.32$, $0.31 + 1/2(0.67 + 0.02) = 0.65$, and $0.31 + 1/2(0.02 - 0.67) = -0.01$. The log transformed data clearly fit much better, suggesting that the better scale to use for the quantitative genetic models is the log transformed scale. In actual practice, the entire set of data is used to estimate μ, a, and d by a method known as least squares, and the goodness of fit of the model to the data can be tested by a chi-square test.

Effects of the scale of measurement are known as **scaling effects**. For example, the a and d values in Table 9.2 are different when calculated for the percent of black coloration x or for the arcsin $(\sqrt{x})$ tabulated values. Since estimates of the additive and dominance values of alleles depend on scaling, so does the heritability. An important point is that the equivalence between the heritability defined by parent-offspring regression and by realized heritability depends on the correct choice of scaling. Only one scaling provides a normal Gaussian distribution of phenotypes and of genetic effects, and that is the appropriate scaling that yields the prediction Equation 9.10.

Change in Gene Frequency

Suppose for the moment that we were practicing artificial selection for increased amount of black coat coloration in the guinea pigs in Table 9.2. Selection for black-coat coloration in a population containing both the c^r (i.e., A) and c^d (i.e., A') alleles would be successful in increasing the allele frequency of A, and the average amount of black coloration among individuals of the next generation would increase. Therefore, in order to calculate the expected increase in black coloration in one generation of selection, we must first calculate the corresponding change in the allele frequency of A. An equation for change in allele frequency with natural selection was derived in Chapter 6, which remains valid for artificial selection if we agree to interpret the "fitness" of an individual as the probability that the individual is included among the group selected as parents of the next generation. With this interpretation of fitness, differences in fitness (i.e., reproductive success) of AA, AA', and $A'A'$ genotypes correspond to the differences in area to the right of the truncation point in Figure 9.11, because only those individuals in the shaded area are allowed to reproduce. The differences in area are easy to calculate if you shift or slide each curve horizontally until its mean coincides with μ^*. The $A'A'$ curve must slide a units to the right, and the AA' and AA curves must slide d and a units to the left. This shifting brings the distributions into coincidence, but it slides the truncation points slightly out of register, as shown in Figure 9.12. The difference in "fitness" between AA and AA', denoted $w_{11} - w_{12}$ (as in Chapter 6), is equal to the small area indicated in Figure 9.12, as is the difference in fitness between AA' and $A'A'$, denoted $w_{12} - w_{22}$. The areas corresponding to $w_{11} - w_{12}$ and $w_{12} - w_{22}$ are approximately rectangles, and the area of a rectangle is the product of the base and the height. The approximation is most accurate when the effect of this one locus on the phenotype is small. Therefore, since Z represents the height of the normal distribution at the point T, we can make the following approximations

$$w_{11} - w_{12} \approx Z[(T - d) - (T - a)] = Z(a - d)$$

$$w_{12} - w_{22} \approx Z[(T + a) - (T - d)] = Z(a + d)$$

9.13

The average fitness $\bar{w}$ of the entire population simply equals B, because B is the proportion of the population saved for breeding. From Chapter 6 we know that

$$\Delta p = pq[p(w_{11} - w_{12}) + q(w_{12} - w_{22})]/\bar{w}$$

where Δp is the change in frequency of the allele A in one generation of selection. Substituting from Equation 9.13 and using $\bar{w} = B$ leads to

$$\Delta p = pq[pZ(a - d) + qZ(a + d)]/B \qquad 9.14$$

or, since $p + q = 1$,

$$\Delta p = (Z/B)pq[a + (q - p)d] \qquad 9.15$$

An equation corresponding to 9.15 could be obtained for any gene affecting the trait, but the values of p, a, and d would differ for each gene. The quantity in square brackets in Equation 9.15 is called the **average excess**. A generalization that accounts for nonrandom mating is found in Falconer (1985).

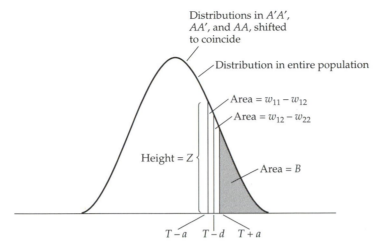

Distributions in $A'A'$, AA', and AA, shifted to coincide

Distribution in entire population

Area = $w_{11} - w_{12}$

Area = $w_{12} - w_{22}$

Height = Z

Area = B

$T - a$ $T - d$ $T + a$

Figure 9.12   Same distribution as in Figures 9.10 and 9.11, but with the distributions of AA, AA', and $A'A'$ shifted laterally to coincide. Shifting the distributions slides the truncation points slightly out of register, so the truncation points for AA, AA', and $A'A'$ become $T - a$, $T - d$, and $T + a$, respectively. The small area that is denoted $w_{11} - w_{12}$ is the difference between the proportions of AA and AA' genotypes that are included among the selected parents, and the area $w_{12} - w_{22}$ is the difference in the proportion of AA' and $A'A'$ genotypes included among the selected parents.

Genetic Model for the Change in Mean Phenotype

Equation 9.15 provides an expression for Δp which can be used to calculate the mean phenotypic value of coat color after one generation of selection. In the next generation, the allele frequencies of A and A' are $p + \Delta p$ and $q - \Delta p$, respectively. With random mating, the mean phenotype in this generation is given by Equation 9.12 as

$$\mu' = (p + \Delta p)^2(\mu^* + a)$$
$$+ 2(p + \Delta p)(q - \Delta p)(\mu^* + d) \qquad\qquad 9.16$$
$$+ (q - \Delta p)^2(\mu^* - a).$$

When the right-hand side of this expression is multiplied out and terms in $(\Delta p)^2$ are ignored because Δp is usually small, then μ' is found to be approximately

$$\mu' \approx \mu + 2[a + (q - p)d]\Delta p \qquad\qquad 9.17$$

The approximation in Equation 9.17 is rather good even for relatively large values of Δp.

Equation 9.17 warrants a little more development since it yields the prediction equation $R = h^2S$ (Equation 9.10) and also provides an expression for h^2 in terms of the parameters a, d, and p that can be interpreted genetically. First, rewrite Equation 9.17 as

$$\mu' - \mu = 2[a + (q-p)d]\Delta p \qquad 9.18$$

Then substitute for Δp from Equation 9.15, which yields

$$\mu' - \mu = (Z/B)2pq[a + (q-p)d]^2 \qquad 9.19$$

Now use the expression for Z/B given in Equation 9.11 to obtain

$$\mu' - \mu = (\mu_S - \mu)2pq[a + (q-p)d]^2/\sigma^2 \qquad 9.20$$

Finally, substitute from Equations 9.8 and 9.9 for the selection differential S and the response R, yielding

$$R = (S)2pq[a + (q-p)d]^2/\sigma^2 \qquad 9.21$$

However, $R = h^2S$ also (Equation 9.10), and so

$$h^2 = 2pq[a + (q-p)d]^2/\sigma^2 \qquad 9.22$$

Equation 9.22 for h^2 is the one we were after, as it defines the heritability in terms of p, q, a, and d—each of which has a genetic meaning.

Equation 9.22 is a valid approximation when a single gene affects the trait in question, and when the effects of that gene are small. However, when many genes affect the trait, the right-hand side of the equation must be replaced by a summation of such terms, one for each gene. That is, for many genes, $R = h^2S$ where

$$h^2 = \Sigma 2pq[a + (q-p)d]^2/\sigma^2 \qquad 9.23$$

in which the summation is over all genes that affect the trait. (However, each gene may have different values of a, d, p, and q.) As will be discussed in more detail later, the quantity

$$\sigma_a^2 = \Sigma 2pq[a + (q-p)d]^2 \qquad 9.24$$

is called the **additive genetic variance** of the trait. Although the individual components in the additive genetic variance are difficult to identify except in contrived examples like the one involving guinea pigs, the collective effects (represented by the summation) can be estimated.

COMPONENTS OF PHENOTYPIC VARIANCE

As Equation 9.24 suggests, the variance of a quantitative trait can be split into various components representing different causes of variation. Similarity between relatives is conveniently expressed in terms of the variance components, but variance partitioning is also of interest in its own right. Since the rate of change of a trait under selection depends on the amount of genetic variation affecting the trait, if there is no genetic variation, there is obviously no response to selection. What is not so obvious is that some components of genetic variation cannot be acted upon by some kinds of selection. In other words, certain populations have ample genetic variation, yet fail to respond to selection. The part of the genetic variation amenable to selection is clarified by partitioning the variance.

Genetic and Environmental Sources of Variation

As shown in Table 9.3, the phenotypic value of any individual can be represented as a sum of three components: (1) the mean μ of the entire population, (2) a deviation from the population mean due to the specific genotype of the individual in question (symbolized as G_1, G_2, and G_3 for AA, AA', and $A'A'$ genotypes, respectively), and (3) a deviation from the population mean due to the specific **microenvironment** of the individual in question. (The environmental deviations are unique to each individual and are represented as E_1, E_2, . . . , E_9.) These microenvironmental effects might be due to random differences in nutrition, temperature, or other external factors, or they might be seen even in an absolutely uniform external environment due to the vagaries of embryonic development. It is important to note that the Gs and Es are not directly observable. Nevertheless, as we shall see, the total variance in phenotypic value can be partitioned into a component due to variation among the Gs and another component due to variation among the Es. The model can be summarized by writing

$$P = \mu + G + E \qquad\qquad 9.25$$

**TABLE 9.3 PHENOTYPES OF VARIOUS GENO-
TYPES AS THE SUM OF μ, G, AND E[a]**

Genotype	Phenotypic Value
AA	$\mu + G_1 + E_1$
AA	$\mu + G_1 + E_2$
AA	$\mu + G_1 + E_3$
AA'	$\mu + G_2 + E_4$
AA'	$\mu + G_2 + E_5$
AA'	$\mu + G_2 + E_6$
$A'A'$	$\mu + G_3 + E_7$
$A'A'$	$\mu + G_3 + E_8$
$A'A'$	$\mu + G_3 + E_9$

[a] μ is the population mean. G is a contribution due to genotype, different for each genotype. E is a contribution due to environment, different for each individual.

where P represents the phenotypic value of any individual and G and E are the genotypic and environmental deviations pertaining to that individual.

To connect the above symbols with actual numbers, we may use Table 9.2 and assume an allele frequency of A of $p = 0.2$. Equation 9.12 then implies that the mean of the population is $\mu = 0.994$. Thus, the respective G_1, G_2, and G_3 deviations for AA, AA', and $A'A'$ genotypes are

$$G_1 = 1.202 - 0.994 = 0.208$$
$$G_2 = 1.059 - 0.994 = 0.065$$
$$G_3 = 0.948 - 0.994 = -0.046$$

For a particular animal of genotype AA whose actual coat color score is, for example, 1.312, the corresponding value of E for the animal would be calculated using Equation 9.25 from the expression $1.312 = 0.994 + 0.208 + E$; thus, for this animal, $E = 0.11$. Similarly, a particular animal of genotype AA' with an actual phenotype of $P = 1.009$ would have a value of E given by $1.009 = 0.994 + 0.065 + E$, or $E = -0.05$. Because the E values are defined as deviations from their mean, the average of Es for any genotype is 0. Likewise, since the Gs are defined as deviations from their mean, the mean of the Gs is 0. This result can be verified in the guinea pig example because

$$(0.2)^2 G_1 + 2(0.2)(0.8)G_2 + (0.8)^2 G_3 = 0$$

Equation 9.25 is appropriate when the effects of genotype and environment are additive—that is, when the deviation of the phenotype of any particular individual from the population mean $(P - \mu)$ can be written as the sum of an effect resulting from the genotype of that individual and a separate effect resulting from the environment of that individual.

PROBLEM 9.4 In Problem 9.3 the values of μ, a, and d were found to be 0.31, 0.67, and 0.02, respectively, for the logarithms of tomato weight. Calculate the additive genetic variance in the F_2 population, the B_1 population and the B_2 population.

ANSWER   In the F_2 population the allele frequency is $p = q = \frac{1}{2}$, so the formula for the additive genetic variance (Equation 9.24) is $\sigma_a^2 = 2pqa^2 = \frac{1}{2}a^2 = 0.224$. In the backcross 1 population B_1, the allele frequencies are $p = \frac{3}{4}$ and $q = \frac{1}{4}$, so, applying Equation 9.24, we get σ_a^2 = 0.173. The backcross 2 population B_2 has allele frequencies $p = \frac{1}{4}$ and $q = \frac{3}{4}$, so $\sigma_a^2 = 0.163$. When the dominance parameter is so small, the additive variance is at a maximum when the allele frequencies are both $\frac{1}{2}$, and the graph of additive variance against allele frequency is symmetric.

To this point the discussion has been restricted to a particular population in a single **macroenvironment**, and the sources of variation have been due to genetic and microenvironmental differences among individuals. A change in macroenvironment is easiest to see in an experimental setting, where, for example, in macroenvironment 1 all of the guinea pigs get twice as much food as in macroenvironment 2. Additivity of genetic and environmental effects is true whenever the ratio of $G_1 : G_2 : G_3$ is the same in each of the relevant environments. For the genotypes in Figure 9.13, for example, if the actual range of environments is the range designated E_1, then the genetic and environmental effects are additive because the ratio $G_1 : G_2 : G_3$ is the same for

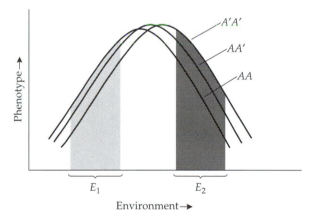

Figure 9.13   The *norm of reaction* is the relation between the phenotype and the environment, and this relation is known to vary from genotype to genotype. Hypothetical norms of reaction for genotypes AA, AA' and $A'A'$ are shown here. In the range of environments denoted E_1, A is very nearly dominant to A'(that is, AA and AA' have nearly the same phenotype). However, in the range E_2, A and A' are very nearly additive (no dominance). The heritability of the trait resulting from this gene differs according to whether the population is reared in E_1 environments or E_2 environments.

any particular environment in E_1. For the same reason, the genetic and environmental effects are additive if the actual range of environments is E_2.

However, if the actual range of environments includes both E_1 and E_2, then the ratio $G_1 : G_2 : G_3$ depends on the particular environment, and therefore the genetic and environmental effects may not be additive. Nonadditivity of genetic and environmental effects is called **genotype-environment interaction**, and in writing Equation 9.25, it appears that there is an assumption that there is no genotype-environment interaction. In actually estimating components of variance, it is not necessary to assume that there is no genotype-environment interaction, because we can explicitly examine the phenotypes when reared in different macroenvironments and directly estimate the magnitude of the interaction. Alternatively, we can arbitrarily define the environmental variance as also including the effects of genotype—environment interaction.

When Equation 9.25 is valid, the total phenotypic variance σ_p^2 in the population equals the mean of $(P - \mu)^2$. However, Equation 9.25 implies that $(P - \mu)^2$ equals $(\mu + G + E - \mu)^2$, which is

$$\sigma_p^2 = (G + E)^2 = G^2 + 2GE + E^2 \qquad 9.26$$

Because G and E are already deviations from their means, the mean of G^2 is the phenotypic variance in the population resulting from differences in genotype, and the mean of E^2 is the phenotypic variance resulting from differences in environment. The mean of G^2 is called the **genotypic variance** and is denoted σ_g^2. The mean of E^2 is called the **environmental variance** and is denoted σ_e^2. The remaining term—the mean of $2GE$—is two times the **genotype-environment covariance**. If the genotypic and environmental deviations are uncorrelated—that is, if there is no systematic association between genotype and environment—then there is said to be no **genotype-environment association** and the mean of $2GE$ equals zero. When there is no genotype-environment association, therefore,

$$\sigma_p^2 = \sigma_g^2 + \sigma_e^2 \qquad 9.27$$

Equation 9.27 is the theoretical foundation for partitioning the variance into genetic and environmental effects. The assumption that genotype-environment association is negligible is frequently a valid assumption in animal and plant breeding where, because breeders have a degree of control not available to, for example, human geneticists, experiments can be intentionally designed in such a way as to minimize genotype-environment association. However, genotype-environment association can occur even in animal and plant breeding. For example, dairy farmers routinely provide more feed supplements to cows that produce more milk; because milk-producing ability is partly due to genotype, this feed regimen will provide superior environments (better feed) to cows that have superior genotypes to begin with, so there will be a genotype-environment association. Similarly the best race horses get the best trainers and the children of the best students often go to the best schools. If one is not careful to correct for such associations, genotype-environment association can inflate the apparent σ_g^2 and possibly give spurious overestimates of heritability.

The biological meaning of Equation 9.27 is shown for the alleles of one gene in Figure 9.14. The solid curves represent the phenotypic distributions in the genotypes AA, AA', and $A'A'$ with means denoted G_1, G_2, and G_3, and the dashed curve represents the phenotypic distribution in the entire population. The total phenotypic variance σ_p^2 is the variance of the dashed distribution; the genotypic variance σ_g^2 is the variance among the Gs (i.e., $\sigma_g^2 = p^2 G_1^2 + 2pq G_2^2 + q^2 G_3^2$, where p is the allele frequency of A); and the environmental variance σ_e^2 is obtained by subtraction: $\sigma_e^2 = \sigma_p^2 - \sigma_g^2$. Although the Gs are not generally known, σ_g^2 must equal zero in a genetically uniform population. The observed variance of a randomly bred population, therefore, provides an estimate of $\sigma_g^2 + \sigma_e^2$, whereas the average observed variance of genetically uniform populations provides an estimate of σ_e^2. The estimate of σ_g^2 is obtained by subtraction, as shown in an example using thorax length in *Drosophila* (Table 9.4). In this case, genetic variation among individuals in the randomly bred population accounts for about $0.180/0.366 = 49.2\%$ of the phenotypic variance. Genetically uniform populations such as inbred lines or crosses

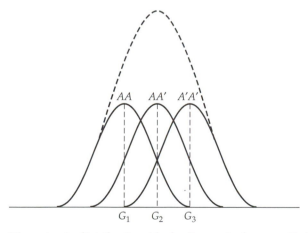

Figure 9.14   Phenotypic distribution (dashed curve) of a quantitative trait in a hypothetical population, showing distributions (solid curves) of three constituent genotypes for two alleles of a gene. The means of AA, AA', and $A'A'$ genotypes are denoted G_1, G_2, and G_3, respectively.

TABLE 9.4 **CALCULATION OF GENOTYPIC VARIANCE (σ_g^2) AND ENVIRONMENTAL VARIANCE (σ_e^2)[a]**

	POPULATIONS	
Variance	*Random-bred*	*Uniform*
Theoretical	$\sigma_g^2 + \sigma_e^2$	σ_e^2
Observed	0.366	0.186

$\sigma_e^2 = 0.186$

$\sigma_g^2 = (\sigma_g^2 + \sigma_e^2) - \sigma_e^2 = 0.366 - 0.186 = 0.180$

Source: Data from Robertson 1957.

[a] Trait is length of thorax in *Drosophila melanogaster* (in units of 10^{-2} mm).

---

between inbreds are not available in human populations, but identical twins are often used instead because of the identical genotypes of the twins.

An example of a naturally occurring organism that exhibits remarkably low levels of genetic variability is the African cheetah (O'Brien et al. 1983, 1987; May 1995). One might suppose that limited genetic variation would result in depressed phenotypic variability as well, but a study of cranial measures by Wayne et al. (1986) revealed that the amount of variability was not appreciably less than that in three other large cats. In fact, there was a significant *increase* in the amount of fluctuating asymmetry (that is, the difference in measurements from the left and the right side of the body). The fluctuating asymmetry is consistent with the notion that genetic homozygosity results in reduced developmental stability—an idea that has considerable empirical support, but so far no good explanation in molecular terms. In any event, reduction of genetic variance, and concomitant high homozygosity, can result in an increase in phenotypic variance due to developmental instability. Since extreme homozygosity may result in phenotypes that are very sensitive to environmental fluctuations, the paradoxical increase in phenotypic variance results from genotype-environment interaction.

Components of Genotypic Variation

So far, the phenotypic variance has been partitioned into the genotypic variance and the environmental variance according to the Equation 9.27. The genotypic variance can be partitioned further into terms that are particularly important for interpreting the resemblance between relatives. The appropriate model is shown in Table 9.5, where the phenotypic means of AA, AA', and $A'A'$ are denoted 1 by $\mu^* + a$, $\mu^* + d$, and $\mu^* - a$, as they were earlier in Figure 9.11. To obtain the G values, the mean of each genotype must be expressed as a deviation from the population mean, which is

TABLE 9.5   EXPRESSIONS FOR POPULATION MEAN AND GENOTYPIC DEVIATIONS

Genotype	Frequency	Mean phenotype	Genotypic deviation from population mean (G)
AA	p^2	$\mu^* + a$	$G_1 = \mu^* + a - \mu = 2q[a + (q - p)d] - 2q^2 d$
AA'	$2pq$	$\mu^* + d$	$G_2 = \mu^* + d - \mu = (q - p)[a + (q - p)d] + 2pqd$
$A'A'$	q^2	$\mu^* - a$	$G_3 = \mu^* - a - \mu = -2p[a + (q - p)d] - 2q^2 d$

$$\text{Population mean } \mu = p^2(\mu^* + a) + 2pq(\mu^* + d) + q^2(\mu^* - a)$$
$$= (p^2 + 2pq + q^2)\mu^* + (p^2 - q^2)a + 2pqd$$
$$= (p + q^2)\mu^* + (p - q)(p + q)a + 2pqd$$
$$= \mu^* + (p - q)a + 2pqd$$

$\mu = \mu^* + (p - q)a + 2pqd$, and the deviations are shown in the last column of Table 9.5. The genotypic variance σ_g^2 is calculated as

$$\sigma_g^2 = p^2 G_1^2 + 2pq G_2^2 + q^2 G_3^2$$
$$= 2pq[a + (q - p)d]^2 + (2pqd)^2$$

9.28

The first term in Equation 9.28 is the *additive genetic variance* σ_a^2 encountered earlier in Equation 9.24. The second term is a new quantity called the **dominance variance**, which is symbolized σ_d^2. From Equation 9.28, therefore,

$$\sigma_g^2 = \sigma_a^2 + \sigma_d^2$$

9.29

which allows us to express the total phenotypic variance as the sum of three terms, namely

$$\sigma_p^2 = \sigma_a^2 + \sigma_d^2 + \sigma_e^2$$

9.30

When Equation 9.22 for heritability is written in terms of variance components rather than p, q, a and d, the equation implies that

$$h^2 = \sigma_a^2 / \sigma_p^2$$

9.31

Equation 9.31 is an important result because it states that the heritability depends only on the additive genetic variance and not on the dominance variance. Therefore, if all the genetic variance in a population results from dominance variance (i. e., $\sigma_a^2 = 0$), then the population cannot respond to individual selection because h^2 equals zero. To say the same thing in another way, the dominance variance σ_d^2 represents that portion of the genetic variance that is not acted upon by individual selection.

Equation 9.31 means that the heritability of a trait is the ratio of the additive genetic variance to the total phenotypic variance. Sometimes the word *heritability* is used in reference to a different variance ratio, namely the ratio of the total genotypic variance to the total phenotypic variance (i.e., σ_g^2/σ_p^2). To avoid confusion, quantitative geneticists distinguish the two types of heritability as follows:

1. The ratio σ_a^2/σ_p^2 is called heritability in the *narrow sense* (This is the variance ratio we have been using all along.)
2. The ratio σ_g^2/σ_p^2 is called heritability in the *broad sense*.

Generally speaking, narrow-sense heritability is the more important with individual selection (or any mode of selection that capitalizes primarily on the additive genetic variance), whereas broad-sense heritability is the more important when selection is practiced among clones (a clone is a group of genetically identical individuals), inbred lines, or varieties. We use the term *heritability* to mean narrow-sense heritability unless otherwise stated.

As emphasized earlier, heritability has no transparent interpretation in simple genetic terms. The same is true of the variance components σ_a^2 and σ_d^2. Even for a single gene, the variance components depend on the particular values of a and d (Figure 9.15), and of course the estimates of heritability must also depend on allele frequency (Figure 9.16). With many genes that act together, σ_a^2 is defined as a summation of the values of $2pq[a + (q - p)d]^2$ for each gene affecting the trait, and σ_d^2 represents a summation of the values of $(2pqd)^2$ for each gene. Furthermore, when the trait is affected by multiple genes, the formula for σ_g^2 in Equation 9.29 must be extended to include an additional term that pertains to interaction among the genes. This interaction term is called the **interaction** variance or the **epistatic** variance and is symbolized σ_i^2. With the interaction variance included, Equation 9.29 becomes

$$\sigma_g^2 = \sigma_a^2 + \sigma_d^2 + \sigma_i^2 \qquad\qquad 9.32$$

The important point to remember about the components of genotypic variance is that they represent the cumulative, statistical effects of all genes affecting the trait. Few inferences about the actual mode of inheritance of the trait are possible from the variance components, particularly concerning the number of genes involved and their individual effects.

PROBLEM 9.5   By definition, a simple Mendelian trait is one that is determined entirely by genotype in the prevailing environment. Therefore, $\sigma_e^2 = 0$ in Equation 9.27, and the broad-sense heritability $\sigma_g^2/\sigma_p^2 = 1$. Show that, for a simple Mendelian recessive, the narrow-sense heritability equals $2q/(1 + q)$, where q is the recessive allele frequency.

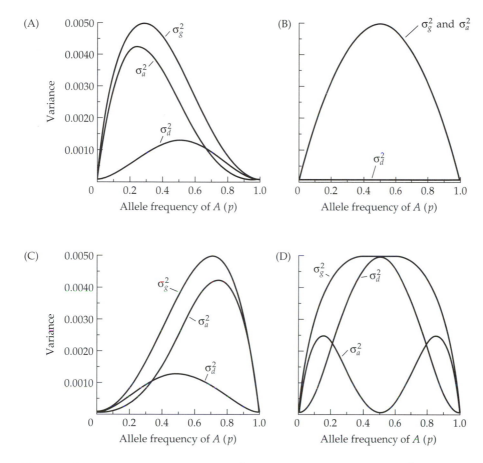

Figure 9.15   Total genetic variance (σ_g^2), additive genetic variance (σ_a^2), and dominance variance (σ_d^2) for a locus with two alleles (A and A') plotted against the frequency of allele A (p). The mean phenotypes of AA, AA' and $A'A'$ are denoted $\mu^* + a$, $\mu^* + d$, and $\mu^* - a$, respectively. In all cases, we have that $\sigma_a^2 = 2pq[a + (q - p)d]^2$, $\sigma_d^2 = (2pqd)^2$, and $\sigma_g^2 = \sigma_a^2 + \sigma_d^2$. (A) $a = d = 0.0701$ (A dominant to A'); (B) $a = 0.1$, $d = 0$ (no dominance), (C) $d = -a = 0.0707$ (A' dominant to A); (D) $a = 0$, $d = 0.141$ (overdominance). For ease of comparison, the values of a and d have been chosen to make the maximum of σ_g^2 equal to 0.005 in each case.

ANSWER   Let the phenotypes of AA, AA', and $A'A'$ be assigned phenotypic values 0, 0, and 1, respectively, so that the A' allele is recessive. In this case, $\mu^* = 1/2$, $a = -1/2$, and $d = -1/2$. The numerator of Equa-

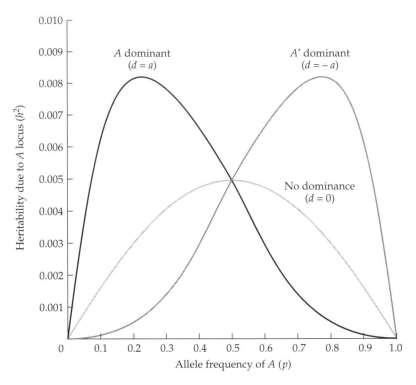

Figure 9.16   Narrow-sense heritability due to a single locus with two alleles (A and A') as a function of p, the allele frequency of A. In general, for one locus, $h^2 = 2pq[a + (q - p)d]^2/\sigma_p^2$, where σ_p^2 is the total phenotypic variance. The curves correspond to $a = 0.1$, and $d = 0.1$ (A dominant), $d = 0$ (no dominance), and $d = -0.1$ (A' dominant).

tion 9.31 is the additive genetic variance, which equals $2pq^3$. Then the mean phenotype equals q^2, and the variance in phenotypic value is $q^2 - (q^2)^2 = q^2(1 - q^2) = q^2(1 + q)(1 - q) = pq^2(1 + q)$. The heritability is the additive variance divided by the phenotypic variance, namely $2pq^3/[pq^2(1 + q)] = 2q/(1 + q)$. When the autosomal recessive trait is rare, $q \approx 0$, and the heritability is approximately equal to the frequency of heterozygous carriers.

COVARIANCE BETWEEN RELATIVES

Components of genetic variation are important because they may be used to express the phenotypic covariance between relatives. Since the distribution of offspring from a given parental genotype depends on the distribution of potential mates, the variance components and estimates of heritability depend not only on allele frequencies but also on the distribution of genotype frequencies. To simplify things, we assume that the trait is determined by one gene with two alleles, and that the population is in Hardy-Weinberg proportions. However, the same results are also true for many genes when the trait is determined by summing the individual allelic effects provided that the population is in multilocus linkage equilibrium.

Table 9.6 displays three genotypes of parents, their genotypic value, and the mean genotypic value of the offspring with random mating. The covariance of the offspring and one parent is calculated by summing the product of the last three columns of Table 9.6 and subtracting the product of the means of the last two columns. After tedious algebra, the covariance of offspring and parent is:

$$\sigma_{OP} = pq \, [a + (q - p)d]^2 = \tfrac{1}{2}\sigma_a^2 \qquad\qquad 9.33$$

This is a remarkably simple result because it says that the covariance in phenotype of parents and offspring is one-half the additive genetic variance. No component of dominance inflates the covariance in this case. Since environmental effects are assumed to be random with respect to genotypic values (there is no genotype–environment correlation), environmental effects also play no role in parent-offspring covariance. We must, however, assume that the *environments* of parents and offspring are uncorrelated for Equation 9.33 to be valid. In order to see the relation between narrow-sense heritability and regression, recall that the regression coefficient is defined as

$$b = \sigma_{xy}/\sigma_x^2 \qquad\qquad 9.34$$

In this case, the regression of offspring on one parent is, from Equation 9.33:

$$b_{OP} = \sigma_{OP}/\sigma_P^2$$
$$= \tfrac{1}{2}\,\sigma_a^2/\sigma_p^2 \qquad\qquad 9.35$$
$$= \tfrac{1}{2}\,h^2$$

TABLE 9.6   DERIVATION OF PARENT–OFFSPRING COVARIANCE

Parent's genotype	Frequency	Genotypic value[a]	Offspring mean genotypic value
AA	p^2	$2q(a - pd)$	$aq + dq(q - p)$
A'A	$2pq$	$a(q - p) + d(1 - 2pq)$	$\tfrac{1}{2}q(q - p) + \tfrac{1}{2}d(q - p)^2$
A'A'	q^2	$-2p(a + qd)$	$-ap - dp(q - p)$

[a] Genotypic values are expressed as deviations from the population mean.

In the regression of offspring on the *midparent* (that is, the average value of the parents), the denominator in Equation 9.35 becomes $\frac{1}{2}\sigma_p^2$ because this is the variance of the mean of the two parents, assuming random mating. Hence the regression coefficient equals $(\frac{1}{2}\sigma_a^2)/(\frac{1}{2}\sigma_p^2)$, and so the regression coefficient of offspring on midparent equals the narrow-sense heritability.

The same reasoning can be followed to obtain covariances between other pairs of relatives, as summarized in Table 9.7. As can be seen, the additive genetic variance can be estimated directly either from parent-offspring covariance or from half-sib covariance. However, full-sib covariance includes a term resulting from dominance. The expressions in Table 9.7 are correct as long as there are no complications such as genotype-environment associations or other nonrandom environmental effects such as full sibs sharing environmental factors common to the whole family but not shared by other families. Since the total variance in phenotypic value σ_p^2 can be estimated directly, once σ_a^2 is estimated from the covariance between relatives, the narrow-sense heritability can be estimated from Equation 9.31. The first three relationships in Table 9.7 are the most useful in quantitative genetics and are commonly used in animal and plant breeding. The other relationships are used mainly in human quantitative genetics.

The genetic covariance between various relatives can also be derived using the concepts of gene identity developed by Cotterman (1940) and extended by Crow and Kimura (1970). In these terms, the generalized covariance for a pair of related individuals is

$$\mathrm{Cov}(x,y) = r\,\sigma_a^2 + u\,\sigma_d^2 \qquad\qquad 9.36$$

where the coefficients r and u are determined from coefficients of coancestry. The coefficient of coancestry, F_{xy}, of two individuals x and y is the inbreeding

---

TABLE 9.7   THEORETICAL COVARIANCE IN PHENOTYPE BETWEEN RELATIVES[a]

Degree of relationship	Covariance
Offspring and one parent	$\sigma_a^2/2$
Offspring and average of parents (midparent)	$\sigma_a^2/2$
Half siblings	$\sigma_a^2/4$
Full siblings	$(\sigma_a^2/2) + (\sigma_d^2/4)$
Monozygotic twins	$\sigma_a^2 + \sigma_d^2$
Nephew and uncle	$\sigma_a^2/4$
First cousins[b]	$\sigma_a^2/8$
Double first cousins	$(\sigma_a^2/4) + (\sigma_d^2/16)$

[a] Variance terms due to interaction between loci (epistasis) have been ignored.

[b] First cousins are the offspring of matings between siblings and unrelated individuals; double first cousins are the offspring of matings between siblings from two different families.

coefficient of a hypothetical offspring of x and y. If individuals A and B are the parents of x, and C and D are the parents of y, then the r and u coefficients in Equation 9.36 are:

$$r = 2F_{xy}$$
$$u = F_{AC}\,F_{BD} + F_{AD}\,F_{BC}$$

9.37

It is an instructive exercise to write out pedigrees for some of the relations in Table 9.7, calculate the coefficients of coancestry, and verify that Equation 9.36 gives the correct result.

Figure 9.17 presents the narrow-sense heritabilities of diverse quantitative traits in farm animals and one important crop plant as estimated from the correlation between relatives. The data are presented merely to show the values of heritability with which breeders typically must deal. It is important to keep in mind that the heritabilities in Figure 9.17 pertain to one population in one type of environment at one particular time. The same trait in a different population or in a different environment might well have a different heritability. Generally speaking, traits that are closely related to fitness (such as calving interval in cattle or eggs per hen in poultry) tend to have rather low heritabilities. Ignoring complications such as antagonistic pleiotropy (discussed later), long-term natural selection is expected to gradually reduce the additive genetic variance until the effect is balanced against the input of new mutations.

For purposes of comparison, Figure 9.18 shows estimated broad-sense heritabilities of a number of quantitative traits in humans. Broad-sense heritabilities vary widely for different traits, as they do in other species. Note the low heritability of fertility, a trait that is obviously closely related to fitness. At the other end of the scale is total fingerprint ridge count, which is apparently not a major component of fitness considering its relatively high broad-sense heritability.

Although it is tempting to think about resemblance between relatives in terms of classical genetic analysis such as Mendel did, there are major differences between the approaches. When the data are measurements of a continuous character in family-structured samples from a population, estimates of statistical components of variance can be obtained. However, these components depend on allele frequencies and environmental conditions, and thus quantities such as heritability are far removed from the basic level of gene action. Direct experimental assessment of variance components (e.g., Mitchell-Olds 1986) shows that different populations do have different heritabilities of many traits. Moreover, a trait with high heritability does not mean that the trait cannot be affected by the environment. For example, phenylketonuria, which is caused by homozygosity for a single defective allele for the enzyme phenylalanine hydroxylase, is a simple Mendelian recessive, yet the phenotype of severe mental retardation can be completely circumvented by a diet low in phenylalanine.

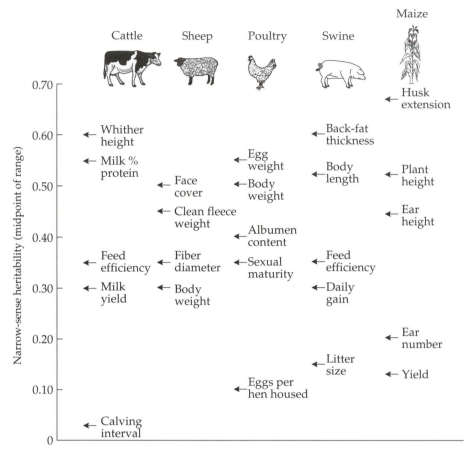

Figure 9.17 Narrow-sense heritabilities for representative traits in plants and animals. Traits closely related to fitness (calving interval, eggs per hen, litter size of swine, yield and ear number of corn) tend to have rather low heritabilities. (Animal data from Pirchner 1969, who gives the range of heritabilities in various studies. The midpoint of the range is plotted here. Corn data from Robinson et al. 1949.)

PROBLEM 9.6 Consider the following hypothetical experiment on plant growth. Seeds were removed from six plants and offspring were grown in two different light conditions. Height of the plant at eight weeks was measured in the parental plants and all the progeny, giving the following data:

Midparent	Offspring with full sunlight	Offspring with 10% full sunlight
1.10	1.12	0.63
1.54	1.53	1.02
1.23	1.22	0.74
1.06	1.04	0.53
1.47	1.43	0.91
1.38	1.32	0.83

Calculate the heritability in both environments.

ANSWER The variance in the midparents can be calculated as $\{\Sigma x_i^2 - [(\Sigma x_i)^2/6]\}/5 = 0.0391$. The covariance between midparents and offspring at full sunlight is $\{\Sigma x_i y_i - [(\Sigma x_i \Sigma y_i)/6]\}/5 = 0.0365$, so the narrow sense heritability is $0.0365/.0391 = 0.93$. At 10% full sunlight, the midparent-offspring covariance is 0.0353, and the heritability is $0.0353/0.0391 = 0.90$. This example illustrates the important principle that a trait may have a very high heritability, yet the phenotypic mean is still strongly altered by a change in the environment. Actually, the heritability itself may also change as one moves from one environment to another.

The distinction between estimation of variance components and knowledge of genetic causes of human differences has particularly important impli-

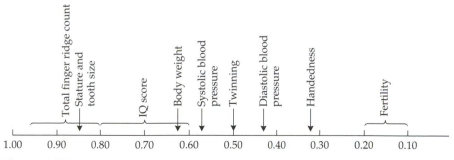

Figure 9.18 Broad-sense heritabilities and ranges of heritabilities of various traits in humans. Uncertainties about the correlation between environments of relatives make such estimates in humans very tentative. (Data from Smith 1975.)

cations for social applicability of human quantitative genetics. Some of the problems are forcefully conveyed by Lewontin (1974) and Feldman and Lewontin (1975). In particular, estimates of heritability within a population, even if they are sound, tell us nothing about the degree to which genetic differences account for the differences in phenotypes between populations (see Problem 9.6). Experimentalists working with organisms that can be manipulated in the field or laboratory can be more rigorous in the assessment of genetic parameters by examining the traits in several environments, and by doing studies that combine classical and quantitative genetic analysis.

Twin Studies and Inferences of Heritability in Humans

Because identical twins are genetically identical, phenotypic differences between identical twins would seem to be a straightforward measure of how much phenotypic variance is caused by environment. Twin studies raise their own unique problems, however, and the results must be interpreted with caution. Before discussing the use of twins in quantitative genetics, we should back up a few steps and first discuss the phenomenon of twinning itself.

Twins are relatively frequent among human births, though the rate of twinning varies from population to population. Among Caucasians in the United States, for example, about one in 88 births results in twins; among Japanese in Japan, the rate is about one in 145 births (Bulmer 1970). Two kinds of twins actually occur. **Identical twins**, often called **monozygotic** or **one-egg** twins, arise from a single zygote that very early in embryonic development splits into two distinct clumps of cells, each clump thereafter undergoing its own embryonic development. Because they arise from a single zygote, identical twins are necessarily genetically identical. The other kind of twins are called **fraternal twins**, **dizygotic** twins, or **two-egg** twins. Fraternal twins arise from a double ovulation in the mother, each egg being fertilized by a different sperm. Because of their mode of origin, fraternal twins are related genetically as siblings. Most of the variation in twinning rates in humans is due to variation in the rate of dizygotic twinning. For example, the rates of monozygotic twinning among Caucasians in the United States and among Japanese in Japan are one in 256 and one in 238, respectively, whereas the respective rates of dizygotic twinning are one in 135 and one in 370 (Bulmer 1970).

For studies in quantitative genetics, identical twins are often compared with same-sex fraternal twins in order to discount the effects of common intrauterine environments. Such an approach is only partially successful, as identical twins often share embryonic membranes *in utero* (the amnion and chorion) that are not usually shared by fraternal twins. Moreover, because identical twins often have astonishingly similar facial features, they may be treated more similarly by parents, teachers, and peers than are fraternal twins. Some of these problems can be overcome by studying twins that are raised apart (in different households), but data of this sort are usually limited (Shields 1962), making estimates of heritability highly imprecise. Even when twins are reared apart, the environments into which they are adopted are generally similar. This effect of correlated environments has the effect of inflating the apparent degree to which traits are genetically determined. In any case, if r_{MZ} and r_{DZ} represent the correlation coefficients of a quantitative trait among monozygotic and dizygotic twins, then $2(r_{MZ} - r_{DZ})$ provides a rough estimate of the broad-sense heritability of the trait. To see where this formula comes from, first look at Table 9.7. The covariance of monozygotic twins in the absence of environmental correlation is $\sigma_a^2 + \sigma_d^2$, so the correlation between monozygotic twins is this covariance divided by the phenotypic variance, or the broad-sense heritability. If monozygotic and dizygotic twins have the same degree of environmental correlation, then subtracting the correlation of one from the other should remove the environmental correlation. The correlation between dizygotic twins is the same as that of full sibs, or $\frac{1}{2}\sigma_a^2 + \frac{1}{2}\sigma_d^2$. Assuming that the phenotypic variance is the same in both types of twins, the expression $2(r_{MZ} - r_{DZ})$ is equal to $[\sigma_a^2 + \frac{1}{2}\sigma_d^2]/\sigma_p^2$, which is not exactly equal to the broad-sense heritability, but it is an approximation (Smith 1975). Even when the mathematically precise estimators are used, the problem of shared environments does not go away.

One human trait that has received an inordinate amount of attention is intelligence. The estimation of heritability of intelligence from twin studies is steeped in controversy. Aside from the necessity to define intelligence as performance on an IQ test, these studies face enormous hurdles in obtaining accurate assessments of causes of patterns of similarity. Ever since the data of Cyril Burt showing a heritability of IQ of 0.771 were cast in doubt, there have been efforts to revive claims of a very high heritability of IQ. Often the language is imprecise, with claims like "about 70% of the variance in IQ was found to be associated with genetic variation" (Bouchard et al. 1990). Even with the best care taken to study only adopted twins reared apart, the problem of correlated environments makes it impossible to obtain an entirely reliable estimate. An important point that is frequently overlooked in this discussion is that a high heritability implies *nothing* about the ability to change a trait by modification of the environment. The question has been raised whether there is any societal good to be gained from knowledge of heritability of IQ; the problem makes for lively reading on both sides (Lewontin et al. 1984; Herrnstein and Murray 1996). Fortunately for evolutionary biologists, organisms that can be reared in controlled conditions do afford the opportunity to obtain meaningful estimates of heritability and other components of quantitative genetic variation, as described in the next section.

EXPERIMENTAL ASSESSMENT OF
GENETIC VARIANCE COMPONENTS

Population biologists typically estimate the heritability and genetic correlations of a trait for the purpose of addressing the genetic constraints on the evolution of the trait. The best experimental approach depends on whether the organisms are sampled directly from a natural population, whether a series of inbred lines is available, and whether laboratory rearing is practical. Because the components of variance (including heritability) are descriptions of a particular population in a particular environment, the ideal would appear to be to use naturally occurring individuals, keeping track of their familial relationships, and to fit the statistical models relating degrees of relationship to expected covariances. In practice, because the natural environment is so variable, it is often preferable to do analyses in a controlled laboratory environment, but this introduces other problems outlined below.

Estimation of genetic variance components in natural populations generally requires a number of restrictive assumptions: (1) diallelic inheritance, (2) no correlation of parental and offspring environments, (3) no linkage or linkage disequilibrium, (4) parents equally inbred, (5) offspring not inbred, (6) samples of relatives drawn at random from a noninbred population, (7) no mutation, migration, selection, and (8) random mating (see Mitchell-Olds and Rutledge 1986 and references therein). In practice, small violations of these assumptions are acceptable, but it is nevertheless a serious challenge to obtain reliable estimates of genetic variance components from samples from natural populations. Most commonly, a full analysis is not carried out in natural populations, but offspring collected from known mothers are studied. One can also estimate a lower bound on the heritability in a natural population by regression of measurements of laboratory-reared offspring on the measurements of parents sampled from nature (Riska et al. 1989).

Once measurements are obtained on individuals with known degrees of relationship, the partitioning of variance into additive, dominance, and environmental components can be done using the standard statistical method of analysis of variance. Many experimental designs permit the effects of various factors on quantitative genetic components to be estimated. For example, assays of variance components could be repeated under several different environmental regimes, and the environmental component in the analysis of variance could be estimated. A second generation of organisms could also be studied and the variance components estimated from parent-offspring covariance. However, the two most common designs are analysis of variance of full-sib families or half-sib families. Using full-sib families alone has the problem that the covariance of full sibs includes both additive and dominance variance. Therefore, when the data are limited to full-sib families, all that can be estimated is the broad-sense heritability. On the other hand, if one studies only half-sibs (whose covariance is $\frac{1}{4}\sigma_a^2$), one can estimate the narrow-sense heritability, but the dominance component may be quite large and remain undetected.

More elaborate designs that feature parents, full sibs and half sibs can give extensive partitioning of variance components. The reliability of the methods can be tested by comparing heritability estimates from either parent-offspring regression or from the covariance of half sibs. One means of using all the data in a single estimate is the method of **maximum likelihood**, a procedure for parameter estimation which solves for parameter values that have the maximum likelihood of obtaining the observed data under a given model. Maximum likelihood methods are preferable to analysis of variance when several traits are being examined to estimate the components of genetic covariance. Analysis of variance becomes extremely cumbersome when sample sizes are not the same at all levels of relationship, and there is no single ideal way to adjust for unequal sample sizes.

The principle behind maximum likelihood is to construct a likelihood function that describes the likelihood of obtaining the observed data given the family structure and a set of unknown parameters to be estimated. The unknown parameters are the magnitudes of the various genetic and environmental variance components. The method then finds the values of the unknown parameters that maximize the likelihood. In practice, the unknown parameters form a variance-covariance matrix, and the computer algorithms entail extensive matrix manipulations. The computer output consists of estimates of heritability that utilize all the data, including parent-offspring, full-sib, half-sib, and any other relationships that are informative. Although no assumption of multivariate normality of the character data is necessary in obtaining the estimates, the estimates have meaning (and the prediction equation is reliable) only if the phenotypes are normally distributed. Testing the statistical significance of the estimates requires normality. Shaw (1987) provides a thorough review of the merits of maximum likelihood methods in quantitative genetics.

Heritability tells us virtually nothing about the actual mode of inheritance of a quantitative trait, useful as the concept may be in predicting response to selection. The heritability of a trait represents the cumulative effect of all genes that affect the trait. Even if a trait is determined by a single gene, heritability depends in a complex manner on the values of p, a, and d, and these individual components cannot be disentangled. (The values of p, a, and d are said to be statistically **confounded**.) With more than one gene, the heritability includes a summation of terms for each gene, and each term has its own particular values of p, q, a, and d. Here, precisely, is the problem: for a quantitative trait determined by, say, 10 diallelic loci, there would be 30 quantities involved in heritability—10 allele frequencies, 10 values of a, and 10 values of *d*. Heritability is but a single number that gives the combined effect of all 30 quantities. It says nothing about any one of them.

It must be emphasized that heritability is a quantity that comes from a mathematical model of reality, and the model has many assumptions. We have assumed that all genes affecting the trait act independently of one another and are unlinked. In actual cases, genes often interact and can be linked. (The model can, however, be extended to incorporate at least partially the effects of linkage and epistasis.) Moreover, the assumption of no correlation of parental and offspring environments is not always easy to test. All in all, while heritability, especially realized heritability, is an indispensable aid to plant and animal breeders, it lends itself to no easy interpretation in simple genetic terms (apart from the statistical description through parent-offspring regression). Another difficulty in interpreting heritability values is that they depend on the range of environments that occur. The denominator σ_p^2 in Equation 9.23 is the total variance in phenotypic value in the population. Because the total variance includes the variance resulting from environmental differences among individuals, increasing the variation in the environment decreases h^2. Exceptionally thoughtful discussions of the concept of heritability and its strengths and limitations are found in Kempthorne (1978) and Jacquard (1983).

Heritability values are determined in part by gene frequencies. Because gene frequencies change during the course of selection, the heritability is also expected to change. In practice, however, the heritability changes sufficiently slowly that over the course of a few generations, it can be regarded as approximately constant. The approximate constancy of heritability has a twofold cause: (1) if a particular gene accounts for only a small proportion of the total phenotypic variance in a quantitative trait, then the gene frequency does not change very rapidly, and (2) the values of *a* and *d* remain nearly constant provided that the environment does not change drastically from one generation to the next. Thus, at least for the first 10 generations or so, heritability usually remains approximately constant and can be used as a constant in the prediction equation (Equation 9.10). To be precise, suppose h^2 is constant and let μ_t and S_t represent the mean of the population and the selection differential in the *t*th generation. Then, over the length of time during which h^2 is approximately constant,

$$\mu_t - \mu_0 = h^2(S_0 + S_1 + \cdots + S_{t-1}) \qquad 9.38$$

The quantity $\mu_t - \mu_0$ is the total response to selection, and $S_0 + S_1 + \cdots + S_{t-1}$ is called the **cumulative selection differential**. During the time in which h^2 is approximately constant, therefore, a plot of μ_t against cumulative selection differential is expected to yield a straight line with slope equal to h^2, as illustrated for a case in mice in Figure 9.19.

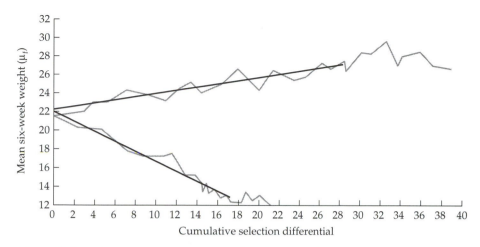

Figure 9.19   Linearity in response against the cumulative selection differential for body weight in mice at age six weeks. Linearity in the up (high-weight) direction continues for about twice as long as it does in the down (low-weight) direction. (After Falconer 1955.)

Indirect Estimation of the Number of Genes Affecting a Quantitative Character

The number of genes that contribute to quantitative traits is not always large. We have already seen an example of seed color in wheat in which the number of genes was three. When the number of genes is relatively small, the number can often be estimated from the means and variances observed in different strains and their hybrids and backcrosses. In the case of two additive alleles of each of three genes, when parental strains are homozygous for all unfavorable or all favorable alleles, then they differ by six units in phenotypic value. The variance in phenotypic value in the F_2 generation equals $\frac{3}{2}$ units^2. If there are n unlinked, additive genes, the difference D in phenotypic value between the means of parental inbred lines is $2n$ units, and the variance σ^2 in the F_2 generation is $n/2$ units^2.

In order to obtain an estimate of the number of genes that is independent of the units of measurement, a ratio is needed to make the units cancel. One possibility, first suggested by Wright (in Castle 1921), is

$$\hat{n} = \frac{D^2}{8\sigma^2}$$ 9.39

With n unlinked, additive genes, $\hat{n} = (2n)^2/[8(n/2)] = n$, as it should. Equation 9.39 is based on the assumptions of complete additivity, equal effects of all genes, no linkage, and fixed differences between parental lines.

When the assumptions are violated, application of the equation usually results in estimates of gene number that are smaller than the actual numbers. For this reason, the quantity estimated in Equation 9.39 is called the **effective** number of genes because it defines a lower limit to the actual number. Figure 9.20 presents the results of a simulation, using a range of 2 to 10 genes to generate sample data to which Equation 9.39 was applied to estimate n. The message is that this method is very approximate and somewhat biased. The statistical properties of the method have been improved (Zeng 1992), but the estimate is still only a rough approximation to the number of genes affecting a trait.

The variance in Equation 9.39 is the **genetic variance**, which is the variance in phenotypic value resulting from genetic differences among individuals. When the environment contributes an amount σ_e^2 to the phenotypic variance, then the variance within the parental inbred lines, or within the F_1 population, equals σ_e^2. This is because the populations are genetically uniform, and the only source of variation in phenotype results from the environment. However, within the F_2 generation, variation in phenotype results in part from genetic variation and in part from environmental variation, and the total variance in phenotypic value equals the summation $\sigma^2 + \sigma_e^2$. Therefore, subtraction of the F_1 variance from the F_2 variance gives an estimate of σ^2 because

$$\sigma^2 = [\sigma^2 + \sigma_e^2] - \sigma_e^2 \qquad\qquad 9.40$$

Further discussion of the genetic variance occurs later in this chapter. Lande (1981) gives several alternative methods of estimating the genetic variance using data from inbreds, hybrids, and backcrosses. Cockerham (1986)

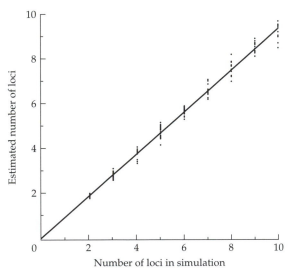

Figure 9.20   Computer-generated samples of an F_2 population having a range of numbers of loci with purely additive effects that influence a trait. For each sample, the number of genes was estimated following Wright's method.

extended the analysis to obtain an unbiased estimation of the difference in parental means, and he combined the data from parentals, F_1, F_2, and backcrosses into a single least-squares estimate.

The scale in which a phenotype is measured is important in using Equation 9.39 because the genes and alleles are assumed to be additive. In the ideal additive case, a plot of the means and variances of inbreds, hybrids, and backcrosses forms a triangle of the type shown in Figure 9.21A. In actual cases, the phenotypes must be measured using a scale that yields an approximate triangle. With fruit in tomatoes, for example, an obvious scale of weight is in grams, but the scale giving the triangle in Figure 9.21B is $x = \log[(\text{weight in grams}) - 0.153]$ (Lande 1981). Using this scale, $D = 1.552$ and $\sigma^2 = 0.0426$, from which Equation 9.39 implies $\hat{n} = (1.552)^2/[8(0.0426)] = 7$ as the effective number of genes.

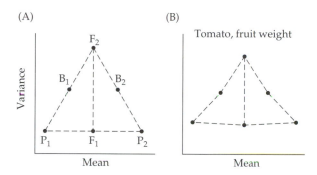

Figure 9.21   (A) Expected triangular relation between means and variance of phenotypic value among inbred parents, P, backcross progeny, B, and hybrids, F, for an ideal quantitative trait determined by unlinked and completely additive genes. (B) Observed relation for fruit weight in tomato, in which the fruit weight is on a logarithmic scale. (After Lande 1981.)

PROBLEM 9.7   In analyzing data on oil content in maize, Lande (1981) found that values of $\log[(\text{percent oil in kernels}) + 1.87]$ approximated a triangular form like those in Figure 9.21. With this scale of measurement, the means of two inbred lines were 0.513 and 1.122, respectively, and the phenotypic variances of the inbred lines, the F_1 generation, and the F_2 generation were 0.00142, 0.00053, 0.00030, and 0.00303, respectively. Use Equations 9.39 and 9.40 to estimate the effective number of genes. Estimate the environmental variance in two different ways: (1) as the mean of the variance in phenotypic value in the parental inbred lines, (2) as equal to the variance in phenotypic value in the F_1 generation.

ANSWER   Using the first estimate, $\hat{\sigma}_e^2 = (0.00142 + 0.00053)/2 = 0.00098$, $\hat{\sigma}^2 = 0.00303 - 0.00098 = 0.00205$, $\hat{D} = 0.513 - 1.122 = -0.609$, and $\hat{n} = (-0.609)^2/8(0.00205) = 22.6$ Using the second estimate, $\hat{\sigma}^2 = 0.00303 - 0.00030 = 0.00273$, and $\hat{n} = 17.0$.

The effective number of genes for oil content in maize kernels calculated in Problem 9.7 bracket those obtained by Lande (1981) using other estimates of the environmental variance. Some of the estimates of number of genes affecting quantitative traits are: tomato fruit weight, 7–11; maize kernel oil content, 17–22, Hawaiian *Drosophila* head size, 6–9, fish eye diameter, 5–7, and human skin color, 4–6. With the notable exception of oil content in maize kernels, the effective number of genes is less than 10. These examples demonstrate that genetic variation in quantitative traits can result from the segregation of a relatively small number of genes. However, the effective number of genes represents a minimum estimate, and the actual number of genes is likely to be larger. In some cases the estimated number of genes is very large. For example, at least 40 genes contribute antigens that are important in the rejection of skin transplants in mice, at least 150 genes contribute to body weight in the mouse and *Tribolium*, and at least 17 genes on the third chromosome of *Drosophila melanogaster* influence sternopleural bristle number (Shrimpton and Robertson 1988). It should be emphasized that the number of genes that affect quantitative traits may differ according to the precise definition of "affect." For example, the overall long-term selection response illustrated in Figure 9.5 may result from a large number of genes, but most of the response that occurs at any one time may result from changes in allele frequency in only a few of them. As we will see in the last section of this chapter, the use of molecular markers to genetically map the position of genes affecting quantitative traits is beginning to give more direct assessments of the number of genes and the distribution of their effects on quantitative characters.

NORM OF REACTION AND PHENOTYPIC PLASTICITY

In considering the influence of environment on the determination of phenotypes, it is tempting to think of environmental effects as random noise added to traits that are basically genetically determined. This view is often misleading. One simple way to analyze the effect of environment on phenotype is to examine the phenotypes of a single genotype in an array of environments; it is possible to do this examination in a number of experimental organisms. The array of phenotypes that results from a given genotype is known as the **norm of reaction**, a term coined by Schmalhausen (1949).

Figure 9.22 shows the norm of reaction of bristle number for 10 strains of *Drosophila pseudoobscura* reared at three temperatures (Gupta and Lewontin 1982). The rank ordering of the strains was found to change at different temperatures, and an astounding 35 to 40% of pairwise comparisons of strains reversed in ranking when one temperature was compared to another. Figure 9.22 shows a greater variation among lines at 14° than at 26°. If we were to do a test of parent-offspring regression at 14° and compare it to the same test at 26°, the former would have a higher heritability. This contrast underscores the fact that heritability is a measure defined in one environment, and it shows that the methods of quantitative genetics using analysis of variance cannot separate the causes of phenotypic differences. In order to detect environmental effects, the norm of reaction must be ascertained by examining phenotypes in a variety of environments.

Referring again to the one-gene model for quantitative traits, the environment can affect the value of heritability because the values of a and d depend on the environment. Figure 9.13 maybe used as an example; it shows the norms of reaction of AA, AA', and $A'A'$ genotypes. If we were dealing with the range of environments denoted E_1 in Figure 9.13, A would be the favored allele and A would be nearly dominant to A'. On the other hand, if we were dealing with the range of environments denoted E_2, A' would be the favored allele and there would be essentially no dominance. Thus, switching a population from E_1 to E_2 would change the values of a and d and substantially alter the heritability of the trait, even though the total phenotypic variance of the population might remain the same.

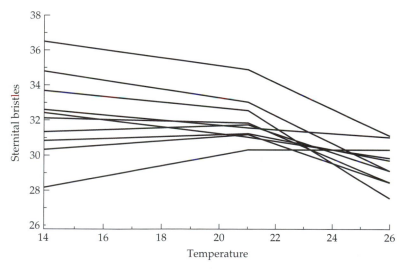

Figure 9.22   By counting the mean number of sternital bristles of a set of genotypes (lines) of *Drosophila* reared at different temperatures, differences in the norm of reaction are apparent from the crossing lines in the plot.

PROBLEM 9.8   Dobzhansky and Spassky (1944) studied the norms of reaction with respect to temperature of the viability of various genotypes containing chromosomes extracted from natural populations of *Drosophila pseudoobscura*. For two such chromosomes (designated *A* and *B*), the following relative viabilities were obtained.

	Genotype		
Temperature (°C)	A/A	A/B	B/B
16.5	0.92	1.00	0.71
25.5	0.32	1.00	0.75

Estimate μ^*, a, and d for these genotypes in populations maintained at 16.5°C and 25.5°C. Then, letting p represent the frequency of the *A* chromosome and q that of the *B* chromosome, assume $p = 0.3$, and estimate the additive genetic variance of viability resulting from these genotypes at both temperatures.

ANSWER   At 16.5° C, $\hat{\mu} = (0.92 + 0.71)/2 = 0.815$, $\hat{a} = 0.92 - 0.815 = 0.105$, $\hat{d} = 1.00 - 0.815 = 0.185$. At 25°C, $\hat{\mu} = 0.535$, $\hat{a} = -0.215$, $\hat{d} = 0.465$. Additive genetic variance for these genotypes is given by Equation 9.24, where $p = 0.3$, and $q = 0.7$. At 16.5°C, $\sigma_a^2 = 0.01346$. At 25.5°C, $\sigma_a^2 = 0.00035$. Note that the additive genetic variance has been decreased by a factor of almost 40, yet all we did was raise the temperature!

The norm of reaction is important in evolutionary genetics because the fate of genetic variation in a population depends on the fitness of the organism, which in turn depends on the environment. In turn, the norm of reaction is itself a property that may be under genetic control and be subject to adaptive evolution by means of natural selection (Schlichting and Pigliucci 1994; Via et al. 1995). In the context of adaptive norms of reaction, evolutionary geneticists often apply the term **phenotypic plasticity**. It is interesting to consider whether natural selection modifies phenotypic plasticity as part of adaptation, especially since extreme phenotypic plasticity is not always favorable. For example, if a plant germinates from a seed in an exceptionally dry year because its genotype has a high level of phenotypic plasticity resulting in the ability to cope very well with dry periods, then in the following season the plant may have insufficient leaf area to compete with its less plastic neighbors that are phenotypically better suited for average moisture. On the other hand, it is also easy to describe a scenario in which the absence of plasticity can prove fatal. A useful way to model the evolution of adaptive phenotypic plasticity is to consider the phenotype in different environments as different traits that may be genetically correlated (Falconer and Mackay 1996; Via and Lande 1985). Further discussion of this model will be deferred until after we have discussed selection of more than one character.

One biological factor that affects the correspondence between genotypes and phenotypes is the chronological age of an organism. A phenotype, such as body weight, often changes with age, and different genotypes can have different age-related growth curves (or other developmental profiles). Consequently, the heritability of traits depends on the age at which individuals are tested. Figure 9.23 shows components of variance of mouse body weight at different ages (Riska et al. 1984). In this study, 2700 mice from 700 full-sib families were measured from days 14 to 70. Variance of all components was maximal at an age of about 20 days (which happens to be the time of maximal growth), and then declined to fairly stable values at age 40 days. Despite

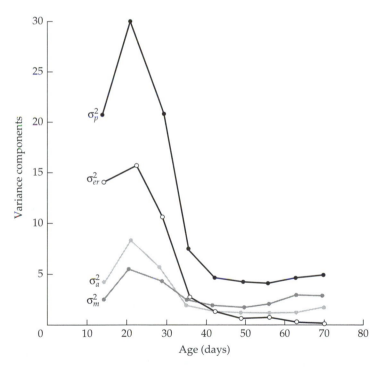

Figure 9.23   Variance components for the logarithm of body weight in a random-bred (genetically variable) strain of mice plotted as a function of age. σ^2_m is the variance due to maternal effects, and $\sigma^2_{er} - \sigma^2_e - \sigma^2_m$. (From Atchley 1984.)

the additive genetic variance decreasing markedly after 20 days of age, the heritability itself actually increased with age, with values of 0.22, 0.27, 0.31, and 0.37 at days 12, 20, 50, and 70. Although there is an assumed increase in environmental variance in many traits in humans, there is no marked tendency for heritability to decrease with age.

THRESHOLD TRAITS AND THE GENETICS OF LIABILITY

Some multifactorial traits do not exhibit continuous variation. Although the variation is discontinuous (individuals either express the trait or not), the trait is nevertheless influenced by multiple genetic factors and also by environment. Such traits are called **threshold** traits. A human example is diabetes, an abnormality in sugar metabolism that affects one or two percent of the Caucasian population. In a sense, diabetes is a continuous trait because the severity of the disease varies from nearly undetectable to extremely severe. On the other hand, diabetes can also be considered a threshold trait because all individuals may be classified according to whether or not they are so severely affected that clinical treatment is required. With such a classification, there are only two phenotypes, "affected" and "not affected," even though there is phenotypic variation within each category. The genetic influence on the trait is shown by the enhanced risk of diabetes in relatives of affected individuals. However, environmental factors such as diet are also important in determining whether high-risk genotypes actually develop the disease. At one time, many threshold traits were "explained" by postulating a simple genetic mechanism (such as a single recessive allele in the case of diabetes) and invoking "incomplete penetrance" to account for the poor fit of pedigree data to a simple Mendelian hypothesis. Now it is preferred, and probably more realistic, to consider threshold traits as true polygenic traits and to calculate heritabilities as for any other quantitative trait. In most cases, however, it is simply not known whether the genetic influence on the trait results from one or a few major genes or is polygenic. Here again, the use of molecular markers to map quantitative trait loci will be able to discriminate between these two possibilities.

The basic idea behind the model of threshold traits is illustrated in Figure 9.24. The normal curve in panel (A) represents the (unobservable) distribution of a hypothetical liability (or risk) toward the threshold trait, measured on a scale such that the mean value is 0 and the variance is 1. It is assumed that individuals whose liability is above a certain threshold (T) actually express the trait. Thus, the shaded area in Figure 9.24A delimits the proportion of individuals in the population who are affected (B_p), and the mean liability among affected individuals is denoted μ_S. Figure 9.24B gives the (again unobservable) distribution of liability among the offspring of affected individuals. The offspring mean is denoted μ', and the proportion of offspring above the threshold is denoted B_o. The setup here is like that in the earlier

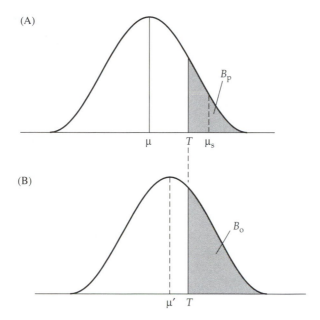

Figure 9.24   (A) Distribution of liability assumed for a threshold trait in a hypothetical population. The shaded area denotes individuals having liability above a critical threshold (T) and consequently affected with the trait. B_p is the frequency of affected individuals in the entire population, μ is the mean liability of individuals in the entire population, and μ_S, the mean liability of affected individuals. (B) Distribution of liability among offspring who have one parent affected with the trait. μ' denotes the mean liability among offspring, and B_o is the proportion of affected offspring.

section of this chapter in which we calculated the regression coefficient of offspring on one parent. In this model it can be shown that the regression coefficient b is given by $b = \mu'/\mu_S$, and the appropriate estimate of the heritability of liability is obtained from the relation

$$h^2 = 2b = 2\mu'/\mu_S \qquad\qquad 9.41$$

The methods for calculating heritability of threshold traits are illustrated in Problem 9.9.

PROBLEM 9.9   For pyloric stenosis, the incidence among males in the general population is $B_p = 0.005$, and the incidence among sons of affected males is $B_o = 0.05$. If liability follows a normal distribution, and 0.005 is the frequency of affected individuals, this represents 2.89 standard deviations above the mean. From these two numbers, infer μ_S and μ' in order to calculate the heritability of liability.

ANSWER   The mean liability of the fathers is $\mu_s = 2.89$. B_p can be obtained from a table of the normal distribution as follows. The tables generally give the value on the x axis in standard deviation units for an observed area in the two tails of the distribution. If the fraction affected is 0.05, and this is the area in one tail, then the area in both tails is 0.10. A probability area of greater than 0.10 is obtained if an observation is more than 2.58 standard deviations from the mean. Thus, $T = 2.58$ is the threshold. Using the same reasoning for the sons, $2B_o = 0.10$, and this area appears in the tail of the normal distribution if an observation is 1.64 standard deviations from the mean. This means that $T - \mu' = 1.64$. We know that $T = 2.58$, so $\mu' = T - 1.64 = 0.94$. From Equation 9.41 we get $h^2 = 2\mu'/\mu_s = 2(0.94)/(2.89) = 0.65$. This is the estimate of the narrow sense heritability of liability to pyloric stenosis.

Twins are frequently used in human quantitative genetics for the study of threshold traits, but twin data are best expressed in terms of concordance. The **concordance** of a trait in a population of twins is the proportion of affected twins that have affected co-twins. For example, suppose that 100 affected individuals are found to be twins and that in 35 cases the co-twin is also affected. The concordance rate is then $35/100 = 35\%$. From the concordance rates for monozygotic and dizygotic twins and the incidence of the trait in the population, the correlations in liability between monozygotic twins (σ_{MZ}) and dizygotic twins (σ_{DZ}) can be calculated (Figure 9.25). The broad-sense heritability is then estimated by $2(\sigma_{MZ} - \sigma_{DZ})$, as discussed earlier. As with quantitative traits in general, twin data are more reliable if the twins are reared apart, but it is seldom possible in practice to obtain a sufficient number of such twin pairs.

CORRELATED RESPONSE AND GENETIC CORRELATION

Genes have pleiotropic effects on phenotype; that is, every gene potentially affects every trait in the organism, either as a primary effect or as a secondary, indirect effect. Therefore, the alleles that are favorable for one quantitative trait may have unfavorable effects on another quantitative trait, and as these alleles are increased in frequency by artificial selection (thereby improving

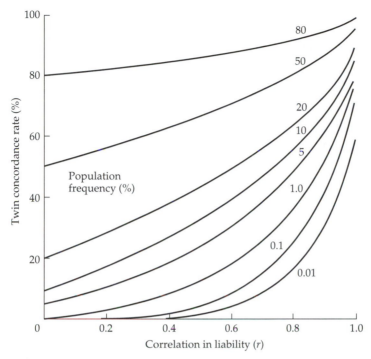

Figure 9.25   Threshold concordance rates expected in monozygotic twins, plotted against the correlation in liability toward the trait and the population incidence of the trait. (From Smith 1975.)

the phenotypic value with respect to the selected quantitative trait), the very same alleles may bring about a deterioration of some other aspect of performance. Pleiotropy is one cause of **correlated response**—a change in phenotypic value of one trait that accompanies response to selection of a different trait. A second possible cause of correlated responses is linkage disequilibrium (Chapter 3)—a favorable allele for one trait that increases in frequency under selection may drag along with it an allele of another, tightly linked gene that has a detrimental effect on an unselected trait.

Correlated responses are quite common in artificial selection and often, but not always, result in a deterioration in reproductive performance. In the case of Leghorn chickens, for example, 12 generations of selection for increased shank length reduced the egg hatchability by nearly half (Lerner 1958). In turkeys, to take another example, there was intense selection during the period 1944–1964 for growth rate, body conformation, and body size, but there was also a steady decline in some aspects of reproductive fitness such as fertility, egg production, and egg hatchability (Nordskog and Giesbrecht 1964). On the other hand, correlated responses can sometimes be useful. For example, selection for larger mature body size often increases litter size in mice and swine. If a trait has a low heritability or is difficult to measure, it is sometimes possible to practice selection for another, correlated trait, obtaining progress in the trait of interest by correlated response. Theoretically, the maximum response to artificial selection occurs when the criterion for selection is determined by a **selection index** (an averaging across several traits) which takes genetic correlations into account. However, the theoretical advantage of index selection is often overridden by practical difficulties in estimating the components of the index and implementing the selection procedure.

From a theoretical point of view, the covariance between two quantitative traits can be partitioned in a manner analogous to the partitioning of the variance for one trait outlined in an earlier section; the covariance can thus be partitioned into an additive covariance, a dominance covariance, an environmental covariance, and so on. The most important theoretical result is that the amount of correlated response with individual selection depends only on the additive covariance, much as the direct response to individual selection depends only on the additive variance. The components of covariance between traits can be estimated from the resemblance between relatives, but often it is preferable to estimate the correlated response by direct observation in a manner analogous to the determination of realized heritability (Falconer and Mackay 1996).

The phenotypic correlation is the correlation one would obtain by measuring two traits, say X and Y, and calculating the correlation coefficient directly. In symbolic terms, the phenotypic correlation is

$$r_\mathrm{p} = \frac{\mathrm{Cov}_p}{\sigma_{PX}\sigma_{PY}} \qquad 9.42$$

where Cov_p is the phenotypic covariance and σ_{PX} and σ_{PY} are the phenotypic standard deviations of characters X and Y. Although correlations do not partition in the same way as variances, the phenotypic covariance can be expressed as the sum

$$\mathrm{Cov}_p = \mathrm{Cov}_a + \mathrm{Cov}_e \qquad 9.43$$

where Cov_a is the additive genetic covariance, and Cov_e is the environmental covariance. The genetic correlation (which is essentially the correlation of additive genetic effects) is defined as

$$r_\mathrm{a} = \frac{\mathrm{Cov}_a}{\sqrt{\sigma_{aX}^2 \sigma_{aY}^2}} \qquad 9.44$$

where Cov_a is the additive genetic covariance, and σ_{aX}^2 and σ_{aY}^2 are the additive genetic variances of the two traits. The genetic covariance is estimated in much the same manner as the additive genetic variance. For example, the additive genetic covariance of two traits between half sibs is $\frac{1}{4}\mathrm{Cov}_a$.

Correlated response (*CR*) occurs when characters other than directly selected traits respond. The magnitude of correlated responses to selection is related to the additive genetic correlation between the selected and correlated characters. The expected response is expressed by the equation

$$CR_Y = ih_X h_Y r_A \sigma_{PY} \qquad\qquad 9.45$$

where X is the directly selected character with narrow-sense heritability h_X^2, i is the intensity of selection, h_Y^2 is the heritability of the correlated character Y, r_A is the genetic correlation, and σ_{PY} is the phenotypic standard deviation of character Y. The **intensity of selection** (*i*) is defined as the selection differential expressed as a multiple of the phenotypic standard deviation. Equation 9.45 says that, all else being equal, a doubling in the genetic correlation will double the magnitude of the correlated response.

It is instructive to consider the results of an artificial selection experiment in more detail. A 23-generation artificial selection experiment for 3-to-9 week weight gain in rats (Baker et al. 1975) resulted in six strains that were characterized for 17 skull traits in order to examine patterns of correlated responses (Atchley et al. 1982). The characters included such measurements as skull length, skull width, interorbital width, braincase depth, mandible width, and jaw length. The experiment consisted of two unselected control groups, two lines selected up and two lines selected down for weight gain, with all lines originating from a common stock. The direct response was remarkable for the magnitude of asymmetry: the males selected for greater weight gain were 3.46 standard deviations larger than the controls, while the down-selected males were 1.30 standard deviations smaller than the controls. Similarly, up-selected females were 1.46 standard deviations larger, and down selected females were 1.02 standard deviations smaller. Asymmetric response is generally attributed to opposing natural selection, and no more specific mechanism can be offered here. Correlated responses showed similar degrees of asymmetry. However, while the replicate lines showed statistical consistency in the magnitudes of *direct* responses, 35 of 51 comparisons between males in replicate lines were significantly different in the *correlated* response.

PROBLEM 9.10 A herd of dairy cattle yields milk with a fat content of 3.4% ± 0.65% and a protein content of 3.3% ± 0.45%. The heritabilities of these traits are 0.60 and 0.70, respectively, and the genetic correlation is 0.55. If selection is practiced for percent protein with a selection intensity of $i = 1.5$, what increase in percent protein and percent fat would be expected? What intensity of selection would produce the same increase in percent fat by direct selection?

ANSWER   Because $i = S/\sigma$, Equation 9.10 can be written as $R = i\sigma h^2$ where i is the intensity of selection. For percent protein, $R = (1.5)(0.45)(0.7) = 0.47$, so expected percent protein = 3.3% + 0.47% = 3.8%. For a correlated response, use Equation 9.45, $CR = (1.5)(0.6)^{1/2}(0.70)^{1/2}(0.55)(0.65) = 0.35$, so expected percent fat = 3.4% + 0.35% = 3.75%. The fat increase corresponds to a direct selection of $i = 0.35/(0.60)(0.65) = 0.90$.

The absence of correlated response can sometimes be very informative. Artificial selection on spacing between vein junctions in the *Drosophila* wing might be expected to result in correlated changes all over the wing. When Weber (1992) did such an artificial selection for a minute (100 cells) region of the wing, he obtained a strong direct response with no significant correlated response in other wing measures. This suggests that the control of morphological development involves many genes, and that independent selection for minute aspects of wing morphology are possible.

As mentioned above, the two primary mechanisms that produce genetic correlation are pleiotropy and linkage. With pleiotropy, genes that affect one character also affect others. These effects may be direct, as when a gene product has two distinct functions, or they may be indirect in the sense that more than one physiological step may connect the two phenotypes. In both cases, replacement of an allele at the relevant gene will affect both phenotypes. Alternatively, linkage disequilibrium can result in genetic correlation even though the genes affecting the two traits are entirely distinct. If there is linkage disequilibrium, then selection of one gene will affect allele frequencies of nearby genes, and the end result is a correlation of changes in two phenotypes. The effects of pleiotropy versus linkage disequilibrium may be distinguished experimentally by attempting to control or eliminate linkage disequilibrium. By following anonymous molecular markers in crosses or selection experiments, it is much easier to distinguish these two causes of genetic correlation, as described later.

Inference of Selection from Phenotypic Data

The results of models for selection on quantitative traits can be used to infer the operation of natural selection, but in this application one must be especially aware of the assumptions of the models; some serious statistical problems must also be overcome. Either by sampling a population at two different times, or by doing a cross-sectional study, changes in the phenotypic distribution resulting from differential mortality of the different phenotypes can be detected (Lande and Arnold 1983; Arnold and Wade 1984). Changes in phenotypic distributions reflect the operation of natural selection, despite the lack of any proven changes in gene frequency. Human birth weight is easily shown to be a character that is under the influence of natural selection, because the mortality rate of very small and very big babies is higher than the mortality of babies close to the population average (Karn and Penrose 1951). Extremes of body weight also have the highest mortality in elderly humans (Harris et al. 1988). Phenotypic selection will result in changes in gene frequency to the extent that the selected phenotypes are heritable, and selection of traits with low heritability will have minimal immediate effects on the genetic composition of the population.

The correspondence between phenotypes and relative fitness can be estimated from the mortality suffered between two times of censusing. A classical example that illustrates the method is a reanalysis by Lande and Arnold (1983) of the data from Bumpus (1899), involving 136 sparrows that had been incapacitated by a severe winter storm. Eight measurements were taken of every bird, and about half of the birds subsequently revived. The phenotypic measures and the fraction of birds that survived provide a way to examine the relation between the phenotypic measures and the fitness. First, the data required transformation to the form of a multivariate Gaussian distribution. After various confounding factors (such as age) were shown to have little effect on mortality, standard methods were used to estimate the partial regression coefficients reflecting the effect of each phenotypic trait in the fraction of birds that survived. One can imagine a mapping from a set of phenotypic measures to a multivariate surface of fitnesses. The partial regression coefficients were standardized to units of phenotypic standard deviation. The weight character was found to have a significant regression in males ($b' = -0.27 \pm 0.09$) and in females ($b' = -0.52 \pm 0.25$). However, the regression coefficients are negative because, contrary to expectation, the smaller birds had a lower mortality. Similar methods were applied more recently by Gibbs and Grant (1987) to a species of Darwin's finches on the Galapagos Island of Daphne Major. Generally, larger body size of birds is favored during dry years, evidently because the most abundant food consists of large, hard seeds. After a prolonged disruption in Pacific Ocean currents known as El Niño, there occurred a year with 10 times the normal rainfall, and a reversal in the selection differential favored small body size. The reversal was consistent with the abundant appearance of small seeds produced by adventitious plant growth.

The Bumpus method for inferring phenotypic selection has faced criticism on a number of grounds. Mitchell-Olds and Shaw (1987) pointed out that, as in other applications of multiple regression, the traits can be intercorrelated (multicolinear), the estimators may be biased when traits are measured with random experimental error, and the estimators are not consistent (do not converge to true values with increasing sample size) when the errors are not identically distributed for all traits. Lande and Arnold (1983) discussed the additional problem of a strongly selected character that is not among those studied. In the end, Mitchell-Olds and Shaw (1987) recommend experimental manipulation to accompany purely observational regression analysis of selection.

The effect of a quantitative character on reproductive fitness can be obtained very directly in some circumstances. When the component of selection is mating success, a simple comparison of phenotypes of mating and nonmating individuals constitutes such a test. For example, Taylor and Kekiç (1988) measured the wing lengths of 55 male *Drosophila melanogaster* that were captured while mating and 55 nonmating males that were caught by random-sweep netting. The mean wing lengths were 1.418 ± 0.012 mm and 1.372 ± 0.013 mm respectively, indicating that the mating males had significantly larger wings. Careful laboratory studies have verified the mating advantage of larger-winged males (Partridge et al. 1995; Wilkinson 1987).

Turelli (1988) emphasized that long-term predictions of responses to natural selection depend critically on the stability of the genetic variance-covariance matrix, defined as a square matrix whose diagonal elements are additive genetic variances of the traits, and the off-diagonal elements are additive genetic covariances. Without direct information on the stability of genetic variance-covariance matrix, **G**, it is difficult to assess the confidence in estimates of selection. Wilkinson et al. (1990) calculated empirical estimates of **G**, the genetic variance-covariance matrix, in *Drosophila* populations selected for 23 generations on thorax length. Statistical tests of the constancy of **G** found that selection for large thorax did not change **G** from the control population, but selection for small thorax did result in significant changes in **G** (Shaw et al. 1995). The change may be in part due to gene frequency changes caused by selection, changes in gene frequency due to random drift, and changes in linkage disequilibrium (the Bulmer effect). Given that significant changes in **G** can be introduced by very short-term laboratory experiments, long-term evolutionary projections based on current patterns of genetic variance and covariance are probably questionable.

EVOLUTION OF QUANTITATIVE TRAITS

Evolutionary quantitative genetics is an application of population genetics in which the distributions of phenotypic variation and covariation and the changes in those distributions are modeled by considering the changes in the underlying genes. The genetic models are intrinsically multilocus; this would make them very cumbersome, except that generally simplifying assumptions are made. This is a challenging field both experimentally and theoretically, and it is an area of very active research at present. In this section some of the basic principles and conclusions of evolutionary quantitative genetics will be considered.

Random Genetic Drift and Phenotypic Evolution

Genetic variation in a finite population always undergoes the process of random genetic drift. If the genetic variation impacts a quantitative character, then the variance in the quantitative character will in turn be affected by random drift. In a discrete population of effective size N_e, the genetic variance changes over successive discrete generations according to

$$E(\sigma_a^2)' = [1 - 1/2N_e]\, E(\sigma_a^2) + \sigma_m^2 \qquad 9.46$$

where σ_m^2 is the increment in genetic variance added each generation by mutation (Clayton and Robertson 1955, 1957; Lande 1979, 1980; Turelli et al. 1988). When the influx of new mutations (inflating the variance) is balanced by the loss of variance due to random genetic drift, the population arrives at the **mutation-drift equilibrium**. The expected genetic variance at the mutation-drift equilibrium is

$$E(\sigma_a^2) = 2N_e\sigma_m^2 \qquad 9.47$$

As in the case of random genetic drift for one gene, the population is expected to arrive at the mutation-drift equilibrium after $4N_e$ generations. Moreover, given two lineages that diverged t generations ago, the expected difference between the mean phenotypes is $2t\sigma_m^2$. Just as the neutral mutation theory predicts that the rate of gene substitution (and hence divergence) is independent of population size and depends only on the neutral mutation rate, the neutral rate of phenotypic divergence depends only on the rate of mutations affecting the phenotype. This result was derived rigorously by Lynch and Hill (1986), who also showed that the rate of divergence is dependent only on the rate of purely additive mutations, and is independent of dominance and epistatic effects. Figure 9.26 shows the increase in within-population variance (from an initial population with no variation), until the population reaches a steady-state balance of mutation and drift. Simulations also verify the nearly linear increase in variance between populations with time. Extending neutral divergence to the case of multiple characters, Lynch and Hill (1986) showed that the neutral divergence of the variance-covariance matrix depends only on the mutational variance-covariance matrix. In a model with n genes, k alleles at each, population size N, and mutation rate μ per locus, Cockerham and Tachida (1987) found that the initial rate of increase of variance between populations depended on N, but the

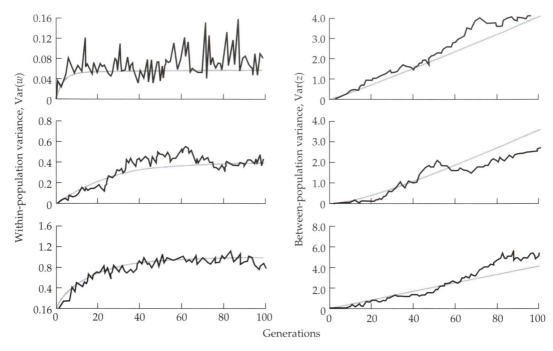

Figure 9.26 Simulations of a model with mutation of genes underlying a quantitative character and random genetic drift in a finite, subdivided population. The jagged lines represent the simulations, while the smooth curves are based on an analytical model. One hundred populations were run in each sample and each case. A mutation rate of 0.001 per locus per generation was used in all runs. (Top) Effective population size = 2, number of loci = 50. (Middle) Population size = 10 with 10 loci. (Bottom) Population size = 10 with 50 loci. The steady-state within population variance increases with population size, but the rate of increase of variance between populations is twice the mutational variance.

asymptotic rate of divergence, and the steady-state variance between populations, depended only on the mutation rates.

Turelli et al. (1988) used the mutation-drift equilibrium as a null hypothesis for devising a statistical test. Evolution has been too rapid to be explained by the neutral model (with 95% confidence) if

$$\frac{\sigma_m^2}{\sigma_p^2} < \frac{(\Delta z / \sigma_p)^2}{2t(1.96)^2}$$

9.48

where σ_m^2 is the mutational variance, σ_p^2 is the phenotypic variance, and Δz is the change in phenotypic mean in time period t. In other words, the mutation-drift equilibrium places a lower bound on the ratio of mutational variance to the total phenotypic variance. If the observed σ_m^2/σ_p^2 is smaller than the critical value given in Equation 9.48, then mutation does not introduce enough variance to account for the observed divergence. The test in Equation 9.48 is only useful on a macroevolutionary time scale, when there is some hope that the diverged populations actually have reached a mutation-drift equilibrium. For situations like Bumpus's sparrows or the influence of the rains caused by El Niño on Darwin's finches, this test is not appropriate, and tests that compare the magnitude of change to that expected by random drift in a population with the given effective size are used (Lande 1976, 1977).

The amount of variance in quantitative traits introduced by mutation each generation is a quantity of considerable interest. Mutational variance can be estimated by two methods, including quantification of the increase in variance among initially identical lines as mutations accumulate over generations, or by response to selection in a population that is initially in mutation-drift equilibrium or is devoid of variation (Lynch 1994). It might be supposed that an artificial selection experiment in a population lacking genetic variation would be doomed to failure, but populations accumulate sufficient mutational variation to produce a response in a few generations (Fry et al. 1995). In both cases, the action of natural selection will bias the estimates of mutational variance, so care is taken to try to stop natural selection. In mutation accumulation experiments in *Drosophila*, this has traditionally been done with balancer chromosomes that prevent recombination and minimize selection by maintaining chromosomes in a heterozygous state. Lynch (1988) and Houle et al. (1996) did comprehensive reviews of experimental estimates of σ_m^2/σ_e^2 and found that this ratio generally falls in the range of 10^{-2} to 10^{-4}. On an absolute scale, the number of deleterious mutations per gamete per generation is very close to 1.0. Interestingly, the estimate of σ_g^2/σ_m^2, the ratio of genetic variance to the mutational variance, is expected to equal the median persistence time of mutant alleles. The estimated persistence times for traits associated with life history averaged around 50 generations, while the persistence time of morphological trait mutations was around 100 generations. In order to see whether this figure is reasonable, we need to consider what the models of mutation-selection balance would predict.

One experiment to quantify mutational variance that gave an estimate orders of magnitude different from those mentioned above was a study of *E. coli* mutation-accumulation lines (Kibota and Lynch 1996). The experiment was done by expanding 50 independent lineages of cells from a single cell, and following each of the 50 lineages for 300 growth cycles (each of about 25 generations). The per-cell rate of deleterious mutation was found to be 0.0002, in contrast to the per-individual deleterious mutation rate in *Drosophila* of about 1.0 per generation. The discrepancy is probably in part caused by the smaller genome size of *E. coli* (about $1/35$ that of *Drosophila*), and the fact that *Drosophila* undergo about 25 cell divisions per organismal generation. In fact, when the *E. coli* mutation rate is scaled up by these two factors, the effective rate of 0.21 mutations per generation is remarkably close to that of *Drosophila*.

Artificial selection experiments can provide information about rates of accumulation of variance through mutation. The continued response seen in the Illinois corn oil experiment in Figure 9.4 is almost certainly due in large part to selection operating on variation introduced subsequent to the start of the study. Enfield (1980) obtained similarly large response to selection on *Tribolium* pupal weight and argued that mutational variance was at least part of the cause. In comparing selection response in large and small populations, Weber (1990a,b) and Weber and Diggins (1990b) were trying to quantify the importance of drift in artificial selection experiments. The striking response they obtained for wing tip height and alcohol vapor tolerance, and the increased response in larger populations, suggested that larger populations are better able to produce the full range of genetic variation for the trait, in part perhaps by mutation, but also by recombination. The importance of having the right combinations of genes for selection response to occur was very clear in a selection experiment for flight speed (Weber 1996). One hundred generations produced a nearly linear increase in flight performance. At that point, selected and control flies were crossed, and the F_1 flies did little better than the controls. Subsequent selection on these hybrids and their descendants recovered performance in just six generations that had taken 75 generations to achieve. It is likely that the secondary selection went so much faster because the F_1 flies had preassembled clusters of alleles conferring stronger flight, so response did not require waiting for rare recombinants to construct them.

Transposable elements are one source of mutations that have received particular attention by evolutionary quantitative geneticists. Because transposable elements may jump with appreciable rates in some species, it was thought that they may contribute a substantial amount of mutational variation. By performing artificial selection in the presence of transposing *P*-elements and controls in the absence of *P*-elements, Mackay (1985) obtained the first evidence in *Drosophila* that *P*-elements may contribute genetic variance that can result in increased selection response (see Problem 9.2). The joint effects of *P*-element insertions on bristle traits and on viability were quantified by Lyman et al. (1996) in a sample of 1094 single-*P*-element insertion lines. The magnitude of effects exceeded those seen in mutation accumulation experiments for spontaneous mutations. They found that most of the variance in bristle number was caused by a few insertions of relatively large effect, and that the elements with the largest phenotypic effect were also the most deleterious. In another study of random *P*-element insertion lines scored for 14 metabolic characters, single insertions were found to have significant effects on several metabolic traits at once, indicating substantial levels of pleiotropy (Clark et al. 1995b). Insertional mutations are not limited to *Drosophila*, and Keightley et al. (1993) have demonstrated that retroviral insertions clearly generate additional genetic variation in body weight in mice. These experiments demonstrate that transposable and retroviral elements may be a significant source of quantitative genetic variation in natural populations.

Mutation-Selection Balance

Stabilizing selection is of great interest for the role it may play in the persistence of heritable genetic variation in multifactorial traits. Intuitively, it seems reasonable to suppose that observed levels of additive genetic variation may result from a balance between stabilizing selection, which tends to reduce genetic variation, and new mutations, which tend to increase it. Deceptively simple when stated verbally, the models become very complex when formulated in mathematical terms, and include such complications as the number of genes, the type of action of alleles and their interactions, linkage between genes, the type and intensity of selection, and the influence of selection on other traits that are related through pleiotropy. **Pleiotropy** is reflected in the widespread tendency of genes to affect several traits simultaneously, which usually results from the fact that complex phenotypic traits are determined by the interactions of the products of many genes during development. The number of genes affecting a trait is relevant to mutation-selection balance because, for a given genetic variance, the selection intensity per locus decreases as the number of loci increases. If the total mutation rate is fixed, the per-locus mutation rate must decrease as the number of loci increases. The challenge is to develop some theory that will describe the relationship between equilibrium additive genetic variance in a population having stabilizing selection balanced against mutation. The parameters will include the effective population size, the mutation rate per locus, the number of loci affecting the trait, the distribution of mutational effects, and the distribution of fitness effects of new mutations.

As you might have surmised, progress in understanding mutation-selection balance in such a complex system has come from making many simplifying assumptions. Following Kimura (1965), Lande (1975), and Turelli (1984), let $p_t(x)$ be the distribution of allelic effects before selection in generation t. In each generation, selection occurs, then mutation, and then reproduction. Assume that phenotypes and underlying genotypic effects have a Gaussian distribution; since selection would be likely to change the distribution, this approximation remains acceptable only under weak selection. Although the model as developed is haploid, it is intended as an approximation for a diploid model assuming additivity of gene effects. Letting $p_t'(x)$ be the density of allelic effects after selection, μ be the mutation rate (the fraction of gametes that are mutated), and $g(x)$ be the density of mutational effects on the phenotype, the recursion for $p(x)$ is

$$p_{t+1}(x) = (1-\mu)p_t'(x) + \mu \int_{-\infty}^{\infty} p_t'(y)g(x-y)dy \qquad 9.49$$

This model, called the Kimura-Lande-Fleming model by Turelli (1984), partitions the phenotypic distribution into two components. A fraction $(1 - \mu)$ of the distribution derives from nonmutant genes, and their distribution is $p_t'(x)$ after selection. Fitness follows a Gaussian distribution with an optimum phenotype coinciding with the mean. Genes that mutate change phenotype x to y with distribution $g(x - y)$, and then undergo selection, resulting in the distribution $p_t'(y)$. Since any phenotype x can mutate to many different values of y, the integral is taken over all values of y. Kimura (1965) showed that, with weak selection, the process in Equation 9.49 comes to a steady state with a Gaussian distribution of allelic effects. This model leads to a paradox. On the one hand, Lande (1975) found an approximate Gaussian equilibrium that held in the case of small mutational effects but required unacceptable high mutation rates. On the other hand, Turelli (1984) introduced the "house-of-cards" view of mutation to this model; it allows individual mutations to have large effects so that the postmutation distribution of allelic effects is $g(y)$, which is independent of x, the premutation effect of an allele. Both models could give reasonable equilibrium distributions of allelic effects, but under vastly different assumptions about numbers of loci affecting traits, mutation rates, and magnitudes of mutational effects. Where in this complex space of parameters does reality lie? It became apparent that only empirical observations would settle the matter.

PROBLEM 9.11   Why does the problem of mutation-selection balance for quantitative characters depend on the number of loci that determine a trait?

ANSWER   The problem of mutation-selection balance for quantitative characters depends on the number of loci that determine a trait in two ways. First, as the number of loci increases, the per-locus selection coefficients must become smaller, and so the per-locus selective effects may fall below $1/2N$ with sufficiently many loci. Second, as the number of loci increases with a fixed mutation rate per locus, the effective number of mutations influencing the trait increases.

As reviewed by Barton and Turelli (1989), there is also a paradox in the results of models and experiments to determine mutational variance. Per-trait mutation rates from mutation-accumulation experiments are on the order of 10^{-2} or 10^{-3}, and per locus mutation rates from classical genetic studies are on the order of 10^{-6}. Logic implies that these two observations are compatible if each trait is determined by 100 to 1000 genes. Yet estimates of classical effective numbers of genes affecting traits typically give a range of 5 to 20 genes. As is often the case in science, such paradoxes may generally best be resolved by breaking out of the standard way of thinking about the problem and taking an altogether different course. The course that is already changing our views of the genetic basis of quantitative traits, namely direct mapping of segments of the genome that affect the traits, is covered in the next section.

QUANTITATIVE TRAIT LOCI

Despite the landmark paper of Fisher (1918) and the impressive work since, there remains a gap between the theory of quantitative traits and the identification of classical Mendelian genes. The good fit of molecular variation to many aspects of the neutral theory contrasts strongly with the appeal of adaptation in morphological evolution. Although the discrepancy is no doubt exaggerated by intuitive notions that something as obvious as horns must be useful, and something as subtle as silent nucleotide substitutions must have negligible effects, it remains an important problem to bring these two levels of population genetics together. One approach is to identify the explicit relations among Mendelian genes and quantitative traits. Let us see how this is being done.

Mapping Genes that Influence Quantitative Characters

Largely because of the density of markers identified by a variety of molecular methods there has been a renewed interest in mapping genes (or blocks of genes) that affect quantitative traits. Observations of linkage between a marker gene and a gene that influences a quantitative trait date back to Sax (1923). Interest in actually mapping "quantitative trait loci" (or QTLs) also has a long history, beginning with Thoday's (1961) studies of bristle traits in *Drosophila*. Reviews of the early work can be found in Thoday (1979). The essence of the experiments on *Drosophila* was to construct lines that differed with respect to a quantitative character (such as "up" and "down" selected lines for bristle number), and to construct a random set of recombinants among them, using multiply marked chromosomes (each bearing several recessive mutations whose phenotype can be readily scored), to infer which chromosomal regions the recombinants carried. The end result of such experiments is not a genetic map in the classical sense, identifying locations of a series of specific genes that determine the quantitative character. The result is rather a statistical description of the character that indicates what fraction of the genetic variance in the parental lines is contributed by each region identified by the flanking markers. When this method is taken to a fine enough genetic scale, it is possible to identify individual genes that affect a trait, and the application of dense sets of molecular markers motivated the original drive to look again at a fine genetic scale (Paterson et al. 1988; Lander and Botstein 1989; Tanksley 1993).

The simplest means for mapping QTLs is to consider a cross between two parental lines that differ widely in phenotype. If the parental lines are fixed for alternate alleles at many loci, the F_1 hybrids will be heterozygous at those loci. Intercrossing the F_1 then produces a highly variable F_2 population, both in terms of the underlying genotypes and in the resulting distributions of phenotypes. Genes that are closely linked may be represented in only one linkage phase in the F_2, so one limitation of this method is that it does not allow very fine resolution of genes, but it is a good place to start.

Suppose we have constructed an F_2 population as described above. This F_2 population is scored for phenotypes and a series of molecular markers that differed between the original parental lines. First consider the case in which a molecular marker indicates the allelic state of a gene that directly affects a quantitative trait. In order to quantify the effects of the gene on the trait, regression methods can be performed as illustrated in Figure 9.27. The assumption in doing regression is that this locus has an effect that will be apparent when averaging across all genotypes at the other loci. Such marginal effects are expected if the genes act additively across loci, but if they interact, the marginal effects may either underestimate or overestimate the importance of a particular gene in determining the trait. If the alleles are labeled A_1 and A_2, and genotypes are indexed by the number of copies of A_2 alleles they possess, then regression of the phenotype on this index will produce a regression coefficient that estimates the value of the additive effect a (Figure 9.27). Similarly, after indexing the homozygotes by 0 and heterozygotes by 1, the regression of the phenotype on this index produces a regression coefficient that estimates the dominance parameter, d. One measure of the significance of each marker on the phenotype might be simply to determine the statistical significance of these regressions in the standard way. As we will see, this is not the best approach.

A shortcoming of the above method is that it assumes that the markers themselves are affecting the quantitative trait. One improvement is to suppose that there is a quantitative trait locus (or QTL) that is not directly observed, but that lies between a pair of markers that are scored (Figure 9.28). The idea is to still use a regression model to relate genotypes at the QTL to observed phenotypes, but now the genotypes are not directly observed. Instead, we infer the probability that each individual has a genotype at the QTL from the genotypes at the flanking markers. For example, if the markers are $A_1A_1 B_1B_1$, the QTL genotype is probably QQ, but a double recombination

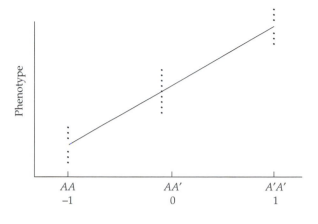

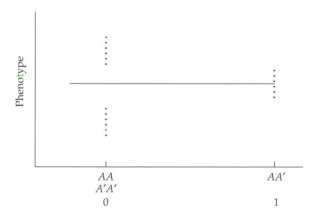

Figure 9.27 Illustration of the use of regression to estimate parameters of QTL expression. Ignoring recombination for now, the genotypes AA, AA' and $A'A'$ are indexed –1, 0 , and 1, and regression of phenotypes in these indices yields an estimate of the additive effect of the locus. Indices 0, 1, and 0 for genotypes AA, AA' and $A'A'$ yield regression estimates of dominance effects.

in either gamete will result in genotype Qq, and a double recombination in both gametes will result in qq. The probabilities of these three events are $1 - r_1 - r_2 + r_1r_2$, $r_1 + r_2 - 2r_1r_2$, and r_1r_2, respectively. One can work out the probabilities of all genotypes by using the gamete frequencies produced by the F_1 individuals as listed in Figure 9.28 to build an 8×8 Punnett square. For each position of the QTL on the map, one fits the model

$$Y = \mu + \Sigma_j \left[\Sigma_i \left(g_{ij} \, x_{ai} \, a_i + g_{ij} \, x_{di} \, d_i \right) \right] + \varepsilon \qquad\qquad 9.50$$

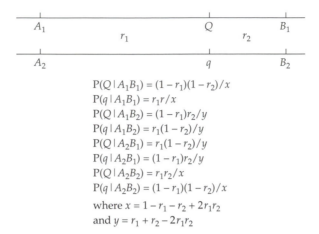

$$P(Q \mid A_1B_1) = (1 - r_1)(1 - r_2)/x$$
$$P(q \mid A_1B_1) = r_1 r_2/x$$
$$P(Q \mid A_1B_2) = (1 - r_1)r_2/y$$
$$P(q \mid A_1B_2) = r_1(1 - r_2)/y$$
$$P(Q \mid A_2B_1) = r_1(1 - r_2)/y$$
$$P(q \mid A_2B_1) = (1 - r_1)r_2/y$$
$$P(Q \mid A_2B_2) = r_1 r_2/x$$
$$P(q \mid A_2B_2) = (1 - r_1)(1 - r_2)/x$$

where $x = 1 - r_1 - r_2 + 2r_1r_2$

and $y = r_1 + r_2 - 2r_1r_2$

Figure 9.28   Composite interval mapping for quantitative trait loci is done by first expressing the probability that each marker locus genotype has a given QTL genotype. The flanking markers are A and B with the QTL in the middle, and recombination frequencies as specified. See text for a description of how the position of the QTL is determined.

This is a multiple regression with phenotype Y expressed as a linear function of grand mean μ, additive terms a_i, dominance terms d_i, and error ε. The x_{ai} are indicator variables with values –1, 0, and 1 for genotypes QQ, Qq, and qq, and x_{di} represents the indicator variables for dominance terms, with x_{di} = 0, 1, 0 for QQ, Qq, and qq. Because the QTL genotype is not actually observed, the probability that marker genotype j had QTL i (g_{ij}) is determined from the markers as described above. In practice, the goodness of fit of this model is assessed by a likelihood ratio, and the position of the QTL in the genome that gives the maximum likelihood is the most likely location for the QTL. Most often there is more than one peak to the likelihood curve, and this may very well reflect the presence of more than one QTL in the genome.

Significance Testing of QTLs

A serious problem with QTL mapping methods is how to decide when a likelihood ratio is significant. In the usual way one does a hypothesis test in statistics, one gets a test statistic whose distribution under the null hypothesis one knows. The null hypothesis is rejected if it has a probability less than 5%. This means one expects to reject the null hypothesis 5% of the time, even if it is true. With QTL mapping, one essentially tests the null hypothesis for thousands of potential locations of the QTL, so on the face of it, false positives would litter the genome. The problem is made even worse by the fact that we do not know precisely what the null distribution of the likelihood ratio test is, and the many tests that are done are not all independent of one another.

Fortunately there is a way out of this morass. Rather than relying on asymptotic theory to get an expected null distribution, we can build an empirical null distribution for the likelihood ratio by randomly permuting (shuffling) the association of marker genotypes and phenotypes (Doerge and Churchill 1996). Just as the Hudson (1992) test of geographic structure gave a null distribution by randomly permuting geographic location, this approach uses all the built-in structure of the data (including variations of sample size, clustered distribution of markers, and so on) to generate a null distribution tailored to the data. When the observed likelihood ratios are tested against this empirical null, then most of the problems outlined above disappear.

Composite Interval Mapping and Other Refinements

The method of interval mapping as described above examines only the pair of markers flanking the putative QTL position in assessing a probability that a particular marker genotype bears a particular pair of QTL alleles. It also considers the phenotype to be determined in a strictly additive fashion, so that the marginal effect of a particular QTL on the phenotype is all that is needed to map the QTL. More recently, these assumptions have been relaxed in the method of **composite interval mapping** (Zeng 1994; Jansen and Stam 1994). This approach is essentially a multiple regression extension of standard interval mapping allowing genotypes at marker loci more distant from the putative QTL site to have an effect on the phenotype. This is done by a statistical approach called partial regression, and in principle it allows there to be some level of epistasis between sets of QTLs in the way they affect a trait. Both simulations and practical application of these methods are very encouraging.

One limitation of using F_2 populations is that blocks of the genome will remain intact. Composite interval mapping algorithms have been generalized to include the crossing designs where the F_2 are intercrossed to give an F_3, which are then intercrossed, and so forth. After a few generations like this, the mixed population is then backcrossed to the two parental populations, and phenotypes and marker genotypes are assessed. Such a design allows much finer resolution because several rounds of genetic recombination have occurred. Even finer mapping can be done by considering "historical" recombination in even longer-running experiments (Xiong and Guo 1997). Other methods for assessing relationships between complex phenotypes and genotypes include nonparametric methods (Kruglyak and Lander 1995a), and, if candidate genes have been identified, one can build a gene tree for each candidate gene and contrasts of phenotypes among clades of the gene tree may be able to distinguish subtle differences in phenotype (Templeton et al. 1995).

PROBLEM 9.12   The third chromosome data of Long et al. (1995) consisted of estimates of line means of sternopleural and abdominal bristles of 84 lines that were also scored for the presence of *roo* transposable elements at 29 positions along the chromosome. Below is a subset of the data, where H means that site on the genome had the same *roo* genotype as the high selected line and L means the line had the same genotype as the low selected line. Determine whether each interval has an effect by a simple *t*-test. The t statistic is

$$t = \frac{\mu_1 - \mu_2}{\sqrt{\dfrac{s_1^2}{n_2} + \dfrac{s_2^2}{n_1}}}$$

where s_1^2 and s_2^2 are the sample variance of the H and L groups and n_1 and n_2 are the sample sizes of the H and L groups respectively. There are $n_1 + n_2 - 2$ degrees of freedom to the *t*-test, and the 5% significance level for t_{11} is 2.201.

Line	Male sternopleural bristle count	Interval 1	2	3
1	16.65	L	L	L
2	18.00	H	L	L
5	16.30	L	L	L
8	18.70	H	H	H
9	16.95	L	L	L
10	16.00	L	L	L
11	16.85	L	L	L
12	16.15	L	L	L
13	18.85	H	H	H
14	20.00	H	H	H
15	19.55	H	L	L
17	15.45	L	L	L
20	18.55	H	H	H

ANSWER   The mean of the lines with an H in interval 1 is $\mu_1 = 18.94$, and the mean of the lines with an L in interval 1 is $\mu_2 = 16.34$. Substituting into the *t*-statistic formula, we get $t = (18.94 - 16.34)/\sqrt{(0.279)/7 + 0.520/6} = 7.33$. This test has $13 - 2 = 11$ degrees of freedom, and the corresponding P value is less than 0.05. We conclude that even with this small subset of the data, there is a QTL near this region that gives the H lines more bristles than the L lines. The *t* statistics for the other two intervals are both 4.07, and are also significant. With so little data, we cannot tell whether the significant effects are all caused by the same QTL or by more than one QTL.

Although the method of scoring F_2 individuals is powerful, it is not essential to perform controlled crosses to make inferences about the effects of marker genes on quantitative characters. This is fortunate because methods for identifying marker genes relevant to polygenic human diseases are thought to have great potential in identifying underlying genes (Lander and Schork 1994). The chances for success in identifying QTLs in humans is much improved by selecting **candidate genes** which may have a functional relation to the trait (Risch and Merikangas 1996). For example, Sing and Davignon (1985) examined a sample of humans for apolipoprotein E (*apoE*) genotype and several quantitative traits relating to serum triglyceride, cholesterol, and low-density lipoproteins. There are three common alleles of *apoE*, yielding six genotypes, the mean phenotypes of which are presented in Table 9.8. About 16% of the total genetic variance in LDL-C (an important carrier of cholesterol) is explained by genotype at the *apoE* locus. The importance of this kind of approach is underscored by the observation that about half of the variation among individuals in serum cholesterol is associated with polygenic variation.

What Have We Learned from Mapping QTLs?

One might think that by mapping QTLs, geneticists would immediately learn where the genes are that affect a character and then be able to identify those genes and quickly isolate them. This has not been the case. Most QTL mapping projects have not been followed to this point. Nevertheless, it is useful to consider what has been learned from the patterns of QTL effects observed. One of the largest QTL mapping studies was that of Stuber et al. (1992), which sought to determine the genetic basis for heterosis or hybrid vigor in maize. The primary competing hypotheses were: (1) hybrid vigor arises from an advantage that comes from being heterozygous for individually important genes, each of which has a heterozygote advantage; (2) heterozygous genotypes for each individual locus are intermediate, but both parentals have low performance because of homozygosity for different recessive deleterious alleles. Stuber et al. found that many regions of the genome had the property that QTL genotypes *QQ* and *qq* were inferior to *Qq*, so it appeared at first as if there were a locus-by-locus heterozygote advantage.

TABLE 9.8   SERUM LIPID LEVELS IN A SAMPLE OF 102 PEOPLE FROM OTTAWA, ADJUSTED FOR AGE, SEX, HEIGHT, AND WEIGHT EFFECTS

	Genotype						Probability (F ratio)	Estimates of the population	
	ε4ε4	ε3ε3	ε2ε2	ε4ε3	ε3ε2	ε4ε2		Mean	Variance
Count	4	63	2	21	10	2	—	—	—
Relative frequency	0.039	0.618	0.020	0.206	0.098	0.020	—	—	—
Variable									
Total cholesterol	180.3	173.8	136.0	183.5	161.4	178.1	0.09	174.16	732.48
HDL-C	53.3	47.3	47.1	47.3	45.7	45.12	0.93	47.32	130.43
LDL-C	102.9	104.2	73.9	112.8	89.5	109.3	0.08	104.00	602.17
VLDL-C	24.0	22.3	15.0	23.3	26.2	23.5	0.83	22.83	127.56
Triglycerides	65.5	74.4	70.6	70.4	79.3	73.5	0.96	73.60	918.01
LDL-Apo B	86.3	83.9	55.8	86.8	78.4	60.5	0.13	83.03	375.72
VLDL-Apo B	8.5	11.4	5.65	10.5	9.2	18.9	0.56	10.91	66.21
LDL-C/total cholesterol	0.56	0.60	0.55	0.61	0.55	0.61	0.26	0.59	0.0055
VLDL-C/total triglycerides	0.37	0.32	0.19	0.33	0.33	0.32	0.79	0.32	0.0200

Source: From Sing and Davignon 1985.

Subsequent work suggests that the regions identified in this mapping effort (which included phenotypic measurements on nearly 100,000 plants and molecular assays done for 76 loci) were still too coarse; the finer mapping suggests that the QTLs are actually blocks of genes in linkage disequilibrium, with each gene tending toward a pattern of deleterious recessive effects.

A similar experiment in hybrid rice initially produced results much more in line with recessive deleterious effects being responsible for the hybrid vigor (Xiao et al. 1995). Rice is predominantly a self-pollinator, while corn outcrosses, so one might have expected rice to have more effectively eliminated recessive deleterious alleles from its genome. Subsequent analysis by Zeng (pers. com.) is finding that both the maize and the rice data show evidence for large amounts of epistasis.

This chapter began with studies of bristle number in *Drosophila* that began in the 1950s. The early work suggested that bristle number was a trait that seemed to fit the classical quantitative genetic model fairly well. Bristle number had an abundance of additive variance and responded well to artificial selection. QTL mapping was applied to bristle number by scoring *roo* elements as the genetic marker in 93 recombinant isogenic lines derived from divergently selected parental lines (Long et al. 1995). The *roo* transposable element is present in high copy number and many sites differed between the two parental lines. Scoring the *roo* elements by in situ hybridization allowed mapping effects to within 4 cM; the results revealed two X-linked QTLs and five third-chromosome QTLs that had a significant effect on bristle number. All seven of these locations happen to coincide with genes known to have large effects on bristle number, such as *achaete-scute*, *hairy*, and *Delta*.

As exemplified by the findings with regard to bristle number, where there are many candidate genes identified for a particular trait, it is often more efficient to start with the candidate genes and ask how much of the population variability they explain. One other candidate gene for bristle number is *scabrous*, and Lai et al. (1995) did one of the most extensive analyses of molecular variation in the *sca* locus and its associations with quantitative variation. The fact that a significant association between molecular variation of *sca* and phenotypic differences in bristle number was found is consistent with the idea that variation at this locus has an effect on bristle number even when averaged over effects of the rest of the genome. Although it is known that outright loss-of-function mutations in *sca* affect bristle number, it is quite another matter to find that quantitative variation in a natural population in the expression of *sca* causes quantitative variation in bristle number. Another idea for testing the role of candidate genes is a quantitative trait complementation test (Mackay and Fry 1996). By crossing lines selected for high and low bristle number to wildtype control lines and to mutant lines having defects in known bristle genes, large differences in bristle traits are sometimes found. The differences were interpreted to mean either that the selected lines had allelic differences at the candidate genes or that they had epistatic interactions with the candidate genes.

The scale at which we look at the genome often colors the way problems are perceived. The Stuber et al. results on hybrid vigor in maize appeared to show heterosis of QTLs at one scale but recessiveness at a finer scale. Quantitative variation in a single gene's expression can be caused by *P*-element insertional mutations all over the genome (Clark et al. 1995b). On the other hand, even molecular variation in the immediate proximity of a structural gene can have complex effects on the gene's expression. The activity of *Adh* varies among lines with the same electrophoretic allele, and it required extremely careful and laborious in vitro mutagenesis and transformation experiments to tease apart the effects of variation at the sites (Stam and Laurie 1996). The conclusion was that more than one nucleotide site affects expression in a way that appears to be epistatic.

The idea of using interspecific hybrid crosses to understand the genetic basis for differences between species was introduced in Chapter 8, and it should be clear that the use of sets of molecular markers make these methods even more powerful. Many plant species originated through ancient hybridization events, and by crossing extant species and following the fate of molecular markers, the genomic composition of the hybrid descendants can be followed. A pair of species of *Helianthus* (sunflower) were thought to have hybridized to give rise to a third species, and all three species can be found today. The two original species were crossed, backcrossed for two generations then selfed for two generations before DNA was extracted and examined at 197 RAPD marker loci (Rieseberg et al. 1995). Many parts of the genome were invariant. More strikingly, the genomic composition of the hybrids was remarkably similar to that of the extant, putatively hybrid origin species of sunflower (Rieseberg et al. 1996). Sometimes the genetic basis for interspecific differences is dominated by one or a few genes of major effect, as was found by Doebley et al. (1995) for the difference in glume architecture between maize and teosinte. Another application of interspecific crosses showed that the genetic basis for interspecific differences in genital arch shape in male *Drosophila* species maps to several genes on more than one chromosome (Liu et al. 1996; Laurie et al. 1997).

When one considers the problem of complex genetic diseases in humans, it becomes clear that the essence of the problem is to understand how genetic variation in a population relates to phenotypic variation for risk of disease. The problem is intrinsically one that requires an understanding of principles of population genetics. The patterns of underlying genetic variation, including differences among populations and linkage disequilibrium, are primary topics of population genetics. As the human genome project nears completion, the rate at which single gene disorders are discovered and characterized will also near exhaustion. It is becoming evident that the next big problem in medical genetics is to understand complex disorders. It seems inescapable that the field of population genetics is about to see a dramatic increase in visibility and importance. We hope this book played a part in inspiring some people to begin thinking about these problems.

SUMMARY

Multifactorial traits are affected by multiple genes and usually by environmental factors as well. Some multifactorial traits, such as height or weight, are continuous in that they demonstrate a continuum of possible phenotypic values. Other multifactorial traits, known as meristic traits, have their phenotypic value determined by enumeration, for example, by counting the number of bristles on a fruit fly sternite. Still other multifactorial traits feature an underlying continuum of liability or risk, and only certain individuals with liabilities above a threshold are affected. These three basic types of multifactorial traits are collectively called quantitative traits.

Many quantitative traits have a distribution of phenotypic values that is approximately normal and described completely in terms of two parameters, the mean μ and the variance σ^2. Traits that are not normally distributed sometimes become normal when measured on an appropriate scale, such as $\ln(x)$ or $\arcsin(\sqrt{x})$, where x represents the original measurement of phenotype. Much of the theory of quantitative genetics is based on the assumption of a normal distribution of phenotypes.

Truncation selection is a method of individual selection in which all individuals whose phenotype lies above a certain value T (the truncation point) are saved and mated randomly among themselves to produce the next generation. If μ denotes the mean phenotype of the original population and μ_S denotes the mean phenotype among selected parents, then the mean phenotype among the progeny (μ') is given by $u' - u = h^2(\mu_S - \mu)$, where h^2 is the heritability of the trait. The quantity $\mu' - \mu$ is usually called the response to selection R, while $\mu_S - \mu$ is called the selection differential S; so the prediction equation for individual selection can be written as $R = h^2 S$. The prediction equation for individual selection can be written equivalently as $R = i\sigma h^2$ where i is called the intensity of selection and $i = S/\sigma$. Intensity of selection is a useful quantity with which to compare diverse breeding programs because i depends only on the proportion of the population saved for breeding. When h^2 is estimated from observed values of R and S, then h^2 is called the realized heritability. Otherwise h^2 can be estimated from the resemblance between relatives, as exemplified by the regression coefficient of offspring on parent. If b denotes the regression coefficient of offspring on a single parent, then $h^2 = 2b$. If b denotes the regression coefficient of offspring on the mean of the parents (midparent), then $h^2 = b$.

In terms of genetic parameters, $h^2 = \Sigma 2pq[a + (q - p)d]^2/\sigma^2$ where summation is carried out across all genes affecting the trait and, for each gene, p is the frequency of the favorable allele, a is the effect of the gene (measured as the average difference between the homozygotes), and d is a measure of dominance (measured as the deviation of the heterozygote from the mean of the homozygotes). The quantity $\Sigma 2pq[a + (q - p)d]^2$ is known as the additive genetic variance. In the above formula for h^2, the symbol σ^2 represents the variance in phenotypic value in the population. Thus, the heritability is the ratio of additive genetic variance to total phenotypic variance.

Resemblance between relatives can also be used to partition the total phenotypic variance of a trait into components due to genotype and environment. The total genotypic variance can be further partitioned. In the absence of genotype-environment interaction (evidenced by parallel norms of reaction) and in the absence of genotype-environment association (evidenced by lack of correlation between genotype and environment), then the total phenotypic variance can be written as the sum of a variance term due to the additive effects of genes (the additive genetic variance), a term due to dominance effects (the dominance variance), a term due to interactions between genes (the epistatic variance), and so forth. The ratio of the additive genetic variance to the total phenotypic variance is the heritability h^2, more precisely called the narrow-sense heritability. This is the heritability that is important in individual selection. Another type of heritability is the ratio of the total genetic variance to the total phenotypic variance; this is called heritability in the broad sense. Heritability in the broad sense is important when selection is practiced between clones, inbred lines, or noninterbreeding varieties. Both types of heritability are specific to a particular trait in a particular population at a particular time, because they depend on the allele frequencies in the population, the environmental variance, and the additive, dominance, and other effects. Furthermore, heritabilities estimate the fraction of the variance in a trait that is due to genetic causes, either in the narrow sense of additive effects or in the broad sense of all genetic effects. The heritability is not informative about the mean phenotypic value of a trait. For example, the mean can be altered dramatically by a change in environment without affecting the heritability.

After long-continued selection for a quantitative trait, response to selection eventually ceases and the population reaches a plateau. A selection limit may be reached even before all the additive genetic variance for the trait has been exhausted, and in many cases, the response to selection ceases because natural selection for fitness-related traits counterbalances the artificial selection for the trait in question. Nevertheless, artificial selection can produce a total response of five or ten or more phenotypic standard deviations, so the trait in the selected population may be well beyond the range of what it was in the original population. This seemingly paradoxical result happens because populations used in selective breeding are typically small enough in number that not every possible genotype can be represented, particularly genotypes that are rare.

Threshold traits are traits that are either present or absent in an individual, but their presence or absence is determined by the value of an underlying quantitative trait known as the liability. The heritability of liability can be calculated from established principles of quantitative genetics when one knows the incidence of the trait in the general population and also the incidence among offspring or other relatives of affected individuals.

The principles of quantitative genetics also have application in understanding the evolution of quantitative traits under natural selection. The models include many parameters and their implications can be investigated by computer simulation or mathematical analysis under simplifying assumptions. At opposite ends of the spectrum are models that assume a large number of genes, each with small effects, and those that assume a small number of genes, each with a relatively large effect. There are methods for estimating the minimum number of genes affecting a quantitative trait, such as Wright's method, but the methods are biased toward underestimates. Another approach is the use of molecular genetic markers to detect linkage with loci that affect the quantitative trait of interest, which are called QTLs or quantitative trait loci. Various methods for detecting QTLs have been devised, but they are all have difficulty distinguishing whether an identified QTL is due to a single allele of large effect or a group of linked alleles, each of small effect. These possibilities can be sorted out by the study of multigeneration pedigrees in which recombination has the opportunity to break up blocks of linked alleles.

PROBLEMS

1. A recent study reports that the heritability of violent behavior in humans is 80%. Many people think that this means that people with the at-risk alleles (if they could be identified) are destined to become violent. What is the fallacy in this argument?

2. The following are the weaning weights (lbs) of lambs in a large flock. Estimate the mean, variance and standard deviation of weight at weaning.

68	79	93	67	73	81	82	81	85	78
72	69	64	82	77	59	68	54	71	57
88	97	69	60	92	62	64	64	90	60

3. Suppose a population of *Drosophila* has a normal distribution of abdominal bristles with a mean of 20 and a standard deviation of 2. What proportion of the population is expected to fall into the following categories of bristle number:
   a. Between 18 and 22.
   b. Greater than 22.
   c. Greater than 24.
   d. Between 20 and 22.
   e. Smaller than 16.

4. Two inbred varieties of tobacco are crossed and give a variance in leaf number in the F_1 generation of 1.5. The variance in the F_2 generation is 6.0. What are the genotypic and environmental variance components and the broad sense heritability?

5. In a population of the flour beetle *Tribolium* the mean weight of pupae is 2000 mg. The phenotypic variance is 40,000 mg^2, and the additive genetic variance is 10,000 mg^2. If individuals with a mean pupa weight two phenotypic standard deviations above the mean are selected, what is the expected average pupa weight among the progeny?

6. If a population of *Drosophila* has a mean number of abdominal bristles of 20 with a narrow sense heritability of 25%, what is the expected bristle number after one generation when the selection differential is four bristles? What is the expected number after 10 generations of equally intense selection?

7. Five generations of selection for decreased plasma cholesterol level in mice decreased the mean from 2.16 mg/100 ml to 2.01 mg/100 ml. The average selection differential was 0.07 mg/100 ml. What is the realized heritability?

8. A quantitative trait has the following mean values for the genotypes shown:

AA	AA'	A'A'
23.8	25.2	19.4

  a. What are the values of a and d?
  b. What allele frequency of A would maximize the mean value of the trait in the entire population?

9. Two strains of rats were selected for increased or decreased pigmentation on the head and back. After 10 generations the high strain had a rating of 3.73 and the low strain a rating of –2.01. The strains were crossed, and the standard deviations in the F_1 and F_2 generations were 0.87 and 0.60, respectively. Estimate the effective number of factors affecting the trait in these strains.

10. Estimate the correlation coefficient in litter size between first and second litters using the following data from 10 females.

First litter	8	9	9	10	10	10	11	11	13	13
Second litter	6	8	12	10	12	12	9	10	12	12

11. How many generations of selection with a selection differential of 20 would be required to increase the average number of eggs laid per hen per year from 180 to 220, given a heritability of 20%?

12. A standard normal distribution is a normal distribution with mean 0 and variance 1. Using this distribution, calculate the mean of the top $\frac{1}{2}$, $\frac{1}{4}$, $\frac{1}{8}$, $\frac{1}{16}$, and $\frac{1}{32}$ of the population. These values are the standardized selection intensities (i) for the proportions saved, which correspond to truncation points of 0.00, 0.68, 1.16, 1.54, and 1.86, respectively.

13. If the selection differential differs in males and females, show that the proper value to use in Equation 9.10 is the mean.

14. Consider a locus with genotypes AA, AA', $A'A'$ whose contribution to a quantitative trait has $a = 0.6$ and $d = 0.2$; another locus with genotypes BB, BB', $B'B'$ contributes to the same trait with $a = 0.4$ and $d = 0$. If the loci are unlinked and additive and the allele frequencies of A and B are 0.5 and 0.7, respectively, calculate the narrow-sense and broad-sense heritability of the trait when the total phenotypic variance is 1.0.

15. A herd of dairy cattle yields milk with a fat content of 3.4% $\pm$ 0.65% (mean $\pm$ standard deviation) and a protein content of 3.3% $\pm$ 0.45%. The heritabilities of these traits are 60 and 70%, respectively, and the genetic correlation is 0.55. If selection is practiced for percent protein with a selection intensity of $i = 1.5$, what increase in percent protein and percent fat would be expected? What intensity of selection would produce the same increase in percent fat by direct selection?

16. Show that a simple Mendelian dominant gene has a narrow sense heritability of $2(1-q)/(2-q)$, where q is the frequency of the dominant allele.

17. For an overdominant locus with two alleles, show that the additive genetic variance at equilibrium equals 0.

18. The intensity of selection i is the mean of the selected parents in a standard normal distribution when B is the proportion saved. It equals the selection differential in units of standard deviation, so $i = (\mu_S - \mu)/\sigma$. Over the range $B = 0.05$ to $B = 0.005$, i is given approximately by $i = 0.8 + 0.41\ln[(1/B) - 1]$ (Simmonds 1977). Calculate i for $B = \frac{1}{2}, \frac{1}{4}, \frac{1}{8}, \frac{1}{16}$, and $\frac{1}{32}$ and compare with the corresponding values in Problem 12.

19. If a normally distributed phenotype X, with truncation point T, is transformed to a standard normal $x = (X - \mu)/\sigma$, with transformed truncation point $t = (T - \mu)/\sigma$, what is the relationship between the ordinate Z of the untransformed distribution at T to the ordinate z of the standardized distribution at the point t?

Suggestions for Further Reading

The following books are suggested for further reading:

Avise, J. C. 1994. *Molecular Markers, Natural History and Evolution*. Chapman and Hall, New York. A thorough treatment of the use of molecular markers to infer past history of migration and phylogeny.

Bulmer, M. G. 1985. *The Mathematical Theory of Quantitative Genetics*. Clarendon Press, Oxford. Along with Falconer and Mackay, this book presents the basics principles of quantitative genetics. Emphasis is on animal breeding, with perhaps more concern for evolutionary questions than the other texts.

Cavalli-Sforza, L. L. 1996. *History and Geography of Human Genes*. Princeton University Press, Princeton, NJ. An examination of the history of human migration as inferred from the legacy of the genes left behind.

Cavalli-Sforza, L. L and W. F. Bodmer. 1971. *The Genetics of Human Populations*. W. H. Freeman and Co., San Francisco. Despite the copyright date, this remains one of the best books on the population genetics of humans, and is full of excellent examples.

Charlesworth, B. 1994. *Evolution in Age-Structured Populations*, Second Edition. Cambridge University Press, Cambridge. With an entire book on this subject, it is not surprising that the treatment is advanced and very thorough.

Crow, J. F. and M. Kimura. 1970. *An Introduction to Population Genetics Theory*. Harper & Row, New York. This is the book from which the previous generation learned population genetics theory, and it remains the best source for many aspects of population genetics theory. It was reprinted in 1978 by Burgess.

Ewens, W. J. 1979. *Mathematical Population Genetics*. Springer-Verlag, Berlin. This is a rather advanced text, and gives a dense and rigorous treatment of the subject for the mathematically inclined.

Falconer, D. S. and T. Mackay. 1996. *Introduction to Quantitative Genetics*. Longman, Essex, England. An update of a classic textbook devoted to quantitative genetics.

Feldman, M. W. and F. B. Christiansen. 1986. *Population Genetics*. Blackwell Scientific Publications, Palo Alto, CA. An elementary text with emphasis on human examples.

Futuyma, D. J. and M. Slatkin. 1983. *Coevolution*. Sinauer Associates, Sunderland, MA. Chapters contributed by many authors survey this intriguing problem which is just as topical today.

Gale, J. S. 1990. *Theoretical Population Genetics*. Unwin Hyman, London. One of the clearest expositions of diffusion methods.

Gillespie, J. H. 1991. *The Causes of Molecular Evolution*. Oxford University Press, Oxford. An excellent survey of problems in contemporary molecular evolution. Offers a serious challenge to the neutral theory.

Golding, B. (ed.). 1994. *Non-Neutral Evolution: Theories and Molecular Data*. Chapman and Hall, New York. A collection of papers dealing with some advanced topics in population genetics.

Hamrick, J. L. (ed.). 1995. *Conservation Genetics: Case Histories from Nature*. Chapman and Hall, New York. How can the principles of population genetics be put to practical use in conservation biology?

Hartl, D. L. 1988. *A Primer of Population Genetics*, Second Edition. Sinauer Associates, Sunderland, MA. An easily digested, thinner treatment than this book

Hillis, D. M., C. Moritz and B. K. Mable (eds.). 1996. *Molecular Systematics*, Second Edition. Sinauer Associates, Sunderland, MA. An introduction to the methods and some of the pitfalls of phylogenetic inference.

Kempthorne, O. 1957. *An Introduction to Genetic Statistics*. John Wiley and Sons, New York. Reprinted in 1969 by Iowa State University Press, Ames. This is a standard reference for the statistical tools needed for estimating parameters of quantitative genetics.

Kimura, M. 1983. *The Neutral Theory of Molecular Evolution*. Cambridge University Press, Cambridge. No student of population genetics should fail to read this book. A concise statement of an extremely influential theory.

Kimura, M. and T. Ohta. 1971. *Theoretical Aspects of Population Genetics*. Princeton University Press, Princeton, NJ. An overview of the application of diffusion methods to a number of classic problems in population genetics.

Kingman, J. F. C. 1980. *Mathematics of Genetic Diversity*. Society for Industrial Applied Mathematics, Philadelphia. A rigorous treatment of selected topics in population genetics theory. Especially good on the sampling theory of neutral genes and the coalescent process of the Wright-Fisher model.

Levine, L. (ed.). 1995. *Genetics of Natural Populations: The Continuing Importance of Theodosius Dobzhansky*. Columbia University Press, New York. Praise for one of the founders of theoretical population genetics.

Lewin, B. 1997. *Genes VI*. John Wiley and Sons, New York. Population geneticists need to learn a lot of molecular biology, and this text provides a good background.

Lewontin, R. C. 1974. *The Genetic Basis of Evolutionary Change*. Columbia University Press, New York. This book remains essential reading for students of population genetics, providing a deep look at the nature of questions that can be addressed by population genetics.

Li, C. C. 1976. *First Course in Population Genetics*. Boxwood Press, Pacific Grove, CA. The style of this book is rather different from any other, and appeals very much to some readers. Li likes to jump right in to numerical examples, and uses them very effectively.

Li, W.-H. 1997. *Molecular Evolution*. Sinauer Associates, Sunderland, MA. An advanced treatise on molecular population genetics.

Manly, B. F. J. 1985. *The Statistics of Natural Selection on Animal Populations*. Chapman and Hall, London. A thorough coverage of the subject with a massive citation listing.

Mather, K. and J. L. Jinks. 1982. *Biometrical Genetics*. Third Edition. Chapman and Hall, London. The serious student of quantitative genetics should read both Falconer and this text. Mather and Jinks focus more on experimental designs for variance partitioning in plants.

Nagylaki, T. 1977. *Selection in One- and Two-Locus Systems*, Lecture Notes in Biomathematics, Volume 15. Springer-Verlag, Berlin. Hard-core approach to the basic models.

Nei, M. 1987. *Molecular Evolutionary Genetics*. Columbia University Press, New York. A rather comprehensive text on the subject, with emphasis on interpretation of molecular data.

Ohta, T. and K. Aoki. 1985. *Population Genetics and Molecular Evolution*. Springer-Verlag, New York. Collection of articles on various aspects of the neutral theory, including its history and role in population genetics.

Provine, W. B. 1986. *Sewall Wright and Evolutionary Biology*. University of Chicago Press, Chicago. A superb scientific biography of one of the leading evolutionary biologists since Darwin.

Provine, W. B. 1987. *The Origins of Theoretical Population Genetics*. University of Chicago Press, Chicago. An engaging early history of theoretical population genetics.

Roughgarden, J. 1995. *Theory of Population Genetics and Evolutionary Ecology: An Introduction*. Prentice Hall, Paramus, NJ. This is the best text available on ecological genetics, with a clear derivation of the diffusion approximation.

Selander, R. K., A. G. Clark and T. S. Whittam. 1991. *Evolution at the Molecular Level*. Sinauer Associates, Sunderland, MA. Contains thirteen review papers by a variety of experts.

Takahata, N. and A. G. Clark (eds.). 1993. *Mechanisms of Molecular Evolution*. Sinauer Associates, Sunderland, MA. A nice set of papers on contemporary molecular population genetics.

Weir, B. S. 1996. *Genetic Data Analysis II*. Sinauer Associates, Sunderland, MA. A detailed examination of methods used in the analysis of population genetic data.

Weir, B. S., E. J. Eisen, M. M. Goodman and G. Namkoong (eds). 1988. *Proceedings of the Second International Conference on Quantitative Genetics*. Sinauer Associates, Sunderland, MA. Fifty-five contributed papers on all aspects of quantitative genetics, with a mix of reviews and original work.

Weiss, K. M. 1993. *Genetic Variation and Human Disease: Principles and Evolutionary Approaches*. Cambridge University Press, Cambridge. An examination of the distribution of genetic variation associated with genetic disorders in human populations.

Wright, S. *Evolution and the Genetics of Populations*. Vol. 1: *Genetic and Biometric Foundations* (1968). Vol. 2: *The Theory of Gene Frequencies* (1969). Vol.3: *Experimental Results and Evolutionary Deductions* (1977). Vol.4: *Variability within and among Natural Populations* (1978). University of Chicago Press, Chicago. A four-volume treatise by one of the founders of the field.

Answers to Chapter-End Problems

CHAPTER 1

1. Some of the possible causes of variation include: (a) variation in size of the flies, (b) variation in primary sequence of G6PD, giving allelic forms with different specific activities, (c) variation in transcription rates, due to either *cis* or *trans* factors, (d) variation in post-transcriptional processing of the message to produce functional mRNA, (e) variation in stability of the mRNA, (f) variation in rates of translation, (g) variation in post-translational processing of the protein, (h) variation in protein stability.

2. One might expect that there would be defects at all the steps giving rise to PKU, but in fact the overwhelming majority of cases are caused by defects in the primary gene sequence. Not all are amino acid changes or nonsense mutations, however. There are many defects at splice junctions, resulting in improper splicing of the mRNA. Mutations in *trans*-acting factors that cause a complete loss of enzyme activity have not been found, presumably because they would cause misexpression of other genes as well, so that defects in *trans*-factors are not identified as PKU.

3. Of the 576 possible single-base mutations, 438 yield a codon that encodes something other than the premutation codon.

4. Many amino acid replacements can be made by a single nucleotide change, such as phenylalanine (UUU) to leucine (CUU). Other replacements require two or even three nucleotide changes. For example, methionine (AUG) codons must be changed at all three sites to encode cystine (UGU or UGC). The rate of mutation from one amino acid to another depends on the number of underlying nucleotide changes that must occur.

5. Pleiotropic effects of mutations result in many examples of apparently disparate phenotypic effects of mutations. Model systems, like *Drosophila*, provide some of the best examples. For example, the gene *maleless* encodes an RNA helicase and is important for normal dosage compensation, increasing the rate of X-chromosome transcription in males only. Mutations in *maleless* result in male death. Other mutations in *maleless* result in a lower temperature at which flies are incapacitated (previously thought to be caused by the a gene, *no-action-potential*, now known to be the same as *maleless*). Examples like this are common in the literature; especially common are the disparate pleiotropic effects seen in human genetic disorders.

6. The males produce gametes that are (e st) and (e^+ st^+) in proportions $\frac{1}{2}$: $\frac{1}{2}$ because they do not undergo meiotic recombination. Females produce gametes with respective frequencies: (e st) 0.315, (e^+ st^+) 0.315, (e st^+) 0.185, and (e^+ st) 0.185. Draw a 2×4 Punnett square to get the frequencies of the eight genotypes of the next generation by multiplying the respective gamete frequencies of males and females. For example, the genotype (e st)/(e st) and (e^+ st^+)/(e^+ st^+) will each have frequency $\frac{1}{2} \times 0.315 = 0.158$, and ($e$ st)/(e^+ st^+) will have frequency 0.315. All four other genotypes have frequency $\frac{1}{2} \times 0.185 = 0.092$.

7. Assuming that the chance of having a son or daughter is each $\frac{1}{2}$, such a stopping rule would have no effect on the sex ratio, which would remain at $\frac{1}{2}$: $\frac{1}{2}$. To show this result, make a table with the counts of sons and daughters of all possible families up to five daughters and one son (and five sons and one daughter). Adding up the fraction of families of each type multiplied by the counts of the two sexes will show that the sum converges to $\frac{1}{2}$.

8. The chance of producing one gamete that is $ABCD$ is $\frac{1}{2} \times \frac{1}{2} \times 1 \times \frac{1}{2} = \frac{1}{8}$. The chance of producing two such gametes is $\frac{1}{8} \times \frac{1}{8} = \frac{1}{64}$, again applying the multiplication rule because the two gametes are independent of one another.

9. Many such problems in probability arise in calculating genetic risk in pedigrees. In this case, both parents have to be heterozygotes in order to have produced an affected offspring. The fraction of offspring of a cross $Aa \times Aa$ that is heterozygous is $\frac{1}{2}$, so this is the chance that the sib is a carrier.

10. The binomial variance is pq/N. For $p = q = \frac{1}{2}$, this is $1/4N$. We would reject the null hypothesis of $p = q = \frac{1}{2}$ if the observed frequency were more deviant than two standard deviations. Solving $2 \times \sqrt{(1/4N)} = 0.05$, we get $N = 400$. This means if we counted only 200 offspring, a sex ratio of 55 : 45 would not be significantly different from 50 : 50.

11. One could calculate the frequency of the A morph in each population and the standard error of the frequency estimates. If the confidence intervals of these frequency estimates overlap sufficiently, then the apparent differences in frequency could be accounted for by sampling effects. More simply, one can do a heterogeneity χ^2 test. In this case, this test is done as a standard 2×2 contingency table. Let the first row be $a = 26$, $b = 28$ and the second row be $c = 10$ and $d = 21$; the χ^2 value is $(ad - bc)^2 N / [(a + b)(c + d)(a + c)(b + d)] = 2.04$. ($N = a + b + c + d$). This is less than the 3.84 critical value for one degree of freedom, and we conclude that it is possible these samples are from the same population.

12. (a) Allele frequency of PGI-$2a$ = $(70 + 19)/114 = 0.78$, of PGI-$2b = 0.22$. (b) Expected genotype frequencies are $0.78^2 = 0.61$, $2(0.78)(0.22) = 0.34$, and $(0.22)^2 = 0.05$, giving expected numbers of 34.8, 19.4, and 2.8.

13. On a log scale, the plot of the human population size versus time curves upward dramatically. This means that our population is growing considerably *faster* than at an exponential growth rate. Think about this as you work Problem 14.

14. If one generation takes 12 days, a year is approximately 30 generations. Each generation the population increases 250-fold, so $250^{30} = 8.67 \times 10^{71}$ is the number of flies. One kilogram is 10^6 milligrams, so the weight will be 8.67×10^{65} kilograms. The mass of the earth is 5.97×10^{24} kg, so apparently flies are not able to keep on breeding at their maximum rate.

CHAPTER 2

1. 80 mm is the mean plus one standard deviation. We know that 68 % of the distribution lies within one standard deviation. This means that 34% are between the mean and plus one standard deviation. Since 50% of the population falls below the mean (the normal distribution is symmetric), this implies that 50% + 34% = 84% of the population is smaller than 80 mm.

2. 90 mm is two standard deviations above the mean. We know that 95% of the distribution lines within two standard deviations of the mean, so 47.5% falls between the mean and plus two standard deviations (between 70 and 90 mm in our example). Because 34% fall between 70 and 80 mm, this means that the difference, or 13.5%, falls between 80 and 90 mm.

3. The sum of the counts is 193, and there are 14 counts, so the sample mean is $193/14 = 13.785$. The sum of squared counts is 2679, so the sample variance is $[(2679 - (193^2/14)]/13 = 1.412$. The sample standard deviation is the square root of the sample variance, or 1.188. The standard error of the mean is the standard deviation divided by the square root of the sample size, or 0.318.

4. The standard error is $\sqrt{(64/100)} = 0.8$. The difference in means of the large sample and the smaller test sample is $60 - 58 = 2$. This difference is $2/0.8 = 2.5$ standard errors. Because we expect 95% of random samples from the same population to have a mean within two standard errors and the observed mean is more deviant than this, we conclude that the sample population is significantly smaller in size than the rest of the species. The actual probability of getting a mean this small (or smaller) by chance is 0.0062.

5. This is an example of the central limit theorem in action. Adding 12 uniform random numbers gives a resulting sum whose distribution is approximately normal.

6. Positive covariance would mean that successive numbers are more similar than they would be had the numbers been independent. This would result in a smaller variance than expected under the central limit theorem.

7. This is most likely an X-linked trait, a hypothesis that can be easily tested if females and sets of their offspring can be examined. The males seem to have 80% F and 20% S alleles, and, if you count up the number of F and S alleles in the females, they too have 80% F alleles and 20% S alleles.

8. Heterozygotes would produce both F and S polypeptide chains, which could be assembled into dimers that are FF, FS, and SS. These three molecules would most likely have different electrophoretic mobility, giving three bands on a protein gel. In the case of tetramers, the heterozygote could produce FFFF, FFFS, FFSS, FSSS, and SSSS. These may or may not all be resolvable on a protein gel, but in the best case, all five bands would be visible.

9. The polymerase chain reaction doubles the amount of a specific fragment of DNA each round (assuming perfect efficiency). This means that 30 rounds of PCR will yield $2^{30} = 1.074 \times 10^9$ copies. In reality PCR is rarely more than 80% efficient.

10. If one starts with a large number of template molecules of DNA, the PCR reaction begins by amplifying many of them. The first round of PCR will have errors, but there will be many different errors, and all will be rare. Subsequent rounds of PCR will have additional errors, but at the end of many rounds of PCR, each individual error will remain rare. Sequencing this mix will give the most common apparent base at each position, and this should be the true base. One can have problems sequencing after PCR reactions if the initial concentration is very low (only a few molecules of template DNA), or if one clones a PCR product. Generally, if PCR products are cloned, careful investigators insist on sequencing more than one to be sure PCR artifacts have not been introduced.

11. Levels of polymorphism at the nucleotide level are often low. In humans, for example, two sequences will differ at an average of around 0.1% of the bases, or around 1 per 1000. Any method of calling variation that has a false positive rate of even one percent could lead to serious misinterpretation of amounts of variation in the population.

12. The average of the pairwise differences is 5.0, so the average heterozygosity per site is $5/1200 = 0.00417$. This per-site heterozygosity is the same as the nucleotide diversity, π.

13. If the DNA types do not match, one can be certain that the sample from the crime scene did not come from the suspect (barring lab errors). This situation is known as exclusion. Such information has resulted in the release of many wrongly accused people. On the other hand, if the DNA types do match, then either the suspect is the source of the blood from the crime scene, or someone else with the same genotype at the scored loci was the source of the blood. The chance that the suspect is the one now depends on how often one gets such a match by chance. Thus the whole problem of matching hinges on population genetic issues.

CHAPTER 3

1. Allele frequency $q = (1/10{,}000)^{1/2} = 0.01$, and $2pq = 2(0.99)(0.01) = 0.0198$ is the frequency of carriers (i.e., about one person in 50).

2. Allele frequency of $d = (170/400)^{1/2} = 0.65$, of $D = 0.35$. Expected genotype frequencies of DD, Dd, and dd are 0.12, 0.46, and 0.42. Fraction of heterozygotes among Rh^+ is $0.46/(0.12 + 0.46) = 0.79$, and expected number of heterozygotes is $0.79 \times 230 = 182$.

3. Allele frequency of $M = (2202 + 1496)/6200 = 0.60$, of $N = 0.40$. Expected numbers are $(0.60)^2(3100) = 1116$, $2(0.60)(0.40)(3100) = 1488$, and $(0.40)^2(3100) = 496$. The $\chi^2 = (1101 - 1116)^2/1116 + (1496 - 1488)^2/1488 + (503 - 496)^2/496 = 0.34$ with one degree of freedom (because there are three classes of data and one parameter estimated from the data), with an associated probability value of about 0.65.

4. Expand $(0.1A_1 + 0.2A_2 + 0.3A_3 + 0.4A_4)^2$ to obtain

$A_1A_1\ (0.1)^2 = 0.01$;

$A_1A_2\ 2(0.1)(0.2) = 0.04$;

$A_1A_3\ 2(0.1)(0.3) = 0.06$;

$A_1A_4\ 2(0.1)(0.4) = 0.08$;

$A_2A_2\ (0.2)^2 = 0.04$;

$A_2A_3\ 2(0.2)(0.3) = 0.12$;

$A_2A_4\ 2(0.2)(0.4) = 0.16$;

$A_3A_3\ (0.3)^2 = 0.09$;

A_3A_4 $2(0.3)(0.4) = 0.24$;

A_4A_4 $(0.4)^2 = 0.16$.

5. If heterozygous parental genotype is Aa, half the gametes contain A and half contain a. The A gamete yields a heterozygous offspring when it unites with an a gamete (probability q) and the a gamete yields a heterozygous offspring when it unites with an A gamete (probability p). Overall probability of heterozygous offspring is $(1/2)q + (1/2)p = (1/2)(q + p) = 1/2$.

6. If the dominant and recessive allele frequencies are p and q, then $D = p^2$, $H = 2pq$, and $R = q^2$, so $DR = p^2q^2$ and $H = (2pq)^2$, confirming $DR = H^2/4$.

7. The frequency of $AA = p^2 = (1 - q)^2 = 1 - 2q + q^2 \cong 1 - 2q$; frequency of Aa heterozygotes $= 2pq = 2(1 - q)q = 2q - 2q^2 \cong 2q$; and frequency of $aa = q^2 \cong 0$.

8. The probability that an individual with the dominant phenotype is heterozygous is $2pq/(p^2 + 2pq) = 2q/(1 + q)$, and among these individuals, the recessive allele frequency is $1/2$; hence, the frequency of recessive allele among dominant phenotypes is $[2q/(1 + q)](1/2) = q/(1 + q)$.

9. Allele frequency $q = 0.05$. Expected frequency of homozygous (color blind) females $= (0.05)^2 = 0.0025$ (about $1/400$); expected frequency of carriers $= 2(0.95)(0.05) = 0.095$ (about $1/10$).

10. Allele frequency $q =$ frequency of affected males. Frequency of carrier females $= 2pq = 2(1 - q)q = 2q - 2q^2 \cong 2q$ when $q^2 \cong 0$. Therefore, the frequency of carrier females is approximately two times the frequency of affected males.

11. Genotype frequencies given by expansion of $(pA + qa)^4$ instead of $(pA + qa)^2$.

12. By the definition in the text, a polymorphic gene is one for which the most common allele has a frequency less than 0.95, making genes 1, 2, and 5 polymorphic in this sample, and giving $P = 3/5 = 60$ percent. Heterozygosity with random mating equals $1 - \Sigma p_i^2$, which for genes 3–5 equals 0.47, 0.11, 0.01, 0, and 0.37, with an average of $H = 0.19$.

13. Normal allele is dominant because of phenotype in F_1. The χ^2 value equals $(88 - 93.75)^2/93.75 + (37 - 31.25)^2/31.25 = 1.41$ with one degree of freedom (since no parameters were estimated from the data). The associated probability level is approximately 0.25, so the hypothesis of a 3:1 ratio cannot be rejected.

14. Expected ratio obtained from the expansion of $[(3/4)D + (1/4)R]^3$, where D and R represent the dominant and recessive phenotypes, respectively, yielding the ratio $27 : 9 : 9 : 9 : 3 : 3 : 3 : 1$ for each 64 progeny. Expectations are $269.6 : 89.9 : 89.9 : 89.9 : 30.0 : 30.0 : 30.0 : 10.0$ among 639 progeny, for a χ^2 of 2.67 with seven degrees of freedom (because there are eight classes of data and no parameters estimated), for which the associated probability value is about 0.92. This is a very good fit, indeed.

15. A_1B_1 (0.3)(0.2) = 0.06;
    A_1B_2 (0.3)(0.3) = 0.09;
    A_1B_3 (0.3)(0.5) = 0.15;
    A_2B_1 (0.7)(0.2) = 0.14;
    A_2B_2 (0.7)(0.3) = 0.21;
    A_2B_2 (0.7)(0.5) = 0.35.

16. (a) Assuming linkage equilibrium, frequencies of gametes A_1B_1, A_1B_2, A_2B_1, and A_2B_2 are p_1q_1, p_1q_2, p_2q_1, and p_2q_2, or (0.7)(0.3) = 0.21, (0.7)(0.7) = 0.49, (0.3)(0.3) = 0.09, (0.3)(0.7) = 0.21, respectively. (b) Theoretical maximum of D equals the smaller of p_1q_2 (0.49) and p_2q_1 (0.09), or 0.09, and 50 percent of 0.09 gives D = 0.045. Gametic frequencies, given in the same order as in (a), are 0.21 + 0.045 = 0.255, 0.49 − 0.045 = 0.445, 0.09 − 0.045 = 0.045, and 0.21 + 0.045 = 0.255.

17. When dominant × dominant matings occur at random, the proportion of homozygous recessive offspring equals the square of the recessive allele frequency among individuals with the dominant phenotype, or $[q/(1 + q)]^2$, because random mating of individuals is equivalent to random union of gametes, just as in the derivation of the Hardy-Weinberg principle. With dominant × recessive matings, the probability of a recessive gamete from the parent with the dominant phenotype is $q/(1 + q)$, and from the parent with the recessive phenotype it is 1; altogether the probability of a homozygous recessive offspring from a dominant × recessive mating is $[q/(1 + q)](1) = q/(1 + q)$, and likewise from a recessive × dominant mating.

CHAPTER 4

1. Average frequency before fusion = $[(q + \varepsilon)^2 + (q − \varepsilon)^2]/2 = q^2 + \varepsilon^2$; after fusion = q^2; difference = ε^2 = variance in allele frequency among subpopulations.

2. Multiply $(1 − F_{IS})(1 − F_{ST}) = 1 − F_{IT}$ and cancel the 1s. The expression says that the probability of autozygosity within the total population equals the probability of autozygosity because of inbreeding within a subpopulation, plus the probability of autozygosity because of random genetic drift, minus the probability of autozygosity for both reasons.

3. Heterozygosities are 0.54, 0.62, 0.66, respectively, average 0.61. Fused population has allele frequencies 0.2, 0.3, 0.5, and heterozygosity 0.62. F_{ST} = 0.02. Maximum F_{ST} occurs when each population is fixed for a different allele, and F_{ST} = 1 for the set of three as well as for each pairwise comparison.

4. Allele frequency is 0.2 in both subpopulations, giving H_S = 0.32. Average H_I = 0.272. F_{IS} = 0.15. (Note from the genotype frequencies that the populations have inbreeding coefficients 0.1 and 0.2, respectively, which

average 0.15). Since the allele frequencies are identical, $H_T = 0.32$ also, so $F_{ST} = 0$. Since $(1 - F_{IT}) = (1 - F_{IS})(1 - F_{ST})$, $F_{IT} = 0.15$.

5. Gametic frequencies are $p_1q_1 + \varepsilon$, $p_1q_2 - \varepsilon$, $p_2q_1 - \varepsilon$, $p_2q_2 + \varepsilon$ in one population, $p_1q_1 - \varepsilon$, $p_1q_2 + \varepsilon$, $p_2q_1 + \varepsilon$, $p_2q_2 - \varepsilon$ in the other; D in fused population equals 0.

6. $p^2 - p^2F + pF = p^2 + p(1-p)F = p^2 + pqF = p(1-q) + pqF = p - pq + pqF = p - pq(1-F)$.

7. Random mating frequencies $\frac{1}{4}$, $\frac{1}{2}$, $\frac{1}{4}$; among offspring of first cousins ($F = \frac{1}{16}$), they are $\frac{17}{64}$, $\frac{30}{64}$, $\frac{17}{64}$. Deficiency of heterozygotes $= (\frac{32}{64} - \frac{30}{64})/(\frac{32}{64}) = \frac{1}{16} = F$.

8. $q = (1/1600)^{1/2} = \frac{1}{40}$, and expected frequency from first-cousin matings is $(1/1600)(\frac{15}{16}) + (\frac{1}{40})(\frac{1}{16}) = 1/465$.

9. Proportions with first-cousin parents are 0.016, 0.022, 0.068, 0.120, and 0.391 for the frequencies given. When $q = 1$, the proportion is 0.01, which says that one percent of the matings are between first cousins.

10. With inbreeding F, mean allele frequency equals $[p^2(1 - F) + pF](1) + [2pq(1 - F)](\frac{1}{2}) = p$. Mean square $MS = [p^2(1 - F) + pF](1) + (2pq(1 - F)](\frac{1}{2})^2$. Variance equals $MS - p^2 = (pq/2)(1 + F)$. When $F = 0$ (random mating), variance $= pq/2$. When $F = 1$ (complete inbreeding), variance $= pq$.

11. $H_1 = 0.62$, $H_2 = 0.48$, $H_S = 0.55$, $H_T = 0.615$, and $F_{ST} = 0.106$. $J_S = [(0.04 + 0.36)(2 + (0.09 + 0)/2 + (0.25 + 0.16)/2] = 0.45$, $J_T = [(0.4)^2 + (0.15)^2 + (0.45)^2] = 0.385$, and $G_{ST} = 0.106$. Note that $J_S = 1 - H_S$; $J_T = 1 - H_T$; so $G_{ST} = F_{ST}$.

12. $F_{ST(1)} = (0.48 - 0.40)/0.48 = 0.167$; $F_{ST(2)} = (0.255 - 0.21)/0.255 = 0.176$; $F_{ST(3)} = (0.495 - 0.49)/0.495 = 0.010$. Average frequencies are $p_1 = 0.4$, $p_2 = 0.15$, $p_3 = 0.45$, and the weighted average $F_{ST} = 0.106$, which equals the G_{ST} calculated in the preceding problem.

13. Because a male contributes the Y chromosome to his sons, paths with two or more consecutive males have probability 0 for the transmission of an X-linked gene.

14. Equilibrium $F = 0.2/(2 - 0.2) = \frac{1}{9}$. Genotype frequencies are $(\frac{1}{9})(\frac{8}{9}) + (\frac{1}{3})(\frac{1}{9})$, $(\frac{4}{9})(\frac{8}{9})$, $(\frac{4}{9})(\frac{8}{9}) + (\frac{2}{3})(\frac{1}{9})$, or 0.135, 0.395, 0.469.

15. $F_A = F_B = 0$; $F_C = F_D = 0$; $F_E = F_F = 2(\frac{1}{2})^3 = \frac{1}{4}$; $F_G = F_H = 2(\frac{1}{2})^3 + 4(\frac{1}{2})^5 = \frac{3}{8}$; $F_I = (\frac{1}{2})^3(1 + F_E) + (\frac{1}{2})^3(1 + F_F) + 4(\frac{1}{2})^5 + 8(\frac{1}{2})^7 = \frac{8}{16}$. Note that the pedigree is of three generations of sib mating.

16. $F = \frac{1}{2}$ in odd-numbered generations (the progeny of selfing), and $F = 0$ in even-numbered generations (the progeny of random mating).

17. $F_0 = 0$, $F_1 = 1$, $F_2 = \frac{1}{4}$, $F_3 = \frac{3}{8}$, $F_4 = \frac{8}{16}$, $F_5 = \frac{19}{32} = 0.59$, so genotype frequencies after five generations are 0.21, 0.17, 0.61; one additional gener-

ation of random mating restores the Hardy-Weinberg frequencies of 0.09, 0.42, 0.49.

18. (a) $F_{ST} = 1 - (1 - 1/100)^{50} = 0.39$; $F_{IS} = \frac{1}{4}$; $F_{IT} = 1 - (0.75)(0.61) = 0.54$. (The exponent in F_{ST} is 50, not 47, because random drift still occurs during the generations of sib mating.) (b) $F_{ST} = 0.39$; $F_{IS} = 0$; $F_{IT} = 0.39$.

19. A_1A_2, A_2A_3, A_1A_3 styles occur in equal frequency at equilibrium, so the probability of any pollen grain landing on a compatible style is $\frac{1}{3}$.

20. Two-way hybrid has genotype A_1A_2 and gametes $\frac{1}{2} A_1 + \frac{1}{2} A_2$, so the probability of identity by descent (F) among offspring is $(\frac{1}{2})^2 + (\frac{1}{2})^2 = \frac{1}{2}$; three-way hybrid has gametes $\frac{1}{4} A_1$, $\frac{1}{4} A_2$, $\frac{1}{2} A_3$, and $F = (\frac{1}{4})^2 + (\frac{1}{4})^2 + (\frac{1}{2})^2 = \frac{3}{8}$; four-way hybrid has gametes $\frac{1}{4} A_1$, $\frac{1}{4} A_2$, $\frac{1}{4} A_3$, $\frac{1}{4} A_4$, and $F = 4(\frac{1}{4})^2 = \frac{1}{4}$.

21. $F_t = (\frac{1}{2})^2(1 + F_{t-2}) + (\frac{1}{2})^3(1 + F_{t-3}) + (\frac{1}{2})^4(1 + F_{t-4}) + \ldots = (\frac{1}{2})^2(1 + F_{t-2}) + (\frac{1}{2})F_{t-1} = \frac{1}{4} + (\frac{1}{2})F_{t-1} + (\frac{1}{4})F_{t-2}$. Therefore, $F_0 = 0$, $F_1 = 0$, $F_2 = \frac{1}{4}$, $F_3 = \frac{3}{8}$, $F_4 = \frac{8}{16}$, $F_5 = \frac{27}{32}$.

22. $F_t = (\frac{1}{2})^2(1) + (\frac{1}{2})^3(1) + (\frac{1}{2})^4(1) + \ldots = (\frac{1}{2})^2 + (\frac{1}{2})F_{t-1}$; therefore, $F_0 = 0$, $F_1 = \frac{1}{4}$, $F_2 = \frac{3}{8}$, $F_3 = 7/16$, $F_4 = \frac{15}{32}$, $F_5 = \frac{31}{64}$. The equilibrium is given by $F = (\frac{1}{4}) + F/2$, or $F = \frac{1}{2}$.

CHAPTER 5

1. Various kinds of mutations anywhere along the gene can result in loss of gene function; once a gene has mutated, only very specific kinds of reverse mutations will restore function.

2. The experiment proved the point because the antibiotic-resistant cells in the colonies on the unselective plate were never themselves exposed to the antibiotic.

3. $P_0 = 11/20 = 0.55$ and $m = -[\ln(P_0)]/N = 1.1 \times 10^{-9}$ per generation.

4. Use the Poisson zero term $P_0 = \exp(-m)$ with $P_0 = 1 - 0.35 = 0.65$, giving $m = -\ln P_0 = 0.43$ as the average number of lethals per chromosome.

5. Dominant lethals $(2 \text{ to } 10 \times 10^{-2})/(5 \times 10^{-4}) = 40$ to 200 rads; recessive visibles $(8 \times 10^{-6})/(7 \times 10^{-8}) = 114$ rads; reciprocal translocations $(2 \text{ to } 5 \times 10^{-4})/(1 \text{ to } 2 \times 10^{-5}) = 10 - 50$ rads. (Incidentally, humans appear to be somewhat less radiation sensitive than mice.)

6. With $m = 0$, $p_t = p_0(1 - \mu)^t$. For 10, 100, 1000, and 10,000 generations, p_t equals 0.99995, 0.9995, 0.995, and 0.95, respectively. Note that the approximation $p_t = 1 - \mu t$ is very accurate in this case.

7. Use $p_t = p_0(1 - \mu)^t$ with $p_0 = 1$, $\mu = 0.01$, and $p_t = 0.90$. Then $t = \ln(0.90)/\ln(0.99) = 10.5$ generations.

8. Use $q_t = q_0 + \mu t$. For $t = 0$ to 12 generations, each four-generation interval increases q by 2×10^{-6}, so $\mu = (2 \times 10^{-6})/4 = 5 \times 10^{-7}$ per generation. (b) For

$t = 12$ to 24, each interval increases q by 0.04×10^{-6}, so $\mu = (0.04 \times 10^{-6})/4 = 10^{-8}$ per generation. The novel metabolite decreases the mutation rate; such substances are called antimutagens.

9. (a) Equilibrium $p = \mu/(\mu + v) = \frac{1}{11}$; (b) $\frac{1}{101}$; (c) $\frac{1}{2}$; (d) $\frac{1}{11}$.

10. The equation for reversible mutation implies that $(p_t - p)/(p_0 - p) = (1 - \mu - v)^t$, and halfway to equilibrium means that $p_t - p = (\frac{1}{2})(p_0 - p)$, so $(1 - \mu - v)^t$ equals $\frac{1}{2}$. Therefore $t \ln(1 - \mu - v) = \ln(\frac{1}{2})$ or, using the approximation, $t \cong \ln2/(\mu + v) \cong 0.7/(\mu + v)$. For the values of μ and v specified, $t \cong 6.4 \times 10^4$ generations.

11. $q_1 = q_0 + \mu_0$; $q_2 = q_1 + \mu_1 = q_0 + \mu_0 + \mu_1$; and so on to obtain $q_t = q_0 + \Sigma \mu_i$. Can write this as $q_0 + \mu t$ if μ is interpreted as the arithmetic mean of the mutation rates $(\Sigma\mu_i)/t$.

12. The effective number of alleles equals the reciprocal of homozygosity (sums of squares of allele frequencies). These examples have $n_e = 2.8$ and 8.

13. Rare alleles contribute negligibly to the homozygosity; hence the effective number of alleles is essentially independent of the number of rare ones.

14. Use the relation $F = 1/[4N(m + \mu) + 1]$ with $N = 50$, $\mu = 10^{-5}$ and $m = 10^{-3}$ to obtain $F = 0.83$, so $H = 1 - F = 0.17$.

15. The expected number of neutral alleles in a sample of size two equals $1 + x$ when $\theta = x/(1 + x)$. Since $H = \theta/(1 + \theta)$, the heterozygosity is $H = x$.

16. Use $F_t = 1 - [1 - 1/(2N)]^t$ with $N = 50$ and $t = 200$, so $F_{200} = 0.866$.

17. Use $F_t = 1 - [1 - 1/(2N)]^t \cong 1 - \exp(-t/2N)$ with $F_t = \frac{1}{4}$. Thus $t \cong -2N\ln(\frac{3}{4}) \cong 0.6N$ generations.

18. Use the equation $p_t = \bar{p} + (p_0 - \bar{p})(1 - m)^t$ with $p_t = 0.5$, $p_0 = 0.2$, $\bar{p} = 0.8$ and $m = 0.01$. Then $t = 69$ generations.

19. $(1 - m)^{10} = 0.6$, so allele frequencies are 0.32, 0.44, 0.56, and 0.68.

20. Using $p_t - p = (p_0 - p)(1 - m)^t$, note that $\Sigma(p_t - p)^2 = \Sigma(p_0 - p)^2(1 - m)^{2t}$, or $s_t = s_0(1 - m)^2t$.

21. Use $F = 1/(4Nm + 1)$ with $1/(4Nm + 1) < 0.05$, so $m > 4.75/N$.

CHAPTER 6

1. $p_1 = p_0/(1 - q_0s_0)$; $q_1 = q_0s_0/(1 - q_0s_0)$; and so on give $(p_1/q_1) = (p_0/q_0)s_0$; $p_2/q_2 = (p_0/q_0)(s_0s_1)$; . . . ; $p_n/q_n = (p_0/q_0)(s_0s_1 \cdots s_{n-1})$. To use the formula with s, set $s = (s_0s_1 \cdots s_{n-1})^{1/n}$, so s is the geometric mean of the selection coefficients.

2. Use the general equation for two allele viability selection $p' = (p^2w_{11} + pqw_{12})/\bar{w}$, to get $p_1 = 0.736$, $p_2 = 0.768$, and $p_3 = 0.796$.

3. Defining the fitnesses as $1 - s : 1 : 1 - t$, the equilibrium allele frequency is $\hat{p} = t/(s + t)$, where (a) $s = 0.7$, $t = 0.3$; (b) $s = 0.07$, $t = 0.03$; (c) $s = 0.007$, $t = 0.003$. In all cases $p = 0.3$.

4. $\bar{w} = (0.64)(0.9) + 2(0.8)(0.2)(1) + (0.04)(0.6) = 0.92$. Note that $0.8 = 0.4/(0.4 + 0.1)$, so 0.8 is the equilibrium frequency for this case of overdominance, and in this model the equilibrium occurs at maximum $\bar{w}$.

5. $q' = pq(1 - h)/[p^2 + 2pq(1 - h)] = q(1 - h)/(1 - qh) \cong q(1 - h)$ with $q'/q = 0.99$ given, so $h = 0.01$.

6. $t = 4[-2.2 + 11.5] = 37.2$ generations. Note that the equation can be written as $t = (2/s)[\ln(p_t/p_0) - \ln(q_t/q_0)]$, and with $p_t/p_0 << 1$, $\ln(q_t/q_0) \cong 0$. For a large range of values of p_t/p_0 and s, t ranges from 5 to 50. This explains in part why significant pesticide resistance evolves so rapidly in diverse species.

7. For the haploid, $p' = p/(1 - qs)$. For the diploid, $p' = [p^2 + pq(1 - s)]/[p^2 + 2pq(1 - s) + q^2(1 - s)^2] = p[p + q(1 - s)]/[p + q(1 - s)]^2 = p/[p + q(1 - s)] = p/(1 - qs)$.

8. Interchange p and q and change the sign of s to obtain $\ln(q_t/p_t) + 1/p_t = [\ln(q_0/p_0) + 1/p_0] - st$, or $\ln(p_t/q_t) - 1/p_t = [\ln(p_0/q_0) - 1/p_0] + st$, where p_t is now the allele frequency of the favored recessive and s is the selection coefficient against individuals with the dominant allele.

9. $d\Delta p/dp = (\frac{1}{2} - p)(1 - p) - p(\frac{1}{2} - p) - p(1 - p)$. This value is positive for $p = 0$ and $p = 1$ (indicating unstable equilibria) and negative for $p = \frac{1}{2}$ (indicating local stability). The latter point is also globally stable.

10. For a recessive lethal, $w_{11} = w_{12} = 1$, $w_{22} = 0$. Then $q' = pq/(p^2 + pq) = q/(1 + q)$, or $1/q_n = 1 + 1/q_{n-1}$, which implies $1/q_n = n + 1/q_0$, or $q_n = q_0/(1 + nq_0)$. For $q_n = q_0/2$, $n = 1/q_0$ generations.

11. Into the equation $\hat{q} = \mu/hs$ substitute $hs = \frac{1}{2}$, hence $q = 18 \times 10^{-5}$; the expected frequency of affected individuals is $2pq = 36 \times 10^{-5}$.

12. From the equation $\hat{q} = \sqrt{(\mu/s)} = (4 \times 10^{-6}/0.2) = 4.5 \times 10^{-3}$, thus, $q = 4 \times 10^{-6}/(0.05)(0.2) = 4.0 \times 10^{-4}$. The five percent heterozygous effect reduces the equilibrium frequency by more than an order of magnitude.

13. From the equation $\hat{q} = \sqrt{(\mu/s)} = \sqrt{(10^{-6}/0.6)} = 1.3 \times 10^{-3}$; this would be 1.0×10^{-3} if homozygotes did not reproduce.

14. Frequency of heterozygotes $\cong 2q$, frequency of heterozygotes from new mutations $\cong 2\mu$, ratio $= \mu/q$, and since $q = \mu/h$, the desired ratio is h.

15. The given condition is $(1/0.9) + (v_{12}/1) > 2$, or $v_{12} > 0.89$.

16. Substituting into the given expression for equilibrium, $w_{11} = 1$, $w_{12} = 1$, $v_1 = 1 - s$, $v_2 = 1$, we get $\hat{p} = 1 - s$ is the frequency of A, so $\hat{q} = s$ is the frequency of a.

17. Set $p = 1 - 0.125 = 0.875$ and $k = 0.75$. Then w_{12} satisfies $-2(0.25)w_{12}/(1 - 2w_{12}) = 0.875$, or $w_{12} = 0.7$. Actually, segregation distorter is active only in

males, and a model that takes this into account yields $w_{12} = 0.79$ in males and $w_{12} = 1$ in females (Hartl 1970).

18. The reason is that homozygotes for some alleles may be superior to heterozygotes containing other alleles.

19. The equilibrium allele frequency for this overdominant case is: $\hat{p} = (w_{12} - w_{22})/(2w_{12} - w_{11} - w_{22}) = 0.25$. The mean fitness at this point is obtained by substituting into $\bar{w} = p^2 w_{11} + 2pq w_{12} + q^2 w_{22} = 0.909$. The marginal fitness of the A'' allele is 0.8, because all genotypes bearing this allele have a fitness of 0.8. Because the marginal fitness of the A'' allele is less than the mean fitness, A'' does not increase in frequency.

20. Fitnesses are $A_1 A_2$, 0.7; $A_1 A_3$, 0.6; $A_1 A_4$, 0.5; $A_2 A_3$, 0.5; $A_2 A_4$, 0.4; $A_3 A_4$, 0.3; and $p_i = \frac{1}{4}$; $w = (\Sigma A_i A_i$ fitnesses $+ 2\Sigma A_i A_j$ fitnesses$) = 0.5$.

CHAPTER 7

1. The expected heterozygosity in a finite population decreases according to $H_t = (1 - 1/2N)H_{t-1}$. In this case, we expect H to go from 0.5 to 0.495 in one generation. Since H fell considerably below this, it would appear that there was substantial inbreeding. A χ^2 test could be done to formally test the statistical significance.

2. Since $H_t = H_0\, e^{-(t/2N)}$, we can substitute and take logs to get $\ln(0.05) = -t/[2(20)]$. Solving, we get $t = 59.9$ generations. For a population that is 10 times the size, the time to decrease to 5% of the original heterozygosity is 10 times as long, or 599 generations.

3. For an autosomal gene, the 24 cats represent 48 copies of the gene, so the probability of ultimate fixation is $\frac{1}{48}$ and the probability of ultimate loss is $\frac{47}{48}$. For an X-linked gene, the 24 cats represent 36 copies of the gene, so the probability of ultimate fixation is $\frac{1}{36}$ and that of ultimate loss $\frac{35}{36}$.

4. The equation, $\ln(\frac{1}{2}) = -t/2N$, from the boxed problem 7.6, implies that $t = 1.39N$ generations will reduce the average heterozygosity by half. Because the plant is an annual, one generation corresponds to one year, so in this case $t = 50$, giving $N = 36$ as the effective population size.

5. Using Equation 7.12 with $N = 20$ and $t = 8$ yields 0.18. Since self-fertilization cannot occur, use of $N = 20.5$ would be a bit more accurate, but this gives essentially the same answer. The value $F_{ST} = 0.18$ is not much less than the inbreeding coefficient of 0.25 resulting from one generation of brother-sister mating.

6. Dividing and taking logs of both sides of Equation 7.11 yields

$t = [\ln(1 - F_t) - \ln(1 - F_0)]/\ln[1 - (1/2N)]$

From $F = 0.01$ to 0.02 requires $N = 1$ generation, from 0.05 to 0.10 requires 5.4 generations. Using the suggested approximations yields

$-F_t = -t/(2N) - F_0$, and when $F_t = 2F_0$, then $t = 2NF_0$. For $F_0 = 0.01$ or 0.05, the approximation gives $t = 1$ and 5 generations, respectively.

7. $N_f = 100$, and the number of breeding males equals $N_f/5 = 40$. In this situation $N_e = 4N_mN_f/(N_m + N_f) = 4(40)(200)/(40 + 200) = 133$.

8. Here $N_e = 4N_mN_f/(N_m + N_f)$. For 10 cows and one bull, $N_e = 4(1)(10)/(1 + 10) = 3.6$; for 40 cows and one bull, $N_e = 3.9$; for 10 cows and two bulls, $N_e = 6.7$.

9. The appropriate formula is Equation 7.25, which for $N_f = 100$ and $N_m = 10$ yields 37.5 and for $N_f = 10$ and $N_m = 100$ yields 21.4.

10. In the F_1 population, all restriction site differences are heterozygous so that the initial RFLP allele frequencies are $\frac{1}{2}$. Equation 7.15 can be applied to determine the expected fraction of sites that remain heterozygous. With $N = 80$ and $H_0 = 1.00$, for $t = 10$, $H_t = \exp(-10/160) = 0.94 \times 100 = 94$ sites segregating. For $t = 50$, the same approach gives 73 sites segregating.

11. Using Equation 7.38 with $N_e = 50$ and $t = 100$ gives $(1/100)(99/100)^{100} = 0.0037$.

12. Applying Equation 7.34, the expected number of sites segregating in samples of size 10, 20, and 50 are 29.3, 36.0, and 45.0, respectively.

13. Use equation 7.29 to get $F = 1/(1 + 4N\mu) = 1/(1 + 10^6 \times 10^{-6}) = 0.5$. Because $H = 1 - F$, the heterozygosity is also 0.5. For the second part, use the equation in boxed problem 7.9, $F = 1/[4N(\mu + m) + 1]$, and substitute to get $\frac{1}{3} = 1/[10^6(10^{-6} + m) + 1]$, so $m = 10^{-6}$. Very little migration is needed (one migrant every four generations) is needed to make the change in H.

14. The distribution of expected coalescence times is given by Equation 7.39, and the mean is $4N/[k(k-1)]$, where k is the sample size. Substituting, we get $200/[k(k-1)] = 10$, so $k = 5$.

15. Use Equation 7.15 with $H_t = H_0/x$, which yields $(1/x) = \exp(-t/2N)$, or $\ln(1/x) = -t/2N$, which implies that $t = 2N\ln x$. For $x = 2$, the value $t = 1.39N$, agreeing with the value stated in the text.

16. Substitute $H_t = H_0/e$ into Equation 7.15 to obtain $\exp(-1) = \exp(-t/2N)$, whence $t = 2N$.

17. The variance in frequency caused by sampling each cell generation is pq/N. Assuming the stem cells had 50% of each mtDNA type, this is $1/4000$. The 95% confidence interval in the allele frequency after one generation of sampling is $\pm$ two standard deviations, or $+ 0.0316$. In 30 cell generations, it is therefore not unlikely to drift down to the observed frequency of 0.2. Diffusion methods or computer simulation could be used to obtain the probability of the observed result.

CHAPTER 8

1. Apply the Jukes-Cantor formula (Equation 8.15), to get $k = 0.0517$ when $d = 0.05$, and $k = 2.03$ when $d = 0.7$. Note that when there are few differences, k and d are nearly equal. When the proportion of sites that differ nears the limiting value of 0.75, there can be more than one expected substitution per site. Inspection of Equation 8.16 shows that the variance in the estimate of k increases as d increases.

2. The analog to Equation 8.15 if there were six different nucleotides is $k = -(5/6) \ln(1 - 6d/5)$. Substitute $d = 0.2$ and get $k = 0.229$. Note that for four-base DNA, application of Equation 8.15 gives $k = 0.233$. The compensation for potentially missed back mutations is greater for four-base than for six-base DNA.

3. If you assume that most differences are synonymous (third position), then reading frame begins with position 1 at the far left. The sequences code for 20 amino acids, of which two are different. Altogether there are 60 nucleotides (13 different), 40 nonsynonymous sites (two different), 20 synonymous sites (11 different).

   a. Use Equation 8.5 with $D = 2/20$, $\hat{K} = 0.105$.
   b. Use Equation 8.15 with $d = 13/60$, $\hat{k} = 0.256$.
   c. Use Equation 8.15 with $d = 2/40$, $\hat{k} = 0.052$.
   d. Use Equation 8.15 with $d = 11/20$, $\hat{k} = 0.991$.

4. With $\hat{D} = 0.33$ in Equation 7.21, $\hat{k} = 0.43$, and $t = 0.43/(2 \times 0.01/\text{yr}) = 21.5$ yr. The date of divergence is approximately $1983 - 21.5 = 1961$ or 1962.

5. Values of $\hat{k}$ from Equation 8.15 are

	HIV2	VISNA	MMLV
HIV1	0.45	0.95	1.31
HIV2		0.89	1.37
VISNA			1.37

Data indicate that HIV1 and HIV2 are most closely related to each other, more distantly to VISNA, and still more distantly to MMLV. HIV1, HIV2, and VISNA are all about equally distant from VISNA.

6. The selective constraints are probably weak, as 5×10^{-9} approximates the rate of synonymous substitution.

7. Most likely the protein is undergoing very rapid amino acid replacement because of natural selection.

8. Compensatory substitutions might be expected, i.e., an $A \rightarrow G$ substitution at a site matched with a $T \rightarrow C$ substitution at the position with which it pairs. This is the pattern actually observed.

9. $d = \frac{3}{4}$, which means that the nucleotide differences are as great as expected in comparing two random sequences.

10. $0.005 \times 10^{-6} = 5 \times 10^{-9}$ substitutions along a lineage. The rate of divergence between two lineages is twice this value, or 1% per million years.

11. The proportion of synonymous sites that substituted is $34/310 = 0.11$, and the proportion of nonsynonymous sites that substituted is $16/633 = 0.025$. Using these as the observed divergence values (d) in Equation 8.15, the estimates of the number of substitutions per synonymous and nonsynonymous site are 0.118 and 0.026, respectively.

12. (a) $3N/4$; (b) $N/4$; (c) $N/4$.

13. The level of intrapopulation variation would differ between the two scenarios. With a high rate of substitution, there would be more intrapopulation variation, and less interpopulation differentiation than in the case with low rates of substitution and migration. It is very difficult to obtain reliable estimates of the neutral mutation rate and population size because the models nearly always confound these factors. Mutation and migration are similarly confounded in models of subdivided populations. The neutral mutation rate can be estimated if one is willing to compare to different species and date the divergence with the fossil record.

14. The digits on the branches of the tree indicate the site position along the sequence where each substitution occurred:

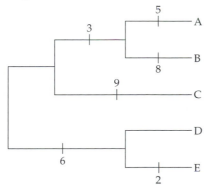

15. (a) Set $4N = 10 \times 2\ln(2N)$ and iterate. Convergence is very rapid for any starting value, and $N = 17.9$; (b) $N = 323.6$.

16. Set $f = c_2$ and solve for $\lambda = (n-1)/4N$. Then $c_1 = c_2 = 1/(1 + 4N\mu] \cong 1$ because $4N\mu \ll 1$.

17. $c_1 = c_2 = 1/n$; $f \cong 1$.

18. When $t = $ half life, $e^{-ht} = \frac{1}{2}$, so $t = -\ln(\frac{1}{2})/h$.

19. Distribution of persistence times is a standard exponential distribution with parameter h, $P(t) = h \exp(-ht)$, and the mean of an exponential is the reciprocal of the exponential parameter.

CHAPTER 9

1. Heritability is a quantitative measure of variance partitioning. It says nothing about the ability of a trait to be altered by environmental effects. High heritability does not mean that a trait cannot be modified by medical intervention or education. This point is especially important in interpreting heritability of abilities and behaviors in humans.

2. $\hat{\mu} = 2206/30 = 73.5$, $\hat{s}^2 = [(166,206/30) - (2206/30)^2](30/29) = 137.64$, $\hat{s} = 11.7$.

3. The range 18–22 is one standard deviation, so a proportion $1 - 0.317 = 0.683$. (b) > 22 is greater than one standard deviation, so proportion $= 0.317/2 = 0.158$. (c) > 24 proportion $= 0.046/2 = 0.023$. (d) The range is 20–22, proportion $= 0.5 - 0.158 = 0.342$. (e) < 16 proportion $= 0.046/2 = 0.023$.

4. $s^2 = 1.5$, $s^2 = 6.0 - 1.5 = 4.5$, and the broad sense heritability $= 4.5/6.0 = 75\%$.

5. Phenotypic standard deviation $= 200$ mg, so $S = 400$ mg; $h^2 = 10,000/40,000 = 25\%$, so expected mean $= 2000 + (0.25)(400) = 2100$.

6. After one generation, $\mu' = 20 + 0.25(4) = 21$; after 10 generations (using Equation 9.38), $\mu' = 20 + 0.25(4)(10) = 30$ bristles.

7. $h^2 = $ Response/cumulative selection differential $= 0.15/(0.07 \times 5) = 0.43$.

8. (a) $a = (23.8 - 19.4)/2 = 2.2$; $d = [2(25.2) - 23.8 - 19.4]/2 = 3.6$. (b) Write values as if they were relative fitnesses, i.e., $1 - 0.056$, 1, $1 - 0.230$ for AA, AA', $A'A'$. Since maximum of mean fitness occurs at $p = 0.230/(0.230 + 0.056) = 0.804$, this value will maximize the mean of the quantitative trait.

9. Using Equation 9.39, $D = 3.73 + 2.01 = 5.74$ and $s^2 = (0.87)^2 - (0.60)^2 = 0.40$, hence $n = (5.74)^2/8(0.40) = 10.3$.

10. Let X = first litter, Y = second litter. Then $\Sigma X = 104$, $\Sigma X^2 = 1106$, $\Sigma Y = 103$, $\Sigma Y^2 = 1101$, $\Sigma XY = 1089$, $\sigma = [1089 - (104)(103)/10]/\{[106 - (104)^2/10][1101 - (103)^2/10]\}^{1/2} = 0.57$.

11. Response/$h^2 = (220 - 180)/0.20 = 200 = $ Cumulative selection differential. Number of generations $= 200/20 = 10$.

12. Selection intensity $= i = [1/\sqrt{2\pi}]\exp(-t^2/2)/B$, where, taking the top half, for example, $B = 1/2$ and $t = 0.0$, so $i = 0.80$. For the other proportions (in order), $i = 1.27$, 1.63, 1.95, 2.26.

13. Use subscripts M and F for males and females. $S_M = m - m_M$, $S_F = m - m_F$. Mean of selected parents is $(m_M + m_M)/2$, so proper total selection differential is $m - (m_M + m_F)/2 = (S_M + S_F)/2$.

14. For A and B loci, $2pq[a + (q - p)d]^2 = 0.18$ and 0.067, respectively, so $\sigma_a^2 = 0.247$, and $h^2 = \sigma_a^2/\sigma^2 = 0.247$. For A and B, $\sigma_d^2 = (2pqd)^2 = 0.01$ and 0, so broad sense heritability $= 0.257$, and 0.247, respectively.

15. Equation 9.10 can be written as $R = ish^2$ where i is the intensity of selection. For percent protein, $R = (1.5)(0.45)(0.7) = 0.47$, so expected percent protein $= 3.3\% + 0.47\% = 3.8\%$. For correlated response use Equation 9.45, $CR = (1.5)(0.60)^{(1/2)} (0.70)^{(1/2)}(0.55)(0.65) = 0.35$, so expected percent fat $= 3.4\% + 0.35\% = 3.75\%$. The fat increase corresponds to a direct selection of $i = 0.35/(0.60)(0.65) = 0.90$.

16. Let the phenotypes of AA, AA', $A'A'$ be $1, 0, 0$ with frequencies p^2, $2pq$, q^2, so A' is dominant with frequency q. Then $a = 1/2$, $d = -1/2$, and $\sigma_a^2 = 2p^3q$. Then $\sigma^2 = p^2 - (p^2)^2 = p^2q(1 + p)$. Therefore $h^2 = 2p^3q/p^2q(1 + p) = 2(1 - q)/(2 - q)$, which is approximately $1 - q$ when q is small.

17. Let fitnesses of AA, AA', $A'A'$ be $1 - s$, 1, $1 - t$, so $a = -(s - t)/2$ and $d = (s + t)/2$. At equilibrium $p = t/(s + t)$ and $q = s/(s + t)$, so $a + (q - p)d = 0$ and $\sigma_a^2 = 0$.

18. For $B = 1/2, 1/4, 1/8, 1/16, 1/32$ approximation gives $i = 0.80, 1.25, 1.60, 1.91, 2.21$, respectively.

19. $Z = [1/\sqrt{(2\pi)}]\exp[-(T - \mu)^2/2\sigma^2]$ and $z = [1/\sqrt{(2\pi)}]\exp(-t^2/2) = [1/\sqrt{(2\pi)}]\exp[-(T - \mu)^2/2\sigma^2]$. Therefore, $z = \sigma Z$.

Bibliography

Adams, J. and R.H. Ward. 1973. Admixture studies and the detection of selection. Science 180:1137–1143.

Aguadé, M., N. Miyashita and C.H. Langley. 1992. Polymorphism and divergence in the *Mst26A* male accessory gland gene region in *Drosophila*. Genetics 132:755–770.

Ajioka, J.W. and D.L. Hartl. 1989. Population dynamics of transposable elements. pp. 939–958. *In* D.E. Berg and M.M. Howe (eds.), *Mobile DNA*. American Society for Microbiology, Washington, DC.

Akashi, H. 1995. Inferring weak selection from patterns of polymorphism and divergence at silent sites in *Drosophila* DNA. Genetics 139:1067–1076.

Aoki, K. 1981. Algebra of inclusive fitness. Evolution 35:659–663.

Aquadro, C.F., D.J. Begun and E.C. Kindahl. 1994. Selection, recombination and DNA polymorphism in *Drosophila*. pp. 46–56. *In* B. Golding, (ed.), *Non-Neutral Evolution: Theories and Molecular Data*. Chapman and Hall, New York.

Arnold, S.J. and M.J. Wade. 1984. On the measurement of natural and sexual selection: Applications. Evolution 38:720–734.

Asmussen, M.A., J. Arnold and J.C. Avise. 1987. Definition and properties of disequilibrium statistics for associations between nuclear and cytoplasmic genotypes. Genetics 115:755–768.

Atchley, W.R., J.J. Rutledge and D.E. Cowley. 1982. A multivariate statistical analysis of direct and correlated response to selection in the rat. Evolution 36:677–698.

Avise, J.C. 1986. Mitochondrial DNA and the evolutionary genetics of higher animals. Philos. Trans. R. Soc. Lond. B Biol. Sci. 312:325–342.

Avise, J.C. 1994. *Molecular Markers, Natural History and Evolution*. Chapman and Hall, New York.

Avise, J.C., C. Giblin-Davidson, J. Laerm, J.C. Patton and R.A. Lansman. 1979. Mitochondrial DNA clones and matriarchal phylogeny within and among geographic populations of the pocket gopher, *Geomys pinetis*. Proc. Natl. Acad. Sci. USA 76:6694–6698.

Ayala, F.J. and M.L. Tracy. 1974. Genetic differentiation within and between species of the *Drosophila willistoni* group. Proc. Natl. Acad. Sci. USA 71:999–1003.

Ayala, F.J., B.S.W. Chang and D.L. Hartl. 1993. Molecular evolution of the *Rh3* gene in *Drosophila*. Genetica 92:23–32.

Ayala, F.J., D.E. Krane and D.L. Hartl. 1994. Genetic variation in *IncI1-ColIb* plasmids. J. Mol. Evol. 39:129–133.

Bafna, V. and P. Pevzner. 1996. Genome rearrangements and sorting by reversals. SIAM J. Computing 25:272.

Baker, R.L., A.B. Chapman and R.T. Wardell. 1975. Direct response to selection for post-weaning weight gain in the rat. Genetics 80:171–189.

Baird, M., I. Ballazs, A. Giusti, L. Miyazaki, L. Nicholas, K. Wexler, E. Kanter, J. Glasssberg, F. Allen, P. Rubenstein and L. Sussman. 1986. Allele frequency distribution of two highly polymorphic DNA sequences in three ethnic groups and its applicability to the determination of paternity. Am. J. Hum. Gen. 39:489–501.

Ballard, J.W.O. and M. Kreitman. 1994. Unraveling selection in the mitochondrial genome of *Drosophila*. Genetics 138: 757–772.

Barton, N.H. and M. Turelli. 1989. Evolutionary quantitative genetics: How little do we know? Annu. Rev. Genet. 23:337–370.

Begun, D.J. and C.F. Aquadro. 1993. African and North American populations of *Drosophila melanogaster* are very different at the DNA level. Nature 365:548–550.

Benveniste, R.E. 1985. The contributions of retroviruses to the study of mammalian evolution. pp. 359–417. *In* R.J. MacIntyre (ed.), *Molecular Evolutionary Genetics*. Plenum Press, New York.

Berg, D.E. and M.M. Howe (eds.). 1989. *Mobile DNA*. American Society for Microbiology, Washington, DC.

Bergman, A., D.B. Goldstein, K.E. Holsinger and M.W. Feldman. 1995. Population structure, fitness surfaces, and linkage in the shifting balance process. Genet. Res. 66:85–92.

Bermingham, E., T. Lamb and J.C. Avise. 1986. Size polymorphism and heteroplasmy in the mitochondrial DNA of lower vertebrates. J. Hered. 77:249–252.

Berry, A. and M. Kreitman. 1993. Molecular analysis of an allozyme cline: Alcohol dehydrogenase in *Drosophila melanogaster* on the East Coast of North America. Genetics 134:869–893.

Birky, C.W., T. Maruyama and P. Fuerst. 1983. An approach to population and evolutionary genetic theory for genes in mitochondria and chloroplasts, and some results. Genetics 103:513–527.

Bishop, J.A. and L.M. Cook. 1975. Moths, melanism and clean air. Sci. Am. 232:90–99.

Bonnell, M.L. and R.K. Selander. 1974. Elephant seals: Genetic variation and near extinction. Science 184:908–909.

Bouchard, T.J., D.T. Lykken, M. McGue, M.L. Segal and A. Tellegen. 1990. Sources of human psychological differences: The Minnesota study of twins reared apart. Science 250:223–228.

Boursot, P., H. Yonekawa and F. Bonhomme. 1987. Heteroplasmy in mice with deletion of a large coding region of mitochondrial DNA. Mol. Biol. Evol. 4:46–55.

Bowcock, A.M., J.R. Kidd, J.L. Mountain, J.M. Hebert, L. Carotenuto, K.K. Kidd and L.L. Cavalli-Sforza. 1991. Drift, admixture, and selection in human evolution: A study with DNA polymorphisms. Proc. Natl. Acad. Sci. USA 88:839–843.

Bowcock, A.M., A. Ruiz-Linares, J. Tomfohrde. E. Minch, J.R. Kidd and L.L. Cavalli-Sforza. 1994. High resolution of human evolutionary trees with polymorphic microsatellites. Nature 368:455–457.

Braverman, J.M., R.R. Hudson, N.L. Kaplan, C.H. Langley and W. Stephan. 1995. The hitchhiking effect on the site frequency spectrum of DNA polymorphisms. Genetics 140:783–796.

Britten, R.J. 1986. Rates of DNA sequence evolution differ between taxonomic groups. Science 231:1393–1398.

Brown, W.M. 1980. Polymorphism in mitochondrial DNA of humans as revealed by restriction endonuclease analysis. Proc. Natl. Acad. Sci. USA 77:3605–3609.

Brown, W.M. 1985. The mitochondrial genome of animals. pp. 95–130. *In* R.J. MacIntyre (ed.), *Molecular Evolutionary Genetics*. Plenum Press, New York.

Brown, W.M., M. George and A.C. Wilson. 1979. Rapid evolution of animal mitochondrial DNA. Proc. Natl. Acad. Sci. USA 76:1967–1971.

Bryan, G.J., J.W. Jacobson and D.L. Hartl. 1987. Heritable somatic excision of a *Drosophila* transposon. Science 235:1636–1638.

Buckler, E.S., A. Ippolito and T.P. Holtsford. 1997. The evolution of ribosomal DNA: Divergent paralogues and phylogenetic implications. Genetics 145:821–832.

Bulmer, M. 1994. *Theoretical Evolutionary Ecology*. Sinauer Associates, Sunderland, MA.

Bulmer, M.G. 1970. *The Biology of Twinning in Man*. Oxford University Press, London.

Bumpus, H.C. 1899. The elimination of the unfit as illustrated by the introduced sparrow *Passer domesticus*. Woods Hole Mar. Biol. Sta. Biol. Lectures 6:209–226.

Buonagurio, D.A., S. Nakada, J.D. Parvin, M. Krystal, P. Palese and W.M. Fitch. 1986. Evolution of human influenza A viruses over 50 years: Rapid uniform rate of change in NS genes. Science 232:980–982.

Buri, P. 1956. Gene frequency in small populations of mutant *Drosophila*. Evolution 10:367–402.

Cann, R.L., M. Stoneking and A.C. Wilson. 1987. Mitochondrial DNA and human evolution. Nature 325:31–36.

Castle, W.E. 1921. An improved method of estimating the number of genetic factors concerned in cases of blending inheritance. Science 54:223.

Cavalli-Sforza, L.L. 1974. The genetics of human populations. Sci. Am. 231:81–89.

Cavalli-Sforza, L.L. and W.F. Bodmer. 1971. *The Genetics of Human Populations*. W.H. Freeman, San Francisco.

Cavalli-Sforza, L.L. and M.W. Feldman. 1978. Darwinian kin selection and "altruism." Theor. Popul. Biol. 14:268–280

Chakraborty, R. and M. Nei. 1977. Bottleneck effects on average heterozygosity and genetic distance with the stepwise mutation model. Evolution 31:347–356.

Charlesworth, B. 1980. *Evolution in Age-Structured Populations*. Cambridge University Press, Cambridge.

Charlesworth, B. and D.L. Hartl. 1978. Population dynamics of the segregation distorter polymorphism of *Drosophila melanogaster*. Genetics 89:171–192.

Charlesworth, B., M.T. Morgan and D. Charlesworth. 1993. The effect of deleterious mutations on neutral molecular variation. Genetics 134:1289–1303.

Charlesworth, B., P. Sniegowski and W. Stephan. 1994. The evolutionary dynamics of repetitive DNA in eukaryotes. Nature 371:215–220.

Charlesworth, D., B. Charlesworth and M.T. Morgan. 1995. The pattern of neutral molecular variation under the background selection model. Genetics 141:1619–1632.

Christiansen, F.B. and O. Frydenberg. 1974. Geographical patterns of four polymorphisms in *Zoarces viviparus* as evidence of selection. Genetics 77:765–770.

Clark, A.G. 1984. Natural selection with nuclear and cytoplasmic transmission. I. A deterministic model. Genetics 107:679–701.

Clark, A.G. 1988. Deterministic theory of heteroplasmy. Evolution 42:621–626.

Clark, A.G. and M.W. Feldman. 1986. A numerical simulation of the one-locus, multiple-allele fertility model. Genetics 113:161–176.

Clark, A.G. and C.M.S. Lanigan. 1993. Prospects for estimating nucleotide divergence with RAPDs. Mol. Biol. Evol. 10:1096–1111.

Clark, A.G., L. Wang and T. Hulleberg. 1995a. Spontaneous mutation rate of modifiers of metabolism in *Drosophila*. Genetics 139:767–779.

Clark, A.G., L. Wang. and T. Hulleberg. 1995b. *P*-element induced variation in metabolic regulation in *Drosophila*. Genetics 139:337–348.

Clark, A.G., B.G. Leicht and S.V. Muse. 1996. Length variation and secondary structure of introns in the *Mlc1* gene in six species of *Drosophila*. Mol. Biol. Evol. 13:471–482.

Clayton, G.A. and A. Robertson. 1955. Mutation and quantitative variation. Am. Nat. 89:151–158.

Clayton, G.A. and A. Robertson. 1957. An experimental check on quantitative genetical theory. II. Long-term effects of selection. J. Genet. 55:152–170.

Clegg, M.T., R.W. Allard and A.L. Kahler. 1972. Is the gene the unit of selection? Evidence from two experimental plant populations. Proc. Natl. Acad. Sci. USA 69:2474–2478.

Clegg, M.T., M.P. Cummings and M.L. Durbin. The evolution of plant nuclear genes. Proc. Natl. Acad. Sci. USA, *in press*.

Cockerham, C.C. 1986. Modifications in estimating the number of genes for a quantitative character. Genetics 114:659–664.

Cockerham, C.C. and H. Tachida. 1987. Evolution and maintenance of quantitative genetic variation by mutations. Proc. Natl. Acad. Sci. USA 84:6205–6209.

Coen, E., Strachan, T. and G. Dover. 1982. Dynamics of concerted evolution of ribosomal DNA and histone gene families in the melanogaster species subgroup of *Drosophila*. J. Mol. Biol. 158:17–35.

Cohen, J.E. 1995. Unexpected dominance of high frequencies in chaotic nonlinear population models. Nature 378:610–612.

Cook, L.M. 1965. Inheritance of shell size in the snail *Arianta arbustorum*. Evolution 19:86–94.

Cook, L.M., G.S. Mani and M.E. Varley. 1986. Postindustrial melanism in the peppered moth. Science 231:622–613.

Cotterman, C.W. 1940. A calculus for statistico-genetics. Ph.D. thesis, Ohio State University, Columbus, Ohio.

Coyne, J.A. 1985. The genetic basis of Haldane's rule. Nature 314:7346–738.

Coyne, J.A. 1996. Genetics of differences in pheromonal hydrocarbons between *D. melanogaster* and *D. simulans*. Genetics 143:353–364.

Coyne, J.A., B. Charlesworth and H.A. Orr. 1991. Haldane's rule revisited. Evolution 45:1710–1714.

Coyne, J.A., N.H. Barton and M. Turelli. 1997. A critique of Sewall Wright's shifting balance theory of evolution. Evolution 51:643–671.

Cross, S.R.H. and A.J. Birley. 1986. Restriction endonuclease map variation in the *Adh* region in populations of *Drosophila melanogaster*. Biochem. Genet. 24:415–433.

Crow, J.F. 1985. The neutrality-selection controversy in the history of evolution and population genetics. pp. 1–18. *In* T. Ohta and K. Aoki (eds.), *Population Genetics and Molecular Evolution*. Japanese Scientific Societies Press, Tokyo.

Crow, J.F. and K. Aoki. 1982. Group selection for a polygenic behavioral trait: A differential proliferation model. Proc. Natl. Acad. Sci. USA 79:2628–2631.

Crow, J.F. and M. Kimura. 1970. *An Introduction to Population Genetic Theory*. Harper & Row, New York.

Curie-Cohen, M. 1982. Estimates of inbreeding in a natural population: A comparison of sampling properties. Genetics 100:339–358.

Curtsinger, J.W. 1984. Evolutionary landscapes for complex selection. Evolution 38:359–367.

Daniels, G.R. and P.L. Deininger. 1985. Integration site preferences of the Alu family and similar repetitive DNA sequences. Nucleic Acids Res. 13:8939–8954.

Darwin, C. 1871. *The Descent of Man and Selection in Relation to Sex*. Appleton, New York.

Dayhoff, M.O. 1972. *Atlas of Protein Sequence and Structure*, Volume 5. National Biomedical Research Foundation, Silver Spring, MD.

Demers, G.W., K. Brech and R.C. Hardison. 1986. Long interspersed *L1* repeats in rabbit DNA are homologous to *L1* repeats of rodents and primates in an open reading frame region. Mol. Biol. Evol. 3:179–190.

Densmore, L.D., J.W. Wright and W.M. Brown. 1985. Length variation and heteroplasmy are frequent in mitochondrial DNA from parthenogenetic and bisexual lizards (genus *Cnemidophorus*). Genetics 110:689–707.

Dickerson, R.E. 1971. The structure of cytochrome *c* and the rates of molecular evolution. J. Mol. Evol. 1:26–45.

Dobzhansky, T. and B. Spassky. 1944. Genetics of natural populations. XI. Manifestation of genetic variants of *Drosophila pseudoobscura* in different environments. Genetics 20:270–290.

Doebley, J., A. Stec and C. Gustus. 1995. *teosinte branched* 1 and the origin of maize: Evidence for epistasis and the evolution of dominance. Genetics 141:333–346.

Doerge, R.W. and G.A. Churchill. 1996. Permutation tests for multiple loci affecting a quantitative character. Genetics 142:285–294.

Doolittle, R.F. 1985. The genealogy of some recently evolved vertebrate proteins. Trends Biochem. Sci. 10:233–237.

DuBose, R.F., D.E. Dykhuizen and D.L. Hartl. 1988. Genetic exchange among natural isolates of bacteria: Recombination within the *phoA* gene of *Escherichia coli*. Proc. Natl. Acad. Sci. USA 85:7036–7040.

DuMouchel, W.H. and W.W. Anderson. 1968. The analysis of selection in experimental populations. Genetics 58:435–449.

Dudley, J.W. 1977. 76 generations of selection for oil and protein percentage in maize. pp. 459–473. *In* E. Pollak, O. Kempthorne and T.B. Bailey Jr. (eds.), *International Conference on Quantitative Genetics*. Iowa State University Press, Ames, IA.

Dudley, J.W. and J.R. Lambert. 1992. Ninety generations of selection for oil and protein content in maize. Maydica 37:1–7.

Easteal, S. 1985. Generation time and the rate of molecular evolution. Mol. Biol. Evol. 2:450–453.

Ehrlich, J., D. Sankoff and J.H. Nadeau. Synteny conservation and chromosome rearrangements during mammalian evolution. Genetics, *in press*.

Ehrman, L. 1970. The mating advantage of rare males in *Drosophila*. Proc. Natl. Acad. Sci. USA 515: 345–348.

Endo, T., K. Ikeo and T. Gojobori. 1996. Large scale search for genes on which positive selection may operate. Mol. Biol. Evol. 13:685–690.

Enfield, F.D. 1980. Long term effects of selection: The limits to response. pp. 69–86. *In* A. Robertson (ed.), *Selection Experiments in Laboratory and Domestic Animals*. Commonwealth Agricultural Bureau, Slough, England.

Epling, C. and T. Dobzhansky. 1942. Genetics of natural populations, VI: Microgeographic races in *Linanthus parryae*. Genetics 27:317–332.

Ewens, W.J. 1979. *Mathematical Population Genetics*. Springer-Verlag, New York.

Ewens, W.J. 1982. On the concept of the effective population size. Theor. Popul. Biol. 21:373–378.

Falconer, D.S. 1985. A note on Fisher's "average effect" and "average excess." Genet. Res. 46: 337–347.

Falconer, D.S. and T.F.C. Mackay. 1996. *Introduction to Quantitative Genetics*, Fourth Edition. Longman, Essex, England.

Feldman, M.W. and R.C. Lewontin. 1975. The heritability hang-up. Science 190:1163–1168.

Felsenstein, J. 1978. Cases in which parsimony or compatibility methods will be positively misleading. Syst. Zool. 27:401–410.

Felsenstein, J. 1981. Evolutionary trees from DNA sequences: A maximum likelihood approach. J. Mol. Evol. 17:368–376.

Felsenstein, J. 1982. Numerical methods for inferring evolutionary trees. Quart. Rev. Biol. 57:379–404.

Felsenstein, J. 1985. Confidence limits on phylogenies: An approach using the bootstrap. Evolution 39:783–791.

Ffrench-Constant, R.H., D.P. Mortlock, C.D. Shaffer, R.J. MacIntyre and R.T. Roush. 1991. Molecular cloning and transformation of cyclodiene resistance in *Drosophila*: An invertebrate GABA subtype A receptor locus. Proc. Natl. Acad. Sci. USA 88:7209–7213.

Fisher, R.A. 1918. The correlation between relatives on the supposition of Mendelian inheritance. Trans. Royal Soc. Edinburgh 52:399–433.

Fisher, R.A. 1922. On the dominance ratio. Proc. Roy. Soc. Edin. 42:321–431.

Fisher, R.A. 1930. *The Genetical Theory of Natural Selection*. Clarendon, Oxford.

Fisher, R.A. 1958. *The Genetical Theory of Natural Selection*, Second Edition, Dover Press, New York.

Fitch, W.M. and E. Margoliash. 1967. A method for estimating the number of invariant amino acid coding positions in a gene using cytochrome *c* as a model case. Biochem. Genet. 1:65–71.

Ford, E.B. and P.M. Sheppard. 1969. The *medionigra* polymorphism of *Panaxia dominula*. Heredity 24:561–569.

Fry, J.D., K.A. DeRonde and T.F.C. Mackay. 1995. Polygenic mutation in *Drosophila melanogaster*: Genetic analysis of selection lines. Genetics 139:1293–1307.

Fu, Y. and W.-H. Li. 1993. Statistical tests of neutrality of mutations. Genetics 133:693–709.

Fuerst, P.A., R. Chakraborty and M. Nei. 1977. Statistical studies on protein polymorphism in natural populations. I. Distribution of single locus heterozygosity. Genetics 86:455–483.

Galton, F. 1889. Natural Inheritance. MacMillan, London.

Gaut, B.S. and M.T. Clegg. 1993. Molecular evolution of the *Adh1* locus in the genus *Zea*. Proc. Natl. Acad. Sci. USA 90:5095–5099.

Gibbs, K.L. and P.R. Grant. 1987. Oscillating selection on Darwin's finches. Nature 327:511–513.

Gillespie, J.H. 1986. Variability of evolutionary rates of DNA. Genetics 113:1077–1091.

Gillespie, J.H. 1989. Lineage effects and the index of dispersion of molecular evolution. Mol. Biol. Evol. 6:636–648.

Gillespie, J.H. 1991. *The Causes of Molecular Evolution*. Oxford University Press, Oxford.

Gillespie, J.H. and C.H. Langley. 1974. A general model to account for enzyme variation in natural populations. Genetics 76:837–848.

Gingerich, P.D. 1986. Temporal scaling of molecular evolution in primates and other mammals. Mol. Biol. Evol. 3:205–221.

Glover, D.M. and D.S. Hogness. 1977. A novel arrangement of the 18S and 28S sequences in a repeating unit of *Drosophila melanogaster* rDNA. Cell 10:167–176.

Golding, G.B. 1983. Estimates of DNA and protein sequence divergence: An examination of some assumptions. Mol. Biol. Evol. 1:125–142.

Golding, G.B. (ed.). 1994. *Non-Neutral Evolution*. Chapman and Hall, New York.

Golding, G.B., C.F. Aquadro and C.H. Langley. 1986. Sequence evolution within populations under multiple types of mutation. Proc. Natl. Acad. Sci. USA 83:427–431.

Goldman, N. 1993. Statistical tests of models of DNA substitution. J. Mol. Evol. 36:182–198.

Goldman, N. and Z. Yang. 1994. A codon-based model of nucleotide substitution for protein-coding DNA sequences. Mol. Biol. Evol. 11:725–736.

Goodman, M., B.F. Koop, J. Czelusniak and M.L. Weiss. 1984. The γ-globin gene: Its long evolutionary history in the β-globin gene family of mammals. J. Mol. Biol. 180:803–823.

Gray, P.W. and D.V. Goeddel. 1983. Cloning and expression of murine immune interferon cDNA. Proc. Natl. Acad. Sci. USA 80:5842–5846.

Grossman, A.I., L.G. Koreneva and L.E. Ulitskaya. 1970. Variation of the *alcohol dehydrogenase* (*ADH*) locus in natural populations of *Drosophila melanogaster*. Genetika 6:91–96 (in Russian).

Grun, P. 1976. *Cytoplasmic Genetics and Evolution*. Columbia University Press, New York.

Gupta, A.P. and R.C. Lewontin. 1982. A study of reaction norms in natural populations of *Drosophila pseudoobscura*. Evolution 36:934–948.

Haldane, J.B. S. 1956. The estimation of viabilities. J. Genet. 54:294–296.

Hale, L.R. and R.S. Singh. 1986. Extensive variation and heteroplasmy in size of mitochondrial DNA among geographic populations of *Drosophila melanogaster*. Proc. Natl. Acad. Sci. USA 83:8813–8817.

Hamilton, W.D. 1964. The genetical evolution of social behaviour I. J. Theor. Biol. 7:1–16.

Hammer, M.F. and L.M. Silver. 1993. Phylogenetic analysis of the α-globin pseudogene-4 (*Hba-ps4*) locus in the house mouse species complex reveals a stepwise evolution of *t* haplotypes. Mol. Biol. Evol. 10:971–1001.

Hardies, S.C., M.H. Edgell, C.A. Hutchinson III. 1984. Evolution of the mammalian β-globin gene cluster. J. Biol. Chem. 259:3748–3756.

Hardies, S.C., S.L. Martin, C.F. Voliva, C.A. Hutchison III and M.H. Edgell. 1986. An analysis of replacement and synonymous changes in the rodent *L1* repeat family. Mol. Biol. Evol. 3:109–125.

Hardison, R.C. 1984. Comparison of the β-like globin gene families of rabbits and humans indicates that the gene cluster 5′—γ—β-3′ predates the mammalian radiation. Mol. Biol. Evol. 1:390–410.

Harris, H. 1966. Enzyme polymorphisms in man. Proc. R. Soc. Lond. B Biol. Sci. 164:298–310.

Harris, H., D.A. Hopkinson and Y.H. Edwards. 1977. Polymorphism and the subunit structure of enzymes. A contribution to the neutralist-selectionist controversy. Proc. Natl. Acad. Sci. USA 74:698–701.

Harris, T., E.F. Cook, R. Garrison et al. 1988. Body mass index and mortality among nonsmoking older persons: The Framingham Heart Study. JAMA 259:1520–1524.

Harrison, R.G., D.M. Rand and W.C. Wheeler. 1987. Mitochondrial DNA variation in field crickets across a narrow hybrid zone. Mol. Biol. Evol. 4:144–158.

Hartl, D.L. 1970. A mathematical model for recessive lethal segregation distorters with differential viabilities in the sexes. Genetics 66:147–163.

Hartl, D.L. 1979. Four volume treatise on population biology. BioScience 29:179–180.

Hartl, D.L. 1989. Transposable element *mariner* in *Drosophila* species. pp. 531–536. *In* D.E. Berg and M.M. Howe (eds.), *Mobile DNA*. American Society for Microbiology, Washington, DC.

Hartl, D.L. 1994. *Genetics*, Third Edition. Jones and Bartlett, Boston, MA.

Hartl, D.L. and D.E. Dykhuizen. 1981. Potential for selection among nearly neutral allozymes of *6-phosphogluconate dehydrogenase* in *Escherichia coli*. Proc. Natl. Acad. Sci. USA 78:6344–6348.

Hartl, D.L. and S.A. Sawyer. 1988. Why do unrelated insertion sequences occur together in the genome of *Escherichia coli*? Genetics 118:537–541.

Hartl, D.L., E.N. Moriyama and S.A. Sawyer. 1994. Selection intensity for codon bias. Genetics 138:227–234.

Hasegawa, M. and S. Horai. 1991. Time of the deepest root for polymorphism in human mitochondrial DNA. J. Mol. Evol. 32:37–42.

Hastings, A. 1985. Four simultaneously stable polymorphic equilibria in two-locus, two-allele models. Genetics 109:255–261.

Hauswirth, W.W. and P.J. Laipis. 1982. Mitochondrial DNA polymorphism in a maternal lineage of Holstein cows. Proc. Natl. Acad. Sci. USA 79:4686–4690.

Haymer, D.S. and D.L. Hartl. 1982. The experimental assessment of fitness in *Drosophila*: Comparative measures of competitive reproductive success. Genetics 102:455–466.

Hedrick, P.W. and C.C. Cockerham. 1986. Partial inbreeding: Equilibrium heterozygosity and the heterozygosity paradox. Evolution 40:856–861.

Herrin, G., Jr. 1993. Probability of matching RFLP patterns from unrelated individuals. Am. J. Hum. Genet. 52:491–497.

Herrnstein, R.J. and C. Murray. 1996. *The Bell Curve, Intelligence and Class Structure in American Life*. Simon and Schuster, New York.

Hey, J. 1997. Mitochondrial and nuclear genes present conflicting portraits of human origins. Mol. Biol. Evol. 14:166–172.

Hey, J. and R.M. Kliman. 1993. Population genetics and phylogenetics of DNA sequence variation at multiple loci within the *Drosophila melanogaster* species complex. Mol. Biol. Evol. 10:804–822.

Hillis, D.M. 1996. Inferring complex phylogenies. Nature 383:130–131.

Hillis, D.M., J.P. Huelsenbeck and C.W. Cunningham. 1994. Application and accuracy of molecular phylogenies. Science 264:671–676.

Hiraizumi, Y., L. Sandler, and J.F. Crow. 1960. Meiotic drive in natural populations of *Drosophila melanogaster*. III. Populational implications of the segregation-distorter locus. Evolution 14:433–444.

Holgate, P. 1966. A mathematical study of the founder principle of evolutionary genetics. J. Appl. Prob. 3:115–128.

Holland, J., K. Spindler, F. Horodyski, E. Grabau, S. Nichol and S. VandePol. 1982. Rapid evolution of RNA genomes. Science 215:1577–1585.

Hood, L., M. Kronenberg and T. Hunkapiller. 1985. T cell antigen receptors and the immunoglobulin supergene family. Cell 40:225–229.

Houle, D., B. Morikawa and M. Lynch. 1996. Comparing mutational variabilities. Genetics 143:1467–1483.

Hudson, R.R. 1987. Estimating the recombination parameter of a finite population without selection. Genet. Res. 50:245–250.

Hudson, R.R. 1990. Gene genealogies and the coalescent process. Oxford Surveys Evol. Biol. 7:1–44.

Hudson, R.R. 1993. The how and why of generating gene genealogies. pp. 23–36. *In* N. Takahata and A.G. Clark (eds.), *Mechanisms of Molecular Evolution*. Sinauer Associates, Sunderland, MA.

Hudson, R.R. and N.L. Kaplan. 1995. Deleterious background selection with recombination. Genetics 141:1605–1617.

Hudson, R.R., M. Kreitman and M. Aguadé. 1987. A test of neutral molecular evolution based on nucleotide data. Genetics 116:153–159.

Hudson, R.R., D.D. Boos and N.L. Kaplan. 1992. A statistical test for detecting geographic subdivision. Mol. Biol. Evol. 9:138–151.

Hudson, R.R., M. Slatkin and W.P. Maddison. 1992. Estimation of levels of gene flow from DNA sequence data. Genetics 132:583–589.

Hudson, R.R., K. Bailey, D. Skarecky, J. Kwiatowski and A.J. Ayala. 1994. Evidence for positive selection in the superoxide dismutase region of *D. melanogaster*. Genetics 136:1329–1340.

Hughes, A.L. and M. Nei. 1988. Pattern of nucleotide substitution at major histocompatibility complex class I loci reveals overdominant selection. Nature 335:167–170.

Hunkapiller, T. and L. Hood. 1986. The growing immunoglobulin supergene family. Nature 323:15–16.

Iltis, H. 1932. *Life of Mendel* (E. and C. Paul, trans.). Norton, New York.

Ikemura, T. 1985. Codon usage and tRNA content in unicellular and multicellular organisms. Mol. Biol. Evol. 2:13–34.

Ioerger, T.R., A.G. Clark and T.-H. Kao. 1991. Polymorphism at the self-incompatibility locus in Solanaceae predates speciation. Proc. Natl. Acad. Sci. USA 87:9732–9735.

Isaac, A.M., G. Sargent and J. Cooke. 1997. Control of vertebrate left-right asymmetry by a *snail*-related zinc finger gene. Science 275:1301–1304

Iwasa, Y. and A. Pomiankowski. 1995. Continual change in mate preferences. Nature 377:420–422.

Jacquard, A. 1983. Heritability: One word, three concepts. Biometrics 39:465–477.

Jansen, R.C. 1993. Interval mapping of multiple quantitative trait loci. Genetics 135:205–211.

Jansen, R.C. and P. Stam. 1994. High resolution of quantitative traits into multiple loci via interval mapping. Genetics 136:1447–1455.

Johannsen, W. 1909. *Elemente der exackten Erblichkeit-slehre*. Fischer, Jena.

Jones, D.T., W.R. Taylor and J.M. Thornton. 1992. The rapid generation of mutation data matrices from protein sequences. Comp. Appl. Biosci. 8:275–282.

Jukes, T.H. and C.R. Cantor. 1969. Evolution of protein molecules. pp. 21–132. *In* H.N. Munro (ed.), *Mammalian Protein Metabolism III*. Academic Press, New York.

Kaplan, N.L. and J.F.Y. Brookfield. 1983. Transposable elements in Mendelian populations. III. Statistical results. Genetics 104:485–495.

Karlin, S. and D. Carmelli. 1975. Numerical studies of two-locus selection models with general viabilities. Theor. Popul. Biol. 7:364–398.

Karn, M.N. and L.S. Penrose. 1951. Birth weight and gestation time in relation to maternal age, parity and infant survival. Ann. Eugen. 16:147–164.

Kazazian, H.H., C. Wong, H. Youssoufian, A.F. Scott, D.G. Phillips and S.E. Antonarakis. 1988. Haemophilia A resulting from de novo insertion of *L1* sequences represents a novel mechanism for mutation in man. Nature 332:164–166.

Keightley, P.D., M.J. Evans and W.G. Hill. 1993. Effects of multiple retrovirus insertions on quantitative traits of mice. Genetics 135:1099–1106.

Keith, T.P., L.D. Brooks, R.C. Lewontin, J.C. Martinez-Cruzado and D.L. Rigby. 1985. Nearly identical distributions of *xanthine dehydrogenase* in two populations of *Drosophila pseudoobscura*. Mol. Biol. Evol. 2:206–216.

Kempthorne, O. 1978. Logical, epistemological and statistical aspects of nature-nurture data interpretation. Biometrics 34:1–23.

Kettlewell, H.B.D. 1956. Further selection experiments on industrial melanism in the *Lepidoptera*. Heredity 10:287–301.

Kettlewell, H.B.D. 1973. *The Evolution of Melanism: The Study of a Recurring Necessity*. Clarendon, Oxford.

Kibota, T.T. and M. Lynch. 1996. Estimate of the genomic mutation rate deleterious to overall fitness in *E. coli*. Nature 381:694–696.

Kimura, M. 1955. Solution of a process of random genetic drift with a continuous model. Proc. Natl. Acad. Sci. USA 41:144–150.

Kimura, M. 1964. *Diffusion Models in Population Genetics*. Methuen, London.

Kimura, M. 1965. A stochastic model concerning the maintenance of genetic variability in quantitative characters. Proc. Natl. Acad. Sci. USA 54:731–736.

Kimura, M. 1968. Evolutionary rate at the molecular level. Nature 217:624–626.

Kimura, M. 1969. The number of heterozygous nucleotide sites maintained in a finite population due to steady flux of mutations. Genetics 61:893–903.

Kimura, M. 1971. Theoretical foundation of population genetics at the molecular level. Theor. Popul. Biol. 2:174–208.

Kimura, M. 1976. Population genetics and molecular evolution. Johns Hopkins Med. J. 138:253–261.

Kimura, M. 1980a. A simple method for estimating evolutionary rate of base substitutions through comparative studies of nucleotide sequences. J. Mol. Evol. 16:111–120.

Kimura, M. 1980b. Average time until fixation of a mutant allele in a finite population under continued mutation pressure: Studies by analytical, numerical, and pseudosampling methods. Proc. Natl. Acad. Sci. USA 77:522–526.

Kimura, M. 1983. *The Neutral Theory of Molecular Evolution*. Cambridge University Press, Cambridge.

Kimura, M. 1985. The role of compensatory neutral mutations in molecular evolution. J. Genet. 64:7–19.

Kimura, M. 1986. DNA and the neutral theory. Philos. Trans. R. Soc. Lond. B Biol. Sci. 312:343–354.

Kimura, M. and J.F. Crow. 1963. The measurement of effective population numbers. Evolution 17:279–288.

Kimura, M. and T. Ohta. 1969. The average number of generations until fixation of a mutant gene in a finite population. Genetics 61:763–771.

Kimura, M. and T. Ohta. 1971. *Theoretical Aspects of Population Genetics*. Princeton University Press, Princeton, NJ.

King, J.L. and T.H. Jukes. 1969. Non-Darwinian evolution: Random fixation of selectively neutral mutations. Science 164:788–798.

Kingman, J.F.C. 1980. *Mathematics of Genetic Diversity*. Society for Industrial and Applied Mathematics, Philadelphia.

Kingman, J.F.C. 1982. On the genealogy of large populations. J. Appl. Prob. 19A:27–43.

Kirby, D.A., S.V. Muse and W. Stephan. 1995. Maintenance of pre-mRNA secondary structure by epistatic selection. Proc. Natl. Acad. Sci. USA 92:9047–9051.

Kirkpatrick, M. and N. Barton. 1995. Déjà vu all over again. Nature 377:388–389.

Koop, B.F. and L. Hood. 1994. Striking sequence similarity over almost 100 kb of human and mouse T-cell receptor DNA. Nat. Genet. 7:48–53.

Krane, D.E., R.W. Allen, S.A. Sawyer, D.A. Petrov and D.L. Hartl. 1992. Genetic differences at four DNA typing loci in Finnish, Italian and mixed Caucasian populations. Proc. Natl. Acad. Sci. USA 89:10583–10587.

Kreitman, M. 1983. Nucleotide polymorphism at the alcohol dehydrogenase locus of *Drosophila melanogaster*. Nature 304:412–417.

Kreitman, M.E. and M. Aguadé. 1986. Excess poly-morphism at the *Adh* locus in *Drosophila melanogaster*. Genetics 114:93–110.

Kreitman, M. and R.R. Hudson. 1991. Inferring the evolutionary histories of the *Adh* and *Adh-dup* loci in *Drosophila melanogaster* from patterns of polymorphism and divergence. Genetics 127:565–582.

Kruglyak, L. and E.S. Lander. 1995a. A nonparametric approach for mapping quantitative trait loci. Genetics 139:1421–1428.

Kruglyak, L. and E.S. Lander. 1995b. Complete multi-point sib-pair analysis of qualitative and quanti-tative traits. Am. J. Hum. Gen. 57:439–454.

Lai, C., R.F. Lyman, A.D. Long, C.H. Langley and T.F.C. Mackay. 1995. Naturally occurring varia-tion in bristle number and DNA polymorphisms at the scabrous locus in D. melanogaster. Science 266:1697–1702.

Lande, R. 1975. The maintenance of genetic variabili-ty by mutation in a polygenic character with linked loci. Genet. Res. 26:221–234.

Lande, R. 1976. Natural selection and random genetic drift in phenotypic evolution. Evolution 30:314–334.

Lande, R. 1977. The influence of the mating system on the maintenance of genetic variability in polygenic characters. Genetics 86:485–498.

Lande, R. 1979. Quantitative genetic analysis of mul-tivariate evolution, applied to brain:body size allometry. Evolution 33:402–416.

Lande, R. 1980. The genetic covariance between char-acters maintained by pleiotropic mutations. Genetics 94:203–215.

Lande, R. 1981. The minimum number of genes con-tributing to quantitative variation between and within populations. Genetics 99:541–553.

Lande, R. and S.J. Arnold. 1983. The measurement of selection on correlated characters. Evolution 37:1210–1226.

Lande, R. and S.J. Arnold. 1985. Evolution of mating preferences and sexual dimorphism. J. Theor. Biol. 117:651–664.

Lande, R. and R. Thompson. 1990. Efficiency of marker-assisted selection in the improvement of quantitative traits. Genetics 124:743–756.

Lander, E.S. and D. Botstein. 1989. Mapping Mendelian factors underlying quantitative traits using RFLP linkage maps. Genetics 121:185–199.

Lander, E.S. and N.J. Schork. 1994. Genetic dissection of complex traits. Science 265:2037–2048.

Landsmann, J., E.S. Dennis, T.J.V. Higgins, C.A. Appleby, A.A. Korti and W.J. Peacock. 1986. Common evolutionary origin of legume and non-legume plant hemoglobins. Nature 324:166–168.

Langley, C.H. and W.M. Fitch. 1974. An examination of the constancy of the rate of molecular evolu-tion. J. Mol. Evol. 3:161–177.

Langley, C.H., J.F.Y. Brookfield and N. Kaplan. 1983. Transposable elements in Mendelian popula-tions. I. A theory. Genetics 104:457–471.

Laurie, C.C., J.R. True, J. Liu and J.M. Mercer. 1997. An introgression analysis of quantitative trait loci that contribute to a morphological differ-ence between *D. simulans* and *D. mauritiana*. Genetics 145:339–348.

Lawlor, D.A., F.E. Ward, P.D. Ennis, A.P. Jackson and P. Parham. 1988. HLA-A and B polymorphisms predate the divergence of humans and chim-panzees. Nature 335:268–271.

Leicht, B.G., S.V. Muse, M. Hanczyc and A.G. Clark. 1995. Constraints on intron evolution in the gene encoding the myosin alkali light chain in *Drosophila*. Genetics 139:299–308.

Lerner, I.M. 1958. *The Genetic Basis of Speciation*. John Wiley and Sons, New York.

Levin, D.A. 1978. Genetic variation in annual *Phlox*: Self-compatible versus self-incompatible spe-cies. Evolution 32:245–263.

Levings, C.S. III 1983. The plant mitochondrial genome and its mutants. Cell 32:659–661.

Levy, M. and D.A. Levin. 1975. Genetic heterozygosi-ty and variation in permanent translocation het-erozygotes of the Oenothera biennis complex. Genetics 79:493–512.

Lewontin, R.C. 1974. The analysis of variance and the analysis of causes. Am. J. Hum. Gen. 26:400–411.

Lewontin, R.C., L.R. Ginzburg and S.D. Tuljapurkar. 1978. Heterosis as an explanation for large amounts of genetic polymorphism. Genetics 88:149–170.

Lewontin, R.C., S. Rose and L.J. Kamin. 1984. *Not in Our Genes*. Pantheon Books, New York.

Li, W.-H. and C.-I. Wu. 1987. Rates of nucleotide sub-stitution are evidently higher in rodents than in man. Mol. Biol. Evol. 4:74–77.

Li, W.-H., C.-C. Luo and C.-I Wu. 1985a. Evolution of DNA sequences. pp. 1–94. *In* R.J. MacIntyre (ed.), *Molecular Evolutionary Genetics*. Plenum Press, New York.

Li, W.-H., C.-I Wu and C.-C. Luo. 1985b. A new method for estimating synonymous an nonsyn-onymous rates of nucleotide substitution con-sidering the relative likelihood of nucleotide and codon changes. Mol. Biol. Evol. 2:150–174.

Li, W.-H., M. Tanimura and P.M. Sharp. 1988. Rates and dates of divergence between AIDS virus nucleotide sequences. Mol. Biol. Evol. 5:313-330.

Liu, J., J.M. Mercer, L.F. Stam, G.C. Gibson, Z.-B. Zeng and C.C. Laurie. 1996. Genetic analysis of a morphological shape difference in the male genitalia of *D. simulans* and *D. mauritiana*. Genetics 142:1129–1145.

Lohe, A.R., E.N. Moriyama, D.-A. Lidholm and D.L. Hartl. 1995. Horizontal transmission, vertical degeneration, and stochastic loss of *mariner*-like transposable elements. Mol. Biol. Evol. 12:62–72.

Long, A.D., S.L. Mullaney, L.A. Reid, J.D. Fry, C.H. Langley and T.F.C. Mackay. 1995. High resolution mapping of genetic factors affecting abdominal bristle number in *D. melanogaster*. Genetics 139:1273–1291.

Long, A.D., S.L. Mullaney, T.F.C. Mackay and C.H. Langley. 1996. Genetic interactions between naturally occurring alleles at quantitative trait loci and mutant alleles at candidate loci affecting bristle number in *Drosophila melanogaster*. Genetics 144:1497–1510.

Lyman, R.F., F. Lawrence, S.V. Nuzhdin and T.F.C. Mackay. 1996. Effects of single *P*-element insertions on bristle number and viability in *D. melanogaster*. Genetics 143:277–292.

Lynch, M. 1988. The rate of polygenic mutation. Genet. Res. 51:137–148.

Lynch, M. 1994. Neutral models of phenotypic evolution. pp. 86–108. *In* L. Real (ed.), *Ecological Genetics*. Princeton University Press, Princeton, NJ.

Lynch, M. and W.G. Hill. 1986. Phenotypic evolution by neutral mutation. Evolution 40:915–935.

Mackay, T.F.C. 1985. Transposable element induced response to artificial selection in *Drosophila melanogaster*. Genetics 111:351–374.

Mackay, T.F.C. and J.D. Fry. 1996. Polygenic mutation in *Drosophila melanogaster*: Genetic interactions between selection lines and candidate quantitative trait loci. Genetics 144:671–688.

Mackay, T.F.C., R.F. Lyman and M.S. Jackson. 1992a. Effects of *P* element insertions on quantitative traits in *Drosophila melanogaster*. Genetics 130: 315–332.

Mackay, T.F.C., R.F. Lyman, M.S. Jackson, C. Terzian and W.G. Hill. 1992b. Polygenic mutation in *Drosophila melanogaster*: Estimates from divergences among inbred strains. Evolution 46: 300–316.

Manly, B.F.J. 1985. *The Statistics of Natural Selection on Animal Populations*. Chapman and Hall, London.

Margot, J.B., G.W. Demers and R.C. Hardison. 1988. Complete nucleotide sequence of the rabbit β-like globin gene cluster: Analysis of intergenic sequences and comparison with the human β-like globin gene cluster. J. Mol. Biol. 205:15–40.

Maruyama, K. and D.L. Hartl. 1991. Evolution of the transposable element *mariner* in *Drosophila* species. Genetics 128:319–329.

May, R.M. (ed.). 1981. *Theoretical Ecology*. Blackwell, Oxford.

May, R.M. 1985. Evolution of pesticide resistance. Nature 315:12–13.

May, R.M. 1995. The cheetah controversy. Nature 374:309–310.

Maynard Smith, J. and K.C. Sondhi. 1961. The genetics of a pattern. Genetics 45:1039–1050.

McCarrey, J.R. and K. Thomas. 1987. Human testis-specific PGK gene lacks introns and possesses characteristics of a processed gene. Nature 326:501–505.

McDonald, J.H. and M. Kreitman. 1991. Adaptive protein evolution at the *Adh* locus in *Drosophila*. Nature 351:652–654.

McLellan, T. and L.S. Inouye. 1986. The sensitivity of isoelectric focusing and electrophoresis in the detection of sequence differences in proteins. Biochem. Genet. 24:571–577.

McVean, G.T. and L.D. Hurst. 1997. Evidence for a selectively favorable reduction in the mutation rate of the X chromosome. Nature 386:388–392.

Merbs, S.L. and J. Nathans. 1992. Absorption spectra of human cone pigments. Nature 356:433–435.

Milkman, R. and M.M. Bridges. 1990. Molecular evolution of the *Escherichia coli* chromosome. III. Clonal frames. Genetics 126:505–517.

Milkman, R. and M.M. Bridges. 1993. Molecular evolution of the *Escherichia coli* chromosome. IV. Sequence comparisons. Genetics 133:455–468.

Miller, R.H. 1988. Human immunodeficiency virus may encode a novel protein on the genomic DNA plus strand. Science 239:1420–1422.

Mitchell-Olds, T. 1986. Quantitative genetics of survival and growth in *Impatiens capensis*. Evolution 40:107–116.

Mitchell-Olds, T. and J.J. Rutledge. 1986. Quantitative genetics in natural populations: A review of the theory. Am. Nat. 127:379–402.

Mitchell-Olds, T. and R.G. Shaw. 1987. Regression analysis of natural selection: Statistical inference and biological interpretation. Evolution 41:1149–1161.

Miyata, T., H. Hayashida, K. Kuma, K. Mitsuyasu and T. Tasunaga. 1987. Male-driven molecular evolution: A model and nucleotide sequence analysis. Cold Spring Harbor Symp. Quant. Biol. 52:863–867.

Monnerot, M., J.-C. Mounolou and M. Solignac. 1984. Intra-individual length heterogeneity of *Rana esculenta* mitochondrial DNA. Biol. Cell 52:213–218.

Montgomery, E.A. and C.H. Langley. 1983. Transposable elements in Mendelian populations. II. Distribution of three *copia*-like elements in a natural population of *Drosophila melanogaster*. Genetics 104:473–483.

Moriyama, E. and D.L Hartl. 1993. Codon usage bias and base composition of nuclear genes in *Drosophila*. Genetics 134:847–858.

Morse, B., P.G. Rothberg, V.J. South, J.M. Spandorfer and S.M. Astrin. 1988. Insertional mutagenesis of the *myc* locus in a human breast carcinoma. Nature 333:87–90.

Mourant, A.E., A.C. Kopec and K. Domaniewska-Sobczak. 1976. *The Distribution of Human Blood Groups and other Polymorphisms*, Second Edition. Oxford University Press, New York.

Mukai, T., T.K. Watanabe and O. Yamaguchi. 1974. The genetic structure of natural populations of *Drosophila melanogaster*. XII. Linkage disequilibrium in a large local population. Genetics 77:771–793.

Muse, S.V. and B.S. Gaut. 1994. A likelihood approach for comparing synonymous and nonsynonymous nucleotide substitution rates with application to the chloroplast genome. Mol. Biol. Evol. 11:715–724.

Muse, S.V. and B.S. Weir. 1992. Testing for equality of evolutionary rates. Genetics 132:269–276.

Myrianthopoulos, N.C. and S.M. Aronson. 1966. Population dynamics of Tay-Sachs disease. I. Reproductive fitness and selection. Am. J. Hum. Genet. 18:313–327.

NRC (National Research Council). 1992. *DNA Technology in Forensic Science*. National Academy Press, Washington, DC.

NRC (National Research Council). 1996. *DNA Forensic Science: An Update*. National Academy Press, Washington, DC.

Nathans, J., T.P. Piantanida, R.L. Eddy, T.B. Shows and D.S. Hogness. 1986. Molecular genetics of inherited variation in human color vision. Science 232:203–210.

Neel, J.V. and E.A. Thompson. 1978. Founder effect and the number of private polymorphisms observed in Amerindian tribes. Proc. Natl. Acad. Sci. USA 75:1904–1908.

Nei, M. 1975. *Molecular Population Genetics and Evolution*. American Elsevier, New York.

Nei, M. 1986. Definition and estimation of fixation indices. Evolution 40:643–645.

Nei, M. 1987. *Molecular Evolutionary Genetics*. Columbia University Press, New York.

Nei, M. 1996. Phylogenetic analysis in molecular evolutionary genetics. Annu. Rev. Genet. 30:371–403.

Nei, M. and R.K. Chesser. 1983. Estimation of fixation indices and gene diversities. Ann. Hum. Genet. 47:253–259.

Nei, M. and T. Gojobori. 1986. Simple methods for estimating the numbers of synonymous and nonsynonymous nucleotide substitutions. Mol. Biol. Evol. 3:418–426.

Nei, M. and D. Graur. 1984. Extent of protein polymorphism and the neutral mutation theory. Evol. Biol. 17:73–118.

Nei, M. and A.K. Roychoudhury. 1982. Genetic relationship and evolution of human races. Evol. Biol. 14:1–59.

Nei, M. and A.K. Roychoudhury. 1993. Evolutionary relationships of human populations on a global scale. Mol. Biol. Evol. 10:927–943.

Nei, M., T. Maruyama and R. Chakraborty. 1975. The bottleneck effect and genetic variability in populations. Evolution 29:1–10.

Nei, M., P.A. Fuerst and R. Chakraborty. 1976. Testing the neutral mutation hypothesis by distribution of single locus heterozygosity. Nature 262:491–493.

Nevo, E.. 1978. Genetic variation in natural populations: Patterns and theory. Theor. Popul. Biol. 13:121–177.

Nichols, B.P. and C. Yanofsky. 1979. Nucleotide sequences of *trpA* of *Salmonella typhimurium* and *Escherichia coli*: An evolutionary comparison. Proc. Natl. Acad. Sci. USA 76:5244–5248.

Nilsson-Ehle, H. 1909. Kreuzungsuntersuchungen an Hafer und Weizen. Lunds. Univ. Aarskr. NF 5,2:1–22.

Nordskog, A.W. and F.G. Giesbrecht. 1964. Regression in egg production in the domestic fowl when selection is relaxed. Genetics 50:407–416.

Novick, A. 1955. Mutagens and antimutagens. Brookhaven Symp. Biol. 8:201–215.

O'Brien, S.J., D.E. Wildt, D. Goldman, C.R. Merril and M. Bush. 1983. The cheetah is depauperate in genetic variation. Science 221:459–461.

O'Brien, S.J., M.E. Roelke, L. Marker, A. Newman, C.A. Winkler, D. Meltzer, L. Colly, J.F. Evermann, M. Bush and D.E. Wildt. 1985. Genetic basis for species vulnerability in the cheetah. Science 227:1428–1434.

O'Brien, S.J., D.E. Wildt, M. Bush, T.M. Caro, C. FitzGibbon, I. Aggundey and R.E. Leakey. 1987. East African cheetahs: Evidence for two population bottlenecks. Proc. Natl. Acad. Sci. USA 84:508–511.

Oakeshott, J.G., J.B. Gibson, P.R. Anderson, W.R. Knibb, D.G. Anderson and R.K. Chambers. 1982. *Alcohol dehydrogenase* and *glycerol-3-phosphate dehydrogenase* clines in *Drosophila melanogaster* on different continents. Evolution 36:86–96.

Ohta, T. 1973. Slightly deleterious mutant substitutions in evolution. Nature 246:96–98.

Ohta, T. 1982. Allelic and nonallelic homology of a supergene family. Proc. Natl. Acad. Sci. USA 79:3251–3254.

Ohta, T. and M. Kimura. 1971. On the constancy of the evolutionary rate of cistrons. J. Mol. Evol. 1:18–25.

Ota, T. and M. Nei. 1994. Divergent evolution and evolution by the birth-and-death process in the immunoglobulin VH gene family. Mol. Biol. Evol. 11:469–482.

Pace, N.R., D.K. Smith, G.J. Olsen and B.D. James. 1989. Phylogenetic comparative analysis and the secondary structure of ribonuclease P RNA: A review. Gene 82:65–75.

Palopoli, M.F. and C.-I Wu. 1994. Genetics of hybrid male sterility between *Drosophila* sibling species: A complex web of epistasis is revealed in interspecific studies. Genetics 138:329–341.

Partridge, L., B. Barrie, N.H. Barton, K. Fowler and V. French. 1995. Rapid laboratory evolution of adult life history traits in *Drosophila melanogaster* in response to temperature. Evolution 49:538–544.

Paterson, A.H., E.S. Lander, J.D. Hewitt, S. Peterson, S.E. Lincoln and S.D. Tanksley. 1988. Resolution of quantitative traits into Mendelian factors by using a complete linkage map of restriction fragment length polymorphisms. Nature 335:721–726.

Pearl, R. 1927. The growth of populations. Quart. Rev. Biol. 2:532–548.

Peetz, E.W., G. Thomson and P.W. Hedrick. 1986. Charge changes in protein evolution. Mol. Biol. Evol. 3:84–94.

Pesole, G., E. Sbisa, G. Preparata and C. Saccone. 1992. The evolution of the mitochondrial D-loop region and the origin of modern man. Mol. Biol. Evol. 9:587–598.

Pipkin, S.B., E. Franklin-Springer, S. Law and S. Lubega. 1976. New studies of the alcohol dehydrogenase cline in *D. melanogaster* from Mexico. J. Hered. 67:258–266.

Pirchner F. 1969. *Population Genetics in Animal Breeding*. W. H. Freeman, San Francisco.

Policansky, D. and E. Zouros. 1977. Gene differences between the sex ratio and standard gene arrangements of the X chromosome in *Drosophila persimilis*. Genetics 85:507–511.

Powell, J.R., A. Caccone, G.D. Amato and C. Yoon. 1986. Rates of nucleotide substitution in *Drosophila* mitochondrial DNA and nuclear DNA are similar. Proc. Natl. Acad. Sci. USA 83:9090–9093.

Powers, L. 1951. Gene analysis by the partitioning method when interactions of genes are involved. Bot. Gaz. 113:1–23.

Prakash, S. 1977. Gene polymorphism in natural populations of *Drosophila persimilis* Genetics 85:513–520.

Provine, W.B. 1986. *Sewall Wright and Evolutionary Biology*. University of Chicago Press, Chicago.

Race, R.R. and R. Sanger. 1975. Blood Groups in Man, Sixth Edition. J.B. Lippincott, Philadelphia.

Raju, N.B. 1994. Ascomycete spore killers: Chromosomal elements that distort genetic ratios among products of meiosis. Mycologia 86:461–473.

Rand, D.M. and R.G. Harrison. 1986. Mitochondrial DNA transmission genetics in crickets. Genetics 114:955–970.

Rand, D.M. and L.M. Kann. 1996. Excess amino acid polymorphism to mitochondrial DNA: Con-

trasts among genes from *Drosophila*, mice, and humans. Mol. Biol. Evol. 13:735–748.

Raymond, M., A. Callaghan, P. Fort and N. Pasteur. 1991. Worldwide migration of amplified insecticide resistance genes in mosquitoes. Nature 350:151–153.

Reed, T.E. and J.V. Neel. 1959. Huntington's chorea in Michigan. Am. J. Hum. Genet. 11:107–136.

Relethford, J.H. 1995. Genetics and modern human origins. Evol. Anthropol. 4:53–63.

Rieseberg, L.H., C.R. Linder and G.J. Seiler. 1995. Chromosomal and genic barriers to introgression in *Helianthus*. Genetics 141:1163–1171.

Rieseberg, L.H., B. Sinervo, C.R. Linder, M.C. Unger and D.M. Arias. 1996. Role of gene interactions in hybrid speciation: Evidence from ancient and experimental hybrids. Science 272:741–744.

Riley, M.A. 1993. Positive selection for colicin diversity in bacteria. Mol. Biol. Evol. 10:1048–1059.

Risch, N. and K. Merikangas. 1996. The future of genetic studies of complex human diseases. Science 273:1516–1517.

Riska, B., W.R. Atchley and J.J. Rutledge. 1984. A genetic analysis of targeted growth in mice. Genetics 107:79–101.

Riska, B., T. Prout and M. Turelli. 1989. Laboratory estimates of heritabilities and genetic correlations in nature. Genetics 123:865–871.

Robertson, H.M. 1993. The *mariner* transposable element is widespread in insects. Nature 362:241–245.

Robertson, H.M. and E.G. MacLeod. 1993. Five major subfamilies of *mariner* transposable elements in insects, including the Mediterranean fruit fly, and related arthropods. Insect Mol. Biol. 2:125–139.

Robinson, H.F., R.E. Comstock and P.H. Harvey. 1949. Estimates of heritability and degree dominance in corn. Agron. J. 41: 353–359.

Rogers, A.R. and H. Harpending. 1992. Population growth makes waves in the distribution of pairwise genetic differences. Mol. Biol. Evol. 9:552–569.

Rohlf, F.J. and M.C. Wooten. 1988. Evaluation of the restricted maximum-likelihood method for estimating phylogenetic trees using simulated allele-frequency data. Evolution 42:581–595.

Roughgarden, J. 1979. *Theory of Population Genetics and Evolutionary Ecology: An Introduction*. Macmillan, New York.

Roughgarden, J. 1996. *Theory of Population Genetics and Evolutionary Ecology: An Introduction*. Prentice Hall, New York.

Rowen, L., B.F. Koop and L. Hood. 1996. The complete 685-kb DNA sequence of the human β-T cell receptor locus. Science 272:1755–1762.

Rubin, G.M. 1983. Dispersed repetitive DNAs in *Drosophila*. pp. 329–361. *In* J.A. Shapiro (ed.), *Mobile Genetic Elements*. Academic Press, New York.

Ruvolo, M.S. Zehr, M. von Dornum, D. Chang and J. Lin. 1993. Mitochondrial COII sequences and modern human origins. Mol. Biol. Evol. 10:1115–1135.

Rzhetsky, A. and M. Nei. 1992. A simple method for estimating and testing minimum-evolution trees. Mol. Biol. Evol. 9:945–967.

Saitou, N. and M. Nei. 1987. The neighbor-joining method: A new method for reconstructing phylogenetic trees. Mol. Biol. Evol. 4:406–425.

Sakoyama, Y., K.-J. Hong, S.M. Byun, H. Hisajima, S. Ueda, Y. Yaoita, H. Hayashida, T. Miyata and T. Honjo. 1987. Nucleotide sequences of immunoglobulin epsilon genes of chimpanzee and orangutan: DNA molecular clock and hominoid evolution. Proc. Natl. Acad. Sci. USA 84:1080–1084.

Satta, Y., C. O'hUigin, N. Takahata and J. Klein. 1993. The synonymous substitution rate of the major histocompatibility complex in primates. Proc. Natl. Acad. Sci. USA 90:7480–7484.

Sawyer, S. and D.L. Hartl. 1986. Distribution of transposable elements in prokaryotes. Theor. Popul. Biol. 30:1–17.

Sawyer, S.A. 1989. Statistical tests for detecting gene conversion. Mol. Biol. Evol. 6:526–538.

Sawyer, S.A., D.E. Dykhuizen, R.F. DuBose, L. Green, T. Mutangadura Mhlanga, D.F. Wolczyk and D.L. Hartl. 1987. Distribution and abundance of insertion sequences among natural isolates of *Escherichia coli*. Genetics 115:51–63.

Sax, K. 1923. The association of size differences with seed coat pattern and pigmentation in *Phaseolus vulgarus*. Genetics 8:552–560.

Schaeffer, S.W. and E.L. Miller. 1991. Nucleotide sequence analysis of *Adh* genes estimates the time of geographic isolation of the Bogota population of *Drosophila pseudoobscura*. Proc. Natl. Acad. Sci. USA 88:6097–6101.

Scharloo, W. 1987. Constraints in selection response. pp. 125–150, *In* V. Loeschcke (ed.), *Genetic Constraints on Adaptive Evolution*. Springer-Verlag, Berlin.

Schlichting, C.D. and M. Pigliucci. 1994. Gene regulation, quantitative genetics and the evolution of reaction norms. Evol. Ecol. 8:1–15.

Schlötterer, C., C. Vogl and D. Tautz. Polymorphism and locus-specific effects of polymorphism at microsatellite loci in natural *Drosophila melanogaster* populations. Genetics, *in press*.

Schmalhausen, I.I. 1949. *Factors of Evolution: The Theory of Stabilizing Selection*. Blakiston, Philadelphia.

Selander, R.K., D.A. Caugant and T.S. Whittam. 1987. Genetic structure and variation in natural populations of *Escherichia coil*. pp. 1625–1648. *In* J.L. Ingraham, K. Brooks Low, B. Magasanik, M. Schaechter and H.E. Umbarger (eds.), *Escherichia coli and Salmonella typhimurium: Cellular and Molecular Biology*. American Society for Microbiology, Washington, DC.

Shaw, F.H., R.G. Shaw, G.S. Wilkinson and M. Turelli. 1995. Changes in genetic variances and covariances: G whiz! Evolution 49:1260–1267.

Shaw, R.G. 1987. Maximum likelihood approaches applied to quantitative genetics of natural populations. Evolution 41:812–826.

Shen, S., J.L. Slightom and O. Smithies. 1981. A history of the fetal globin gene duplication. Cell 26: 191–203.

Shields, J. 1962. *Monozygotic Twins Brought Up Apart and Brought Up Together*. Oxford, London.

Shimmin, L.C., B. H.-J. Chang and W.-H. Li. 1993. Male-driven evolution of DNA sequences. Nature 362:745–747.

Shrimpton, A.E. and A. Robertson. 1988. The isolation of polygenic factors controlling bristle score in *Drosophila melanogaster*. II. Distribution of third chromosome bristle effects within chromosome sections. Genetics 118:445–459.

Simmonds, N.W. 1977. Approximations for *i*, intensity of selection. Heredity 34:413–414.

Simmons, M.F. and J.F. Crow. 1977. Mutations affecting fitness in *Drosophila* populations. Annu. Rev. Genet. 11:49–78.

Simon, M., J. Zieg, M. Silverman, G. Mandel and R. Doolittle. 1980. Phase variation: Evolution of a controlling element. Science 209:1370–1374.

Simonsen, K.I., G.A. Churchill and C.F. Aquadro. 1995. Properties of statistical tests of neutrality for DNA polymorphism data. Genetics 141:413–429.

Sing, C.F. and J. Davignon. 1985. Role of apolipoprotein E polymorphism in determining normal plasma lipid and lipoprotein variation. Am. J. Hum. Genet. 37:268–285.

Singer, M.F. 1982. SINEs and LINEs: Highly repeated short and long interspersed sequences in mammalian genomes. Cell 28:433–434.

Singh, R.S. and L.R. Rhomberg. 1987. A comprehensive study of genetic variation in natural populations of *Drosophila melanogaster*. II. Estimates of heterozygosity and patterns of geographic differentiation. Genetics 117:255–271.

Singh, R.S., R.C. Lewontin and A.A. Felton. 1976. Genetic heterogeneity within electrophoretic "alleles" of *xanthine dehydrogenase* in *Drosophila pseudoobscura*. Genetics 84:609–629.

Skibinski, D.O., M. Woodwark and R.D. Ward. 1993. A quantitative test of the neutral theory using pooled allozyme data. Genetics 135:233–248.

Skibinski, D.O.F., C. Gallagher and C.M. Beynon. 1994. Sex-limited mitochondrial DNA transmission in the marine mussel *Mytilus edulis*. Genetics 138:801–810.

Slagel, V., E. Flemington, V. Traina-Dorge, H. Bradshaw and P. Deininger. 1987. Clustering and subfamily relationships of the *Alu* family in the human genome. Mol. Biol. Evol. 4:19–29.

Slatkin, M. 1985. Rare alleles as indicators of gene flow. Evolution 39:53–65.

Slatkin, M. and R.R. Hudson. 1991. Pairwise comparisons of mitochondrial DNA sequences in stable and exponentially growing populations. Genetics 129:555–562.

Slatkin, M. and W.P. Maddison. 1989. A cladistic measure of gene flow inferred from the phylogenies of alleles. Genetics 123:615–619.

Slatkin, M. and W.P. Maddison. 1990. Detecting isolation by distance using phylogenies of genes. Genetics 126:249–260.

Slightom, J.L., A.E. Blechl and O. Smithies. 1980. Human fetal $^G\gamma$ and $^A\gamma$ globin genes: Complete nucleotide sequences suggest that DNA can be exchanged between these duplicated genes. Cell 21:627–638.

Smit. A.F.A. 1996 The origin of interspersed repeats in the human genome. Curr. Opin. Genet. Dev. 6:743–748.

Smith, C. 1975. Quantitative inheritance. pp. 382–441. *In* G. Fraser and O. Mayo (eds.), *Textbook of Human Genetics*. Blackwell, Oxford.

Solignac, M.M. Monnerot and J.C. Mounolou. 1983. Mitochondrial DNA heteroplasmy in *Drosophila mauritiana*. Proc. Natl. Acad. Sci. USA 80:6942–6946.

Solignac, M., J. Genermont, M. Monnerot and J.-C. Mounolou. 1984. Genetics of mitochondria in *Drosophila*: mtDNA inheritance in heteroplasmic strains of *D. mauritiana*. Mol. Gen. Genet. 197:183–188.

Solignac, M., J. Genermont, M. Monnerot and J.-C. Mounolou. 1987. *Drosophila* mitochondrial genetics: Evolution of heteroplasmy through germ line cell divisions. Genetics 117:687–696.

Sourdis, J. and M. Nei. 1988. Relative efficiencies of the maximum parsimony and distance-matrix methods in obtaining the correct phylogenetic tree. Mol. Biol. Evol. 5:298–311.

Spencer, H.G. and R.W. Marks. 1988. The maintenance of single-locus polymorphism. I. Numerical studies of a viability selection model. Genetics 120:605–613.

Sprague, G.F. 1978. Introductory remarks to the session on the history of hybrid corn. pp. 11–12. *In* D.B. Walden (ed.), *Maize Breeding and Genetics*. John Wiley and Sons, New York.

Stam, L.F. and C.C. Laurie. 1996. Molecular dissection of a major gene effect on a quantitative trait: The level of alcohol dehydrogenase expression in *D. melanogaster*. Genetics 144: 1559–1564.

Stephens, J.C. 1985. Statistical methods of DNA sequence analysis: Detection of intragenic recombination or gene conversion. Mol. Biol. Evol. 2:539–556.

Stewart, D.T., C. Saavedra, R.R. Stanwood, A.O. Ball and E. Zouros. 1995. Male and female mitochondrial DNA lineages in the blue mussel (*Mytilus edulis*) species group. Mol. Biol. Evol. 12:735–747.

Stocker, B.A.D. 1949. Measurements of rate of mutation of flagellar antigenic phase in *Salmonella typhimurium*. J. Hyg. 47:398–412.

Stuber, C.W., S.E. Lincoln, D.W. Wolff, T. Helentjaris and E.S. Lander. 1992. Identification of genetic factors contributing to heterosis in a hybrid from two elite maize inbred lines using molecular markers. Genetics 132:823–839.

Südhof, T.C., J.L. Goldstein, M.S. Brown and D.W. Russell. 1985. The LDL receptor gene: A mosaic of exons shared with different proteins. Science 228:815–822.

Sved, J.A. 1975. Fitness of third chromosome homozygotes in *Drosophila melanogaster*. Genet. Res. Camb. 25:197–200.

Sved, J.A. and B.D.H. Latter. 1977. Migration and mutation in stochastic models of gene frequency change. J. Math. Biol. 5:61–73.

Tajima, F. 1983. Evolutionary relationship of DNA sequences in finite populations. Genetics 105:437–460.

Tajima, F. 1989. Statistical method for testing the neutral mutation hypothesis by DNA polymorphism. Genetics 123:585–595.

Tajima, F. 1993. Simple methods for testing the molecular evolutionary clock hypothesis. Genetics 135:599–607.

Takahata, N. 1983. Population genetics of extranuclear genomes under the neutral mutation hypothesis. Genet. Res. 42:235–256.

Takahata, N. 1984. A model of extranuclear genomes and the substitution rate under within-generation selection. Genet. Res. 44:109–116.

Takahata, N. 1987. On the overdispersed molecular clock. Genetics 116:169–179.

Takezaki, N., A. Rzhetsky and M. Nei. 1995. Phylogenetic test of the molecular clock and linearized trees. Mol. Biol. Evol. 12:823–833.

Tanksley, D.S. 1993. Mapping polygenes. Annu. Rev. Genet. 27:205–233.

Taylor, C.E. and V. Kekiç. 1988. Sexual selection in a natural population of *Drosophila melanogaster*. Evolution 42:197–199.

Teissier, G. 1942. Persistence d'un gène léthal dans une population de *Drosophiles*. Compt. Rend. Acad. Sci. 214:327–330.

Templeton, A.R. 1982. Adaptation and the integration of evolutionary forces. pp. 15–31. *In* R. Milkman (ed.), *Perspectives on Evolution*. Sinauer Associates, Sunderland, MA.

Templeton, A.R. 1993. The "Eve" hypothesis: A genetic critique and reanalysis. Am. Anthropol. 95: 51–72.

Templeton, A.R., E. Routman and C.A. Phillips. 1995. Separating population structure from population history: A cladistic analysis of the geographical distribution of mtDNA haplotypes in the tiger salamander, *Ambystoma tigrinum*. Genetics 140:767–782.

Thoday, J.M. 1961. Location of polygenes. Nature 191:368–370.

Thoday, J.M. 1979. Polygene mapping: Uses and limitations. pp. 219–234. *In* J.N. Thompson and J.M. Thoday (eds.), *Quantitative Genetic Variation*. Academic Press, New York.

Tishkoff, S.A., E. Dietzsch, W. Speed, A.J. Pakstis, J.R. Kidd, K. Cheung, B. Bonné-Tamier, A.S. Santachiara-Benerecetti, P. Moral, M. Krings, S. Paabo, E. Watson, N. Risch, T. Jenkins and K.K. Kidd. 1996. Global patterns of linkage disequilibrium at the CD4 locus and modern human origins. Science 271:1380–1387.

True, J.R., B.S. Weir and C.C. Laurie. 1996. A genome-wide survey of hybrid incompatibility factors by the introgression of marked segments of *D. mauritiana* chromosomes into *D. simulans*. Genetics 142:819–837.

Turelli, M. 1984. Heritable genetic variation via mutation-selection balance: Lerch's zeta meets the abdominal bristle. Theor. Popul. Biol. 25:138–193.

Turelli, M. 1988. Phenotypic evolution, constant covariances and the maintenance of additive variance. Evolution 42:1342–1347.

Turelli, M., J.H. Gillespie and R. Lande. 1988. Rate tests for selection on quantitative characters during macroevolution and microevolution. Evolution 42:1085–1089.

Turelli, M., A.A. Hoffman and S.W. McKechnie. 1992. Dynamics of cytoplasmic incompatibility and mtDNA variation in natural *Drosophila simulans* populations. Genetics 132:713–723.

Unseld, M., J.R. Marienfeld, P. Brnadt and A. Brennicke. 1997. The mitochondrial genome of Arabidopsis thaliana contains 57 genes in 366,924 nucleotides. Nat. Genet. 15:57–65.

Uyenoyama, M.K. 1995. A generalized least-squares estimate of the origin of sporophytic self-incompatibility. Genetics 139:975–992.

Uyenoyama, M. and M.W. Feldman. 1980. Theories of kin and group selection: A population genetics perspective. Theor. Popul. Biol. 17:380–414.

Via, S. and R. Lande. 1985. Genotype-environment interaction and the evolution of phenotypic plasticity. Evolution 39:505–522.

Via, S., R. Gomulkiewicz, G. De Jong, S.M. Scheiner, C.D. Schlichting and P.H. van Tienderen. 1995. Adaptive phenotypic plasticity: Consensus and controversy. TREE 10:212–217.

Vigilant, L.M. Stoneking, H. Harpending, K. Hawkens and A.C. Wilson, 1991. African populations and the evolution of human mitochondrial DNA. Science 253:1503–1507.

Vogel, F. and A.G. Motulsky. 1986. *Human Genetics*, Second Edition. Springer-Verlag, New York.

Wade, M.J. and C.J. Goodnight. 1991. Wright's shifting balance theory: An experimental study. Science 253:1015–1018.

Wahlund, S. 1928. Composition of populations from the perspective of the theory of heredity. Hereditas 11:65–105 (in German).

Wakeley, J. 1993. Substitution rate variation among sites in hypervariable region 1 of human mitochondrial DNA. J. Mol. Evol. 37:613–623.

Walsh, J.B. 1988. Sequence-dependent gene conversion: Can duplicated genes diverge fast enough to escape conversion? Genetics 117:543–557.

Walsh, J.B. 1995. How often do duplicated genes evolve new functions? Genetics 139:421–428.

Watanabe, S., S. Kondo and E. Matsunaga (eds.). 1975. *Human Adaptability, Volume 2: Anthropological and Genetic Studies on the Japanese*. University of Tokyo Press, Tokyo.

Watterson, G.A. 1975. On the number of segregating sites in genetical models without recombination. Theor. Popul. Biol. 7:256–276.

Wayne, R.K., W.S. Modi and S.J. O'Brien. 1986. Morphological variability and asymmetry in the cheetah (*Acinonyx jubatus*), a genetically uniform species. Evolution 40:78–85.

Weber, K.E. 1990a. Artificial selection on wing allometry in *Drosophila melanogaster*. Genetics 126: 975–989.

Weber, K.E. 1990b. Increased selection response in larger populations. I. Selection for wing-tip height in *Drosophila melanogaster* at three population sizes. Genetics 125:579–584.

Weber, K.E. 1992. How small are the smallest selectable domains of form? Genetics 130:345–353.

Weber, K.E. 1996. Large genetic change at small fitness cost in large populations of *D. melanogaster* selected for wind tunnel flight: Rethinking fitness surfaces. Genetics 144:205–213.

Weber, K.E. and L.T. Diggins. 1990. Increased selection response in larger populations. II. Selection for ethanol vapor resistance in *Drosophila melanogaster* at two population sizes. Genetics 125: 585–597.

Weir, B.S. and C.C. Cockerham. 1984. Estimating *F* statistics for the analysis of population structure. Evolution 38:1358–1370.

Wellauer, P and I.B. Dawid. 1979. Isolation and sequence organization of human ribosomal DNA. J. Mol. Biol. 128:289–303.

Whittam, T.S., H. Ochman and R.K. Selander. 1983. Multilocus genetic structure in natural populations of *Escherichia coli*. Proc. Natl. Acad. Sci. USA 80:1751–1755.

Wilkinson, G. 1987. Equilibrium analysis of sexual selection in *Drosophila melanogaster*. Evolution 41:11–21.

Wilkinson, G.S., K. Fowler and L. Partridge. 1990. Resistance of genetic correlation structure to directional selection on *Drosophila melanogaster*. Evolution 44:1990–2003.

Williams, S.M., R. DeSalle and C. Strobeck. 1985. Homogenization of geographical variants at the nontranscribed spacer of rDNA in *Drosophila mercatorum*. Mol. Biol. Evol. 2:338–346.

Wilson, D.S. 1983. The group selection controversy: History and current status. Ann. Rev. Ecol. Syst. 14:159–187.

Woese, C.R. 1981. Archaebacteria. Sci. Am. 244:98–122.

Wolfe, K.H., W.-H. Li and P.M. Sharp. 1987. Rates of nucleotide substitution vary greatly among plant mitochondrial, chloroplast and nuclear DNAs. Proc. Natl. Acad. Sci. USA 84:9054–9058.

Worton, R.G., J. Sutherland, J.E. Sylvester, H.F. Willard, S. Bodrug, I. Dube, C. Duff, V. Kean, P.N. Ray and R.D. Schmickel. 1988. Human ribosomal RNA genes: Orientation of the tandem array and conservation of the 5′ end. Science 239:64–68.

Wright, S. 1921. Systems of mating. Genetics 6:111–178.

Wright, S. 1931. Evolution in Mendelian populations. Genetics 16:97–159.

Wright, S. 1943a. Analysis of local variability of flower color in *Lynanthus parryae*. Genetics 28:139–156.

Wright, S. 1943b. Isolation by distance. Genetics 28:114–138.

Wright, S. 1969. *Evolution and the Genetics of Populations. Volume 2. The Theory of Gene Frequencies.* University of Chicago Press, Chicago.

Wright, S. 1977. *Evolution and the Genetics of Populations. Volume 3. Experimental Results and Evolutionary Deductions.* University of Chicago Press, Chicago.

Wright, S. 1978. *Evolution and the Genetics of Populations. Volume 4. Variability within and among Natural Populations.* University of Chicago Press, Chicago.

Wu, C.-I. and W.-H. Li. 1985. Evidence for higher rates of nucleotide substitution in rodents than in man. Proc. Natl. Acad. Sci USA 82:1741–1745.

Xiao, J., J. Li, L. Yuan and S.D. Tanksley. 1995. Dominance is the major genetic basis of heterosis in rice as revealed by QTL analysis using molecular markers. Genetics 140:745–754.

Xiong, M. and S.-W. Guo. 1997. Fine-scale mapping of quantitative trait loci using historical recombination. Genetics 145:1201–1218.

Yang, Z. 1996a. Maximum likelihood models for combined analyses of multiple sequence data. J. Mol. Evol. 42:587–596.

Yang, Z. 1996b. Among site rate variation and its impact on phylogenetic analysis. TREE 11:367–371.

Yokoyama, S., L. Chung and T. Gojobori. 1988. Molecular evolution of the human immunodeficiency and related viruses. Mol. Biol. Evol. 5:237–251.

Zeng, Z.-B. 1992. Correcting the bias of Wright's estimates of the number of genes affecting a quantitative character: A further improved method. Genetics 131:987–1001.

Zeng, Z.-B. 1994. Precision mapping of quantitative trait loci. Genetics 136:1457–1468.

Zhivotovsky, L.A. and M.W. Feldman. 1995. Microsatellite variability and genetic distances. Proc. Natl. Acad. Sci. USA 92:11549–11552.

Zuckerkandl, E. and L. Pauling. 1962. Molecular disease, evolution, and genetic heterogeneity. pp. 189–225. *In* M. Kasha and B. Pullman (eds.), *Horizons in Biochemistry*. Academic Press, New York.

Author Index

Subject Index

ABOUT THE BOOK

Editor: Andrew D. Sinauer
Project Editor: Nan Sinauer
Production Manager: Christopher Small
Electronic Bookbuilder: Greg Martel, Precision Graphics
Illustration Program: Precision Graphics
Cover Design: Mary Portner
Copy Editor: Roberta Lewis
Cover Manufacturer: Henry N. Sawyer Company, Inc.
Book Manufacturer: Best Book Manufacturers, Inc.